Advances in Material Research and Technology

This Series covers the advances and developments in a wide range of materials such as energy materials, optoelectronic materials, minerals, composites, alloys and compounds, polymers, green materials, semiconductors, polymers, glasses, nanomaterials, magnetic materials, superconducting materials, high temperature materials, environmental materials, Piezoelectric Materials, ceramics, and fibers.

Shadia J. Ikhmayies
Editor

Advances in Biomaterials Research

Springer

Editor
Shadia J. Ikhmayies
Department of Physics
University of Jordan
Amman, Jordan

ISSN 2662-4761 ISSN 2662-477X (electronic)
Advances in Material Research and Technology
ISBN 978-3-031-97480-9 ISBN 978-3-031-97481-6 (eBook)
https://doi.org/10.1007/978-3-031-97481-6

This Springer imprint is published by the registered company Springer Nature Switzerland AG
The registered company address is: Gewerbestrasse 11, 6330 Cham, Switzerland

Preface

A biomaterial is any natural, synthetic, or hybrid material that is used in medical devices or implants, which is placed in contact with the human biological system to replace or restore deteriorating bodily functions. Biomaterials include natural materials found within the body, such as bone, connective tissue, and cartilage. Synthetic biomaterials include glass, polymers, ceramics, titanium alloys, cobalt-chromium alloys, composites, and stainless steel. Because biomaterials are placed in close and continuous contact with the recipient's body fluids or tissues, they must have special properties that can be tailored to meet the needs of a specific application. A biomaterial must be biocompatible, non-carcinogenic, corrosion resistant, and have low toxicity and corrosiveness. Most of the biomaterials stimulate the formation of cells and perform the antimicrobial functions, as well as preventing the harmful transformation of certain cells. But any compounds that are released from the biomaterial may result in positive or negative responses from the biological systems, ranging from increasing wound healing and improving biological functions to toxicity and activation of the immune response. Biomaterials are classified based on their structure, chemical composition, and properties related to biological applications. Based on their physical properties, biomaterials are classified into three categories: bioceramics, biopolymers, and biometals. They can also be classified at the organ and system level, based on the parts of the body they treat. Biomaterials incorporated by Nanoparticles provide a composite material that exhibits new or improved properties and offers therapeutic potential.

The rapid advances in the field of biomaterials to meet the growing needs in healthcare and medical practices are evident. This progress has enabled the production and engineering of biomaterials to be used in the manufacture of medical devices and implants, such as hip joints, bone plates, heart valves, and cardiac pacemakers. Moreover, these advances led to the use of biomaterials in disease treatment, drug delivery to cells, cancer immunotherapy, cell regeneration, antimicrobial treatment, and much more. Various types of biomaterials are used in these applications, such as lipid-based biomaterials, inorganic biomaterials, and polymer-based biomaterials. However, there are a lot of challenges related to the properties of biomaterials such as biosafety, biocompatibility, toxicity, pollution of environment, and the disposal of large amounts of waste resulting from the extensive use of non-biodegradable synthetic polymeric biomaterials. Therefore, emerging biomaterials which are highly biocompatible with the human body, besides being environmentally friendly will have tremendous potential in medical applications.

This book covers fundamentals, reviews, and research findings of various biomaterials, making it a recommended reference for graduate and undergraduate students and early researchers in various disciplines of biomaterials science and engineering. It covers design, synthetic methods, characterization, and some applications, such as drug and gene delivery, tissue engineering, orthopedic and central nervous system applications, and 3D organ printing. It also covers and discusses many important topics, such as multifunctional and multimodal biomaterials, bionanoparticles, antibacterial/bactericidal activity, and applications such as 3D and 4D printing of biomaterials, electrochemical biosensors for heavy metal detection, and artificial organs. Readers of this book will find rich material on biopolymers, bioactive glasses, glass-ceramics, and their uses in delivery systems, tissue engineering, as well as therapeutic uses, bone healing, and other medical applications. The book includes 11 chapters, briefly described in the following paragraphs:

Chapter one, "Biomaterials: Fundamentals, Processing, and Applications" by Mohammad Hossein Ehsani and Sanaz Alamdari reviews the fundamentals of structure, processing, and properties of biomaterials, and focuses on the new methodologies for developing or designing new biomaterials, especially the modern green synthesis techniques. The chapter also reviews recent developments in nanostructured biopolymers used for the design of biomaterials for delivery systems and tissue engineering. Advanced 3D bioprinting techniques for constructing innovative scaffolds in medicine and industry are also reviewed in this chapter. Chapter two, "Biomaterials Today and Tomorrow" by Esmat M. A. Hamzawy et al. discusses the recent advances in biomaterial research, and focuses on material design and fabrication. It shines light on some of the potential applications of biomaterials such as tissue engineering, orthopedic applications, central nervous system applications, regenerative medicine, and gene delivery. It also presents the challenges and future views to provide developments of novel biomaterials, 3D printing of organs, and drug delivery, Chapter three, "New Trends in the Design of Biomaterials for 3D and 4D Printing" by Rubén del Olmo et al. provides an overview of the state-of-the-art, current, and future perspectives in the field of biomaterials additive manufacturing (AM) with their actual technological limitations and biovalidation strategies, i.e., from the manufacturing process technology to the direct biomedical applications. Chapter four, "Application of Biomaterials for Tissue Engineering by Abu Hashem et al., focuses on the latest developments in biomaterials research and their applications in the field of tissue engineering, as well as their classification, properties, and their preparation techniques. Furthermore, the prospects, obstacles, potential solutions, and strategies for future developments are discussed in this chapter.

Chapter five "Progress in Multifunctional and Multimodal Biomaterials for Therapeutic Applications" by Sérgio R. S. Veloso et al. provides a discussion on the most recent and relevant progress in biomaterials for therapeutic applications, starting from hydrogels and liposomes to the combination of these materials with composites towards smart multifunctional materials. Chapter six "Inorganic Particles Incorporated into Electrospun Nanofibers: Bioactivity and Antibacterial Properties" by Verônica Ribeiro dos Santos Borges et al. presents an overview of obtaining nanofibers by the electrospinning technique, and exploring the methods for overcoming the biological limitations of nanofibers by incorporation of inorganic particles such as bioactive glass, glass ceramics, calcium phosphate, and metal nanoparticles into polymeric nanofibers by encapsulation and electrospray deposition to promote bioactivity and antimicrobial activity. Chapter seven "Selected VI-Element Based Glasses, Glass-Ceramics and Nanoparticles: Synthesis and Antibacterial/Bactericidal Activity" by Dariush Souri provides a review of tellurate transition metal oxide containing glasses (TMOGs) (especially ternary silvered telluro-vanadates), their corresponding glass ceramics (GCs), bare, shelled, and doped ZnSe NPs. The antibacterial/bactericidal potential of these materials which was detected by employing serial dilution and/or agar disk-diffusion\ bioassays are also discussed. Chapter eight "Bioactive Glass and Glass-Ceramics for Bone Healing and Medical Applications" by Patrick Munyensanga et al. provides an overview of current research on bioactive glasses, glassceramics, and related composites, including their fabrication techniques and processing methods. It also discusses the factors influencing the preparation of bioactive glass and glass-ceramics, highlighting how modifications of their composites contribute to the formation of micro- and macroscale structures.

Chapter nine, "Advances in Microbial-Based Electrochemical Biosensors for Heavy Metal Detection", by Ab Rahman Marlinda et al. focuses on the selection of appropriate mechanisms for heavy metal detoxification from microbial cells, as well as additional uses of microbial cells such as bacteria, yeast, and microalgae in biosensors for heavy metal detection. The advantages, limitations, and challenges of the biosensing components of biosensor technology which provides an invaluable and essential overview for individuals interested in microbial sensing technologies are discussed in this chapter too. Chapter ten, "Therapeutic Effects of Metallic Ions in the Physiological Environment: A Review on Recent Developments" by Shaher Bano Zaidi et al., summarizes the recent studies on ionic substitutions in bio ceramics (BCs) that are commonly found in human bodies. Furthermore, it discusses BCs for joint replacement, that are synthesized from different precursors and their effect on bone regeneration and remodeling. The challenges associated with the toxicity of inorganic ions and their therapeutic implications are discussed in detail. Recent advances in BCs doped with inorganic ions in bone tissue regeneration and associated therapeutic implications are highlighted. Finally, conclusions are drawn with future research perspectives and directions based on the identified challenges, and chapter eleven, "Advancements and Applications of Biomaterials in Medicine, Artificial Organs, and Biology" by Wardha Zafar et al., reviews and discusses some of the biomaterials currently in use or are under study for biomedical applications including their use in surgical devices across a variety of therapeutic areas. Moreover, the development of scaffolds for tissue engineering, tissue replacement, and augmentation of body tissue, providing essential strategies for better interventions in a wide range of medical conditions is also discussed.

Amman, Jordan Shadia J. Ikhmayies

Contents

Biomaterials: Fundamentals, Processing, and Applications

Mohammad Hossein Ehsani and Sanaz Alamdari

Abstract Recent advancements in the fields of material science and biology have allowed for significant progress in the production of novel biomaterials. In the following years, it is envisaged that the discovery of novel biocompatible materials and refinement of existing material and composition processes would widen the range of biomaterials applications. Biomaterials research to achieve superior qualities in physical and biological performance demands enhanced understanding across different disciplines and the creation of novel design techniques. Additive manufacturing, nanotechnology, and tissue engineering all benefit from biomaterials in commonly available natural items. This chapter focuses on the most current discoveries in biomaterials research and development, allowing readers to benefit from the most recent breakthroughs; Also, hydrogel-based 3D bio-printed scaffolds are mentioned with an emphasis on improved manufacturing processes from the standpoint of tissue engineering.

Keywords Biomaterials · Biopolymers · 3D bioprinting · Green synthesize · Biocompatible

1 Introduction

The first batch of bioactive glass implants (amorphous silicate-based materials) in rat femurs was produced over 40 years ago for bone repairs [1]. Six weeks later, the glass implants had formed an intense, live interfacial contact with the bone. With this discovery, researchers showed that unique materials can be produced and processed to have specific bioactive behaviors with tissues. This paradigm shift has benefitted millions of patients globally. The subject of biomaterials is constantly expanding

M. H. Ehsani (✉)
Faculty of Physics, Semnan University, Semnan, Iran
e-mail: ehsani@semnan.ac.ir

S. Alamdari
Department of Nanotechnology, Faculty of New Sciences and Technologies, Semnan University, Semnan, Iran

S. J. Ikhmayies (ed.), *Advances in Biomaterials Research*, Advances in Material Research and Technology, https://doi.org/10.1007/978-3-031-97481-6_1

because of an older population and rising average weight. Biomaterials are widely used to repair or replace a biological structure's shape and function to enhance human health and lifespan [2]. Recent advances in material science and biology have aided in the creation of novel biomaterials. Existing biomaterials will be refined, or new biocompatible materials will be developed in the coming years. To increase biomaterials' physical and biological performance, researchers must expand their knowledge of different fields and establish novel design approaches. A biomaterial is any material used to produce devices that safely, reliably, economically, and biologically replace human parts or functions. It is a natural or synthetic material that replaces or functions near living tissue. They make a biomolecular and spatial environment that is important for cell growth and vascularization that promotes human health by restoring the function of natural living tissues and organs in the body. According to Fig. 1, biomaterials can be categorized into synthetic, natural, and semisynthetic [3]. Natural biomaterials are protein and polysaccharide-based materials that exhibit bioactivity, biocompatibility, tunable degradation, mechanical kinetics, and structural similarity to biological tissue. They also aid in cell adhesion, proliferation, differentiation, and function. Natural biopolymers are frequently treated in aqueous solutions. These biomaterials are mechanically weak and have variable compositions. Synthetic biomaterials such as polylactic acid (PLA) and copolymers such as poly (lactic-co-glycolic acid) (PLGA) replicate human extracellular matrix systems that modify the functionalities of the biomaterials. Changing the lactide/glycolide ratio and polymerization conditions can alter these materials' tensile strength, mechanical modulus, and degradation characteristics [3, 4].

Lack of physiologically functioning domains prevents immune response. Synthetic biomaterials have well-defined compositions and customizable mechanical, and chemical features. In contrast, natural biomaterials have a less well-defined composition, but exhibit intrinsic bioactivity due to natural extracellular matrix motifs. As a result, natural biomaterials that have been adequately chosen and processed tend to be more biocompatible than synthetic equivalents. Semisynthetic biomaterials were a new type of materials that was introduced. (Natural polymers are combined with synthetic polymers to produce them.) Synthetic polymers are generally combined with natural polymers' ideal mechanical and thermal properties to produce a frequently biodegradable material and as biocompatible as natural polymers. In this way, semisynthetic polymers can outperform natural and synthetic carriers in terms of performance. Therefore, a growing trend in recent years has been to blend natural and synthetic polymers to improve particle qualities [4]. Since natural biomaterials are recognized by cells as having a higher biological affinity, they have stronger interactions with cells and so function better in biological systems.

The use of nanoscale materials or devices may open up new avenues for biomaterials application. Nanotechnology may aid in developing medications with more exact dosages, aimed at particular targets, or that have timed interactions with other substances. This book chapter highlights the most recent research and practical applications in new biomaterials for industrial applications. The essential aspects of structure, processing, characteristics, use, and methodologies for developing or designing new biomaterials, notably recently green synthesis technologies, are given in sections

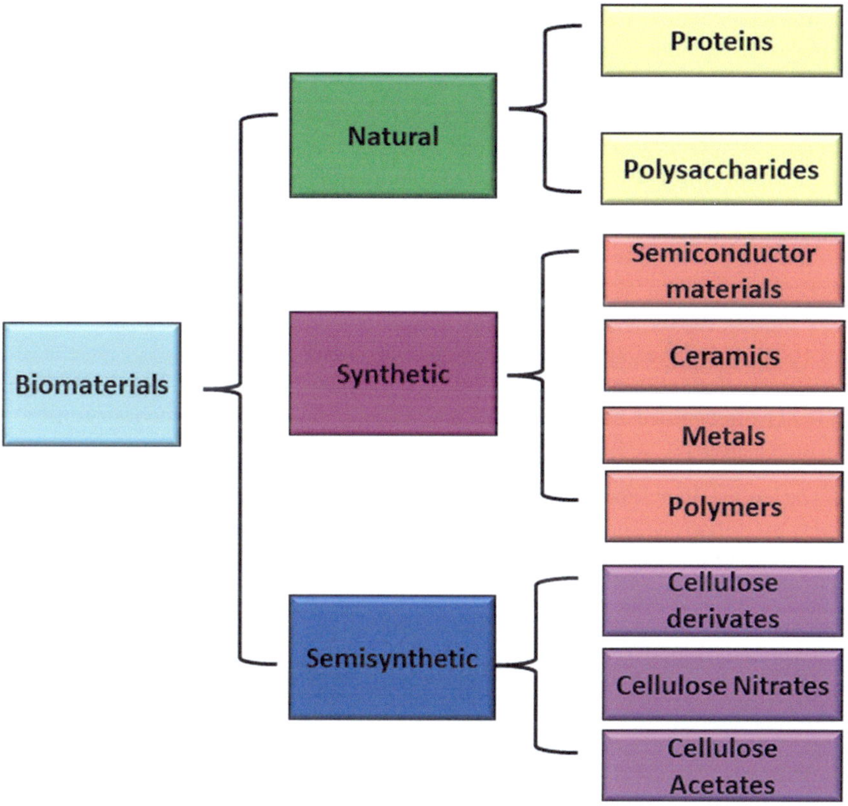

Fig. 1 Classification of biomaterials

one to five. Then, an overview of recent advancements and related research in nanostructured biopolymers will be given to design biomaterials for delivery systems and tissue engineering. An analysis of advanced extrusion 3D bioprinting techniques to construct innovative scaffolds in medicine and industry, and the possibilities of novel approaches to fabricate heterogeneous functional structures will be reviewed at the end of the chapter.

2 Fundamentals of Biomaterials and Biocompatibility

Biomaterials are employed in the human body as heart valves, blood vessel stents, shoulder, knee, hip, elbow, ear, and dental implant. On the other side, as people live longer, the need for reconstructive surgery and longer-lasting implants has increased. Medical equipment, such as menstruation cups, stents, and ventilators, are made from biocompatible materials. Immunity is the body's defense mechanism against pathogens, including viruses, parasites, bacteria, and fungi that can infect the body. Using a non-biocompatible substance in a healthcare application may cause this type of reaction. A hazardous reaction can occur in the short-term, long-term, or when mixed with other elements in the human body. Therefore, biocompatibility testing is essential for all medical equipment, regardless of classification. Three reasons make the biocompatibility of biomaterials a very important thing to think about:

1. This is because these materials stay close to human tissues for a long time and can't be removed by the person who has them.
2. Biocompatibility is the ability of a material to perform its intended function without causing local or systemic side effects in the recipient of the material. This is a complex process, which depends on many factors.
3. The biocompatibility of fixed prosthodontic materials, such as ceramics is typically disregarded because many practitioners feel that its biocompatibility is unquestionable if the material is on the market.

The United States Food and Drug Administration (FDA) must authorize all biomaterials used in vivo. The FDA may provide a Premarket Approval (PMA) for using a substance if it is substantially identical to one that was in use before the FDA Act of 1976. It must undergo a set of "biocompatibility" testing to be considered safe for human consumption [5]. The term "biocompatibility" refers to a material's capacity to work well with a certain host reaction in a given application (see Fig. 2). For example:

1. The material must not be toxic and carcinogenic.
2. Must not leach or release soluble components into a living system.
3. It is better to have the right mechanical strength and density.

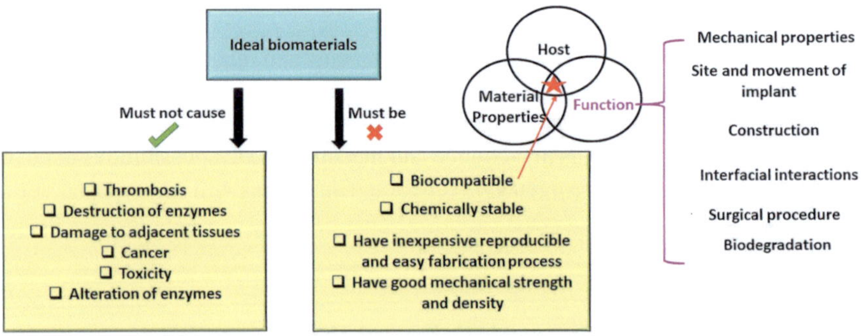

Fig. 2 The features of ideal biomaterials and factors that influence the functional biocompatibility of an implantable device

4. Used material must be biocompatible and chemically stable.
5. Have a relatively inexpensive, reproducible, and easy fabrication process on a large scale.

Biocompatibility standards often are Acute systemic toxicity, Cytotoxicity, Hemolysis, Intravenous toxicity, Mutagenicity, Oral toxicity, Pyrogenicity, and Sensitization. Briefly, as a biomaterial, it must be biocompatible, which means it must not cause an allergic response when used. The material must also have superior mechanical qualities, osseointegration, corrosion resistance, and wear resistance. Implant materials must be non-toxic and produce no inflammatory or allergic responses in the human body. The body's response to an implant determines its biocompatibility [6]. To determine a material's biocompatibility, it is important to consider the host reaction to the material and its degradation in its natural environment. Meet the functional needs of biocompatible material uses are:

1. Be able to sustain a load for months when used to immobilize a fracture, decrease water loss when used to treat a wound, and have an appropriate refractive index when used as a vitreous replacement in the eye.
2. Get a good host response. If desired, this allows for total inertness and a precise biological activity that benefits the receiver. The host response is the tissue's reaction to the implant, which governs the patient's physiological performance after implantation and is governed by the material's properties, particularly its anatomical chemical stability.
3. The application location must also be considered. Both functional and physiological factors influence the biocompatibility requirements of the material.

So, searching for new materials that can last a lifetime without failure or revision surgery with an extended lifespan, good corrosion resistance in body environment, high strength and low Young's modulus, high flexibility, great biocompatibility, and no cytotoxicity is required. Table 1 shows 14 famous biocompatible biomaterials. Additional criteria for biocompatibility certification include where a substance is used, how long it is exposed for, and whether or not it is combined with other substances. For instance, some biocompatible silicon is only acceptable for short-term use. The materials mentioned above cannot be presumed biocompatible without further testing; hence, they cannot be utilized in all medical applications.

F. Campos et al. reported a new delivery method using tissue-like biomaterials that regulate magnetic-responsive nanoparticles (MagNP-OH) [6]. The biocompatibility and functioning of artificial tissues implanted subcutaneously in Wistar rats were compared to injected magnetic-responsive nanoparticles (NPs). Their results showed that MagNP-OH meets the biosafety and biocompatibility standards for potential clinical application, and can be included in fibrin-agarose biomaterials to produce new biomaterials with superior biomechanical characteristics, biosafety, and biocompatibility [6]. According to the transmission electron microscope (TEM) images in Fig. 3; to assess each biomaterial's ultrastructural biocompatibility, it was found that none of the research groups showed evidence of necrosis or cell changes and supported the materials' outstanding biocompatibility [6].

Table 1 Biocompatible materials

Biocompatible material	Composition
Alumina (a ceramic biomaterial)	Al_2O_3
Bioglass (ceramic biomaterial)	Composed of SiO_2, CaO, Na_2O, and P_2O_5
Cobalt-chromium alloy (metal)	CoCr
Hydroxyapatite (ceramic biomaterial)	$Ca_5(PO_4)_3(OH)$
Polyvinylchloride (PVC) (polymer)	$(C_2H_3Cl)_n$
Polyethylene (PE) (polymer)	$(C_2H_4)_n$
Polypropylene (PP) (polymer)	$(C_3H_6)_n$
Polytetrafluoroethylene (PTFE) (polymer)	$(C_2F_4)_n$
Polymethylmethacrylate (PMMA) (polymer)	$(C_5O_2H_8)_n$
Stainless steel	It is a mixture of iron, chromium, and in some cases nickel, and other metals
Trimethylcarbonate (polymer)	$C_4H_6O_3$
TMC NAD-lactide (polymer)	$CH_3[C_6H_8O_4]_m[C_4H_6O_3]_nCH_3$
Titanium and titanium alloys (metal)	Aluminum, vanadium, molybdenum, niobium, tantalum, zirconium, and manganese are among the metals that may be found in titanium alloys
Zirconia (ceramic biomaterial)	ZrO_2

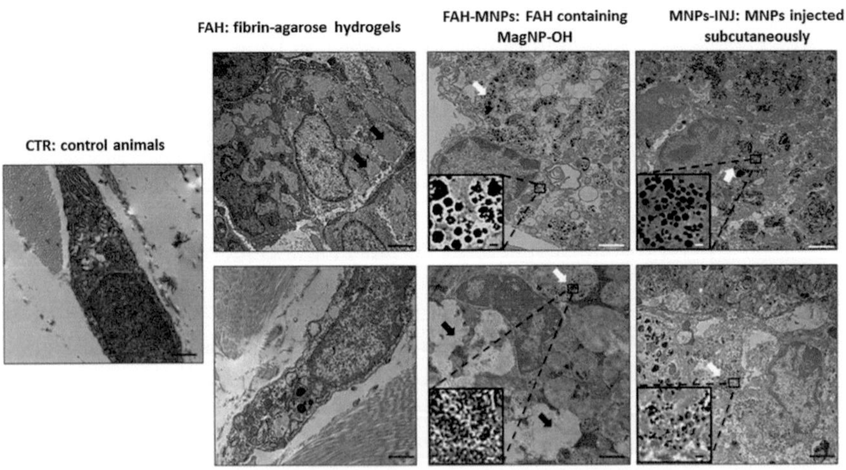

Fig. 3 TEM images of ultrastructural analysis of materials grafted in vivo for 12 weeks. Abbreviations in the figure: CTR: control animals; FAH: fibrin-agarose hydrogels; FAH-MNPs: FAH containing MagNP-OH; MNPs-INJ: MNPs injected subcutaneously. Black arrow: phagosome containing agarose; white arrow: MNPs. The inserted images correspond to higher magnifications of the same images. Scale bars: 300 μm (large images) and 20 μm (inserts). Reprinted from Campos et al. [6], with permission from Elsevier. Copyright (2021)

To utilize them as medical devices, they must be biocompatible. This is determined by the intended usage, location, and interaction length with tissues. The technique used to assess biocompatibility is equally important. So, clinicians must know these techniques. In vitro, animal, and clinical testing determine biocompatibility [7].

3 Processing and Applications of Biomaterials in the Food and Medicine Industry

Many pharmacological, industrial, and medical uses for biomaterials exist, as indicated in Fig. 4. Biomaterials in various types (ceramic, polymer, composite, natural, and metal) have a great place in medical applications. For example, magnetic nanoparticles such as Au-Fe_3O_4, Fe_3O_4, and their core shelled composites with a superparamagnetic behavior are the most promising candidates in cancer therapy using hyperthermia [8, 9].

Scaffolds are vital in tissue engineering because they facilitate cell proliferation and differentiation. The scaffold's design also determines the final organ's shape. Numerous in vitro and in vivo investigations have examined the biocompatibility of 3D printed scaffolds and their interactions with cells [7]. So, in tissue engineering, a 3D matrix, or "scaffold," must be very porous, so that cells can move through it and guide their growth and regeneration in 3D structures. This is especially important in the field of bone tissue engineering and regeneration, because the bone is a 3D composite structure that is very complex. It also changes how bone tissue looks depending on where it is in the body. In other words, the type of configuration and the

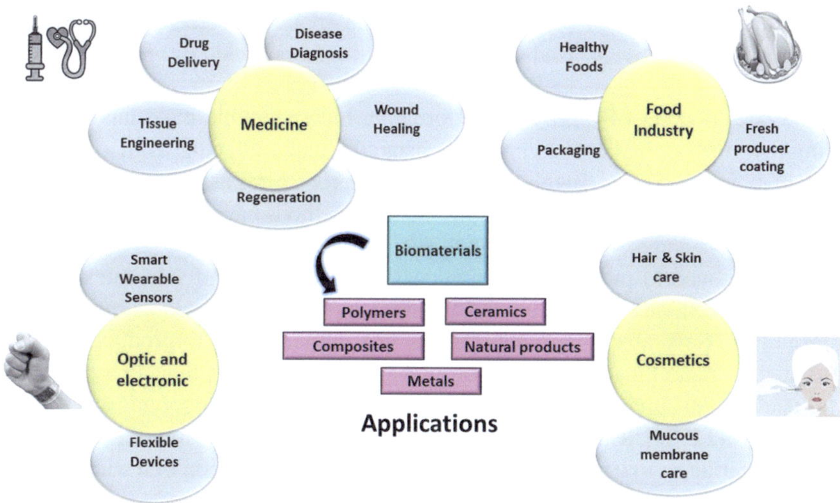

Fig. 4 The applications of biomaterials in various fields

type of biomaterial that should be used depend on what kind of site they will be used for, what kind of mechanical loads there are, and how quickly they should be incorporated. Ideally, the scaffold should be porous enough to allow cells to penetrate, tissues to grow, rapid vascular invasion, and nutrients to get to them. Besides that, the matrix should be designed to help new bones form anatomically correct shapes. Its degradation kinetics should be so that the biodegradable scaffold retains its physical properties (e.g., strength) for at least 6 months [1, 3]. Implants, biomedical devices, and tissue engineering scaffolds rely heavily on biomaterials to cure illnesses, restore body functions, build healthy tissues, or regenerate sick ones [10]. Also, in cosmetics, plastic surgery and its methodologies have received several developments and applications in recent years, with associated plastic cosmetic biomaterials and the use of new technology emerging at a rapid pace. As people's living conditions have improved and their desire for "beauty" has risen, the safety of plastic cosmetic materials has been given increasing attention [11]. Biomaterials used in plastic surgery must meet the following criteria to be approved for use [12, 13]; For example, they should not have the following characteristics:

1. Allergic reaction.
2. Pyrogenic source.
3. Carcinogenic or teratogenic effect.
4. Microbial survival matrix.
5. Definitive ability to integrate with the host material.
6. Inflammatory or foreign body reaction.
7. Antigenic or tissue-borne diseases.

Mucilage derived from chia seed (MCHs) has recently been found to have high cytocompatibility on the L929 fibroblast cell line, as well as potential mucoadhesive properties for use in the food, cosmetic, and pharmaceutical sectors [14]. Snail slime, a potential component in cosmetics as an antioxidant, sunscreen, and tyrosinase inhibitor, was effectively used to produce gold nanoparticles in another research [15].

The development of innovative, affordable, biocompatible materials is critical to improve the population's living circumstances, and wellbeing as the number of implant recipients grows. In this perspective, the procedures used to produce biomaterials must be cheap, quick, and simple. Several methods have been used to produce novel bioactive and biocompatible materials [16]. For example, sol–gel methods are widely employed to produce bioactive coatings, powders, and substrates that can modulate protein and cell incorporation and behavior and be used as implants and sensors [17]. Several studies have used the sol–gel technique to produce biomaterials such as nano bioactive glass, porous bioactive glass, and bioactive glass [18–21]. The sol–gel approach offered nanoscale combinations of inorganic and organic compounds [22–24]. This simple method allows for the production of homogenous hybrid materials at low temperatures, allowing for the integration of a wide range of chemicals [23].

Due to its tunable mechanical characteristics and shape processing, Poly Lactic Acid (PLA) coated Poly glycolide acid (PGA) is ideal for building cartilage with a complicated shape in pioneering tissue engineering experiments from the past. On the other side, patients' microtia ears have been suggested as a suitable cell source, because of the ability of microtia chondrocytes to produce elastic cartilage, and because they can be removed without harming good cartilage [24]. To produce patient-specific ear cartilage in vitro, enlarged microtia chondrocytes, biodegradable compound scaffold, and in vitro culture technique were used [24, 25]. Given the well-established in vitro engineering technique and the appropriate surgical procedure, the results mark a substantial milestone in the clinical translation of tissue-engineered human ear-shaped cartilage [25] (see Fig. 5).

Biomaterials' potential applications as food stabilizers, pathogen control agents, sensors, and edible packaging materials are noticeable. Today COVID-19 causes a lot of infections and deaths because it spreads so quickly. During this crisis, it's important to use healthy drugs and fruits. To keep perishable fruits fresh and beneficial for a longer time, antibacterial nanostructured coatings that are safe, environmentally friendly, durable, and edible are needed. Most of the food and vegetables we eat daily can become contaminated with microbes during processing, supply, and transportation. In these cases, viruses already in the environment can spread quickly through surfaces and spread the disease between people. While developing biomaterial-based food packaging, maintaining food safety and quality is crucial. It extends the shelf life of the food while also providing protection and quality control

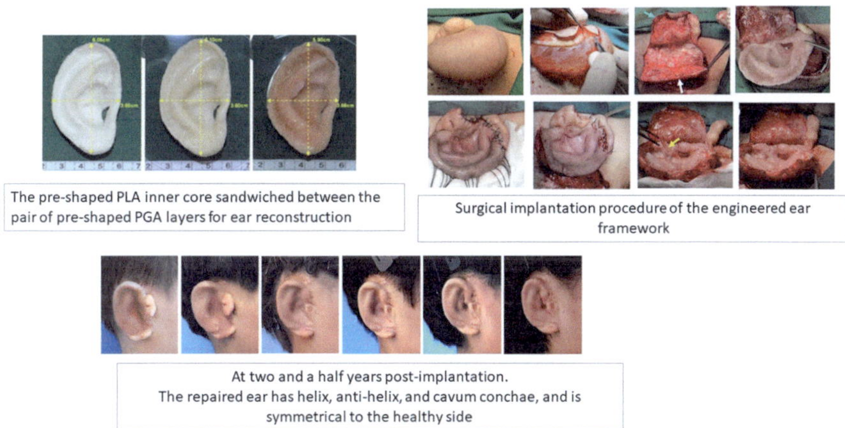

Fig. 5 Examples of the materials used in plastic and cosmetic surgery; Left: Cross view of chimeric, engineered cartilage with outer semipermeable cartilage with inner opaque Polycaprolactone (PCL) lattices and opaque tissue between the PCL bar; Right: Surgical implantation procedure of the engineered ear framework; After 12 weeks of dermal expansion, abundant capillaries and thin kidney-shaped skin capsules were formed. After removing the expander, the expanded skin flap and retro auricular neurovascular fascia flap were dissected, respectively; Down: Operated sample shows peanut-like structures; Two and a half years after transplantation, the reconstructed ear shows typical auricular features with helix, antihelix, and concha space, roughly symmetrical to the healthy side. Reprinted from Zhou et al. [25], under the terms and conditions of the Creative Commons (CCBY-NC-NDlicense) license. http://creativecommons.org/licenses/by-nc-nd/4.0/

under international standards. Nano-biotechnology helps to build greatly improved packaging by producing multifunctional biomaterials for prospective applications that give the customers more brilliant materials. Nano-biocomposites may aid in developing innovative food packaging materials with an improved barrier, mechanical strength, antibacterial, and antioxidant capabilities. Active bioplastics now used in the food packaging industry include starch-derived bioplastics, polylactic acid, and polyhydroxybutyrate [26]. Biomaterials with innovative functions are gaining popularity over plastic and paper carry bags, films, and composites [27]. Consumers now only want sophisticated, active bio-derived components to ensure food quality. Biomaterials have several features that make them valid for manufacturing high-end food items [28].

Moisture/gas barrier, solubility in water or lipid, pH dependence, physicochemical qualities, and non-toxicity are all required for biodegradable or edible foods. These features rely on the biomaterial, derivatization, and processing. Various plasticizers, property boosters, texturing agents, and cross-linking agents are added during biocomposite synthesis to improve certain qualities. The development of bio-derived products relied on polymer cohesion and adherence to surfaces. Cohesion/adhesion affects product properties, including resistance, permeability, porosity, and flexibility. It enhances porosity while reducing gas and solute barrier characteristics. The excessive solvent evaporation, chilling, or other pretreatments required for industrial-scale preparation may cause premature immobilization of biocomposites. Polysaccharides and polypeptides are gaining popularity due to their unique physicochemical and biological properties. The two kinds of biomaterials are utilized in the food business. The advancements in biomaterials enable faster pathogen diagnosis and good barrier protection while designing food packaging. Biomaterials would increase physiochemical properties, barrier stability, and may be biodegradability. Nano-biocomposites will have a significant influence on food packaging in the future. Due to their structural diversity, polyhydroxyalkanoates (PHAs) and natural waxes are widely used in packaging [29]. Starch polymers also have limited mechanical characteristics and are very moisture sensitive.

Recently, the development of novel biomaterials to suppress microbial growth has been required to assure food quality and safety [29]. Antimicrobial thin films applied to a product's surface are being used to preserve foods. According to several reports, antibacterial coatings and films can significantly reduce the growth of hazardous germs [30]. Coma et al. found that adding these beneficial bacteria to the cheese surface inhibits the development of *L. monocytogenes* and *L. innocua* [31]. The antibacterial methylcellulose/chitosan (3:2) films were disclosed by Chen et al. [32]. In recent years, several natural antimicrobial thin-film polymers have been employed, including proteins such as soy and wheat gluten as well as polysaccharides such as chitosan, starch, Hydroxypropyl cellulose (HPC), and Carboxymethyl cellulose (CMC) [33].

Biosensors are compact devices designed to detect, record, and transmit information about biological processes [34, 35]. The shelf life of packaged foods is influenced by both intrinsic and extrinsic factors [36]. Intrinsic factors are inherent to the food itself, while extrinsic factors include storage temperature, relative humidity, and gas composition. Effective food packaging provides protection against air, water vapor, UV radiation, and chemical and microbial contamination.

Chitosan and gum Arabic were combined with non-toxic zinc oxide nanoparticles in a recent study to produce a film-forming coating for edibles that is safe to eat. After 7 days of storage at room temperature, the physicochemical attributes of avocados, including their look and brightness; weight; firmness; and lowering sugar content, dramatically improved after coating with the ZnO/chitosan/gum Arabic edible film (see Fig. 6) [37].

As it was mentioned before, chitosan is a non-toxic, biocompatible, biodegradable polymer with excellent intrinsic properties obtained by deacetylation of chitin. It is also safe and known as FDA-approved material extracted from shrimp, fungi, and crabs [38]. Chitosan is widely used in the medical, agricultural, textile, and wastewater treatment industries due to its uniqueness. Several antibacterial non-toxic inorganic solid compounds have been successfully used as additives in food-preserved coatings [39]. Mixing nanofillers such as TiO_2, ZnO, and Ag NPs into biopolymers enhances mechanical properties and speeds up other activities, including antibacterial activity, biosensing, and oxygen scavenging [39]. Among several non-toxic and antibacterial compounds that have been used as additives in packaging, ZnO

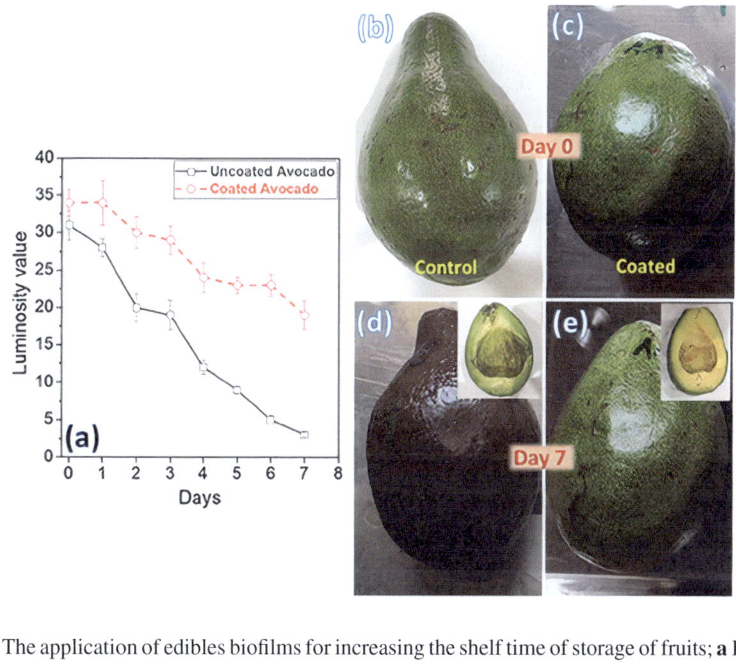

Fig. 6 The application of edibles biofilms for increasing the shelf time of storage of fruits; **a** Brightness parameters during storage of uncoated and coated avocado; Appearance of **b**, **d** uncoated and **c**, **e** coated fruit after 7 days of storage; Reprinted from Le et al. [37], with permission from Elsevier. Copyright (2021)

has been considered a safe agent and vital element for health in the food industry [4, 40–42].

In addition to its antibacterial capabilities, nanoscale ZnO is an excellent candidate for UV-protection because of its potential application for UV preservation (see Fig. 7). The long-term effects of UV exposure on ZnO nanostructured thin films have been explored exactly. Sol–gel-derived ZnO thin films were tested for UV-protection in our work, which involved exposing the films for different periods of time. The findings of the experiments show that the synthesized samples can be used in the medical area, food preserving with desirable stability, and UV-protection factor of +50.

Due to the worldwide outbreak of the coronavirus and the pollution generated by chemical plastic packaging, there is a growing recognition of the importance of environmentally friendly biofilms and antibacterial coatings. A biodegradable hybrid film was created by combining green-synthesized zinc oxide nanoparticles (ZnO NPs) with a chitosan (CS) matrix. The film was made utilizing a straightforward casting technique [4]. The ZnO nanoparticles (NPs) were produced using an extract derived from wild *Mentha pulegium*. The composite films had a transparent, smooth, and uniform surface. The flexible bio-based hybrid films showed notable antibacterial and antioxidant activities, emitted bright visible light at a wavelength of 480 nm, and had the ability to block UV rays. When stored at a temperature of 23 °C, the ZnO/CS films shown the ability to increase the shelf life of fruits by up to eight days (see Fig. 8). Additionally, these films effectively served as a barrier against oxygen and water. The ZnO/CS film, which is capable of being broken down by natural processes, is expected [4] (see Fig. 8).

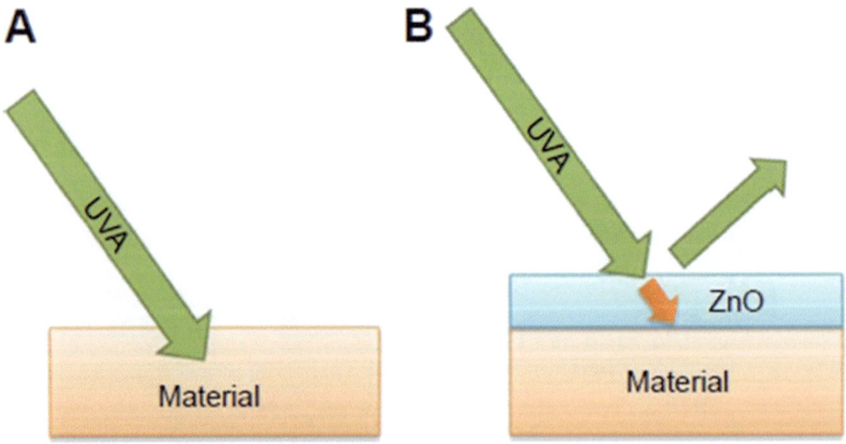

Fig. 7 The effect of coatings on UV shielding. Reproduced from Ghamsari et al. [41], under the terms of the Creative Commons attribution-noncommercial license (CC-BY-NC). https://creativecommons.org/licenses/by-nc/3.0/

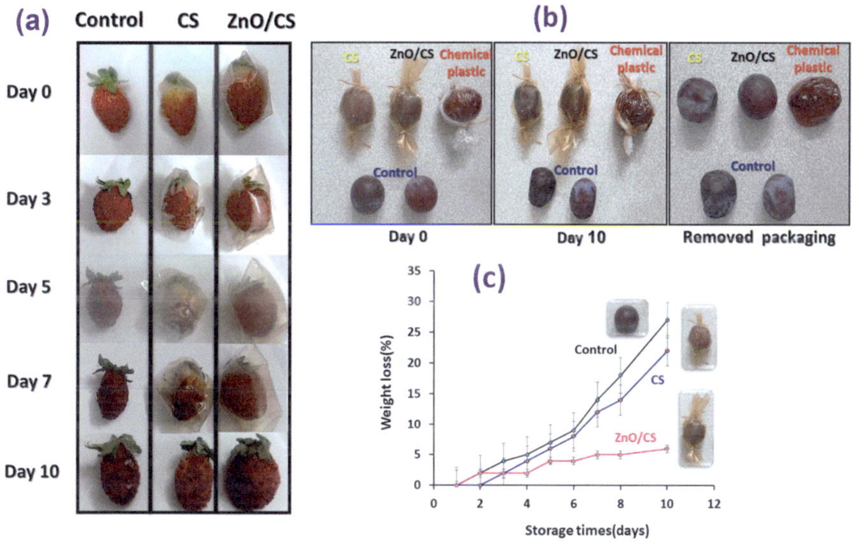

Fig. 8 Multifunctional food packaging film. Reprinted from Alamdari et al. [4], with permission from Elsevier. Copyright (2020)

Regarding electronics, the amount of information about bio-nanocomposites is continually growing. It may be possible to make new materials by combining carbon-based materials like graphene with biopolymers. A new type of electronic device can be made that is biodegradable, flexible, and easy to work with (see Figs. 8, 9, and 10) [4]. In some cases, these new devices can be easily connected to human tissues. There has been a lot of interest in developing materials with fluorescence or plasmonic properties that can be used for biosensors and implantable devices. Most of the time, the biopolymer matrix acts as passive support for the nanoparticles, which makes them biocompatible. However, some biopolymers have built-in optical properties that can be used with fillers to improve their properties or add new ones. Also, Au NPs and Au nanorods (NRs) were combined with biopolymers to make plasmonic/photonic hybrid films. In a recent study, photoluminescent and transparent biocomposite films based on chitosan/cellulose nanocrystals for anti-counterfeiting applications were synthesized by casting solvent (see Fig. 9) [43].

A graphene oxide and silk fiber solution were vacuum-filtered and then chemically reduced to produce graphene/silk composite electrodes in another investigation. Fibers showed a resistance of 2.2 M Ω cm^{-1} compared to neat silk-based composites, which is a substantial increase [44].

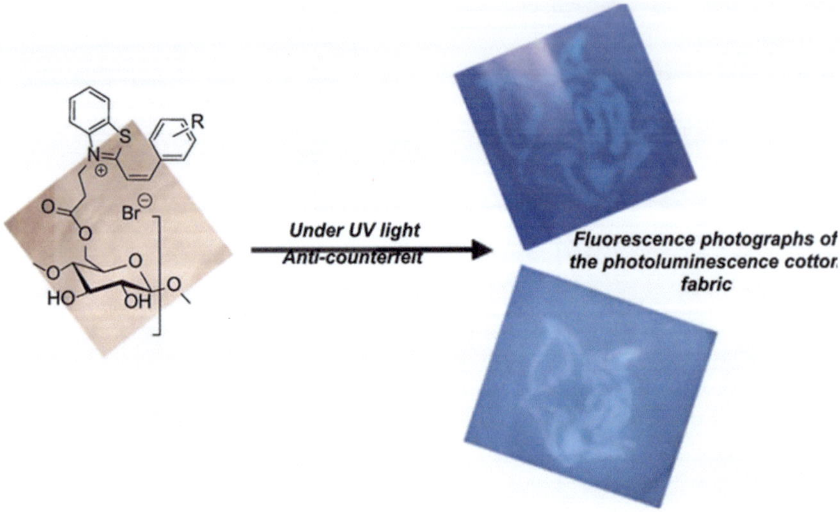

Fig. 9 Photoluminescent biocomposite films of chitosan-based on styrylbenzothiazolium-g-cellulose nanocrystal for anti-counterfeiting applications. Reprinted from Fardioui et al. [43], with permission from Elsevier. Copyright (2021)

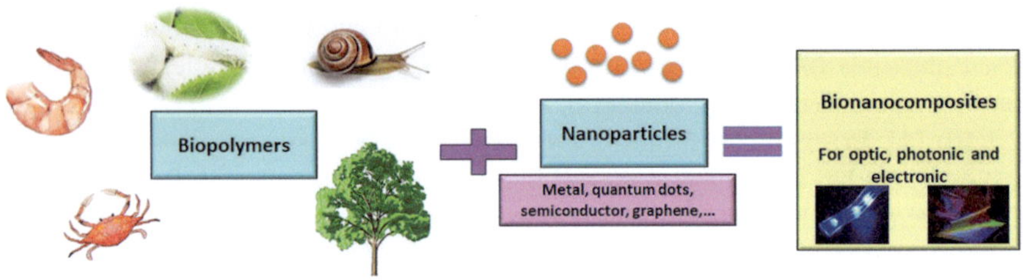

Fig. 10 The applications of biopolymers in electronic

4 Green Trends in Using Plant Extracts for the Synthesis of Biogenic Metal Nanoparticles Toward Environmental and Pharmaceutical Advances

Engineered tissues and organs may be produced by combining the concepts of physiologically active chemicals with scaffolds and other engineering techniques. Natural surfactants derived from plants are an excellent source for green synthesizing materials for various points. Environmentally friendly procedures and materials are used in green chemistry to decrease hazardous and harmful scientific research. Thus, it utilizes safe energy or reduces waste materials using a variety of experimental and industrial synthetic approaches. Environmental and human health benefits are the primary motivations for using eco-friendly practices and technology. Green synthesized biomaterials have attracted a lot of attention since they can be produced in huge quantities with the same physical, chemical, and structural qualities for a wide variety of tissue types, hence reducing the negative consequences of conventional techniques. Natural materials have been used to produce polymer-based biomaterials, such as polysaccharides and proteins, for tissue regeneration.

When developing tissue engineering, chitosan, an acceptable biocompatible and biodegradable polysaccharide with varied chemical activities, is a promising material. Collagen and hyaluronic acid (HA), protein-based biomaterials sourced from natural resources, have also been employed in various scaffolds for repairing or enhancement. Natural materials' biocompatibility is based on the body's ability to absorb them, and there are no long-term repercussions beyond the time of replacement. Trauma, injury, and disease-related bone deterioration have been a major issue in medical research for a long time. Tissue engineering has a distinct role in applied sciences because it has the potential to raise and alter the expectations of people. As an alternative, we'll look at current research using safe and environmentally friendly materials to help preserve or repair injured bone tissue to its pre-disaster state. Hydroxyapatite (HAP) is a naturally occurring calcium apatite with the chemical formula $Ca_{10}(PO_4)_6(OH)_2$ [45].

HAP powder is white. Human bone is made up of modified HAP (bone mineral) to the tune of 50% by volume and 70% by weight. Carbonated calcium-deficient HAP is a dental component. Another example of green chemistry is manufacturing nanostructured HAP from recovered eggshell biowaste. This technique uses eggshell biowaste for calcium and ammonium dihydrogen orthophosphate for phosphate. The approach uses commercially available calcium hydroxide and ammonium dihydrogen orthophosphate as calcium and phosphate supplies. For example, a bamboo fiber/nHAP/poly (lactic-co-glycolic acid) (PLGA) composite was produced by freeze-drying [46].

Moreover, in vitro studies using a comparable body system indicated that bamboo fiber scaffolds accelerated bamboo fiber nHAP/PLGA degradation and were cytocompatible. Because nHAP/PLGA has improved qualities in tissue engineering, bamboo fiber/nHAP/PLGA may be a contender for bone scaffold regeneration. Also, recently combining cellulose fibers with a polysiloxane matrix, we generated multifunctional antibacterial and UV protective cotton surfaces using green in situ biosynthesis of silver nanoparticles (AgNPs) using sumac leaf extract to reduce and stabilize the agent. It was a two-step procedure. First, the siloxane-treated cotton samples were submerged in an $AgNO_3$ solution. The results were outstanding with AgNPs and aromatic phenolic components of sumac leaf extract, along with UV-protection and antibacterial action (see Fig. 11) [47].

Novel alternatives to chemical techniques for synthesizing NPs have been presented in the literature, including various plant extracts. In the production of metal and metal oxide NPs, such as gold (Au), silver (Ag), and titanium dioxide (TiO_2) NPs, citrus fruit extracts or peels have proven effective. As mentioned, due to its unique features and use in a wide range of fields, including medicines, cosmetics, photonics, and photocatalysis, zinc oxide (ZnO) has attracted considerable attention

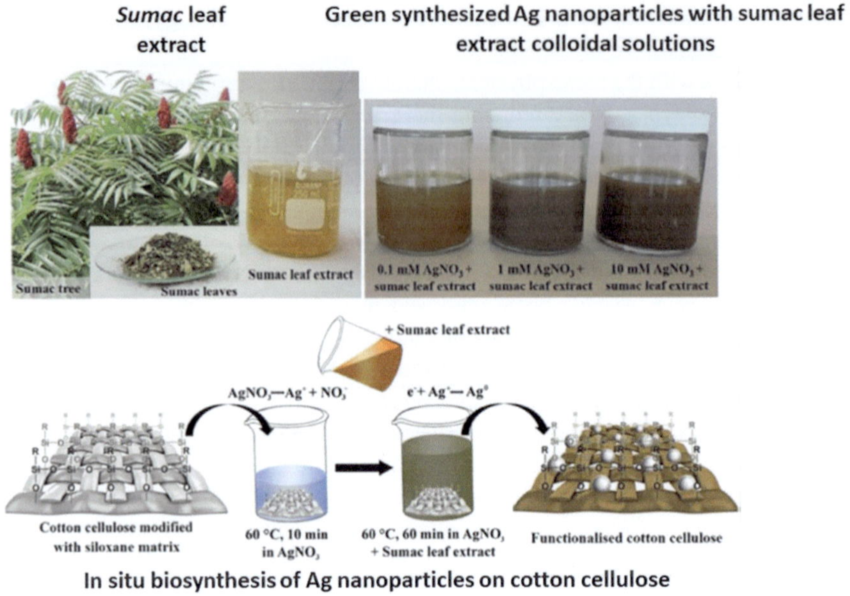

Fig. 11 Brief scheme of green-synthesized AgNPs/cotton cellulose. Reproduced from Štular et al. [47], under the terms of the Creative Commons CC-BY license. https://creativecommons.org/licenses/by/4.0/

from researchers across the world. Sol–gel techniques, co-precipitation, laser vaporization, microemulsion, and ball milling may all be used to produce ZnO nanoparticles (NPs), which can be used in many applications [48–50]. In a recent study, ZnO NPs were prepared successfully using *Sambucus ebulus (S. ebulus)*, leaf extract, and their physicochemical properties were investigated (see Fig. 12).

Green synthesized NPs exhibited outstanding UV absorption and significant yellow-orange emission at room temperature when exposed to UV irradiation. The prepared nanoparticles had strong antibacterial activity against various microorganisms and could photocatalytically degrade methylene blue dye pollution. The acquired results showed that biosynthesized ZnO NPs had fascinating properties that may be useful in various future applications. The *Sambucus ebulus* is a natural plant that thrives in most of South America, Asia, North Africa, and the United States [51]. *S. ebulus* leaf extract has many carbonyls, carboxyl, and hydroxyl groups in polyphenols and flavonoids, which help reduce metal ions and stabilize nanoparticles [51]. Because they may serve as antioxidants, anti-inflammatory, and anticancer agents, these substances can positively impact human health. Due to its ease of collection, natural origin, and outstanding biological capabilities, it may be a suitable substitute for the pricey current chemical surfactants.

As previously stated, *S. ebulus* includes a high concentration of natural antioxidants such as flavonoids, lectins, anthocyanins, vitamin C, and phenolic compounds, all of which may be used as stabilizing agents [51].

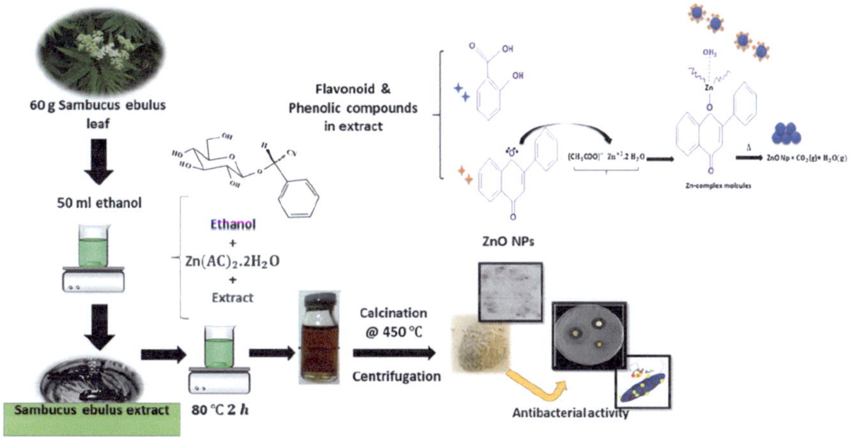

Fig. 12 Brief scheme of green synthesis of ZnO using Sambucus ebulus. Reproduced from Alamdari et al. [40] under the terms and conditions of the Creative Commons Attribution (CC BY) license. http://creativecommons.org/licenses/by/4.0/

The development of new green-synthesized biomaterials with sustainable, low-energy, or low-cost processes for the production of various tissues and health products, particularly in the medical, treatment, cosmetic, and food industries and the reduction of toxic material consumption are both critical. For example, new green components and composites, together with improvements and replacement of organs or tissues, would be a huge benefit to many kidney disease sufferers.

5 Advances in Nanostructured Biopolymer-Based Biomaterials

A new class of materials known as "biopolymers and bio-nano composites" has emerged due to recent advancements in nanotechnology and an increasing need for environmentally friendly products. Biopolymers are combined with synthetic components with at least one nanoscale dimension in these hybrid materials. A biopolymer can be classified as directly extracted from biomass, synthesized from bio-derived monomers, or generated by microorganisms (see Fig. 13) [52].

It is possible to make biopolymers from natural molecules such as polylactic acid (PLA), polyhydroxyalkanoates (PHA), silk, cellulose, chitosan, and synthetic compounds such as these. Biopolymers have several advantages over petroleum-based plastics, including their ability to degrade naturally and interact with nanofillers and multiscale assemblies. Petroleum-based plastic harms our health and the health of the environment in which we live. It can devastate natural environments. It contaminates the groundwater we drink and irrigate the crops we eat. Barnacles, tubeworms,

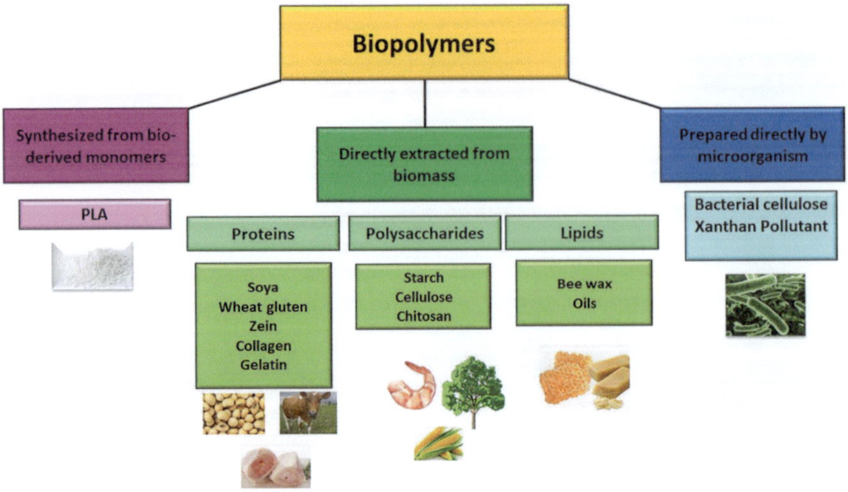

Fig. 13 Classification of biopolymer types

and algae are examples of invasive organisms that they may carry. Worst of all, it may take up to 1,000 years for the plastic to decompose, and even when it does, it has no constructive effect on the ecosystem. Due to the non-biodegradable nature of petroleum-based plastic, environmental contamination produced by traditional plastic packaging has lately become increasingly severe. Searching for a multifunctional alternative to reduce conventional polymers in the plastics sector has become a top priority. There has been a pressing need to focus on studying and developing innovative bio-based materials that are biocompatible and biodegradable to replace petroleum-based plastics.

Polysaccharides and polypeptides are two types of famous bioagents for producing biopolymers. Polysaccharides include alginate, cellulose, starch, glycogen, and chitin. Bacteria and fungi also produce additional biopolymers. Some polysaccharides, including alginates and cellulose, can be produced by microorganisms and employed in bio-packaging formulations due to their gel-forming capacity.

There are several polysaccharides having thin-film forming capabilities [53, 54]. In general, polysaccharide-based films have high gas permeability and low moisture barrier capabilities due to their hydrophilic features and polymer chain structure [53]. It can make biodegradable products from natural polymers. Polypeptides used in food include soy, maize, gluten, casein, collagen, and gelatin [54]. The major determinants of degradation are physiochemical and biological characteristics. Packaging, biodegradable bags, carry bags, and other molded food products may be made from these starch films. Cassava, pectin, lignin, hemicellulose, and waxy components make up natural lignocellulosic fibers [55]. The unique functional characteristics of cellulose give several specific aspects during derivatization [56]. As the most common polysaccharide on Earth, cellulose possesses a wide range of beneficial Physicochemical qualities, including cheap cost, wide availability, rich surface chemistry, and great biodegradability and biocompatibility well as strong thermal and mechanical durability [57, 58]. It has a broad range of uses in textile, packaging, paper, and lightweight composites because of its unique qualities. A variety of cellulosic materials have been used as an aid medium to reduce or stabilize silver nanoparticles, including nanoporous cellulose gel, microfibrils of cellulose, and bacterial cellulose. It is possible to combine cellulose with various inorganic fillers, metals, and oxides to enhance the functional properties of the finished product because of the presence of hydrophilic hydroxyl groups and active sites [59]. Chitin and chitosan are other famous biopolymers mostly found in crustaceans and fungi. Chitosan is a safe food ingredient for human consumption. The rate of chitosan degradation is heavily influenced by crystal structure and acetylation. A wide spectrum of biomaterials with outstanding physicochemical characteristics may be made from it. Gums are polysaccharides that can gel in liquids depending on pH, charge density, and counterion. Glycine alginic acid is an anionic polymer discovered in a kind of brown algae [56–59]. Alginate is a glycosidically linked binary linear copolymer of a-guluronic acid and b-mannuronic acid. This polysaccharide has been extensively studied for its biodegradability [59]. Many animal and plant proteins are used as biopolymers. Water-soluble albumin and salt soluble globulin may be isolated from wheat proteins during gluten processing. Gliadin acts as a filler that dilutes glutenin interactions [60]. Soybeans are used to make edible oils and other items [61]. Soy protein helps emulsification, foaming, adhesiveness, solubility, cohesion, and dough formation. They are more elastic, transparent, and homogeneous than other protein-based films. Like cellulose in plants, collagen can be used to make bioplastics [62]. Casein is a milk protein that is widely utilized in food production. Whey proteins are another protein derived from milk serum by changing the pH during casein processing [63]. Whey proteins can control oxygen, moisture, lipid, carbon dioxide, taste, aroma transfer, and biodegradability [64]. Other proteins such as elastin and keratin have also been employed as biopolymers in various applications [65]. Peanut, pistachio, sunflower, and barley proteins are among the many plant proteins utilized in food production. When used in edible food packaging, they can cause moderate allergic responses and other health issues. Pretreatments with physical and thermal agents can reduce allergenicity [66]. As a result of their distinctive structural characteristics, fascinating physical-chemical properties, and many potential uses, cellulose and chitosan are two of the most researched natural biopolymers. Combining these and other nanoparticles may produce a broad range of fascinating functional nanocomposites with unique features. Recent research has focused on cellulose-based inorganics, such as cellulose/gold, cellulose/Ag, and cellulose/calcium phosphate composites, for diverse uses [67]. On the other side, silver, zinc, copper, and some other inorganic metals and semiconductors have antimicrobial properties with minimal toxicity [68]. It has been discovered that silver nanoparticles (Ag NPs) have antibacterial properties due to their ability to affect deoxyribonucleic acid (DNA) replication and the breakdown of the proton driving force across cell membranes, among other factors. In addition, it does not damage human cells. Cellulose molecules help assemble and stabilize metal oxide nanoparticles. Cellulose is said to help metal oxides be more stable and function better, allowing them to be used in more applications. The cellulose fibers have several surfaces OH groups and important porosity features, making them ideal cellulose/silver composite models.

Composites of cellulose/silver cellulose fibers combine the advantages of both materials. Silver nanoparticle dispersion over cellulose fibers is problematic because of the aggregation of silver nanomaterials. The composites' performance was lowered because of this aggregation method. For the production or immobilization of silver nanoparticles on cellulose, hydrolysis, and sonochemistry have been devised [69]. In one or two processes, metal oxides may be added to cellulose to produce monodisperse nanoparticles and then connect them to cellulose fibers, which is the best way to do it. When silver nanoparticles are combined with cellulosic materials and electrostatically bound, this approach helps to produce antimicrobial nanoparticle materials that are all of the same sizes. Because of its simplicity, convenience, and high yield, in situ formation is superior to alternative preparation procedures. There are various ways to produce medicinal biofilms, as shown in Fig. 14. Intermolecular interactions between biopolymers and solution components are at the heart of the physicochemical process of film formation [52]. Early identification of polymer type is crucial to the film formation process because the biopolymer's ability produces a continuous crystal or amorphous structure [52]. New hydrophilic hydrogen bonds and Van der Waals forces are formed when polysaccharides undergo electrolytic and ion cross-linking [52]. Conformational denaturation, electrostatic charges, and amphiphilicity distinguish protein-based biopolymers from other film-forming materials [52]. Protein films are produced by introducing molecular connections between peptide chains, such as ionic and hydrogen bonds, Van der Waals forces, hydrophobic interactions, covalent polar and nonpolar bonds-disulfide bridges, and crosslinking, into the protein. It involves Van der Waals and electrostatic forces, hydrophobic and hydrogen bonds, excluded volume effects, or chemical binding, as in the Maillard reaction involving amino acids and sugars such as fructose, glucose, etc. [70]. The biopolymer is dissolved, dispersed, and emulsified in a solvent to form films and coatings. Water, ethanol, related mixes, and organic acids are used as solvents for food films and coatings. The dry procedure involves heating a film-forming compound with minimal solvent concentration to make a film [71]. Both procedures use a single biopolymer or a mix of biopolymers [72]. The dry process includes extrusion, injection molding, compression molding, co-extrusion, blown-film extrusion, and others [73, 74]. Electrospinning is a useful method for generating fibers from polymer solutions. The approach involves stretching a polymer droplet with a certain voltage. Electrostatic forces shape and stretch a charged solution jet to produce a regular fiber. Cellulose and its derivatives are beneficial in creating hydrogels and polymer-based composite materials for 3D printing. 3D printing includes stacking raw materials to produce goods directly from a 3D digital model.

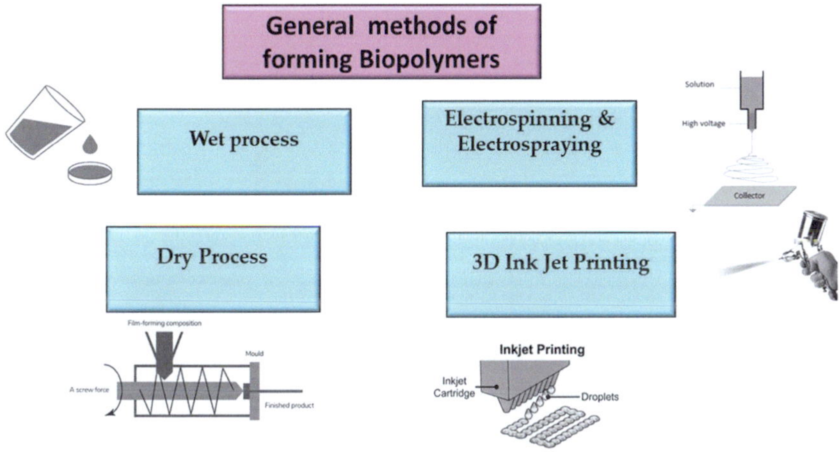

Fig. 14 General film formation routes of biopolymers

6 Recent Progress in 3D Bioprinting of Biomaterials

Tissue engineering (TE) is an interdisciplinary field involving biological science and materials engineering concepts to heal, maintain, and improve tissue function. The cell scaffold matures into a functional construct by extracting cells from a patient and seeding onto or integrating them into a tissue scaffold [75]. It may be inserted inside the patient to aid in tissue repair and healing. Tissue scaffolds are often designed as functional constructions based on their composition. Three-dimensional (3D) bioprinting has recently gained prominence among other approaches for generating tissue scaffolds to bridge the gap between designed and natural tissues. Its worldwide market, estimated at $ 487 million in 2014 [76], $ 1.76 billion in 2022, and is expected to grow at a CAGR of 13.87% from 2023 to 2030 [76]. 3D bioinks (mostly biomaterials, live cells, and/or bioactive chemicals) are printed as dynamic structures with scaffold functionalities to regenerate target tissues [76, 77]. Current 3D bioprinting technologies for engineering functional human tissues and organs that recapitulate their native prototypes may be classified into four major approaches: droplet-based, extrusion-based, laser-induced forward transfer, and stereolithography bioprinting. Extrusion-based bioprinting (EBB) is a technology that uses continuous strands or fibers of biomaterials to build 3D scaffold structures layer-by-layer [78]. Despite new bioprinting techniques being developed, EBB remains the most widely used method in research and industry to manufacture 3D cell-laden scaffolds because of its low cost, accessibility, and ability to reproduce tissue complexity [78, 79]. Since hydrogels are well-known for their composition, their structure is comparable to many human tissues' extracellular matrix (ECM). They can be easily made utilizing relatively moderate settings and aqueous chemistries. They represent an ideal environment for scaffold construction [78–82]. The dispensing head can deposit bioink onto its printing stage through pneumatic, piston- or screw-based processes for EBB

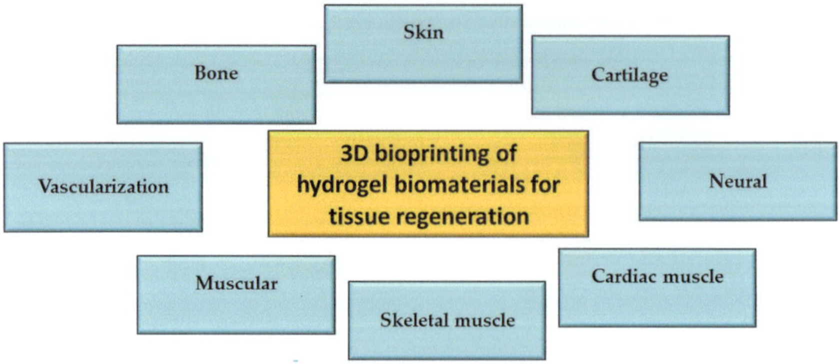

Fig. 15 Applications of 3D bioprinting in tissue regeneration

systems. For TE, modern EBB manufacturing techniques have many applications [78–82] (see Fig. 15).

One of the body's essential organs is the skin that protects the body from viruses and irritants, as well as UV damage and any other hazardous substances that may enter the skin. The fact that it is in close contact with the outside world renders it vulnerable to many ailments. Researchers have been producing a variety of wound dressings and natural product-based skin replacements dependent on the size, extent, and depth of the wound [83]. To be considered ideal, a bioprinted skin must possess several distinct properties, including the ability to transfer nutrients and eliminate wound exudates, biocompatibility, desired mechanical properties, proper surface chemistry, and a high porosity network with interconnected pores. Human plasma and original skin biopsies were used in a study to produce fibrin-based bilayer dermal constructions. Studies conducted in the lab and on animals showed that the 3D-bioprinted skin constructions were strikingly identical to natural human skin [82] (see Fig. 16). The combination of EBB and electrospinning in TE of bone has also been researched following the hybrid bioprinting methodologies. To produce a 3D composite scaffold, PCL/gelatin distributed nanofibers were infused into the meshes of the PCL construct [82–85]. When combined with electrospinning equipment, a 3D printing method produced a scaffold with vertical strands and a thin nanofiber sheet, which was then used to insert cells. Using the cell-laden alginate struts, an even distribution of cells was achieved across the nanofiber layers [86].

TE has made significant progress in replicating the anatomy of cartilage tissue using 3D bioprinting. Chondrocytes' low proliferation rate and cartilage's inability to self-repair owing to avascularity and its functional and structural complexity may induce cartilage loss as a result of trauma or infection. The zonal complexity of natural tissue, encompassing zones with various cell morphologies and arrangements, ECM arrangements, components, and distribution, makes the production of functional articular cartilage difficult. When it comes to imitating articular cartilage tissue, 3D bioprinting has made significant progress, and EBB remains the most common and cost-effective way of dispensing. A novel bio-cartilage was reported

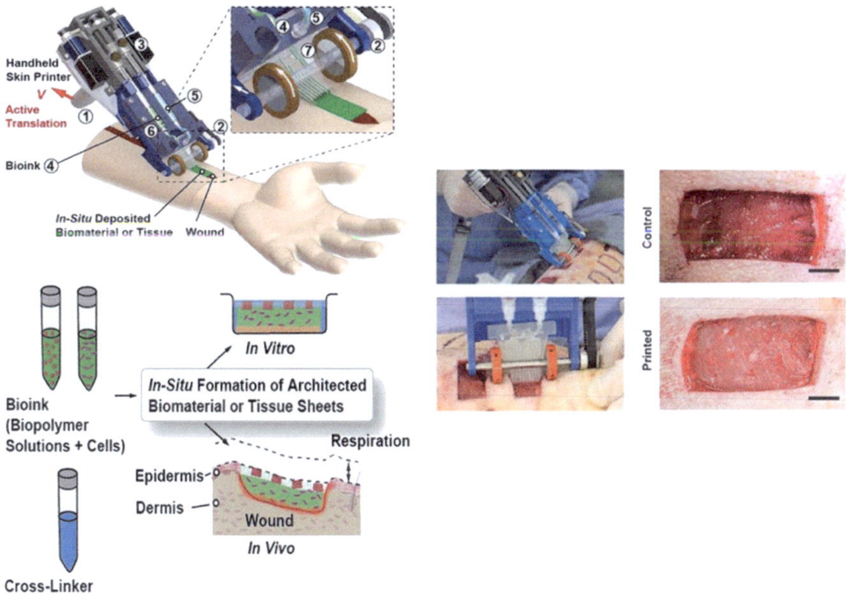

Fig. 16 (Left) In situ production of skin sheets in pig and mouse wound models as direct therapy using specific skin cells in bioinks and employing a handheld bioprinter to generate skin cell-laden sheets with controllable thickness, width, and composition via incorporating dermal and epidermal cells into various cross-linkable hydrogels; (Right) Such portable 3D printers could be revolutionary in the general healthcare market, as patients would not have to wait for lab-grown cell skin grafts. Additionally, the technology can be used for emergencies such as burns and real-time emergency care. Reproduced from Askari et al. [82], under the terms of the Creative Commons (CC BY) Attribution 3.0 Unported Licence. https://creativecommons.org/licenses/by/3.0/

by Kesti et al. to produce cartilage-specific bioinks that may be used to make cartilage transplants [87]. This bioink stimulates chondrocyte proliferation and efficient deposition of cartilage matrix proteins, according to magnetic resonance imaging (MRI) and histological evaluations after 8 weeks in vitro [87].

For every successful cornea transplant, there are others in need. Around 10 million people worldwide need corneal surgery [80, 81]. Corneal scarring already blinds 4.9 million individuals. The cornea is the eye's first light-letting layer. Millions of people are blind due to a lack of donor corneas. For a decade, scientists have been trying collagen gel artificial corneas. The gel must be twisted to fit the eye and focus light on restoring vision. Live corneal cell gel can now self-assemble like origami paper [80, 81]. The same approach may be used to produce human organs, helping millions of transplant patients. The live cells in collagen acted as micro-engines, restricting it. They can shape a 1-inch piece of tissue into a cornea. Some substances rendered corneal collagen gels less flexible. Scientists may utilize this information to compress the gel mixture in different places to generate a shape. Like a cornea, the gel curled when one segment constricted more than the other. This method may be used to produce artificial tissues from organs that include contractile cells [81]. 3D printing, which is being used to develop new artificial organs, may support this idea. 3D & 4D printing of artificial livers, kidneys, hearts, and other organs may be possible in the future.

The musculoskeletal system is composed mainly of dynamic bone tissue, which is inflexible and harder than other connective tissues due to the substantial mineralization of the ECM [88–90]. A worldwide clinical issue, bone tissue healing generates significant morbidity in trauma patients and is a primary socio-economic concern [91]. Autogenous bone grafts from intra- or extra-oral locations are still the gold standard for bone replacement, although they have poor graft amount, donor site morbidity, and infection. Many researchers have tried and failed to find therapeutic ways for fabricating human bone as a highly organized and vascularized tissue. Consequently, bone tissue engineering (BTE) is rapidly evolving as a viable alternative to bone transplantation. Cell and osteon in growth in vivo requires a combination of macropores that allow for cell-to-cell and cell-to-scaffold connections and micropores that will enable for cell–scaffold interactions in vitro [88–90]. To promote angiogenesis in vivo, an increase in macroporosity has been reported [91] In contrast, a degree of microporosity has been proven to improve cell–scaffold interactions, leading to an increase in osteogenic activity. Gelatin/carboxymethyl chitin/HA was used to produce a 3D bioactive scaffold, which was then lyophilized in a cryogenic setting [91]; The inclusion of bulk and surface porosity may lead to an increase in water uptake ratio, cell retention capacity, cell infiltration, adhesion, proliferation, alkaline phosphatase (ALP) level, and mineralization, according to obtained results [91].

Apart from EBB, 4D bioprinting, which integrates time with 3D bioprinting, has emerged as the next-generation solution of TE since it allows the construction of complex and functional structures. The addition of the fourth dimension, "time" to 4D bioprinting, allows for more control over the evolution of 3D printed biomaterials and bioniks, as well as the synthesis of biomimetic tissues from the printed constructions [88–91] (see Figs. 17 and 18). Using stimuli-responsive materials, 4D bioprinting may produce dynamic 3D-patterned living constructs that change shape under different inputs. 4D bioprinting has demonstrated tremendous promise to transform tissue engineering, medication delivery, and other sectors. But it is still in its infancy, and clinical applications are still far away. With advances in materials science, printing technology, software, and numerical modeling, 4D bioprinting may become a reality [89, 90]. 4D printing has limitless potential and is already influencing a wide range of industries. Although 4D printing relies heavily on 3D printing, its material may alter over time to adapt to the environment and future uses of bio-responsive and self-folding materials. Our goal is that this technology will progressively increase what we can do with bio-synthetic organs and tissues, offering hope to the millions of individuals waiting for transplants [88].

Among various materials, ZnO's biocompatibility, and antibacterial properties are used in wound dressings, medication delivery systems, and medical imaging. ZnO speeds wound healing, targets medication delivery, and improves imaging diagnostics, improving healthcare outcomes [92]. The diversity of these uses suggests it might improve medical treatments and diagnostics [93]. Beyond traditional usage,

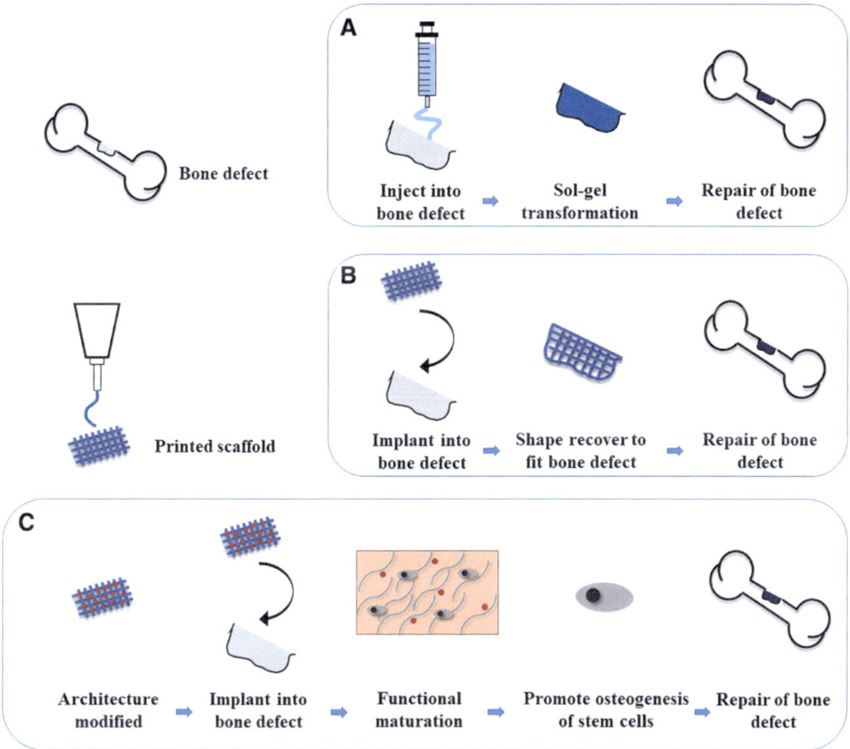

Fig. 17 Applications of 4D printing in medicine: **a** The hydrogel can be injected into the uneven defect region and changed into a gel, **b** Personalized bone defect healing is possible due to a shape memory scaffold that changes size to fill the vacant area, **c** 4D bone tissue printing based on the creation of a biomimetic microenvironment. Reprinted from Wan et al. [88], with permission from Elsevier. Copyright (2020)

Fig. 18 Scheme of future 4D printing

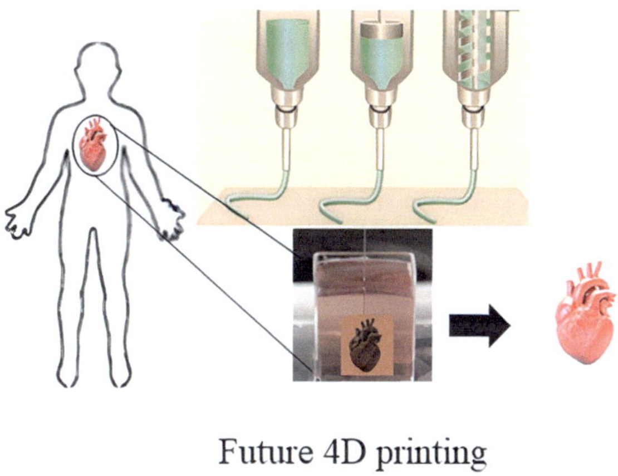

ZnO is improving 3D printing and 4D applications. 3D printing uses ZnO-based materials to construct complex structures like scaffolds for tissue engineering and biomedical implants. ZnO is suitable for complicated medical devices and prosthetics because its material characteristics may be accurately controlled. ZnO adds dynamic responses to temperature, humidity, and light to 4D printing. ZnO-based constructions may change and morph over time, enabling smart biomedical devices, responsive drug delivery systems, and adaptive implants that improve patient care and treatment results. Despite these advances, safety issues with increasing ZnO nanoparticle concentrations persist. These concerns must be addressed via ongoing research to enable safe and sustainable ZnO use in many applications [94]. To effectively utilize ZnO's potential to shape science and technology, biocompatibility, sustainable manufacturing, and multifunctional applications should be prioritized.

7 Conclusions

Briefly, it was mentioned that biomaterials of various types (ceramic, polymer, composite, natural, and metal) have a great place in the medicine, food, and cosmetic industries. Several methods have been used to produce novel bioactive and biocompatible materials. Natural surfactants derived from plants are an excellent source of green synthesizing materials for various points. Environmentally friendly procedures and materials are used in green chemistry to decrease scientific research's hazardous and harmful effects. Thus, it utilizes safe energy or reduces waste materials using a variety of experimental and industrial synthetic approaches. Environmental and human health benefits are the primary motivations for using eco-friendly practices and technology. Green synthesized biopolymers have attracted a lot of attention since they can be produced in huge quantities with the same physical, chemical, and structural qualities for various tissue types and packaging, reducing conventional techniques' negative consequences. On the other side, 3D bioprinting has gained prominence among different approaches for generating tissue scaffolds to bridge the gap between designed and natural tissues to increase the hope between patients who need organ transplantation, such as skin and cartilage tissue. Soon, 4D printing as a new way to acquire external stimuli-responsive structures will overcome some of the constraints of 3D bioprinting technologies in manufacturing high-resolution constructs that produce a great revolution in medicine and creating artificial organs.

References

1. D. Ege, K. Zheng, A.R. Boccaccini, ACS Appl. Bio Mater. **5**(8), 3608 (2022)
2. J. Nath, K. Sharma, S. Kumar, V. Kumar, R. Sehgal, Polymeric and natural composites, in *Advances in Material Research and Technology*, ed. by M.S. Hasnain, A.K. Nayak, S. Alkahtani (Springer, Cham, 2022), pp. 109–150
3. C. Gutierrez Cisneros, V. Bloemen, A. Mignon. Polymers, **13**, 760 (2021)

4. S. Alamdari, O. Mirzaee, F. Nasiri Jahroodi, M.J. Tafreshi, M. Sasani Ghamsari, S. Salmani Shik, M.H. Majles Ara, K.Y. Lee, H.H. Park, Surf. Interfaces **34**, 102349 (2022)
5. L. Gautam, A. Jain, P. Shrivastava, S. Vyas, S.P. Vyas, Polymeric and natural composites, in *Advances in Material Research and Technology*, ed. by M.S. Hasnain, A.K. Nayak, S. Alkahtani (Springer, Cham, 2022), pp. 1–24
6. F. Campos, A.B. Bonhome-Espinosa, R. Carmona, J.D.G. Durán, P. Kuzhir, M. Alaminos, M.T. López-López, I.A. Rodriguez, V. Carriel, Mater. Sci. Eng. C **118**, 111476 (2021)
7. M. Pickup David, J. Speight Richard, C. Knowles Jonathan, E. Smith Mark, J. Newport Robert, Mater. Res. Bull. **43**(2), 333–342 (2008)
8. Z. Salimi, M.H. Ehsani, A.S. Dezfuli, Z. Alamzadeh, J. Supercond. Novel Magn. **35**, 215 (2021)
9. S. Alamdari, M. Sasani Ghamsari, H. Afarideh, A. Mohammadi, S. Geranmayeh, M.J. Tafreshi, M.H. Ehsani, M.H. Majles Ara, Opt. Mater. **92**, 243 (2019)
10. S.V. Murphy, A. Atala, Adv. Healthc. Mater. **3**(5), 803 (2014)
11. W. Peng, Z. Peng, O. Tang, Materials (Basel) **13**, 4108 (2020)
12. K. Amin, R. Moscalu, A. Imere, R. Murphy, S. Barr, Y. Tan, R. Wong, P. Sorooshian, F. Zhang, J. Stone, J. Fildes, A. Reid, J. Wong, Nanomedicine **14**, 2679 (2019)
13. D.A. Banyard, J.M. Bourgeois, A.D. Widgerow, G.R. Evans, Plast. Reconstr. Surg. **135**, 1740 (2015)
14. I.F.S. da Ramos, L.M. Magalhães, C.O. Pessoa, P.M.P. Ferreira, M.S. dos Rizzo, J.A. Osajima, E.C. Silva-Filho, C. Nunes, F. Raposo, M.A. Coimbra, A. B. Ribeiro, M.P. Costa, Ind. Crops Prod. **171**, 113981 (2021)
15. V. Rizzi, J. Gubitosa, P. Fini, S. Nuzzo, A. Agostiano, P. Cosma, J. Photochem. Photobiol. B. **224**, 167 (2021)
16. M. Mittal, D. Mittal, N.K. Aggarwal, Environ. Sci. Pollut. Res. **29**, 11039 (2022)
17. D.M. Pickup, R.J. Speight, J.C. Knowles, M.E. Smith, R.J. Newport, Mater. Res. Bull. **43**, 333 (2008)
18. B. Saravanakumar, M. Rajkumar, V. Rajendran, Mater. Lett. **65**, 31 (2011)
19. B. Lei, S. Alamdari, M.S. Ghamsari, M.J. Tafreshi, Mater. Lett. **64**, 2293 (2010)
20. A. Prasad, A. Maha Lakshmi, S. Murimadugula, P. Venkateswara Rao, S. Chitra, G. Perumal, M. Doble, K. Kumari, M. Özcan, N. Madaboosi, P. Syam Prasad, Ceram. Int. **50**(17B), 30699–30711 (2024)
21. T. Kokubo, H.M. Kim, M. Kawashita, Biomaterials **24**, 2161 (2003)
22. S. Alamdari, M.S. Ghamsari, M.J. Tafreshi, Ceram. Int. **46**, 4484 (2020)
23. S. Alamdari, M. Sasani Ghamsari, M.H. Majles Ara, B. Efafi, Mater. Lett. **158**, 202 (2017)
24. S. Alamdari, M.J. Tafreshi, M.S. Ghamsari, Appl. Phys. A **125**, 165 (2019)
25. G. Zhou, EBioMedicine **28**, 287 (2018)
26. R. Reshmy, E. Philip, A. Madhavan, R. Sindhu, A. Pugazhendhi, P. Binod, R. Sirohi, M.K. Awasthi, A. Tarafdar, A. Pandey, Environ. Pollut. **283**, 117071 (2021)
27. J. Liu, Q. Ru, Y. Ding, Food Res. Int. **49**, 170 (2011)
28. S.E. Naleway, V. Thomas, D. Restrepo, H.C. Schniepp, JOM **72**, 1432 (2020)
29. M.M. Kabir, H. Wang, K.T. Lau, F. Cardona, Compos. B **43**, 2883 (2012)
30. E.A. Abdel-Razik, J. Polym. Sci. Part A **27**, 343 (1989)
31. V. Coma, A. Martial-Gros, S. Garreau, A. Copinet, F. Salin, A. Deschamps, Food Microbiol. **67**, 1162 (2002)
32. M. Chen, G.H. Yeh, B.H. Chiang, J. Food Process. Preserv. **20**, 379 (1996)
33. Ó. Ramos, S. Silva, J. Soares, J. Fernandes, M. Poças, M. Pintado, F. Malcata, Food Res. Int. **45**, 351 (2012)
34. M. Pereda, A.G. Ponce, N.E. Marcovich, R.A. Ruseckalte, Food Hydrocoll. **25**, 1372 (2011)
35. D. Zhai, B. Liu, Y. Shi, L. Pan, Y. Wang, W. Li, R. Zhang, G. Yu, ACS Nano **7**(4), 3540 (2013)
36. S. Alamdari, M. Hemmati, M.J. Tafreshi, M. Ehsani, Prog. Phys. Appl. Mater. **3**, 119–123 (2023)
37. K.H. Le, M.D.-B. Nguyen, L.D. Tran, H.P.N. Thi, C.V. Tran, K.V. Tran, H.P.N. Thi, N.D. Thi, Y.S. Yoon, D.D. Nguyen, D.D. La, Prog. Org. Coat. **158**, 106339 (2021)

38. A. Rouhani Shirvan, M. Shakeri, A. Bashari, in *The Textile Institute Book Series*, ed. by Shahid-ul-Islam, B.S. Butola (Woodhead Publishing, 2019), pp. 107–133
39. N. Noshirvani, B. Ghanbarzadeh, R.R. Mokarram, M. Hashemi, V. Coma, Int. J. Biol. Macromol. **99**, 530 (2017)
40. S. Alamdari, M. Sasani Ghamsari, C. Lee, W. Han, H.-H. Park, M.J. Tafreshi, H. Afarideh, M.H.M. Ara, Appl. Sci. **10**, 3620 (2020)
41. M. Sasani Ghamsari, S. Alamdari, W. Han, H.H. Park, Int. J. Nanomed. **12**, 207 (2016)
42. S. Alamdari, S. Taheri, E. Heydari, M.H. Majles Ara, Opt. Quant. Electron. **56**, 443 (2024)
43. M. Fardioui, M. El Mehdi Mekhzoum, A. El Kacem Qaiss, R. Bouhfid, Int. J. Biol. Macromol. **184**, 981 (2021)
44. B. Liang, L. Fang, Y. Hu, G. Yang, Q. Zhu, X. Ye, Nanoscale **6**, 4264 (2014)
45. A. Saatchi, A. Razaghian Arani, A. Moghanian, M. Mozafari, Ceram. Int. **47**, 260 (2021)
46. L. Jiang, Y. Li, C. Xiong, S. Su, H. Ding, ACS Appl. Mater. Interfaces **9**(5), 4890 (2017)
47. D. Štular, E. Savio, B. Simončič, M. Šobak, I. Jerman, I. Poljanšek, A. Ferri, B. Tomšič, Appl. Surf. Sci. **563**, 150361 (2021)
48. M.V. Sujitha, S. Kannan, Acta A Mol. Biomol. Spectrosc. **102**, 15 (2013)
49. S. Kaviya, J. Santhanalakshmi, B. Viswanathan, J. Muthumary, K. Srinivasan, Acta A Mol. Biomol. Spectrosc. **79**, 594 (2011)
50. A.M. Amanulla, R. Sundaram, Mater. Today Proc. **8**, 323 (2019)
51. M. Shokrzadeh, S.S. Saravi, J. Med. Plants Res. **4**, 95 (2010)
52. A. Lisitsyn, A. Semenova, V. Nasonova, E. Polishchuk, N. Revutskaya, I. Kozyrev, E. Kotenkova, Polymers **13**, 1592 (2021)
53. V. Coma, Polimeros **23**, 287 (2013)
54. J. Liu, Q. Ru, Y. Ding, Food Res. Int. **49**, 170 (2013)
55. R. Reshmy, A. Madhavan, E. Philip, S.A. Paul, R. Sindhu, P. Binod, A. Pugazhendhi, R. Sirohi, Environ. Technol. Innov. **21**, 101335 (2021)
56. I. Mladenović-Ranisavljević, G. Babić, M. Vuković, D. Voza, Water **13**, 3537 (2021)
57. R. Reshmy, E. Philip, S. Paul, A. Madhavan, S. Raveendran, B. Parameswaran, A. Pandey, Biomass Convers. Biorefin. **9**, 1 (2020)
58. H. Charreau, E. Cavallo, M.L. Foresti, Carbohydr. Polym. **237**, 116039 (2020)
59. B. Ghanbarzadeh, H. Almasi, A. Entezami, Innov. Food Sci. Emerg. Technol. **11**, 697 (2010)
60. N.A. El-Wakil, E.A. Hassan, R.E. Abou-Zeid, A. Dufresne, Carbohydr. Polym. **124**, 337 (2015)
61. J. Liu, Q. Ru, Y. Ding, Food Res. Int. **49**, 170 (2012)
62. Q. Zhang, H. Lu, N. Kawazoe, G. Chen, Acta Biomater. **10**, 2005 (2014)
63. A. Cherkashina, A. Rassokha, I. Ryshchenko, O.B. Komarova, Eastern-Eur. J. Enterp. Technol. **2**, 56 (2020)
64. M. Ozdemir, J.D. Floros, J. Food Eng. **84**, 185 (2008)
65. V. Ferraro, M. Anton, V. Santé-Lhoutellier, Trends Food Sci. Technol. **1**, 1 (2016)
66. M. Yasir Khalid, A. Al Rashid, Z. Ullah Arif, W. Ahmed, H. Arshad, J. Mater. Res. Technol. **14**, 2601 (2021)
67. A. Salama, Int. J. Biol. Macromol. **127**, 606 (2019)
68. N. Valinezhad, A. Farhad Talebi, S. Alamdari, Int. J. Biol. Macromol. **241**,124503 (2023)
69. S.M. Maliyekkal, K.P. Lisha, T.A. Pradeep, J. Hazard. Mater. **181**, 986 (2010)
70. H.J. Qazi, A. Ye, A. Acevedo-Fani, H. Singh, Food Hydrocoll. **117**, 106692 (2021)
71. P. Umaraw, A.K, A. Verma, Crit. Rev. Food Sci. Nutr. **57**, 1270 (2017)
72. J. Ju, Y. Xie, Y. Guo, Y. Cheng, H. Qian, W. Yao, Crit. Rev. Food Sci. Nutr. **59**, 2467 (2019)
73. T. Anukiruthika, P. Sethupathy, A. Wilson, K. Kashampur, J.A. Moses, Food Sci. Food Saf. **19**, 1156 (2020)
74. J. Jeya Jeevahan, M. Chandrasekaran, S.P. Venkatesan, V. Sriram, G. Britto Joseph, G. Mageshwaran, R.B. Durairaj, Trends Food Sci. Technol. **100**, 210 (2020)
75. A. Eltom, G. Zhong, A. Muhammad, Adv. Mater. Sci. Eng. **2019**, 1 (2019)
76. Grand View Research, Market Analysis Report. https://www.grandviewresearch.com/industry-analysis/scaffold-technology-market. Accessed at 23 Oct 2023

77. P.S. Gungor-Ozkerim, I. Inci, Y.S. Zhang, A. Khademhosseini, M.R. Dokmeci, Biomater. Sci. **6**, 915 (2018)
78. S. Kyle, Z.M. Jessop, A. Al-Sabah, I.S. Whitaker, Adv. Healthc. Mater. **6**(16), 1700264 (2017)
79. S. Madiedo-Podvrsan, J.P. Belaïdi, S. Desbouis, L. Simonetti, Y. Ben-Khalifa, C. Collin-Djangone, J. Soeur, M. Rielland, Sci. Rep. **11**, 6217 (2021)
80. A.K. Pandey, N. Mudgil, Y. Wadgav, S.S. Mishra, AIMS Public Health **24**, 186 (2021)
81. M. Khalili, M. Asadi, H. Kahroba, M.R. Soleyman, H. Andre, E. Alizadeh, J. Cell. Physiol. **236**, 3275 (2021)
82. M. Askari, M.A. Naniz, M. Kouhi, A. Saberi, A. Zolfagharian, M. Bodaghi, Biomater. Sci. **9**, 535 (2021)
83. J. Igielska-Kalwat, E. Kilian-Pięta, S. Połoczańska-Godek, Polymers **14**, 749 (2022)
84. J.L. Jorcano, Biofabrication **9**, 1 (2017)
85. G.H. Kim, J.G. Son, S. Park, W.D. Kim, Rapid Commun. **29**, 1577 (2008)
86. N. Hakimi, R. Cheng, L. Leng, M. Sotoudehfar, P.Q. Ba, N. Bakhtyar, S. Amini-Nik, M.G. Jeschke, A. Günther, Lab Chip **18**, 1440 (2018)
87. M. Kesti, C. Eberhardt, G. Pagliccia, D. Kenkel, D. Grande, A. Boss, M. Zenobi-Wong, Adv. Funct. Mater. **25**, 7406 (2015)
88. Z. Wan, P. Zhang, Y. Liu, L. Lv, Y. Zhou, Acta Biomater. **101**, 26 (2020)
89. H.A. Alshahrani, J. Sci. Adv. Mater. Dev. **6**, 167 (2021)
90. M. Ramezani, Z. Mohd Ripin, J. Funct. Biomater. **14**, 347 (2023)
91. D. Gupta, A.K. Singh, A. Dravid, J. Bellare, ACS Appl. Mater. Interfaces **11**, 20437 (2019)
92. S. Alamdari, O. Mirzaee, M.J. Tafreshi, R. Riedel, Compos. B Eng. **265**, 110934 (2023)
93. M. Aliannezhadi, S.Z. Mirsanaee, M. Jamali et al., Sci. Rep. **14**, 2035 (2024)
94. M. Hosseinpour, O. Mirzaee, S. Alamdari, J.L. Menéndez, H. Abdoos, J. Environ. Manage. **345**, 118586 (2023)

Biomaterials Today and Tomorrow

Esmat M. A. Hamzawy, Gehan T. El-Bassyouni, Sayed H. Kenawy, Mostafa Mabrouk, Samah S. A. Eldera, and Areg Omar

Abstract Biomaterials are components that have been contrived to intermingle with biological systems for medical interventions, which could be either curative (repair, augment, treat, or replace a tissue function of the body) or diagnostic. Majority of biomaterials stimulate the formation of cells, perform anti-microbial functions, or prevent the harmful transformation of some cells. This chapter will discuss recent advances in biomaterial research, focusing on material design and fabrication. Some of the potential applications of biomaterials will be discussed including tissue engineering (TE), orthopedic application, central nervous system applications, regenerative medicine, and gene delivery. Moreover, this chapter will explore the advances at multi-scale levels (macro-, micro-, and nano-scale). Finally, challenges and future views will be presented to provide development of novel biomaterials that will have a role in immune modulation, viral control, 3D printing of organs, as well as, reducing the side effects of drug delivery to human cells and increasing the efficiency and performance of drugs inside the body.

Keywords Biomaterials · Tissue engineering (TE) · Regenerative medicine · Orthopedic application · Cartilage repair · Immune modulation · Central nervous system · Delivery vectors

E. M. A. Hamzawy (✉)
Glass Research Department, Advanced Materials Technology and Mineral Resources Research Institute, National Research Centre (NRC), Cairo, Egypt
e-mail: ehamzawy9@gmail.com

G. T. El-Bassyouni · S. H. Kenawy · M. Mabrouk
Refractories, Ceramics and Building Materials Department, Advanced Materials Technology and Mineral Resources Research Institute, National Research Centre (NRC), Cairo, Egypt

S. S. A. Eldera · A. Omar
Physics Department, Faculty of Science, Al Azhar University (Girls Branch), Cairo, Egypt

S. J. Ikhmayies (ed.), *Advances in Biomaterials Research*, Advances in Material Research and Technology, https://doi.org/10.1007/978-3-031-97481-6_2

1 Introduction

These are ingredients contrived to interrelate with biological structures used for medical commitments, either healing (repair, augment, treat, or replace a tissue function of the body) or diagnostic materials that are normally known as biomaterials [1]. Biomaterial science includes components of biology, chemistry, medicine, and tissue engineering (TE). Most of biomaterials are application—specific materials, which imply that not all biomaterials are suitable for all the biomedical applications [2]. For each tissue there are different remarks that can be followed to judge the potential ability of the investigated materials. For instance, in bone tissue regeneration it is fundamental to test the ability of potential implanted materials to attract precipitations of various calcium phosphates on their surfaces upon soaking in simulated fluids such as simulated body fluid (SBF) and phosphate buffer saline (PBS), which is called the bioactivity feature. Bioactivity could also be identified in another meaning as the capability of an engineered biomaterial to prompt a biological reaction that stimulates bone cells to generate the defected site starting with producing the bone inorganic component (Hydroxyapatite $Ca_{10}(PO_4)_6(OH)_2$) [3]. It was confirmed that various inorganic materials are very promising for the above mentioned feature (bioactivity). In this regard, ceramics and bioactive glass have demonstrated impressive findings, as they were found to enhance the ability of implanted materials to attach well with the nearby tissue in either osteoproductive or osteoconductive roles [4]. In addition, bone growth is often stimulated by inorganic/polymer composite materials while dispersing into adjacent body fluid [5–7]. Therefore, ideal synthetic biomaterials for bone substitutes must be biocompatible and designed to simulate regeneration of the extracellular matrix of the natural bone. Such materials should afford bone regeneration environment, assist complete bone regeneration, tolerate dynamic mechanical loading, and degrade along with the formation of new bone [8]. Another important category belongs to biomaterials is biopolymers, which are frequently used as biocompatible materials, especially in regeneration of soft tissues, because of their flexible characteristics along with their important application as supportive matrices of biological molecules. However, they should go through a series of structured tests to confirm biocompatibility, which is very necessary permit to be licensed for their practice in biomedical field. The principal intention of biocompatibility tests is to estimate the acute and chronic toxicity of materials and define the possible unfavorable properties that might take place upon utilization. Therefore, these investigations are extremely necessary for an assumed material that could be applicable at tissue regeneration field, especially delicate tissues like central nervous and blood circulating systems [9]. The obtained features of those materials are highly dependent on their origin; therefore, the following section will introduce a brief illustration of the biomaterials categorization according to their origin.

1.1 Fabrication and Design of Natural Biomaterials Structures

Biomaterials are categorized into natural and synthetic biomaterials. Naturally derived biomaterials have been validated to display numerous advantages compared to synthetic ones [10]. The naturally derived biomaterials can be sorted into various groups, comprising polysaccharide-based (glucose, cellulose, chitin/chitosan), protein-based (silk, collagen, gelatin), and decellularized tissue-derived biomaterials (liver, blood vessels, decellularized heart valves). Their wide spread as implanted materials is owed to their interconnectivity, microstructure and characteristic bioactivity that simulates the formation of extracellular matrix (ECM), assisting the adhesion, cell infiltration, conveyance of oxygen and nutrients, differentiation and lastly renovating the function of defective tissues or organs and their configuration.

The mechanical properties, microstructure, cellular activity, and bio-stability of the biomaterials can be finely adjusted for specific applications through mixing of natural and artificial biopolymers using chemical or physical treatments. This would result in promotion of the degradation rate, essential mechanical support of the ECM to imitate the microenvironment for supporting cellular activity [11]. Figure 1 represents the evolving stages of the implanted materials to be suitable for tissue reconstruction.

Ordinary biomaterials are extensively designed to meet the needs of several tissue engineering (TE) and regeneration applications to enterprise the three-dimensional organs [12]. Conversely, the highest complications linked to biomaterials are the instable biodegradation rate and mechanical properties frequency that frequently restricts their medical uses. Several approaches were utilized to overcome these restrictions. The bio-stability and mechanical properties of natural biomaterials can be modified by combining various synthetic and natural polymers (Polylactic acid PLA, Polycaprolactone PCL, Polyethylene glycol PEG, poly(glycolic acid) PGA. etc.) as well as chemical and physical treatments [13]. The micro-pores of natural biomaterials are also important factor that allows the migration, proliferation, cell infiltration, and adhesion, thus result in repairing of the function and structure of the damaged tissues. Permeable natural biomaterials assist the transfer of nutrients and oxygen to endorse tissue rejuvenation. Furthermore, they perform a significant function in the transport of bioactive molecules, nanoparticles, cells, growth factors and medications to enhance the bioactivity of biomaterials. Pointing out the improvement of natural biomaterials for different tissue applications is worthy noted. In tissue engineering (TE), in spite of the wide development of natural biomaterials, many issues are still needed to be covered for the successful management. For instance, in natural biomaterials, the precise pore sizes are desired for the proliferation, the exact cell type infiltration and definite configurations, bio-stability of natural biomaterials and mechanical properties are required for representing the tissue ECM. Available natural biomaterials may not totally fulfill these needs to exceptionally facilitate regeneration, cell infiltration, and 3D organ tissue engineering, due to their intrinsic disadvantages. Consequently, the confidence in the mixture of synthetic components

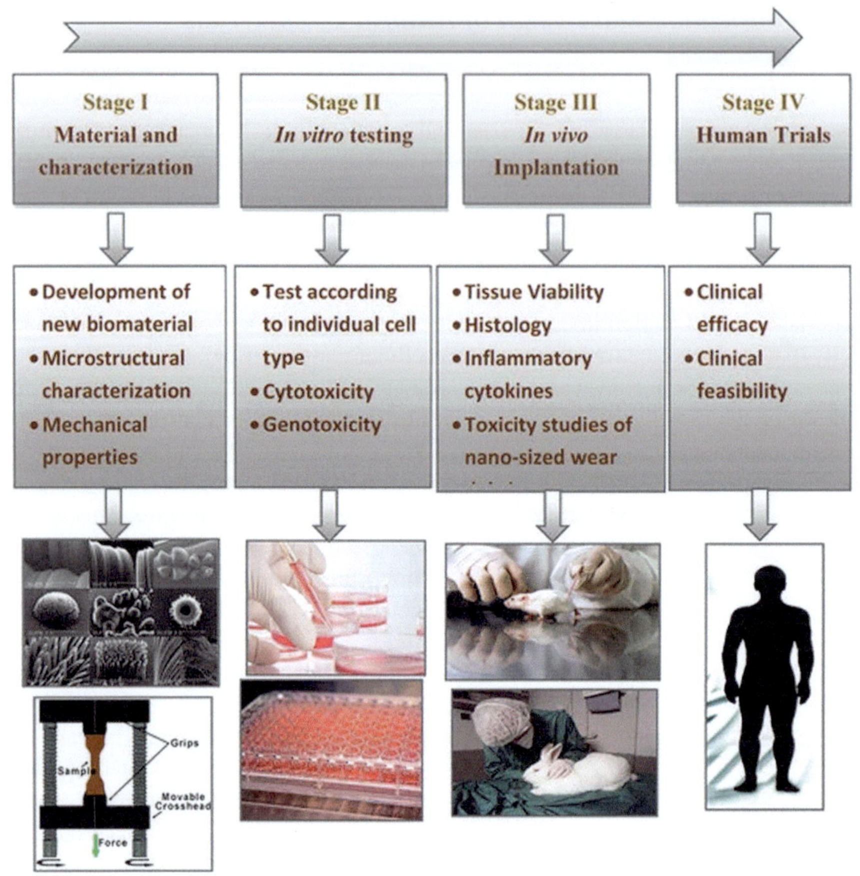

Fig. 1 Illustration of the implantable materials main stages. Reprinted from Velu et al. [11], under the terms of the Creative Commons CC BY license. https://creativecommons.org/licenses/by/4.0/

and natural materials, to pronounce the use of chemical and physical treatments and to create complex materials that would be a useful technique for biomedical applications [14].

1.2 Biomedical Application of Biomaterials

Biomaterials have several health applications such as artificial ligaments and tendons, *cancer therapy*, bone plates, joint replacements, contact lenses, surgical sutures for wound healing, clips, *nerve regeneration*, and in breast implants. Moreover, these materials have some non-medical applications such as the assay of blood proteins in laboratories and the growth of cells in the culture medium [15] (Fig. 2).

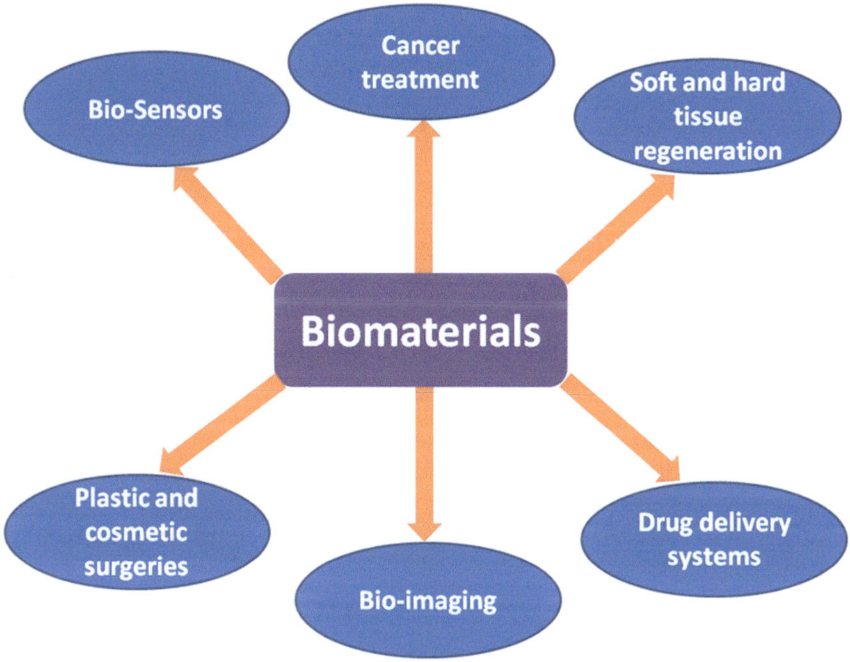

Fig. 2 Applications of biomaterials in health care field

Effective tissue regeneration approaches focus on the use of innovative biomaterials, to regulate the cell behavior and stimulate its regeneration and structure. Moreover, biomaterials are important in material stiffness, surface topography, porosity, biomaterial chemistry, and degradation properties as they have a significant role in monitoring tissue regeneration. Progressive biomaterial structures and tactics are established to emphasize the distribution of bioactive elements (e.g., small molecules, proteins, and peptides) to impact cell regeneration and performance [16]. Utilizing major constituents of TE alongside with signaling molecules and cells, biomaterials have been recognized. Biomaterial research has considerably extended the ways to lead chemical, physical, and biological parameters (Fig. 3). The physical parameters include porosity, surface topography, and material rigidity, were confirmed to control the behavior of cells related to growth and adhesion. On the other side, chemical parameters impact numerous features of cell behavior [17]. Both factors can convey bioactivity and degradability in order to synchronize replacement and restoration of the defect via *neo*tissue (newly formed tissue). Scaffolds of biomaterials are joined through signaling molecules in different behaviors to affect cell behaviors through small drug molecules or growth factors (GFs). Imminent investigation motivated on the relationship between several parameters and their synergistic properties on many tissues [18]. Additional studies could similarly figure out the response of

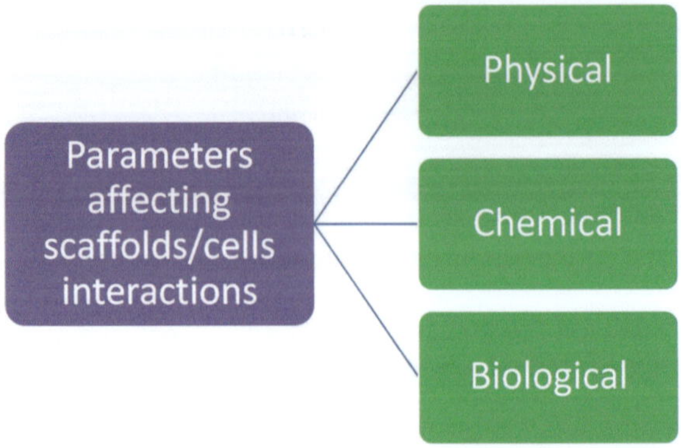

Fig. 3 The parameters affecting on scaffolds/cells interactions

the cell-/tissue-specific to the cited parameters in order to improve additional specified biomaterials adjusted for specific tissue regeneration, along with the provided parameters by the utilized biomaterials [19]. Moreover, research has shed light on the ability of external cues in the form of magnetic or electrical stimulation on the tissues regeneration.

2 Advanced Developments and Engineering of Biomaterials

2.1 *Recent Developments in Regenerative Medicine and Biomaterials*

The usage of biomaterials by means of stem cell-based regenerative medicine, assured the renovation of damaged tissues [20]. In regenerative medicine, mesenchymal stem cells (MSCs) have been described as an operational treatment for primary ovarian insufficiency (POI). Studies of molecular biology established the effectiveness of MSCs in restoring ovarian function, which discovered the genetic modulations, signaling pathways, and paracrine messaging proteins which drive the cell's mechanisms of action [21]. Scientists have connected biocompatible materials with regenerative therapies to offer efficient mechanical support, and persist cellular delivery which could renovate ovarian function and recover the development of follicle [22]. For POI patients to deploy the therapeutic product, the highest concern includes the insensitivity of the procedures. Such skills could be useful to supply biomaterials and stem cells toward the ovary. Another problem with regenerative medicine and stem cell therapy is the potential for immunogenicity and tumorigenesis. Nevertheless, the two obstacles could be addressed by acellular treatments. The use of secretome and exosomes with biomaterials carriers is a favorable acellular strategy.

2.2 Hepatic Tissue Engineering and Biomaterials

Liver toxicity is one of the reasons of acute liver failure and also a major factor of drug designs, causing huge financial losses for pharmaceutical enterprises. Progresses in tissue engineering (TE) and material science have been newly observed, generating progress in biomaterials that have supported complex organs in tissue engineering (TE) [23]. Liver shows a frequent, multicellular structural design, in which the main parenchymal cells (hepatocytes), making 80–85% of total liver cells arranged in cords sandwiched by ECM. Liver has an outstanding capability to entirely regenerate once two-thirds of the liver is partially hepactectomied. In the course of the in vivo liver regeneration, hepatocytes proliferate forming clusters of hepatocyte (Fig. 4). More developed biomaterials and protocols for accomplishing a well-organized vascularization of the developing hepatocytes are obligatory to generate liver tissues for various uses. Invented vascular tissues and liver must be subjected to energetic culture conditions. A crucial mechanical factor that affects phenotype and cell survival is fluid shear stress which originates the mechanosensitive gene expression [24].

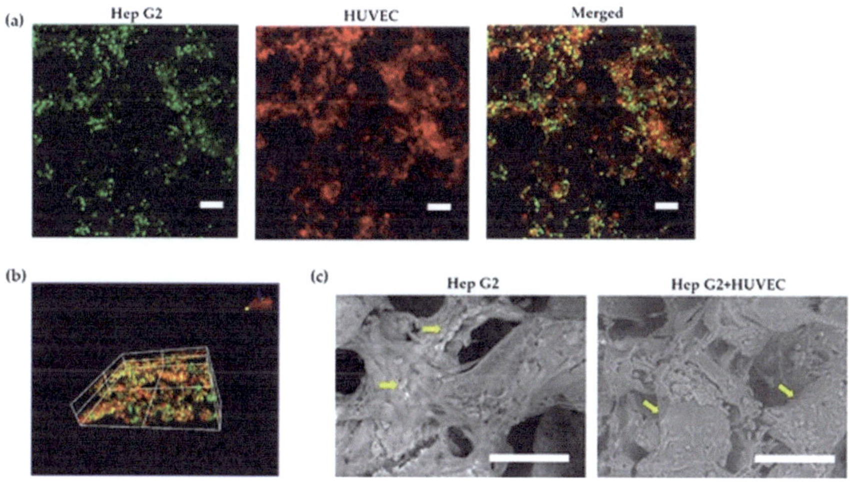

Fig. 4 **a** Live imaging of Hep G2 and human umbilical vein endothelial cells (HUVEC) after 10 days of perfusion culture (scale bars 100 μm, magnification 4×); **b** volume z-stacked view of the Hep G2 and HUVEC cultured on the scaffold; **c** SEM images of Hep G2-covered scaffold and co-culture of Hep G2 and HUVEC at day 10. The yellow arrows show the morphology of Hep G2 and HUVEC grown on the scaffold (scale bar 100 μm, magnification 200×). Reprinted from Li et al. [24], under the terms of the Creative Commons CC BY license. https://creativecommons.org/licenses/by/4.0/

2.3 Skeletal Muscle Engineering

In order to improve the redevelopment of blood vessels, muscle, and neuromuscular junctions (NMJs), the arena of biomaterials grounded on skeletal muscle engineering has been mostly progressed. Progresses in the three-dimensional (3D) printing technology and the usage of ECM-derived polymers are influencing the future of the skeletal muscle engineering [25]. Through innovations in scaffold production techniques, science of biomaterials and editions of decellularized tissues, remarkable translational research is continuing and evolving the field of skeletal muscle engineering [26]. Electrospinning and 3D printing skills are joined to generate composite, multimaterial constructs with impressive architecture and physicochemical possessions that were inaccessible a decade ago. Decellularized skeletal muscle and hybrid biomaterials have been employed in new ways to endorse myogenesis in addition to angiogenesis and neurogenesis, progressions that are necessary to the regeneration of muscle [27]. To enable integration with the defect site and afford waste interchange and nutrient to the regenerated tissue, and for the formation of NMJs, vasculature is needed to re-innervate regrown tissue. Pro-neurogenic and pro-angiogenic growth factors contribute in angiogenic, myogenic and neurogenic crosstalk, triggering synergistic effects. Therapies of skeletal muscle combine tactics to stimulate these pathways. A complex process in skeletal muscle regeneration rely on a number of signaling molecules, cell types, physicochemical properties and architectural features to be fruitful. Nevertheless, it has not ever been so nearby to translate therapies into the clinic in order to restore muscles structurally and functionally. Biomaterials must retain favorable properties to effectively direct muscle regeneration for skeletal muscle tissue engineering.

The material structure impact other physicochemical properties comprising, degradation kinetics, and tensile modulus which are imperative for assistant of skeletal muscle regeneration. Several tissue engineering (TE) applications, constructing an interconnected porous scaffold is valuable because it guarantees cells to proliferate and migrate in order to plug the full depth of the matrix. It similarly provides the delivery of needed nutrients and the removal of waste to help the ingrowth of vasculature [28]. The methodologies established to redevelop skeletal muscle, being categorized into three main groups: in vivo, in vitro, besides in situ muscle engineering. Such tactics have restrictions and benefits. In vitro skeletal muscle engineering characteristically comprises differentiating and growing cells into myotubes. Before implantation, further conditioning of the tissue construct, often take place in a bioreactor. Cells nominated for these systems to accomplish blood vessel and NMJs formation consist of multipotent cells and/or co-culture of cell types [29].

2.4 Cardiovascular Tissue Engineering

Heart disease treatments include methods ranging from medications to surgical intrusions [30]. Different types of biomaterials were used in cardiovascular diseases whether natural such as heart transplantations, donor valve, porcine or bovine heart valves, and natural polymers. Synthetic biomaterials used in cardiovascular diseases include metals, synthetic polymers, and composite ceramics [31]. However, natural grafts have many limitations, as immune complications, rareness of donor tissue, anticoagulation therapy, and restricted resilience. Likewise, traditional synthetic materials have many drawbacks as prompting vessel irritation, and causing inflammation at implantation site. Therefore, there is a major need to find new strategies to overcome these barriers and to discover novel biomaterials that can be used in cardiovascular tissue engineering. The major cardiovascular tissue engineering challenge is to discover three-dimensional (3D) materials that simulate native tissues and can repair cardiac tissues and heart valves [32].

Cardiovascular tissues engineering has different challenges than those of other tissues. It requires materials with proper mechanical properties (elasticity, strength, and durability). Furthermore, these substrates should display excellent hemodynamics, porosity, and hemocompatibility. Moreover, they must have the ability for tissue renovation and growth, have no immunogenicity and degradation resistance [33]. Researchers tried to find new strategies to overcome the existing drawbacks in cardiovascular tissue engineering. These trials included many approaches such as the use of stem cells, carbon nanotubes, and magnetic nanoparticles in addition to the uncommon eco-friendly materials [32, 34, 35].

Demurtas et al. [35] reported that stem cells therapy is a promising approach to restore injured myocardial tissues. However, additional experimental studies are required to conquer immunological and tumorigenicity issues. Alagarsamy et al. [34] summarized the use of carbon nanotubes (CNTs) and their derivatives in drug delivery, immunomodulation, biosensors, and heart tissue engineering (see Table 1). They indicated that carbon nanotubes have an effective role in cardiovascular diseases. However, more studies should be done to overcome their drawbacks. Future studies should handle their rout of administration, ideal therapeutic dose, as well as developing facilities for large-scale production of carbon nanotube materials. In parallel, another survey illustrated the promising application of magnetic nanoparticles in identification and treatment of cardiovascular diseases [36]. Morrison et al. [32] explored the use of unconventional material, as silk, paper, and plant leaves that have been used for cardiovascular tissue engineering. These materials are potential materials for tissue engineering (TE) as they provide cost effective alternative for commercial available biomaterial. Plant leaves have many advantages as they are naturally porous, vascular, and eco-friendly, however there in vivo biocompatibility and functionality are still unknown. Also; silk based materials can exhibit biocompatible and biodegradable features that support growth of heart tissues. Paper was also used and had many advantages such as porosity, durability, and low cost, but

Table 1 Examples of CNTs and their derivative for cardiovascular applications [34]

CNT composite	Field of application	In vitro/in vivo validation	Significant outcomes	Refs.
GCE[a]-coated poly styrene/carboxylated-MWCNT[b]/whiskered nanofibers/HRP-anti-cTnI[c]	Biosensor	Human serum and PBS solution	The linear detection limit of cTnI was between 0.5 and 100 ng/mL	[37]
Chitosan/polyvinyl alcohol/MWCNTs nanofibers	Tissue engineering (TE)	Somatic stem cells	Improved tensile and electrical properties and increased cardiac gene expressions (troponin I, CX43[d])	[38]
MWCNTs/dipyridamole	Drug delivery	Varied pH solutions (hydrochloric acid and PBS)	Improved dissolution rate and controlled release of the drug	[39]
Tissue-engineered blood vessels/C-SWCNT[e]/resveratrol	Immune modulation	Vascular smooth muscle cells (VSMCs)/SD[f] rats	Polarized the shift of M1 inflammatory macrophage to M2 reparative macrophages along with preserved contractile phenotype of VSMCs	[40]

[a] GCE: Glassy carbon electrode
[b] MWCNT: Multi-walled carbon nanotubes
[c] HRP-anti-cTnI: Horseradish peroxidase-anti-cardiac Troponin I
[d] CX43: Connexin 43
[e] SWCNT: Single-walled carbon nanotubes
[f] SD: Sprague Dawley

its problems are lack of biodegradability in vivo as well as the reduction of the mechanical properties in wet environment [32].

2.5 Recent Developments in Dental Biomaterials

All classes of biomaterials have been used in dental applications. Each type of biomaterials has many advantages, and shortcomings as illustrated by Patel et al. [41], Nanotechnology has showed great ability in the treatment of dental related diseases. It was used to increase the antimicrobial activity and drug delivery in oral cavity; however, it is not ideal due to its high toxicity [34]. In dentistry, biomaterials face many environmental challenges: For example, the oral cavity has hard and soft tissues, with great number of microorganisms as well as hot and cold material exposure. Thus, the material should not support microorganisms' adhesion and growth, yet the material needs to attach to biological soft and hard tissues. Moreover, the material should be biocompatible, non-allergic, non-inflammatory, and non-toxic, have proper mechanical properties, remain sterile and easy to be shaped

[42]. In this section, we will briefly illustrate the structure of human tooth and highlight different biomaterials and their promising current and future contributions in dental applications. Biomaterials in general and bioactive glass in particular have been used in many dental applications. The main dental applications include maintenance of endosseous ridge, pulp capping, treatment of dentine hypersensitivity and reconstruction of periodontal defects and maxillary sinus floor.

2.5.1 Maintenance of Endosseous Ridge

The removal of tooth leaves behind cavities that cause resorption of the jaw alveolar bone. To avoid alveolar bone resorption, bioglass was shaped in the form of solid cone, and implanted into extracted tooth site, thus it was firmly bonded to bone by bioactive bond, and then after few weeks' dentures were implanted [43]. Recently, a novel dental implant was developed from metal and nano-micro glass that successfully form bone on the alveolar ridge and increase the vertical bone at rats 6 weeks after surgery [44].

2.5.2 Reconstruction of Periodontal Osseous Defects

Early in 2005 Radin et al. [45] created a cylinder-shaped defect in the iliac crest of New Zealand rabbit and implanted $90SiO_2$–$5CaO$–$5P_2O_5$ xerogel granules through the opening in the cortical bone. Two weeks later, widespread trabecular bone growth was identified in the defect packed with granules. The growth happened in different residences; at the borders of the defect, opening of the defect, and in nearby the cortical bone, thus providing a promising tissue response to silica xerogel implantation. They recommended that xerogels have possible controlled release materials for healing of bone infections and as transporters for controlled delivery of drugs.

Bioactive glass particles (BAGs) have been used in many studies to treat periodontal defects [46]. Granules of bioactive glass have been used to regenerate bone around root of tooth and fill bone cavities at implant site. Polymeric materials have been used as drug delivery systems that have been potentially used against periodontitis diseases [47]. Recently, a new article illustrated that electrospun polymeric membranes have effective role in periodontal tissue regeneration. The chemistry of BAG is imitating the natural hard tissues composition, and has a bioactive role in the regeneration. The apparent antimicrobial properties motivated the research of BAGs in clinical application in dentistry. Numerous reconstruction applications have been reported in the last two decades [48].

2.5.3 Treatment of Dentine Hypersensitivity

Dentine hypersensitivity is triggered by the exposure of dentine due to inflammation of gum surrounding the teeth. Dentine contains tubules that reach the tooth pulp and nerve ending, which causes severe pain upon contact with cold or hot drinks. Tooth paste contains bioactive glass has been used in this condition. Granules of bioglass (BG) are attached to the dentine while brushing teeth. As a result, it causes hydroxyapatite precipitation over the dentine, the tubules are closed preventing hot or cold drinks from reaching the nerve endings, thus, decrease pain [49]. Additionally, toothpaste containing bioactive glass nanoparticles was used; the results indicated that the dentine tubules still closed after 24 h of using toothpaste [50]. Toothpaste including calcium fluoride (CaF_2) and bioactive glass was also studied. The tooth paste induces fluroapatite, which has greater resistance against acid attack than hydroxyapatite [51, 52]. Khan et al. [53] prepared a novel paste containing fluoride bioactive glass and zinc oxide nanoparticles that has ability to seal dentin tubules effectively. They concluded that dentifrices encompassing bioactive glass could assist as a substitute in dentin sensitivity management [53].

2.5.4 Reconstruction of Maxillary Sinus Floor

Clinical trials have been used for sinus floor augmentation. Such trials showed that in sinus floor treatment the combination of bioactive glass with autogenous bone in patients with severe chronic sinusis was suitable [54]. Other recent studies showed the potential role of polymer ceramic composite or polymer based hydrogel on sinus floor augmentation [55].

2.5.5 Dental Pulp Therapy

Dental pulp risk may be due to dental caries, or acute trauma. Biomaterials used in dental pulp therapy stimulate the inflamed tissue around the pulp to repair the dentine (dentine bridge formation) that isolates the pulpal chamber from irritation area, consequently, protecting its contents. Calcium hydroxide, bioceramics, polymer–tricalciumphosphate composite, hydrogels and glass ionomers were sucssessfully used in vital pulp theray. Future studies in dentistry should devolop new smart nanomaterial that can be used in different dental applications, including diagnosis, prevention and treatment [56]. Likewise, the combination of smart nanoparticles with protein and gene delivery represents other future trend in dentistry. However, the real challenge is to use these materials while overcoming their severe problems [34, 57–59].

2.6 Wound Healing

Skin is the largest organ in the human body that acts as a barrier between body and the surrounding environment. It protects the essential organs and defends the body against harmful agents. After injury, the integrity of the skin is compromised and the organism begins the wound healing process to restore the structure and functions of this organ. Wound healing is a highly interactive and dynamic biological process involving extracellular matrix molecules, soluble mediators, various resident cells (fibroblasts and keratinocytes) and infiltration of leukocyte subtypes which, together, work to re-establish the integrity of the damaged tissue and replace the missing one. To improve patient's chances of survival and reduce the loss of vital skin functions, this tissue must be covered immediately after it is damaged. this highlights the importance of developing new wound dressings that improve the healing process, making it less painful and, at the same time, contribute to re-establish skin structure and functions in a shorter period of time. In recent years, various dressings have been developed, using natural or synthetic materials, to restore the original properties and structure of the skin. An appropriate biomaterial is a key structural element to allow skin regeneration since a provisional ECM is essential to support the complex repair processes required for skin healing [60, 61].

Polymers offer a versatility that is unmatched by metals and ceramics, and have been extensively used in the production of wound dressings [62]. These polymers are frequently classified based on source (natural and synthetic). Natural materials are obtained from nature, while synthetic materials are produced in the laboratory [63, 64]. One of the most frequently used natural polymer, in wound dressing production, is collagen [65], which can be obtained from a variety of organisms such as bovine, porcine and jellyfish. Poly caprolactone (PCL) is an example of synthetic polymer used for wound dressings. PCL has been utilzed as an outer layer for tiple diabetic wound healing patch that represented impressive enhancements of their phisicomecanical features (see Fig. 5) [66]. Another example of synthetic biomaterial is the Polyurethane that has been extensively studied as a material for wound dressing due to its semipermeability (impermeable to bacteria and water but permeable to gases) creating a moist environment desirable for wound healing [67].

2.7 *Central Nervous System Cellular Repair*

Stroke, traumatic injury, and neurodegenerative diseases are the main causes of central nervous system injury [68]. The central nervous system (CNS) includes the spinal cord and brain which is considered as the most sophisticated area of the nervous system [69]. Cell-based repair is an approach to replace or repair damaged cells or tissues of the CNS, through direct replacement of the injured neurons, or delivery of therapeutic factors that increase angiogenesis and decrease inflammation. Cellular repair for central nervous system (CNS) damage has been applied in preclinical and clinical research but, many obstacles such as poor cell survival, and insufficient sources of donor cells, limit their use in traumatic injury, stork, and neurodegenerative diseases. Nevertheless, advances in CNS cellular repair have been seen by using stem cell, regenerative medicine, and biomaterial with suitable design [68].

Biomaterials used for brain repair have many challenges due to the blood–brain barrier, brain complexity and the inability of brain cells to self-renovation. Previously, researchers used implantable scaffolds for cell based brain repair; however,

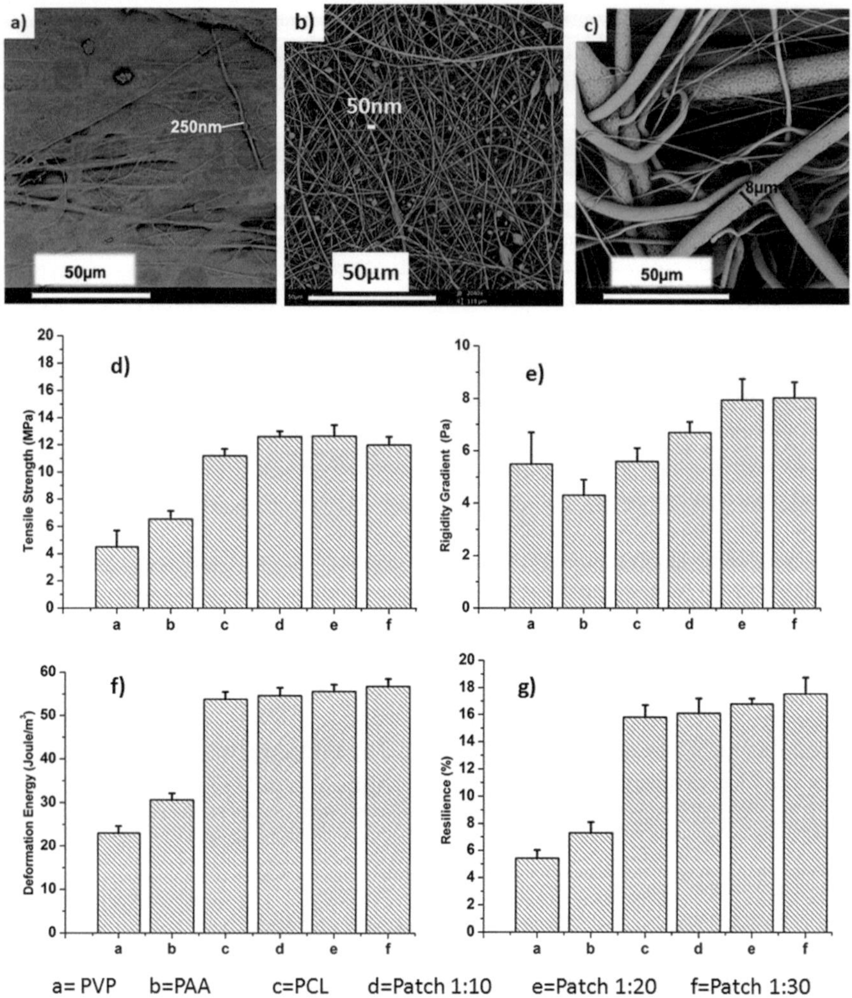

Fig. 5 SEM images of **a** poly (acrylic acid) (PAA) membranous fibers; **b** ciprofloxacin (CFX) loaded polyvinyl pyrrolidone (PVP) fibers (1CFX:10PVP); **c** polycaprolactone (PCL) and mechanical properties; **d** tensile strength; **e** rigidity gradient; **f** deformation energy; **g** resilience (%) of membranous fibers. Reprinted from Mabrouk et al. [66], under the terms of the Creative Commons CC BY license. https://creativecommons.org/licenses/by/4.0/

there is a trouble in accessing the target place. Hence, there was an urgent need to develop injectable scaffolds, known as hydrogels, used directly in brain with minimally invasive surgery. In general, biomaterials used in cellular brain repair should have proper chemical, physical, and biological properties, i.e., they should be chemically and physically stable, have complete degradation without any residues, have long-term biocompatibility, finally should provide a boost structure that looks like the native neural tissues.

2.8 Hydrogels

Hydrogels have several characteristics which make them prime scaffolds to be used in cellular brain therapies. Firstly, hydrogels can act as structural support for the cell attachment, proliferation and differentiation. Moreover, hydrogels can be easily used in intra-cranial delivery owing to their ability to form gels in situ utilizing different stimuli like temperature or pH. Besides, hydrogels can protect the implanted cells from forces applied during injection and host immune response. Also, they can be fabricated from natural or synthetic materials with different designs, shapes, biodegradability, and proper mechanical properties. Finally, hydrogels can also act a reservoir for controlled release of drugs and growth factors [68, 70, 71]. More recently, a subset of hydrogels called cryogels, have displayed to possess several special advantages over traditional hydrogel. Their macroporous structure, not only permits mass fluid transport through the structure but also produces a high surface area to volume ratio for cell growth or drug loading. Moreover, the cryogels are perfect for central nervous system applications as they are very supple and soft, yet also strong, making them a reproducible approach to address neuroscience challenges [72]. Moreover, nanoparticles have the ability to cross blood–brain barrier and deliver biomolecules to the targeted part of the brain. By this means, nanoparticles can develop the pharmacological ingredients, diminish toxicity of active compounds, and increase brain uptake [73]. Therefore, integration of injectable hydrogels with nanoparticles is another current strategy. This process has diverse advantages as the hydrogel saves the cells while nanoparticles offer a persistent long-term release of the biomolecules, to shore cell survival [68].

2.9 Repairing Spinal Cord Lesions

Spinal cord injury harms the spine and is a hopeless disease that forces physical, pecuniary, and psychic encumbrances on patients and their families. In general spinal cord lesions can guide to permanent or temporary failure, leading to paraplegia or tetraplegia, i.e., loss of sense and volitional motor functions. Primary and secondary injuries are the main categories of spinal cord injury. Primary injury leads to cell necrosis, axon disturbance, and vascular loss. This elicits secondary injury that has several phases: acute, sub-acute, moderate, and chronic [74]. Regardless of the major progress in recent years, therapeutic strategies for spinal cord repair are still limited. Researchers should overcome regeneration inhibiting obstacles, which is represented by poor axonal regeneration, poor vascularization in lesion site, abnormal remyelination, scar formation, inflammations and immune problems [75]. Thus, there is a need to combine different therapies and strategies to conquer these difficulties. Several treatment routines, including drugs, enzymes, and growth factors, have been established as cellular therapies for spinal cord repair. Furthermore, cellular based neuro-regeneration is a favorable strategy to treat spinal cord at the local injury site.

Incorporation of biomaterials with stem cells or drugs is also a hopeful method to enhance drug delivery or therapeutic efficacy. Herein, we summarized new strategies for spinal cord lesions repair. Natural and synthetic biomaterials have been used to fabricate implantable bridges including conducting scaffolds (see Fig. 6) (also known nerve guides), which have the ability to cover the damaged spinal cord for more information see [75–77].

3 Future of Biomaterials Research

3.1 Biomaterials and Viral Control

The coronavirus pandemic has showed main deficiencies in our strength to alleviate dispersion of viral disease and afford treatment to patients, causing health and economic crisis. COVID-19 looks like other viruses as SARS and MERS, but it exceeds those in mortality, spread, public and frugal impact [78, 79]. Recently, biomaterials have been widely used as antimicrobial and anti-bacterial agents while their use as antiviral is still little [80]. In this section, we will highlight biomaterials and their possible current and future contributions to deal with coronavirus, whether in prevention, diagnosis, and treatment.

Different metallic nanoparticles, mainly copper, gold, silver and iron have been used as antiviral factors. Copper nanoparticles are used to deactivate Influenza virus and destroy virus complexes [81, 82]. Gold based materials are used in treatment of rheumatoid arthritis and inhibition of influenza A virus and coronaviruses [16, 83]. Silver nanoparticles (AgNPs) have also applied in deactivation of H1N1 influenza and herpes simplex virus type 1 [84–86]. According to the study prepared by Abo-Zeid et al., iron oxide nanoparticles can interact well with the protein receptor binding domain of SARS-CoV-2 which is essential for virus connection to receptors of host cell [87]. In general, metallic NPs can make potent interactions with various viruses leading to viral destruction, and supply suitable way to isolate the bound particles [88]. Consequently, many trials should be made for metallic nanoparticles against coronavirus, taking into account their side effects when used in vivo [89]. In addition to metallic nanoparticles, nanosilica is employed as antigen carriers and in vaccine delivery.

Besides using nanoparticles, we are looking forward to using natural derived materials as cellulose, starch, alginate, lipids and other natural polymers, to confront viruses inside the human body, as they have been used for the same purpose in the food industries. Biomaterials strategies can be also used for virus prevention. Developing new filtration devices able to hold and inactivate SARS-CoV-2 and other viruses is one of the current targets of biomaterials applications. It is essential to design masks that have many advantages as low cost, multiple uses, biodegradability as well as their ability in virus deactivation [90, 91]. Moreover, it is also possible to use biomaterials in detection of coronavirus. Nanomaterials can be used to make

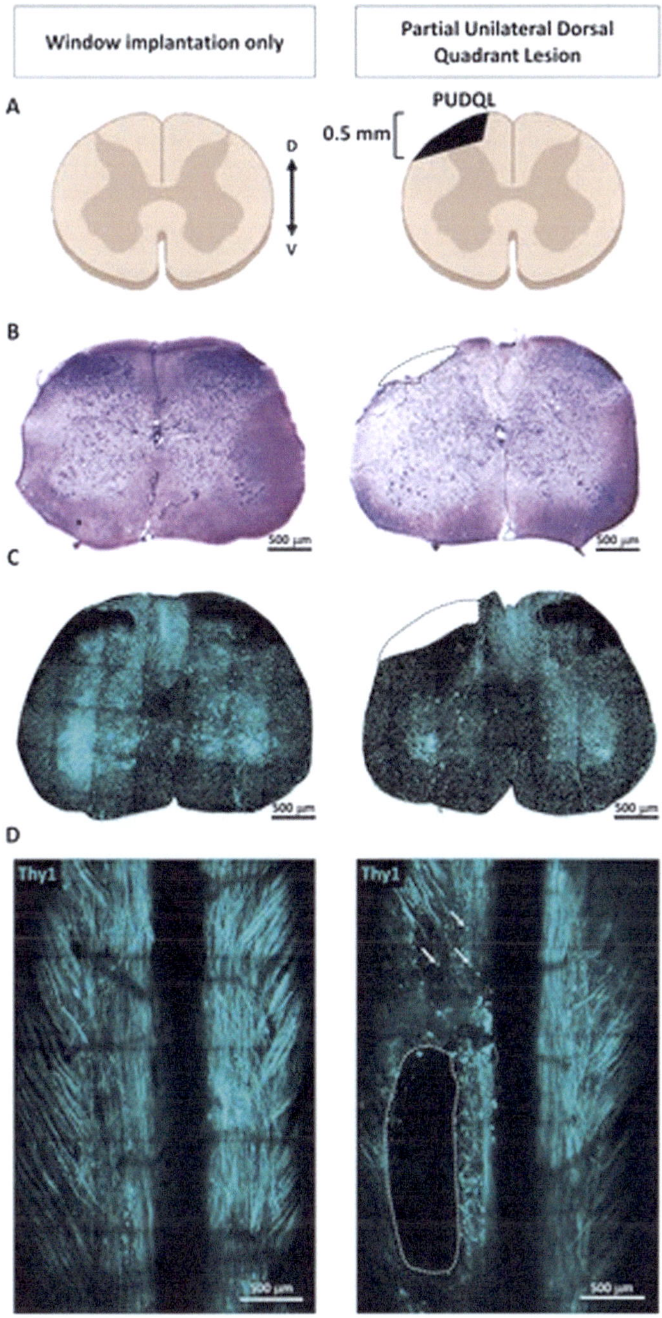
Window implantation only
Partial Unilateral Dorsal Quadrant Lesion
A
PUDQL
0.5 mm
D
V
B
500 µm
500 µm
C
500 µm
500 µm
D
Thy1
Thy1
500 µm
500 µm

◂**Fig. 6** Spinal cord partial unilateral dorsal quadrant lesion (PUDQL) in Thy1-CFP mice. **A** Schematic view of the PUDQL (shadowed area in black). **B** Cresyl violet staining of transversal spinal cord section collected three days (D3) after window implantation or after PUDQL followed by window implantation. **C** Two-photon (2P) fluorescence images of spinal slices consecutive to the one of B). **D** Dorsal view of the same Thy1-CFP spinal cord imaged through the implanted window showing the primary lesion area (dotted lines) and examples of subsequent Wallerian degeneration of axons (arrows) on D3. D: dorsal; V: ventral. Reprinted from El Waly et al. [77], under the terms of the Creative Commons CC BY license. https://creativecommons.org/licenses/by/4.0/

effective and sensitive biosensors that can be used for diagnoses of viruses. Finally, we hope that biomaterials will represent good opportunity to eradicate the Corona pandemic or even future pandemics.

3.2 *Biomaterials and COVID-19*

In December 2019, Wuhan city of China, grieved a new pandemic coronavirus sickness comparable to human coronaviruses. Severe diseases such as Middle East Respiratory Syndrome (MERS) and Severe Acute Respiratory Syndrome (SARS) was recognized and termed COVID-19. This virus has located the world on high panic and on the foundation of the World Health Organization (WHO), nearby 610,000 deaths were recounted, all over the world till July 2020 [92, 93]. The development of COVID-19, the enlargement of new antiviral systems, including vaccines and drugs has newly involved substantial care of scientist.

3.2.1 Carbon Nanostructures for COVID-19

Nanostructured carbon has talented antiviral features, for instance; virus enzyme and viral inhibition activity stalling to fight in contradiction of viruses such as, influenza, human immunodeficiency virus (HIV), herpes simplex virus (HSV) as well as COVID-19 [94]. By means of carbon nanomaterials for antiviral applications at an early stage requires widespread examination. All over the world scientists are exploring innovative materials and biological structures to contest contrary to many diseases that disturb humans and animals. By the rise of novel pathogens as COVID-19 and the occurrence of cancer as one of the important reasons of humanity worldwide, the strength to advance suitable ingredients to report these trials. Carbon nanostructures using their possessions of direct functionalization, ability to package drug, antiviral belongings, and biocompatibility have developed a foremost emphasis of biomedical researchers [95]. Nevertheless, decreasing poisonousness, augmenting biocompatibility, and increasing water solubility have been inspiring for carbon-based biomedical systems [96].

On the other hand, carbon nanomaterials still suffer from their toxic effect on biological systems. Biomedical researchers work on choosing the suitable materials for cancer treatment, antiviral requests, and drug delivery come to be a task. A variety of nanoscale carbon structures is remarkable. In nanoscale dimensions, carbon nanotubes (CNTs), graphene quantum dots (GQDs), graphene, fullerenes, carbon fibers (CFs), carbon dots (CDs), nanodiamonds, amorphous carbon nanostructures, and carbon nano-onions all signify unusual structures and allotropes of carbon (See Fig. 7) [97, 98]. Carbon fibers (CFs) and carbon nanotubes (CNTs) and are categorized as one-, two-, and three-dimensional: one-dimensional (1D) nanomaterials, and graphene as two-dimensional (2D) nanomaterials, nanodiamonds as three-dimensional (3D) nanostructures [99]. New carbon allotrope (fullerenes), have the maximum symmetry among the various carbon nanostructures and consequently reveal a high physical and chemical strength [100]. Researches of the COVID-19 pandemic designate that chloroquine (CQ) may possibly be an effective medication to fight the infection of COVID-19. Density functional theory (DFT) studies indicate that aluminum and silicon doped C_{60} are stable CQ carriers for COVID-19 medicinal owing to the compatible energetics between these species; fullerene acts as the electron acceptor and CQ assists as the electron donor [101].

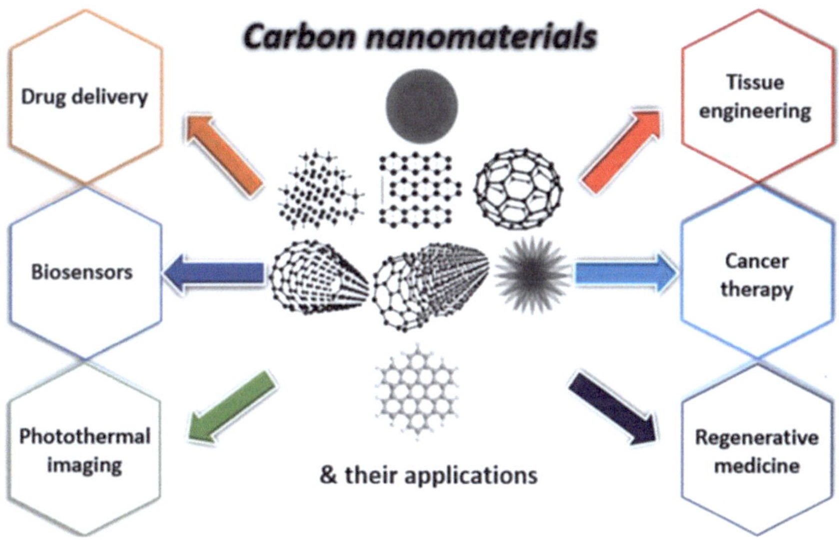

Fig. 7 Carbon nanomaterials and their applications in medicine. Reprinted from Mahor et al. [98], under the terms of the Creative Commons CC BY license. https://creativecommons.org/licenses/by/4.0/

3.2.2 Nanomaterials-Enabled Biosensors for COVID-19 Detection

Several functionalization devices have been developed as shown in Fig. 7 for the detection of Covid-19 [102]. Nanomaterials properties comprising quantum size effects, high surface-to-volume ratio, high adsorption, and reactive capacity as compared to their bulk form are commanding to design biosensing techniques. Biosensors can regulate the virus-mediated infections categorized into four main groups comprising; (i) antibody (ii) nucleic acid, and (iii) antigen-dependent [103]. Consequently, benefits of the various physicochemical properties of nano-dimensions particularly optical, magnetic, electrical, and optmagnetic, efforts to develop nano-enabled biosensing systems planned for the gratitude of virus, liable for respiratory infections [MERS-CoV, SARS-CoV, and severe acute respiratory syndrome coronavirus 2 (SARS-CoV-2)]. The shape and size of nanomaterials can be designed, and, as a result, surface modification/immobilization with several biological species by means of covalent or non-covalent bonding are likely to improve the biosensing characteristics in terms of low-detection limit, high selectivity, sensitivity, and prompt response toward the sample analytes [102]. However, one of the biggest tasks in designing biosensors (e.g., DNA/RNA) is to capture signal of very low magnitude which takes place between the biological species (bio-receptors and analytes). Accordingly, nanomaterials can be used as labels to achieve the significant improvement of signal, high enough to be simply detectable. Classification can be done using metals of gold and silver nanoparticles (Au/Ag) NPs and quantum dots (Cd/Pb) by targeting the DNA/bio-recognizing probe [104]. The synergic effect due to nano-labeling results in significant amplification of the electrochemical signal and, therefore, it is possible to develop ultrasensitive and highly selective labeled-biosensing techniques [104] (Fig. 8).

3.3 3D Printing

Rapid prototyping and additive manufacturing, otherwise called 3D-printing, is a procedure that was first depicted in the year 1986. First inkjet-based procedure was performed in the year 2000, proteins and endothelial cells were printed inside 3D-structures. For this reason, the nozzles of a commercial inkjet printer were altered and supported in programming that engaged specialists to change the nozzle in the necessary x and y positions. In prescription movement, inkjet printers are most as a rule executing this technique. Like the inkjet printers, these printing frameworks store little dabs of the medicine on hydrogel that represent the drug carrier to shape scaffolds which can be used for delivery of drug. As this system requires no high temperature, the stacking of thermosensitive biomolecules just as cells was much appropriate when contrasted with older version of 3D printers [105, 106].

Recently it was confirmed that hydrogels are the most favored biomaterials for tissue regeneration, owed to their adjustable hydrating and mechanical features, accordingly empowers them to imitate the regular tissue itself. PEG, gelatin, and

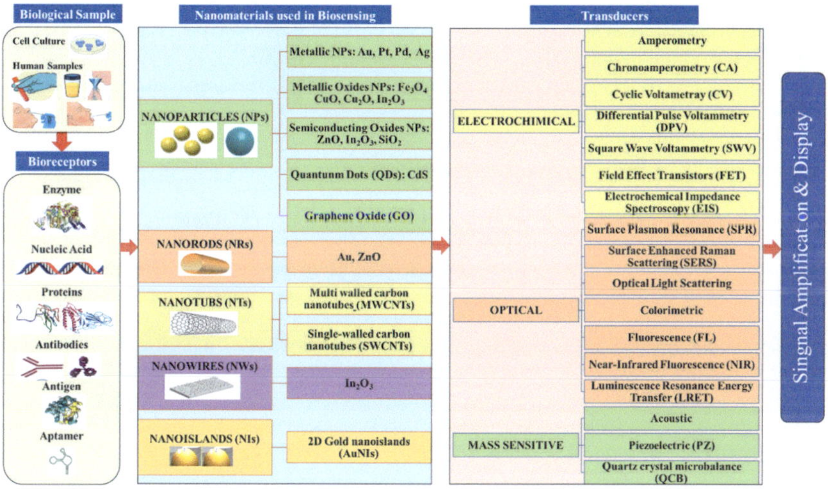

Fig. 8 The schematic diagram of different analytes, bioreceptors for biorecognition elements, transducers with integrated nanostructures as parts of a typical nanobiosensor design for respiratory infections. Reprinted from Ozer et al. [102], under the terms of the Creative Commons License CC BY-NC-ND. http://creativecommons.org/licenses/by-nc-nd/4.0/

fibrin hydrogels just as their derivatives were the most widely recognized hydrogels used from regular and engineered polymers in 3D printing loaded with cells for bone regeneration applications. Extrusion method is the most notable procedure used for cell-loaded hydrogels [107, 108]. A needle siphon stacked up with the hydrogel that moves in z-axis is used and the printing stage moves in y-level for the cell-based printing [107]. Applying voltage on the inkjet head can help to control the speed of the hydrogel extrusion as illustrated in Fig. 9 [108]. These elements are exceptionally fundamental in the 3D printing technique as they determine the pore size and homogeneity of the 3D-scaffolds [103, 109].

In addition, 3D printed scaffolds represent a chance with undiscovered potential. They may be utilized as supported delivery stages for drugs not simply to improve bone recovery, but to target leftover osteosarcoma cells after careful resection and to alleviate diseases after careful implantation of the scaffolds. This may be accomplished by two implications up until now; to straightforwardly upload the medication inside the scaffold (if the medication is appropriate), or to impregnate the platforms with drug-loaded nanoparticles. Methodologies are, for example, labeling the nano-carriers with focusing on mixtures, and refining nano-carriers to react with specific boosts may take into account accuracy in focusing on malignancy cells in adjuvant treatment [110].

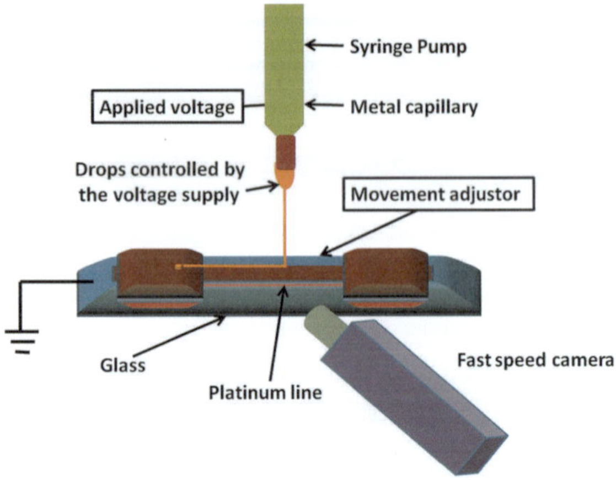

Fig. 9 Schematic diagram of advanced version of which controls the polymer printing rate using voltage supply, Reprinted from Mabrouk et al. [108], with permission from Elsevier. Copyright (2020)

4 Summary

The importance of biomaterials is outlined in that they have a myriad application in both curative and diagnostic measures. They can activate the formation of cells, act as antimicrobial agents, or prevent the harmful conversion of cells. Biomaterials also have an implication in regenerative stem cells, where the restoration of ovarian function is performed as well as the potential applications in stimulating an immune response and preventing tumorigenesis. They also play a role in hepatic tissue engineering where liver failure was treated using complex biomaterials that can reactivate and rearrange 80–85% of the liver cells. In cardiovascular tissue engineering, metals, synthetic polymers, natural polymers, composite ceramics, donor valve, and porcine or bovine heart valves, were used in heart transplantations.

Bioglass is used in the treatment of dental cavities. Xerogel, active glass, and polymers were used as drug delivery systems in the reconstruction of periodontal osseous defects for the treatment of periodontitis diseases. Dentine hypersensitivity is treated using tooth paste that contains bioactive glass that can seals dentin tubules effectively and prevents hot or cold drinks from reaching the nerve endings. In the reconstruction of maxillary sinus floor, bioactive glass with autogenous bone can treat chronic sinusitis. Also the inflamed tissue around the pulp is treated to repair the dentine using biomaterials such as calcium hydroxide, bioceramics, polymer–tricalcium phosphate composite, hydrogels, and glass ionomers.

In wound healing, various dressings have been developed, using natural or synthetic materials, to restore the original properties and structure of the skin. Therefore, polymer-containing metals or ceramics were used in wound dressing production which treat the skin and consequently keep the organs under skin. In the central nervous, advances in cellular repair have been seen using stem cells, regenerative medicine, and biomaterials with suitable design. Hydrogels can be used easily in intra-cranial applications to protect the implanted cells whereas the new cryogels can be used in central nervous system. Nanocarriers also have the ability to cross the blood–brain barrier and deliver biomolecules to the targeted part of the brain. Injectable hydrogels containing nanoparticles save the cells while nanoparticles offer a persistent long-term release of the biomolecules, to shore cell survival. Spinal cord lesions usually cause irreversible damage. However, natural and synthetic biomaterials have been used to fabricate implantable bridges including conducting scaffolds.

Recently some viruses, such as SARS and MERS, caused worldwide health crisis but in the present time, the novel COVID-19 virus has spread all over the world disrupting the economic system as well as causing worldwide shortages in the healthcare systems around the world. Many researcher used carbon nanomaterials and its derivatives against the COVID-19 virus since they found that the hydroxychloroquine may possibly be an effective medication to fight the infection of COVID-19. Gold and silver nanoparticles (Au/Ag NPs) and quantum dots (Cd/Pb) by targeting the DNA/bio-recognizing probe can be used as nanomaterials-enabled biosensors for COVID-19 detection.

All in all, biomaterials lay a solid foundation for the future of regenerative medicine and tissue engineering (TE) as they are the perfect specimens for having properties that facilitate their functions such as biocompatibility and the ability to cross the blood–brain barrier as well as having great drug delivery specifications.

References

1. D.F. Williams, Biomaterials **30**, 5897 (2009)
2. M. Mabrouk, D.B. Das, Z.A. Salem, H.H. Beherei, Molecules **26**, 1 (2021)
3. G.T. El-Bassyouni, S.S. Eldera, S.H. Kenawy, E.M.A. Hamzawy, Heliyon **6**, e04085 (2020)
4. G. Kaur, V. Kumar, F. Baino, J.C. Mauro, G. Pickrell, I. Evans, O. Bretcanu, Mater. Sci. Eng. C. Mater. Biol. Appl. **104**, 109895 (2019)
5. K.M. Tohamy, I.E. Soliman, M. Mabrouk, S. ElShebiney, H.H. Beherei, M.A. Aboelnasr, D.B. Das, Mater. Sci. Eng. C. Mater. Biol. Appl. **93**, 1 (2018)
6. S.A.M. El-Sayed, M. Mabrouk, M.E. Khallaf, B.M. Abd El-Hady, E. El-Meliegy, M.R. Shehata, J. Drug Deliv. Sci. Technol. **57**, 101757 (2020)
7. M. Mabrouk, R. Rajendran, I.E. Soliman, M.M. Ashour, G. Arthanareeswaran, D.B. Das, Pharmaceutics **11**, 294 (2019)
8. V.D.L. Putra, M.J. Song, S. McBride-Gagyi, H. Chang, K. Poole, R. Whan, D. Dean, V. Sansalone, M.L. Knothe Tate, Front. Cell Dev. Biol. **7**, 354 (2020)
9. S.C. Gad, S. Gad-McDonald. *Biomaterials, Medical Devices, and Combination Products*, 1st edn. (CRC Press, 2015), p. 38
10. W.I. Abdel-Fattah, T. Jiang, G.T. El-Bassyouni, C.T. Laurencin, Acta Biomater. **3**, 503 (2007)
11. R. Velu, T. Calais, A. Jayakumar, F. Raspall, Materials (Basel) **13**, 9 (2020)
12. M. Mabrouk, G.T. El-Bassyouni, H.H. Beherei, S.H. Kenawy, E.M.A. Hamzawy, *Advanced 3D-Printed Systems and Nanosystems for Drug Delivery and Tissue Engineering*, ed. by L.C. du Toit, P. Kumar, Y.E. Choonara, V. Pillay (Elsevier, 2020), pp. 83–107
13. G. Nour-Eldeen, M. Abdel-Rasheed, A.M. EL-Rafei, O. Azmy, G.T. El-Bassyouni, Cell Regen. **9**, 7 (2020)
14. D.P. Bhattarai, L.E. Aguilar, C.H. Park, C.S. Kim, Membranes (Basel) **8**, 62 (2018)
15. T. Biswal, S.K. Badjena, D. Pradhan, Mater. Today Proc. **30**, 274 (2020)
16. J. Kim, M. Yeom, T. Lee, H.O. Kim, W. Na, A. Kang, J.W. Lim, G. Park, C. Park, D. Song, S. Haam, J. Nanobiotechnology **18**, 1 (2020)

17. M. Ali, S.L. Payne, Biomater. Res. **25**, 5 (2021)
18. L. Zhu, D. Luo, Y. Liu, Int. J. Oral Sci. **12**, 6 (2020)
19. R. Lian, P. Xie, L. Xiao, Z. Iqbal, S. Zhang, J. Kohn, X. Qu, C. Liu, Y. Li, Front. Mater. **7**, 481 (2021)
20. A. Adawy, G.T. EL-Bassyouni, M. Ibrahim, W.I. Abdel-Fattah, Chapter 2, Bio nano material: the third alternative, in *Nanotechnology Vol. 7: Diagnostics and Therapeutics*, ed. by N.K. Navani, S. Sinha (Studium Press LLC, U.S.A., 2013), pp. 30–33
21. S. Zhang, D. Zhu, X. Mei, Z. Li, J. Li, M. Xie, H.J.W. Lutz, S. Wang, K. Cheng, Bioact. Mater. **6**, 1957 (2021)
22. N. Spears, F. Lopes, A. Stefansdottir, V. Rossi, M. De Felici, R.A. Anderson, F.G. Klinger, Hum. Reprod. Update **25**, 673 (2019)
23. M.P. Nikolova, M.S. Chavali, Bioact. Mater. **4**, 271 (2019)
24. M. Li, R.R. Khadim, M. Nagayama, M. Shinohara, K. Inamura, M. Danoy, M. Nishikawa, K. Furukawa, Y. Sakai, T. Niino, Appl. Sci. **11**, 2473 (2021)
25. M.E. Kalaw, A. Culaba, H. Hinode, W. Kurniawan, S. Gallardo, M.A. Promentilla, Materials (Basel) **9**, 580 (2016)
26. M.A.A. Mahdy, Cell Tissue Res. **375**, 575 (2019)
27. J.M. Fishman, M.W. Lowdell, L. Urbani, T. Ansari, A.J. Burns, M. Turmaine, J. North, P. Sibbons, A.M. Seifalian, K.J. Wood, M.A. Birchall, P. De Coppi, Proc. Natl. Acad. Sci. U. S. A. **110**, 14360 (2013)
28. Q.L. Loh, C. Choong, Tissue Eng. Part B Rev. **19**, 485 (2013)
29. M.J. Rao, M. Srinivasan, R. Rajasekharan, Curr. Genet. **64**, 1071 (2018)
30. M. Leone, F.B. Engel, Clin. Sci. **133**, 1229 (2019)
31. M.T. Lam, J.C. Wu, Expert Rev. Cardiovasc. Ther. **10**, 1039 (2012)
32. E. Morrison, S. Suvarnapathaki, L. Blake, G. Camci-Unal, Curr. Opin. Biomed. Eng. **17**, 100263 (2021)
33. E.S. Fioretta, L. Von Boehmer, S.E. Motta, V. Lintas, S.P. Hoerstrup, M.Y. Emmert, Exp. Gerontol. **117**, 1 (2019)
34. K.N. Alagarsamy, S. Mathan, W. Yan, A. Rafieerad, S. Sekaran, H. Manego, S. Dhingra, Bioact. Mater. **6**, 2261 (2021)
35. J. Demurtas, G.N. Fanelli, S.L. Romano, M. Solari, L. Yang, P. Soysal, G.F.L. Sánchez, I. Grabovac, L. Smith, A. Zorzi, C. Luchini, N. Veronese, Ageing Res. Rev. **67**, 101257 (2021)
36. I. Cicha, C. Alexiou, J. Magn. Magn. Mater. **518**, 167428 (2021)
37. B. Rezaei, A.M. Shoushtari, M. Rabiee, L. Uzun, W.C. Mak, A.P.F. Turner, Talanta **182**, 178 (2018)
38. W. Zhu, H. Huang, Y. Dong, C. Han, X. Sui, B. Jian, Exp. Therap. Med. **17**, 4610 (2019)
39. A. Abedi, B. Bakhshandeh, A. Babaie, J. Mohammadnejad, S. Vahdat, R. Mombeiny, S.R. Moosavi, J. Amini, L. Tayebi, Mater. Chem. Phys. **258**, 123842 (2021)
40. N. Ding, C. Dou, Y. Wang, F. Liu, G. Guan, D. Huo, Y. Li, J. Yang, K. Wei, M. Yang, J. Tan, W. Zeng, C. Zhu, Adv. Healthc. Mater. **7**, 1800026 (2018)
41. E. Patel, P. Pradeep, P. Kumar, Y.E. Choonara, V. Pillay, J. Biomed. Mater. Res. - Part B Appl. Biomater. **108**, 201 (2020)
42. L. Hupa, A. Yli-Urpo, *Bio-Glasses: An Introduction*, 1st edn., ed. by J. Jones, A. Clare (John Wiley Sons Ltd., 2012), p. 161
43. H.E. Skallevold, D. Rokaya, Z. Khurshid, M.S. Zafar, Int. J. Mol. Sci. **20**, 5960 (2019)
44. F. Zhao, Z. Yang, L. Liu, D. Chen, L. Shao, X. Chen, Biomater. Transl. **1**, 82 (2020)
45. S. Radin, G.T. El-Bassyouni, E.J. Vresilovic, E. Schepers, P. Ducheyne, Biomaterials **26**, 1043 (2005)
46. L.L. Hench, R.J. Jones, Bioactive glasses: frontiers and challenges. Front. Bioeng. Biotechnol. **3**, 194 (2015)
47. L.L. Hench, *Handbook of Bioceramics and Biocomposites*, ed. by I.V. Antoniac (Springer Cham, 2016), pp. 1–11
48. X. Liu, X. He, D. Jin, S. Wu, H. Wang, M. Yin, A. Aldalbahi, M. El-Newehy, X. Mo, J. Wu, Acta Biomater. **108**, 207 (2020)

49. L.P.D. Da Cruz, R.G. Hill, X. Chen, D.G. Gillam, Int. J. Dent. **2018**, 5701638 (2018)
50. J.C. Mitchell, L. Musanje, J.L. Ferracane, Dent. Mater. **27**, 386 (2011)
51. V.R. Patel, L. Shettar, S. Thakur, D. Gillam, D.N. Kamala, J. Oral Rehabil. **46**, 1121 (2019)
52. D. Gillam, S.F. Tie, R. Hill, J. Dent. Maxillofac. Res. **3**, 1 (2020)
53. A.S. Khan, I. Farooq, K.M. Alakrawi, H. Khalid, O.W. Saadi, A.S. Hakeem, Med. Princ. Pract. **29**, 338 (2020)
54. J.D. Menezes, R. dos S. Pereira, A.M. de S. Santos, N.B. de Siqueira, F.B.D.J. Boos-Lima, E. Hochuli-Vieira, Oral Maxillofac. Surg. **25**, 541 (2021)
55. J.D. Menezes, R.D.S. Pereira, J.P. Bonardi, G.L. Griza, R. Okamoto, E. Hochuli Vieira, J. Appl. Oral Sci. **26**, e20170296 (2018)
56. S. Cao, J. Han, N. Sharma, B. Msallem, W. Jeong, J. Son, C. Kunz, H.W. Kang, F.M. Thieringer, Materials (Basel) **13**, 1 (2020)
57. J. Ghoddusi, M. Forghani, I. Parisay, Iran. Endod. J. **9**, 15 (2014)
58. H.J. Haugen, P. Basu, M. Sukul, J.F. Mano, J.E. Reseland, Int. J. Mol. Sci. **21**, 3442 (2020)
59. S. Vasiliu, S. Racovita, I.A. Gugoasa, M.-A. Lungan, M. Popa, J. Desbrieres, Int. J. Mol. Sci. **22**, 2585 (2021)
60. M. Mabrouk, Y.E. Choonara, T. Marimuthu, P. Kumar, L.C. du Toit, S. van Vuuren, V. Pillay, Int. J. Pharm. **507**, 41 (2016)
61. M. Mabrouk, D.B. Das, Z.A. Salem, H.H. Beherei, Molecules **26**, 1077 (2021)
62. G.G. Pereira, C.B. Detoni, A.G. Balducci, V. Rondelli, P. Colombo, S.S. Guterres, F. Sonvico, Eur. J. Pharm. Sci. **83**, 203 (2016)
63. D. Akbik, M. Ghadiri, W. Chrzanowski, R. Rohanizadeh, Life Sci. **116**, 1 (2014)
64. A. Oryan, E. Alemzadeh, A. Moshiri, J. Tissue Viability **25**, 98 (2016)
65. A.A. Mahmoud, A.H. Salama, Eur. J. Pharm. Sci. **83**, 155 (2016)
66. M. Mabrouk, P. Kumar, Y.E. Choonara, L.C. Du Toit, V. Pillay, Materials (Basel) **11**, 2128 (2018)
67. A.R. Unnithan, G. Gnanasekaran, Y. Sathishkumar, Y.S. Lee, C.S. Kim, Carbohydr. Polym. **102**, 884 (2014)
68. S. Jarrin, S. Cabré, E. Dowd, Neurochem. Int. **144**, 104971 (2021)
69. Y. Wang, H. Tan, X. Hui, Biomed. Res. Int. **2018** (2018)
70. N. Hlavac, M. Kasper, C.E. Schmidt, Mater. Today Adv. **6**, 100039 (2020)
71. Z.Z. Khaing, R.C. Thomas, S.A. Geissler, C.E. Schmidt, Mater. Today **17**, 332 (2014)
72. D. Eigel, C. Werner, B. Newland, Neurochem. Int. **147**, 105012 (2021)
73. A.R. Khan, M. Liu, M. Khan, G. Zhai, J. Control. Release **268**, 364 (2017)
74. H.J. Jeong, H. Nam, J. Jang, S.J. Lee, Bioengineering **7**, 1 (2020)
75. S. Soares, Y. von Boxberg, F. Nothias, Rev. Neurol. (Paris) **176**, 252 (2020)
76. M. Tsintou, P.P. Wróbel, K. Dalamagkas, *Handbook of Biomaterials Biocompatibility*, ed. by M. Mozafari. *Woodhead Publishing Series in Biomaterials* (Elsevier, 2020), pp. 507–554
77. B. El Waly, V. Escarrat, J. Perez-Sanchez, J. Kaur, F. Pelletier, J.E. Collazos-Castro, F. Debarbieux, Cells **10**, 73 (2021)
78. A.N. Elboraey, H.H. Abo-Almaged, A.A. El-Ashmawy, A.R. Abdou, A.R. Moussa, L.H. Emara, H.M. El-Masry, G.T. El-Bassyouni, M.I. Ramzy, Adv. Pharm. Bull. **11**, 86 (2021)
79. F. Wu, S. Zhao, B. Yu, Y.-M. Chen, W. Wang, Z.-G. Song, Y. Hu, Z.-W. Tao, J.-H. Tian, Y.-Y. Pei et al., Nature **579**, 265 (2020)
80. D. Chakhalian, R.B. Shultz, C.E. Miles, J. Kohn, J. Biomed. Mater. Res. Part A **108**, 1974 (2020)
81. T. Ishida, World Sci. News **99**, 148 (2018)
82. K. Sundberg, V. Champagne, B. McNally, D. Helfritch, R. Sisson, J. Biotechnol. Biomater. **5**, 1 (2015)
83. H.A. Rothan, S. Stone, J. Natekar, P. Kumari, K. Arora, M. Kumar, Virology **547**, 7 (2020)
84. S.K. Kailasa, T.-J. Park, J.V. Rohit, J.R. Koduru, *Nanoparticles Pharmacother*, ed. by A.M. Grumezescu (Elsevier, 2019), pp. 461–484
85. R. Vazquez-Muñoz, B. Borrego, K. Juárez-Moreno, M. García-García, J.D. Mota Morales, N. Bogdanchikova, A. Huerta-Saquero, Toxicol. Lett. **276**, 11 (2017)

86. J.J. Broglie, B. Alston, C. Yang, L. Ma, A.F. Adcock, W. Chen, L. Yang, PLoS ONE **10**, 1 (2015)
87. Y. Abo-Zeid, N.S.M. Ismail, G.R. McLean, N.M. Hamdy, Eur. J. Pharm. Sci. Off. J. Eur. Fed. Pharm. Sci. **153**, 105465 (2020)
88. M. Ghaffari, M. Mollazadeh-Bajestani, F. Moztarzadeh, H. Uludağ, J.G. Hardy, M. Mozafari, Emergent Mater. **4**, 19 (2021)
89. L. Singh, H.G. Kruger, G.E.M. Maguire, T. Govender, R. Parboosing, Ther. Adv. Vaccines **4**, 259 (2017)
90. O. Das, R.E. Neisiany, A.J. Capezza, M.S. Hedenqvist, M. Försth, Q. Xu, L. Jiang, D. Ji, S. Ramakrishna, Sci. Total. Environ. **736**, 139611 (2020)
91. A.L. Allison, E. Ambrose-Dempster, T. Domenech Aparsi, M. Bawn, M. Casas Arredondo, C. Chau, K. Chandler, D. Dobrijevic, H.C. Hailes, P. Lettieri, C. Liu, F. Medda, S. Michie, M. Miodownik, B. Munro, D. Purkiss, J.M. Ward, UCL Open: Environ. **3,** 1 (2021)
92. M. Yang, H. Li, J. Sun, Y. Zhao, D. Tang, Curr. Urol. **14**, 79 (2020)
93. W.K. Boyes, C. van Thriel, Chem. Res. Toxicol. **33**, 1121 (2020)
94. R.Y. Siddiquie, A. Agrawal, S.S. Joshi, Trans. Indian Natl. Acad. Eng. **5**, 343 (2020)
95. Y. Ma, J. Huang, H. Shen, M. Zhang, S. Song, Z. Zhang, *Carbon Nanomaterials for Biomedical Applications*, ed. by M. Zhang, R. Naik, L. Dai (Springer, 2016), pp. 241–267
96. P.R. Riley, R.J. Narayan, Curr. Opin. Biomed. Eng. **17**, 100262 (2021)
97. G.T. El-Bassyouni, Project ID: #7754 (2018) (Scholars' Press, Germany), ISBN: 978-620-2-30686-7
98. A. Mahor, P.P. Singh, P. Bharadwaj, N. Sharma, S. Yadav, J.M. Rosenholm, K.K. Bansal, C **7**, 19 (2021)
99. E. Kai-Hua Chow, M. Gu, J. Xu, *Micro Nano Technologies*, ed. by E.J. Chung, L. Leon, C.B.T.-N. for B.A. Rinaldi (Elsevier, 2020), pp. 223–242
100. S. Bagheri Novir, M.R. Aram, Chem. Phys. Lett. **757**, 137869 (2020)
101. B. Alhalaili, I.N. Popescu, O. Kamoun, F. Alzubi, S. Alawadhia, R. Vidu, Sensors **20**, 6591 (2020)
102. T. Ozer, B.J. Geiss, C.S. Henry, J. Electrochem. Soc. **167**, 37523 (2020)
103. J. Guo, S. Chen, S. Tian, K. Liu, J. Ni, M. Zhao, Y. Kang, X. Ma, J. Guo, Biosens. Bioelectron. **181**, 113160 (2021)
104. Y. Saylan, F. Yilmaz, E. Özgür, A. Derazshamshir, H. Yavuz, A. Denizli, Sensors **17**, 898 (2017)
105. R.R. Jose, M.J. Rodriguez, T.A. Dixon, F. Omenetto, D.L. Kaplan, A.C.S. Biomater, Sci. Eng. **2**(10), 1662 (2016)
106. W.C. Wilson, T. Boland, Anat. Rec. **272**(2), 491 (2003)
107. A.L. Rutz, K.E. Hyland, A.E. Jakus, W.R. Burghardt, R.N. Shah, Adv. Mater. **27**(9), 1607 (2015)
108. M. Mabrouk, H.H. Beherei, D.B. Das, Mater. Sci. Eng., C **110**, 110716 (2020)
109. N. Raja, H.S. Yun, J. Mater. Chem. B **4**(27), 4707 (2016)
110. A. Suleman, P.P.D. Kondiah, M. Mabrouk, Y.E. Choonara, Front. Mater. **8**, 668834 (2021)

New Trends in the Design of Biomaterials for 3D and 4D Printing

Rubén del Olmo, Enrique Martínez-Campos, José Javier Relinque, Alberto Gallardo, Juan Rodríguez-Hernández, Endzhe Matykina, and Ana Santos-Coquillat

Abstract Additive manufacturing (AM), also known as 3D printing, is today a widely used technology in the fabrication of complex-shaped biomaterials from Digital computer-aided (CAD) designs. In this context, 3D fabrication responds to the increasing demand for new scaffolds with tailor-made architectures with the potential for tissue regeneration, including structures specially designed for the vascularization of scaffolds. This chapter aims to provide an overview of the state-of-the-art, current, and future perspectives in the field of biomaterials AM with their actual technological limitations and biovalidation strategies, i.e., from the manufacturing process technology to the direct biomedical applications. Up until now, AM can also improve prosthetics with biometric characteristics with patient-specific anatomical data or facilitate the creation of a new organ (liver, lymphoid organs). Another application is in vitro drug testing (liberation studies, etc.) with 3D-printed models of anatomical parts. So far, the latest advances reveal that material selection and AM printing technology are the most critical factors in achieving the most clinical success rate according to the patient's needs. Moreover, AM of composite materials stands

R. del Olmo
Department of Chemistry and Materials, Research Group of Sustainable Materials Engineering (SUME), Vrije Universiteit Brussel (VUB), Brussels, Belgium

E. Martínez-Campos · J. J. Relinque · A. Gallardo · J. Rodríguez-Hernández
Instituto de Ciencia y Tecnologia de Polimeros (ICTP), CSIC, Madrid, Spain

E. Martínez-Campos · A. Gallardo · J. Rodríguez-Hernández · E. Matykina
Unidad Asociada al ICTP, IQM (CSIC), Grupo de Síntesis Orgánica y Bioevaluación, Instituto Pluridisciplinar (UCM), Madrid, Spain

J. J. Relinque
Departamento de Ciencia de los Materiales e Ing. Met. y Q. I., IMEYMAT, Facultad de Ciencias, Universidad de Cádiz, Puerto Real, Cádiz, Spain

E. Matykina
Departamento de Ingeniería Química y de Materiales, Facultad de Ciencias Químicas, Universidad Complutense, Madrid, Spain

A. Santos-Coquillat (✉)
Université Catholique de Louvain, UCLouvain, Louvain Drug Research Institute, Advanced Drug Delivery and Biomaterials, Brussels, Belgium
e-mail: anamsant@ucm.es

S. J. Ikhmayies (ed.), *Advances in Biomaterials Research*, Advances in Material Research and Technology, https://doi.org/10.1007/978-3-031-97481-6_3

as a promising strategy for many applications, but still there are fundamental relationships between printing, pre-treatment, and post-treatment parameters of each biomaterial for clinical use have to be established. Notwithstanding, although 4D printing has been targeted as the future direction, more studies for its application in the biomedical field have to be implemented.

Keywords Additive manufacturing · 3D biomaterials · Vascularization · Implant · Prosthesis · AM printing technology · Metallic biomaterials · Ceramic biomaterials · Polymeric biomaterials · Tissue engineering · Smart biomaterials · Drug-loaded coatings · Hybrid coatings

List of abbreviations

4DP	4D Printing
5-FU	5-Fluorouracil
ABS	Acrylonitrile Butadiene Styrene
APIs	Active Pharmaceutical Ingredients
AM	Additive Manufacturing
Alg	Alginate
ASTMF	American Society for Testing and Materials
BJ	Binder Jetting
BCP	Biphasic Calcium Phosphate
BAPO	Bisacrylphosphrine Oxide
BSA	Bovine Serum Albumin
TPC	Copolyester Thermoplastic Elastomer
DEF	Diethyl Fumarate
CAD	Digital Computer-Aided
DLP	Digital Light Processing
DIP	Direct Inkjet Printing
DLS	Direct Laser Sintering
DMLS	Direct Metal Laser Sintering
DDS	Drug Delivery Systems
DMEM	Dulbecco's Modified Eagle Medium
EBM	Electron Beam Melting
DED	Energy Direct Energy Deposition
ECM	Extracellular Matrix
FDA	Food and Drug Administration
FDC	Fused Ceramic Deposition
FDM	Fused Deposition Modeling
hMSCs	Human Mesenchymal Stem Cells
hMPCs	Human Muscle Progenitor Cells
HUVECs	Human Umbilical Vein Endothelial Cells
HA	Hydroxyapatite

HPMC	Hydroxypropyl Methylcellulose
HB	Hyperelastic "Bone"
HIF1α	Hypoxia Inducible Factor 1α
LS	Indirect Laser Sintering
Fe	Iron
LCD	Liquid Crystal Display
Mg	Magnesium
MJ	Material Jetting
T_m	Melting Temperature
MAM	Metal Additive Manufacturing
MIG	Metal Inert Gas
MMT	Montmorillonite
MJP	Multi-Jet Printing
1D	One-Dimensional
SPPW	Photopolymer Waveguide
PLGF	Placental Growth Factor
PEO	Plasma Electrolytic Oxidation
PTMC	Poly (Trimethylene Carbonate)
PDLLA	Poly(D,L-Lactide)
PEEK	Poly(Ether-Etherketone)
PEG	Poly(Ethylene Glycol)
PEGDA	Poly(Ethylene Glycol) Diacrylate
PLA	Poly(Lactic Acid)
PLGA	Poly(Lactic-co-Glycolic Acid)
PPF	Poly(Propylene Fumarate)
PTMC	Poly(Trimethylene Carbonate)
PVA	Poly(Vinyl Alcohol)
PA	Polyamide
PCL	Polycaprolactone
PC	Polycarbonate
PDL	Poly-D-Lysine
PEEK	Polyether Ether Ketone
PEI	Polyetherimine
PPSF	Polyphenylsulfone
PP	Polypropylene
PU	Polyurethane
PAVICs	Porcine Aortic Valve Interstitial Cells
PBF	Powder Bed Fusion
RHB	Rhodamine B
RC	Robocasting
SLM	Selective Laser Melting
SLS	Selective Laser Sintering
SMP	Shape Memory Polymers
SL	Sheet Lamination
SBF	Simulated Body Fluid Solution

SOEA	Soybean Oil Epoxidized Acrylate
SCAP	Stem Cells from the Apical Papilla
SL	Stereolithographic
SLA	Stereolithography
SRMs	Stimuli-Responsive Materials
TPS	Thermoplastic Starch
TE	Tissue Engineering
Ti	Titanium
T_g	Transition Temperature
TCP	Tricalcium Phosphate
C3S	Tricalcium Silicate
TIG	Tungsten Inert Gas
2D	Two-Dimensional
TPP	Two-Photon 3D Printing
VEGF	Vascular Endothelial Growth Factor
Zn	Zinc
α-TCP	α-Tricalcium-Phosphate
β-TCP	β-Tricalcium-Phosphate

1 Introduction: Outlook and Clinical Requirements

Over the last several decades, Additive Manufacturing (AM) has been targeted as an **outstanding and flexible technology** by scientists and industrialists for biomaterials manufacturing [1, 2]. AM technology can reduce human body rejection of biomaterials, and it has been widely applied in prosthetics design, scaffolds for tissue engineering, biofabrication by 3D printing of organs, and most recently, drug delivery systems (DDS) [3]. AM allows freedom of design, hybrid materials fabrication, mass customization, and more importantly, reproduction of **complex geometries** [4]. Although the external structure of biomaterials has been described as being of critical importance for successful implantation, AM, in turn, allows for internal modulation of the resulting microstructure. The **internal shape** will influence the biomaterial's mechanical, biological and physical properties, thus eventually promoting improved tissue integration and regeneration [1]. Nevertheless, the level of development and proficiency needed for the clinical use of current 3D printed material has not been reached until recently. In order to achieve these features, further study of AM technologies is required to develop cost-effective designs with adequate core-related properties, and improved surface finishing to fulfill the in vivo performance requirements [5].

Different types of biomaterials are now being used in AM technologies. These include synthetic and natural polymers, metals, bioceramics, and composites, which usually are a synergic combination of the previous ones [6]. Their use is subject to their application and the AM methodology. Selection also relies on the material needs, defining their resolution and speed. Advanced manufacturing technologies allow for the production of **multi-part materials** with different compositions or structural gradients that would not be possible to obtain by old methodologies. For this reason, AM methodologies are being investigated for multiple applications. Among these, patient-specific mold-free **prosthetics** are one of the most investigated to date [7]. New implants are designed following the patient's anatomical features using imaging acquisition and computer-aided design (CAD). Many metallic implants used in hard tissue orthopedics are being cost-effectively designed by AM and daily used in the clinic. However, different challenges need to be overcome, including non-appropriate mechanical properties and low resolution of printed constructs improvement.

Another well-studied field has been the fabrication of scaffolds for **tissue engineering (TE)** [8]. To date in this field, bone and cartilage TE have been studied the most. One of the critical advantages of using AM for this application is related to the possibility of mimicking the tissue structure on the one hand and allowing cells to grow or differentiate on a 3D structure, promoting better osseointegration of the biomaterial on the other hand. Scaffolds that promote cell migration present particular biological and physical properties depending on their use as soft tissue or hard bone. Also, mimicking the extracellular matrix allows the enhancement of proliferation and cell adhesion, providing mechanical support and biochemical cues. However, the main challenge for the printing of these structures is the limited resolution of AM technologies as 20–100 μm cell size cannot be reached without compromising mechanical stability.

Moreover, AM fabricated scaffolds also permit in vitro biomaterial evaluation in a 3D environment previous to in vivo validation in contrast to conventional 2D cultures. These 3D models can be used in vitro to simulate a real organ environment for **disease processes or treatment response studies**. The use of 3D printed microscale tissues to evaluate drug screening has recently attracted attention as patient-specific or genetically compatible models that could provide results closer to in vivo drug response and toxicity [9]. Other in vivo approaches that include drug liberation could also benefit from AM being used as solid dosage forms, implantable or topical DDS. SPRITAM, an AM fabricated pharmaceutical for epilepsy treatment, was Food and Drug Administration (FDA) approved [10] and has been marketed since 2015. The use of AM to develop DDS helps modulate the pharmacokinetic and pharmacogenetic features of personalized medicine, where the characteristics of every patient are taken into consideration.

Additionally, more complex 3D printing technologies began to be developed in the last few years that imply the term **biofabrication**. Biofabrication encompasses the generation of assemblies matching the structure and composition of natural tissues [5]. It can be told apart from other AM technologies because of the incorporation of cells into the inks, making them bioinks [1]. This technology is highly challenging as the printing conditions have to match the physiological conditions for the cells to survive, and cell survival has to be ensured until the implantation process has been completed [3]. Many different hydrogels have already been proposed and employed as part of the bioink [11].

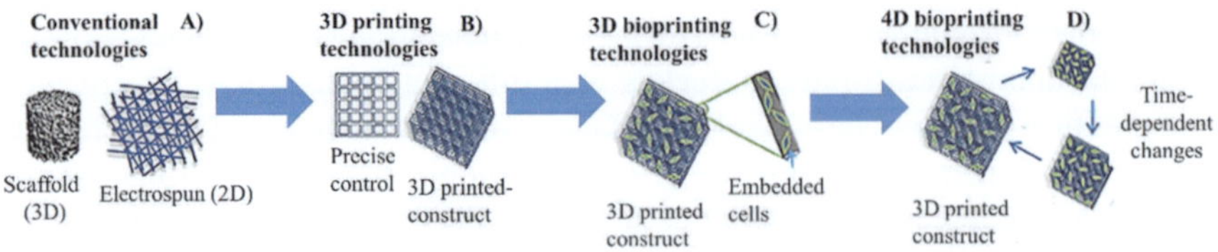

Fig. 1 Emergence of printing technology: **a** 2D/3D conventional, **b** 3D printing, **c** 3D bioprinting, **d** 4D bioprinting. Adapted from Kumar et al. [12], under the terms of the Creative Commons Attribution License (CC BY). https://creativecommons.org/licenses/by/4.0/

To date, AM technologies have been investigated in different areas such as surgery, pharmacology, oncology, and tissue regeneration. Regardless of the promising results of 3D printed biomaterials, the next generation of AM fabricated constructs will comprise smart materials employed for what has been named 4D printing, (i.e., 3D printed structures able to modify their structure as a function of a particular stimulus) and will be addressed in Sect. 3. Figure 1 shows how the technology requirements have impulsed the new strategies in AM of biomaterials.

This visual scheme in Fig. 1 shows that the emergence of 3D and 4D printing technology is a direct consequence of a current premise: To date, there is no perfect conventional 3D additively manufactured material for TE applications. This scenario launches a wide field of research and fundamental knowledge based on new printing technologies, new materials, and their possible synergistic combinations to achieve the maximum success rate from a biological and economic point of view. On that basis, biomaterial printing is one of the most promising breakthroughs and ground-breaking research fields to date, since the number of publications has increased exponentially in the last decade from about 600 to 8320 for the 2002–2022 period (used terms in the Scopus database: "additive manufacturing").

Therefore, this chapter aims to provide an overview of the state-of-the-art, current, and future perspectives in the field of additive manufacturing of biomaterials, i.e., from the manufacturing process technology to the direct biomedical application. Namely, the chapter is divided into seven sections to ascertain the current and future perspectives on the design of biomaterials for 3D and 4D printing:

(i) The current AM technologies used for biomaterial 3D printing from the experimental process to the core-surface level-based properties to fulfill the clinical trials (Sects. 1–5);
(ii) The latest drug-loaded 3D printed medicines and biomaterials as anatomical parts and their biological performance (Sect. 6);
(iii) 4D technology, from the current materials to the future perspective (Sect. 7);

The shared feature of these sections is the systematic discussion of the properties that make the current biomaterials suitable for 3D and 4D printing, their advantages, limitations, and the challenges that remain to date to fulfill the biomedical application.

2 Fabrication Technologies

Overall, the manufacturing method and the printing parameters determine the mechanical, chemical properties, and biocompatibility of the resulting implants. Therefore, the current goal to implant AM technology in biomedicine is to establish the most suitable printing technologies for each type of biomaterials [13, 14].

This is a critical step as, according to the **American Society for Testing and Materials (ASTMF42) committee**, there are seven different AM categories (Table 1): Powder bed fusion (PBF), Material Extrusion, Photopolymerization, Material Jetting (MJ), Sheet lamination, Binder Jetting, and Directed energy deposition (DED) [15].

Within each group, numerous technologies differ from each other in the layer formation (fusion or sintering), raw material input (wire or powder), energy source (laser, electron beam, etc.), scan strategy, process temperature, build chamber atmosphere, etc. [13–15]. As a general overview, Fig. 2 shows the experimental setup and the most widely used feed form for each AM technology listed in Table 1.

2.1 AM Technologies of Polymers

Among all the possible candidates to be used as biomaterials, polymers have been the group of materials more studied in AM to date. This can be explained by their **synthetic versatility** and **adaptability**, as well as the extensive range of physical and chemical properties [21]. Moreover, polymers can be used in different forms and compositions. To illustrate the wide variety of polymeric materials available, examples of polymeric materials employed in FDM include thermoplastics, particle-reinforced polymer composites, polymer matrix composites, fiber-reinforced polymer composites, thermo-responsive polymers, nanocomposites, thermoplastic elastomers composites just to mention few of them [22]. Another key factor of 3D printing techniques is the ability to use different materials in a single manufacturing platform to fabricate **multi-material objects** such as multicomponent bioinks containing two or more biomaterials [22]. The most extensively employed techniques and the materials adapted for each technology are summarized in Table 2. Nowadays, FDM, SLA, SLS, and Bioprinting are the most employed.

Table 2 also summarizes the limitations as well as the benefits of each technique. These reveal that the technique selected for the fabrication of a particular 3D printed part needs to take into consideration aspects such as the resolution required, the type of material, whether cells need to be embedded in the material, the mechanical properties and the printing rate. In addition to the advantages depicted above, printed models employed for pre-surgical planning can minimize the surgery time and blood loss of the patient [24].

Table 1 General classification of current AM technologies [6, 16, 17]

Printing technology	Suitable materials	Biomedical application	Advantages	Limitations
PBF (SLM, MLS, SLS, EBM)[a]	Metal Ceramic Polymers	Long-term and degradable implants: Dental, craniofacial and orthopedic applications	– Suitable for numerous biomaterials – No solvents and support materials are required – Fast printing, high quality, and accurate designs	– Expensive, medium resolution (100–200) μm and usually requires post-processing
Material extrusion (FDM)[b]	Polymers, Hydrogels Bioinks	Scaffolds for soft tissues	– Low cost and highly versatile in terms of freedom of designs and composite materials	– Slow printing rate – Medium resolution (100–200) μm
VAT polymerization (SLA, CLIP, TPP)[c]	Ceramic-polymer matrix	Bone tissue engineering	– Suitable for a wide range of polymers – High printing rate and precision – Optimal surface quality (50–100) μm for polymers	– Limited materials – Brittle ceramic materials – Post-processing is required
Material jetting (MJ)	Photocurable resins and bioinks	Scaffolds for soft tissues	– High resolution (20–100) μm and biocompatibility (cell viability)	– High material waste and low printing rate
Sheet lamination (SL)	Metal Ceramic	Macroscopic models	– Allows design of composite materials and no supporting structure is needed	– High material waste and low printing rate – Delamination and low resolution for accurate geometries (~1 mm)

(continued)

Table 1 (continued)

Printing technology	Suitable materials	Biomedical application	Advantages	Limitations
Binder jetting (BJ)	Metal Polymers	Degradable metallic implants	– Low-cost and high printing rate – Wide range of polymers – On-demand resolution (50–500) μm	– Requires post-processing: Poor adhesion between layers and surface quality – Low mechanical properties
Direct energy deposition (DED)	Metal	Still limited for temporal implants	– Promising for repairing applications – High printing rate – Suitable for composite biomaterials	– Expensive and low printing rate – Low resolution and requires post-processing

[a] Powder bed fusion (PBF), Selective laser melting (SLM), Direct metal laser sintering (DMLS), Selective laser sintering (SLS), Electron beam melting (EBM)
[b] Fused deposition modeling (FDM)
[c] Stereolithography (SLA), Multi-jet printing (MJP), Continuous liquid interface production (CLIP), two-photon 3D printing (TPP)

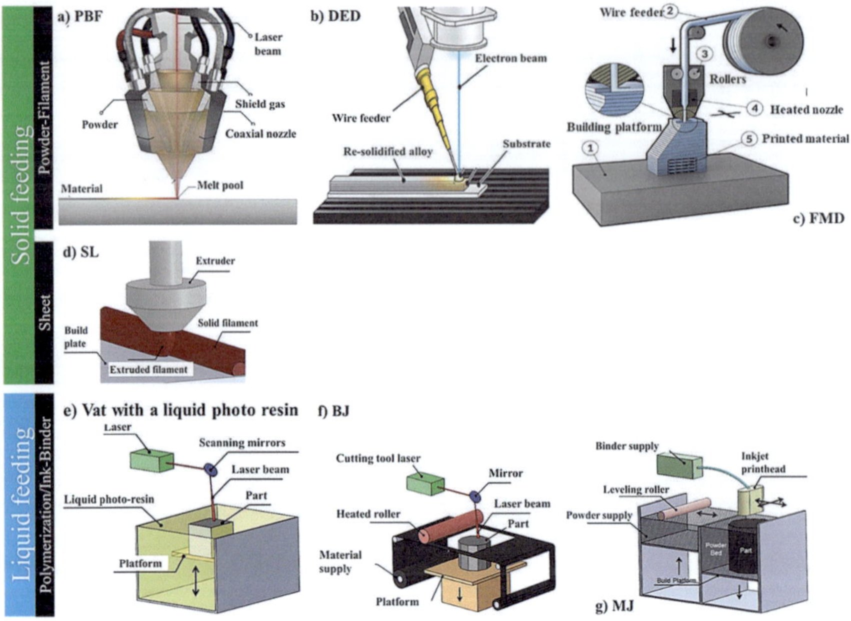

Fig. 2 Schematic illustration of the suitability of AM of three major types of biomaterials (polymers, ceramics, and metals) in their feed-forms and standard, using ASTM processes: **a** Powder bed fusion (PBF), **b** Direct energy deposition (by wire feedstock), **a**, **b** adapted from Selema et al. [18], under the terms and conditions of the Creative Commons Attribution (CC BY) license (http://creativecommons.org/licenses/by/4.0/). **c** Material extrusion (FDM) [19], **d** sheet lamination (SL), **e** VAT photopolymerization, **f** Binder jetting (BJ), **g** Material jetting (MJ). **d–g** Adapted from Razavykia et al. [20], under the terms and conditions of the Creative Commons Attribution (CC BY) license (http://creativecommons.org/licenses/by/4.0/)

2.2 AM Technologies for Ceramic Biomaterials

Overall, additively manufactured ceramic biomaterials have not been used in AM until recently as they present several processing difficulties due to their inherent properties, such as brittleness and hardness [25]. Further, bioceramics melting points are very high, making them incompatible with conventional thermal processing techniques such as FDM, and new phases can be formed when melted (Fig. 3) [5]. For this reason, there is no unique AM technology for ceramic materials. The selection of a specific one depends on the material properties and the targeted application. Furthermore, the final surface finishing and the core-dependent properties are defined by the selected AM technology. On the other hand, the structural dimensions, geometry, surface finishing, and inherent powder properties (e.g., refractive index and melting point) can be tailored by adjusting specific printing parameters of the chosen AM technology [26].

More specifically, **ceramic powders** with a high refractive index and melting point have shown to be **brittle** and **susceptible to thermal shock**, thus producing

Table 2 Additive technologies employed in the fabrication of polymeric 3D printed parts for biomedical applications

Method	Materials	Limitations	Benefits
FDM	Continuous filaments of thermoplastics: ABS, Nylon, PC, PCL, PLA, PLLA, PLGA, TPC, TPS	– Limited selection of materials used in the biomedical field – Thermoplastic polymer used in this method causes shorter lifetime and poor mechanical characteristics – Poor surface quality of printed objects – Not preferred for biomedical implants – Complications with the incorporation of cells or bioactive molecules into filament	– Diverse range of geometries – Complex structures and shapes – Readily available filaments – Time of manufacturing objects – Affordability
SLA	Functional monomers, macromonomers/ oligomers based on PCL, PEGDA, PDL, PLLA, PPF	– Limited number of materials – Limited number of potential materials (toxic substances during photopolymerization)	– High dimensional accuracy – Intricate details – Smooth surface finish
SLS	Polymer powders: PEEK/ HA, PCL, PCL/HA, PDLLA, PEEK, PLGA, PVA	– Using a high-temperature energy source – Expensive technique – Large dimensions of the SLS device – Resolution is dependent upon powder microstructure and the spot size of the laser	– Fast process – Manufacturing large and complex parts (irregular shapes and structures) – Material versatility – Small series in one manufacturing process – High partial accuracy
Binder jetting	Common polymers: Carbohydrates, Polyvinyls, Silica	– Poor adhesion between layers – Poor surface quality – Low mechanical properties	– Enables incorporation of biomolecules and drugs – The cheapest and fastest method – Wide range of used materials – High precision of manufactured objects

ABS (acrylonitrile butadiene styrene), PC (Polycarbonate), PCL (Polycaprolactone), PLA (poly(lactic acid)), PLLA (Poly(L-lactide), PLGA (poly(lactic-co-glycolic acid), TPC (Copolyester thermoplastic elastomer), TPS (Thermoplastic Starch), PEGDA (Poly(ethylene glycol) diacrylate), PDL (poly-D-lysine), PPF (Poly(propylene fumarate)), HA (Hydroxyapatite), PDLLA (Poly(DL-lactide)), PEEK (Polyetheretherketone), PVA (Poly(vinyl alcohol)). Adapted and reproduced with permission from Szymczyk-Ziółkowska et al. [23]. Copyright (2020), Elsevier

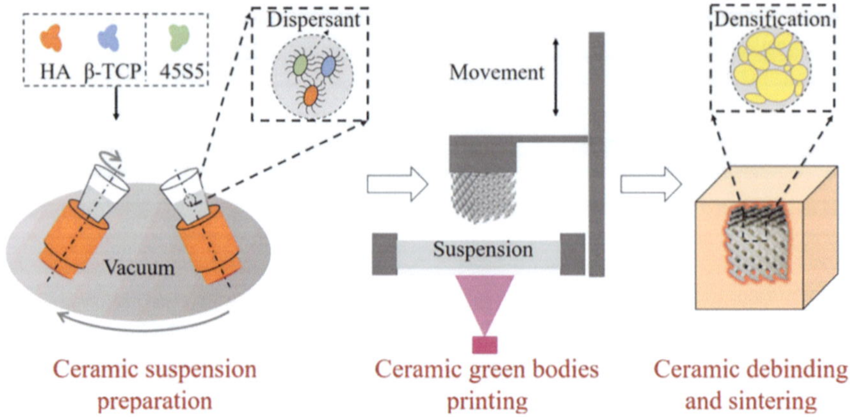

Fig. 3 Schematic illustration of ceramic scaffolds (HA-45S5 bioglass®) prepared by DLP. Reproduced from Hua et al. [34] with permission from Elsevier. Copyright (2021)

defect-containing and cracked surfaces. As a result, complex experimental setups and inert atmospheres are required, thus diminishing the initial advantages of AM [27].

On that basis, AM of ceramic biomaterials can be single- or multi-step processes. Multi-step AM processes are characterized by the need for subsequent thermal post-treatments to obtain the final ceramic components with improved surface finishing. These **multi-step processes** include sheet lamination, extrusion-based technologies (e.g., fused ceramic deposition (FDC), robocasting (RC)), VAT photopolymerization, direct inkjet printing (DIP), binder jetting, and indirect laser sintering (LS), among others [27–29].

On the other hand, direct laser sintering (DLS) and direct energy deposition (DED) processes can produce ceramic implants in a single step.

Among these AM technologies, research and development efforts in biomedicine have focused on Hydroxyapatite-(HA) and Silicon-(Si) based ceramic biomaterials where the residual porosity is to some extent beneficial in some applications, i.e., scaffolds for bone tissue engineering (Sect. 5) [5, 25]. In this scenario, the combination of ceramic powders with polymers and bio-inks has been targeted as the most promising approach to increase the suitability of AM ceramic implants in biomedicine. Namely, **VAT photopolymerization and DLS** technologies have recently been shown to be the most suitable processes for HA- and Si-based implants [25, 28–30].

The VAT photopolymerization encompasses different Lithography-based technologies processes such as Stereolithography (SLA) and digital light processing (DLP) [31–33]. Among them, DLP has been widely employed for the fabrication of 3D printed ceramics such as HA- and Si-based ceramic-biomaterials. In DLP, submicrometric ceramic powders are dispersed in the liquid VAT photopolymer to form a slurry cured by UV light [5, 27, 34]. Then, the liquid photopolymer is cured by UV radiation (Fig. 3). This allows the fabrication of fully dense, and net-shape parts (dimensions: ~1 mm–1 cm) with enhanced geometrical complexity and surface finishing comparable to SL [27].

The first advantage of VAT photopolymerization technologies is their industrial suitability (e.g., Lithoz, Prodways, 3DCeram, and Admatec) [27]. Although the different technologies based on VAT photopolymerization have a similar scientific basis, only the main advantage of DLP over SLA is its higher fabrication speed. In the case of DLP, the entire polymer layer is cured in one step, whereas SLA provides a spot-based photopolymerization method [32]. The main drawback of both technologies is the limited types of ceramic materials that are relatively transparent to UV light (i.e., silica, alumina, and HA) [5, 27].

The DLS method is encompassed within the PBF group, as explained in detail in Sect. 2; Fig. 3. In this case, the ceramic material itself is sintered/fused by the laser beam. Alternatively, the laser beam can induce a chemical reaction from a polymer [33, 35].

The most significant advantage of DLS over most other ceramic AM processes is its one-step nature, thus avoiding the need for thermal post-treatments and sintering processes. However, the thermal stresses during DLS remain a critical problem for ceramic materials, causing cracking and porosity [25, 29, 35]. The DLS process has been used to fabricate HA-silica and tricalcium phosphate (TCP) ceramic implants [36, 37]. Notwithstanding, the research in ceramic-based printing by DLS is still in the early stages [27, 30]. Nevertheless, promising results have been recently achieved in Si-based ceramics, as will be discussed in Sect. 5.

The development of a single-step AM process to produce defect-free and dense ceramic parts is considered the ultimate goal. Given the ambition of such a goal, most research is focused on the development and optimization of viable multi-step additive manufacturing.

2.3 AM Technologies for Metallic Biomaterials

ASTM F2792-12a standard recommends powder bed fusion (PBF), directed energy deposition (DED), and binder-based technologies such as binder jetting and material extrusion for AM of metallic biomaterials [38] (Fig. 4).

So far, for the current well-established metallic biomaterials (i.e., Ti, Fe, Zn, and Mg), **PBF-based technology** (Selective Laser Sintering (SLS), Selective Laser Melting (SLM), Electron Beam Melting (EBM), and Direct Metal Laser Sintering (DMLS)) were shown to be the most adaptable, feasible, and versatile technologies [36, 39]. Except for experimental differences, the overall basis of the PBF process consists of the metallic powder melting/sintering using different energy inputs by adding layer-upon-layer material in a controlled atmosphere.

Binder jetting is similar to PBF (Fig. 2). However, instead of applying thermal energy, an adhesive liquid is dispensed onto the surface of the powder bed, thus binding the powder particles to form the 3D-based component. Polymeric binders, particle suspension inks, metal salt compounds, and organometallic inks can be

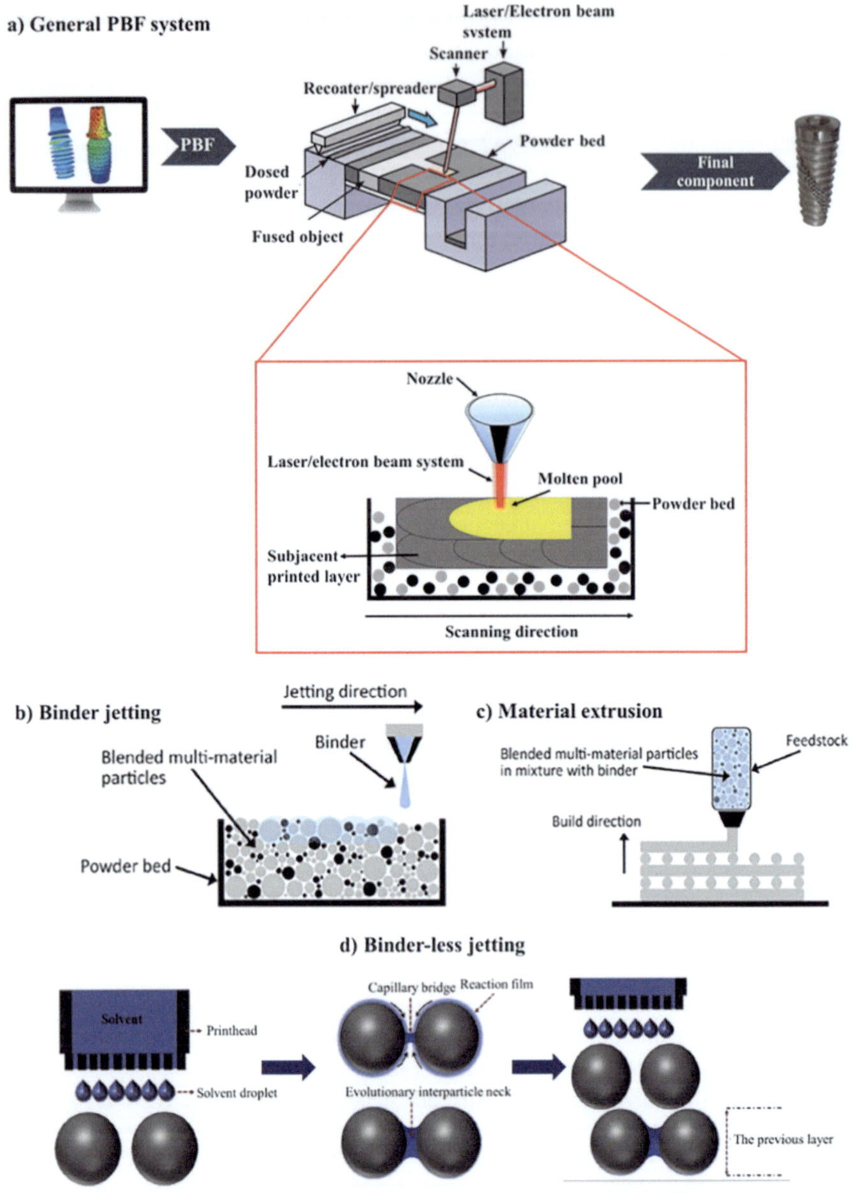

Fig. 4 AM processing techniques for metallic materials: **a** powder bed fusion. Adapted and reprinted with permission from Lowther et al. [47]. Copyright (2019), Elsevier. **b** Binder jetting, **c** material extrusion. **b**, **c** Adapted and reprinted from Putra et al. [42], under the CC BY license (http://creativecommons.org/licenses/by/4.0/), and **d** binder-less jetting. Note that in **a** either laser or electron beam can be used as an energy source; **b**–**d** are variants of binder-based AM printing, but in **d** the solvent, acting as a binder is evaporated during sintering, so de-binding is not required. **d** Adapted and reprinted with permission from Salehi et al. [43]. Copyright (2019), Elsevier

used in this regard. As a consequence of the decomposition of the binder, post-AM sintering causes structural shrinkage [40–42].

In this regard, binder-less jetting has been recently investigated to improve the quality of Mg printing. This process is quite similar to conventional binder jetting, except that, in this case, the capillary forces within the powder act as a binding agent to adhere to the powder particles. This technology requires (i) a single-phase solvent to attach layers of pure Mg powder, and (ii) subsequent heat treatment to sinter and harden the printed implant. So far, the available literature is practically non-existent, recent studies demonstrated that it is possible to achieve printed components with comparable compressive strength and elastic modulus to human bone. Therefore, more research is still needed to address its further application in biomedicine [43].

Extrusion-based AM technology works by continuously feeding metal or ceramic powder-based materials mixed with a polymeric binder through one or several nozzles to build up a 3D structure layer-by-layer. As an advantage, the resulting structures do not have the inherent drawbacks of the laser-based AM technologies described above. The main limitation of this technology lies in the printing resolution, so the design of a suitable multi-material raw material is the critical step for achieving complex structures with multi-material interfaces [42, 44].

Unfortunately, the application of AM is still put on hold for full-scale implementation in the biomedical sector due to a research gap pertaining to uncontrolled surface properties. AM components have an unreliable surface quality, coming from pores, microparticles poorly attached to the 3D printed part, roughness, or other defects and residual stresses, which negatively influence the corrosion resistance and their biocompatibility. Notwithstanding, although these technologies have been recently explored, printing parameters optimization as a function of the alloy composition is still a critical challenge for AM technology [13, 45].

Nevertheless, critical advances have been made over the last several years to overcome these limitations. Namely, in the case of Ti permanent implants, the biomedical industry has always shown great interest in additively manufactured Ti6Al4V alloy as most research has been focused on the direct microstructure-printing parameters relation. More specifically, SLM and EBM have been the most explored techniques for commercially pure titanium (CP-Ti) and its alloys (e.g., Ti–6Al–4V, Ti–Ni, Ti–Nb, Ti–35Zr–28Nb, etc.) [44, 46]. The current trends of additively manufactured Ti alloys for permanent implants are further detailed in Sect. 6.

In SLM, the product is manufactured by melting and fusing the powder material using a laser as the energy source (Fig. 4a). In succession, the powder bed is pushed down a predetermined distance depending on the desired layer thickness (Fig. 4b). Then, the process is repeated cyclically until the total 3D implant/component is printed. Notice that SLM is usually performed in an inert gas atmosphere (He/Ne, N_2) to ensure high purity and minimize the gas contamination and pores in the core material [42, 48].

On the other hand, EBM is another PBF AM method that, unlike SLM, is based on a high-energy electron beam as power input (Fig. 4a). Although SLM and EBM are analogous technologies, EBM is performed in a vacuum atmosphere. Besides, EBM is faster than the SLM method [45, 48]. However, process characteristics depend on the powder composition and the service requirements during the manufactured component´s lifespan. The particularities concerning their advantages and disadvantages in biomedicine are further outlined in Table 3.

For temporary implants (Fe, Mg, and Zn), the main property to take into account is the degradation rate in biological media. This phenomenon depends on the inherent properties of each material. Therefore, the AM technology and the printing parameters will differ as a function of the substrate. In particular, the corrosion rate of these metallic biomaterials can be ranked as follows: Fe < Zn < Mg [42, 52].

In the particular case of Fe and its alloys (e.g., pure Fe, Fe–Mn alloys, and 316L stainless steel), recent studies demonstrated the possibility of processing these materials by numerous AM technologies [53, 54]. Among them, **SLM** has been identified as the most suitable technology as the application of Fe in biomedicine is restricted to the development of **biodegradable porous implants** (as will be detailed in Sect. 5). In addition, processing parameters are similar for the abovementioned Fe-based alloys and stainless steels [52]. Different studies have pointed out that laser power and Fe powder particle size are the most critical parameters to obtain porous implants with different relative densities [55, 56]. Another factor to consider is the cooling rate to avoid high residual stresses and non-refined heterogeneous grains [52, 54, 56].

In the particular case of Mg and Zn, L-PBF technologies have been identified as the most suitable ones. Additionally, Mg can be processed by binder-based material extrusion methods (Sect. 5).

Table 3 General outlook of the current AM technologies for metallic biomaterials (Ti, Fe, Mg, and Zn) [15, 49–51]

AM technology	Advantages	Disadvantages	Application	Remarks
SLM	– Suitable for dense components in one step – No distinction between binder and melted phases – Post-processing is not always required	– Difficult control for composite material – High-quality finishing → ↑ cost of laser source and ↓ scanning speeds – ↑ Residual stress	– Craniofacial – Joint implants – Scaffolds	– Powder size: 10–45 μm – Layer thickness: 20–100 μm – Pre-heating: infrared or resistive heaters at 100–200 °C
EBM	– Time- and cost-effective – High scanning speeds	– Poor surface finishing – Post-processing is usually required	– Knee and hip implants	– Powder size: 45–100 μm – Layer thickness: 50–200 μm – Pre-heating: Electron beam – Chamber temperature 400–100 °C

For Zn-based alloys, its degradation rate is between that of Mg and Fe, which is considered optimal for biodegradable applications. From an experimental perspective, Zn is more volatile than Mg and can be melted in the absence of an inert atmosphere [52]. Nevertheless, the resulting additively manufactured Zn-based implants exhibit severe **spattering and porosity** [57, 58]. Notwithstanding, recent studies reported that pure Zn scaffolds and stents of electropolished L-PBF showed higher hardness and tensile strength than traditional processing methods [59, 60]. However, these **post-treatments** are costly, non-energy-efficient (massive electricity consumption), non-eco-friendly (high chemicals and water consumption), and time-consuming, thus essentially wiping out the initial purpose of AM. Consequently, detailed studies based on the control of the printing parameters, and the atmosphere are needed for Zn-based alloys by AM.

In this scenario, similar observations have been reported in Mg alloys due to their high potential for porous biodegradable scaffolds. However, at present, such implants are still in the early stages of development compared to other biomaterials [61, 62]. Again, the reasons are the inherent properties of Mg. Namely, the low difference between its melting and boiling points, the high flammability of Mg powders, and their high affinity for oxygen make the range of printing parameters optimization quite scarce [61, 63]. In particular, the resulting components exhibit poor interlayer adhesion and numerous surface defects (cracks and pores). Therefore, the number of additively manufactured Mg alloys is still minimal.

3 Additive Manufacturing with Polymers: From Thermoplastics to Bioinks

The following section reviews the state-of-the-art and future clinical application of AM strategies for polymeric materials that have emerged in recent years. This section encompasses the AM of polymers by the most feasible and suitable technologies (discussed in Sect. 2) from thermoplastic polymers to bioinks. Besides, the study of their adaptability and biocompatibility as composite materials to fabricate devices and layered scaffolds is also addressed.

3.1 Thermoplastic and Thermoplastic Elastomer-Based Filaments Employed in FDM

As shown in Sect. 2, in FDM thermoplastic and thermoplastic elastomers are deposited onto a fabrication platform upon melting the material and extrusion through a nozzle [64]. To be employed in this technology, the material must be able to flow through the nozzle during the heating and fusion step and then solidify upon cooling.

(a) *Homo and copolymers for FDM*

Thermoplastic and thermoplastic elastomers that can be melted upon heating and solidified by cooling are the materials employed for FDM (Fig. 2). Moreover, depending on their ability to partially form microstructured domains, these two types of materials can be either **semi-crystalline or amorphous**.

Amorphous thermoplastics and thermoplastic elastomers are more suitable for FDM rather than highly crystalline polymeric material which, as will be described afterward, are more appropriate for **SLS processes** (Fig. 2). The gradual variation of the viscosity as a function of the temperature that occurs in amorphous materials is a non-negligible advantage that significantly improves the extrusion process. As a result, the pressure conditions necessary to extrude these polymers can be easily modulated by changing the nozzle temperature. Overall, crystalline polymers show an abrupt phase transition due to the fusion of their crystalline domains upon heating. This transition changes the physical state of the polymer, i.e., from a solid to a low viscosity liquid. As a result, a homogeneous extrusion process is barely controllable as small temperature changes in the extruder can produce clogging due to polymer solidification [65, 66].

While amorphous polymers are more suitable than semi-crystalline, actually a wide range of both types of polymers is available nowadays for FDM, being the most extensively employed PLA (semi-crystalline) and ABS (amorphous). In particular, PLA is the most commonly used bioplastic derived from the starch of agricultural plants such as corn, sugarcane, sugar beets, and wheat [67]. PLA printing parameters and properties of the printed parts have been largely investigated. For instance, in a recent study by Yang et al. [68] PLA components were printed by FDM to investigate the effect of nozzle diameter, printing temperature, extrusion velocity (flow), printing rate, and layer thickness on tensile strength, surface roughness, and build time of printed parts. In this comprehensive study, several authors stated that both **nozzle diameter and layer thickness** are the critical factors on tensile strength, surface roughness, and build time of printed parts. They also demonstrated that using a larger nozzle diameter, high extrusion rate, printing speed, and larger layer thickness, the resulting 3D printed parts exhibited an increased tensile strength and surface roughness.

Other polymers of interest (also semi-crystalline) are **polycaprolactone** (PCL—FDA approved, biocompatible, and extensively employed for biomedical purposes) or **polyether ether ketone** (PEEK), which are currently being extensively employed in FDM in particular for biomedical purposes. An illustrative example of the use of PCL for biomedical applications was reported by Zein et al. [69]. In particular, they fabricated scaffolds by FDM with different geometrically consistent honeycomb-like patterns with controllable porosity and channel size using a bioresorbable PCL-based polymer. Moreover, they explored the use of variable filament circular cross-sections and their role in the mechanical properties of the 3D printed structures. Another related example was more recently reported by Szojka et al. [70]. In this work, PCL 3D structures for biomedical applications were prepared. The authors designed and produced biomimetic 3D printed meniscus scaffolds, using small porous PCL blocks that mimic the circumferential and radial type I collagen orientation found in the native meniscal tissue. Interestingly, in this work, the model was manually sliced into 37 layers which were customized to have either circumferential or radial fibre orientations. Briefly, even-numbered layers were given a circumferential orientation using Boolean operations with a template of concentric rings (Fig. 5), whereas odd-numbered layers were radially oriented and deposited in a 90° grid pattern. The scaffolds reported may provide a template for biomimetic tissue growth, representing a promising strategy for complete meniscus TE.

In addition to PCL, **PEEK is another excellent candidate** employed in FDM with the potential for different medical applications. PEEK is widely known, not only for its excellent mechanical properties and chemical stability, but also for its **biological stability** and **biocompatibility**, which make this material suitable for a wide myriad of biomedical applications. These include the development of orthopedic implants, scaffolds, artificial organs, surgical guides and templates, prostheses, and drug vessels. In these cases, the manufacturing of PEEK-based medical devices with conventional manufacturing technologies is a rather difficult task due to the existence of **patient specific parameters** that generally demand special specification [71]. Giannatsis et al. [72] reported an illustrative example of the fabrication of orthopedic implants and load-bearing devices, such as hip joints by 3D printing of PEEK. Recent works focused on the creation of fully customized, patient specific, anatomic models, and products to help surgeons in understanding and replacing faulty bones and limbs [73].

In addition to the examples depicted above, some other polymers have been equally employed in FDM that have potential in different biomedical applications. These include **polycarbonate (PC), polyamide (PA/Nylon), polyphenylsulfone (PPSF)** or **high-density polyethylene**, and **thermoplastic elastomers** such as thermoplastic polyurethanes [74, 75].

The polymers depicted above can also be employed in the elaboration of both **blends** (mixtures of two or more different polymeric materials), and **polymer composites** improving the properties of the precursors.

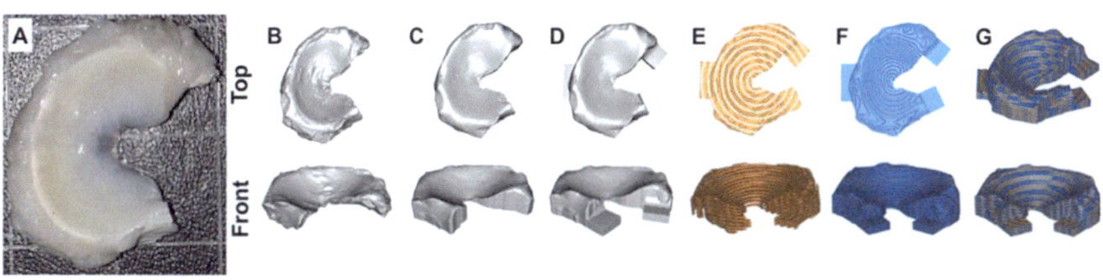

Fig. 5 Sequential process of the meniscus fabrication: from the real sample to the printed part. Fresh meniscus sample (**a**); Raw 3D-scanned meniscus model (**b**); Model processed to remove artifacts (**c**); Model with suture tabs for attachment to the tibia (**d**); separation of layers in circumferentially oriented fibers (**e**), and radially oriented fibers (**f**). Reproduced with permission from Szojka et al. [70]. Copyright (2017), Elsevier

(b) *Multicomponent polymeric materials: use of polymer blends and composites in FDM*

Blending is one of the **easiest and cheapest strategies** to develop novel materials that combine the properties of two or more polymers. An illustrative example of this approach is the commercially available Ultem™ 9085. This material is a blend of PC ($T_g \approx 147$ °C) and polyetherimine (PEI) resin ($T_g \approx 216$ °C). These materials are compliant with the OSU 55/55 standard for aerospace and confined space applications due to their interesting properties as a flame retardant. Compared to PEI, Ultem™ 9085 presents a lower T_g (≈ 186 °C) due to the addition of PC in the mixture; and improved ductility, flowability, and transparency. Besides, Ultem™ 9085 shows good chemical resistance to organic solvents and outstanding mechanical properties including high strength and stiffness. Additionally, it has good chemical resistance to organic solvents [76].

Another interesting example has been recently reported by Kosorn et al. [77]. The authors described the preparation of polymer blend scaffolds of PCL/poly(3-hydroxybutyrate-*co*-3-hydroxyvalerate) (PHBV) fabricated using FDM. The effect of printing parameters, such as filament feeding speed and hatch speed, on filament diameters and pore sizes, were also studied. Various ratios of PCL: PHBV blending including 100/0, 75/25, 50/50, and 25/75 were used. Interestingly, upon cytotoxicity testing, all scaffolds were non-toxic to cells even if they were immersed in phosphate buffer for 150 days. For use as cartilage scaffolds, the proliferation of porcine chondrocytes and total glycosaminoglycans (GAGs) extracted from the cultures on the surface-treated scaffolds with different blending ratios were investigated. The scaffold with a blending ratio of 25/75 exhibited the highest proliferation and total GAGs.

Another type of multi-component material is the **micro/nanoscale polymer composite**, which involves the use of different types of fillers such as **microparticles** and **ceramic powders, fibers,** or **even piezoelectric materials**. Micro and nanoscale composites have been largely employed to improve the properties of the original polymers, typically improving their mechanical properties (elastic modulus, flexural modulus, compressive strength, compressive modulus, etc.) [78] or providing the base material with other additional characteristics. Indeed, the filler employed can improve the functional properties of the material such as its anti-bacterial behavior. Table 4 presents some examples of polymer-based composites containing ceramics as the main materials used for bone scaffold fabrication using FDM.

3.2 Biocompatible Resins for 3D Printing by Photocuring

Most 3D printing based on photocuring biocompatible resins are included in the **vat polymerization category** (Fig. 2), including different techniques, such as SLA, DLP, liquid crystal display (LCD), multi-jet printing (MJP), continuous liquid interface production (CLIP), and two-photon 3D printing (TPP), among others [79]. The

Table 4 Previous studies on FDM filaments and bone scaffold fabrication for biomedical applications

Materials	Application	Ceramics-polymer composition	Mechanical strength
HA, PCL, MMT (Montmorillonite)	Fabrication of FDM filament	HA:PCL:MMT (0–10%:88–100%:0–4%)	Elastic modulus: 155.6 MPa (10:92:2) Flexural modulus: 185.10 MPa (10:88:4)
HA, PCL	Bone scaffold	30% HA:70% PCL	Compressive strength: 15.43 ± 1.28 MPa Elastic modulus: 80.16 ± 3.18 MPa
HA, TCP, DCB, PCL	Bone scaffold	30% HA/TCP/DCB:70% PCL	Compressive modulus: 338 MPa (PCL-HA), 253 MPa (PCL-TCP), 241 MPa (PCL-DCB)
HA, β-TCP, PLGA	Bone scaffold	75% β-TCP:25% PLGA HA used as a coating material	–
β-TCP, PCL	Bone scaffold	20% β-TCP:80% PCL	–
HA, nylon	Fabrication of FDM filament	50% HA:50% nylon	–
HA, PP (polypropylene)	Bone scaffold	20.5% HA:51.7% PP:13.4% wax:14.4% plasticizer	Tensile strength: 7.0 MPa Tensile modulus: 733 MPa
TCP, PCL	Bone scaffold	20% TCP:80% PCL	–
HA, PCL	Bone scaffold	15% HA:85% PCL	–
PLA, TCP	Fabrication of FDM filament	2.5% TCP:97.5% PLA 5% TCP:95% PLA	–
PA, HA, zirconium oxide (ZrO_2)	Fabrication of FDM filament	85% PA:5% HA:15% ZrO_2	Tensile strength: 22.78 MPa Tensile modulus: 761.93 MPa
PA, β-TCP, ZrO_2	Fabrication of FDM filament	PA:ZrO_2:β-TCP (60–70%:15%:15–25%)	Tensile strength: 24.25 MPa (70:15:15) Flexural strength: 29.06 MPa (70:15:15) Tensile strength: 23.20 MPa (65:15:20) Flexural strength: 28.83 MPa (65:15:20) Tensile strength: 22.57 MPa (60:15:25) Flexural strength: 30.03 MPa (60:15:25)

(continued)

Table 4 (continued)

Materials	Application	Ceramics-polymer composition	Mechanical strength
PLA, HA	Bone scaffold	50% PLA:50% HA	Stiffness: 124.04 ± 25.21 MPa
PVA, β-TCP	Bone scaffold	5–20% TCP:80–95% PVA	–

References are included in the original work. Adapted with permission from Mohd Pu'ad et al. [78]. Copyright (2020), Elsevier

main advantages of these technologies are the **high printing rate, high precision, and the surface quality** of the printed objects which present a rather smooth aspect. However, a major limitation, still under investigation, is the performance of photo-cured material, which is usually **too brittle**, **tends to deform, shows poor weather resistance**, and, in general, presents rather **low biocompatibility**. As a result, up to date, the 3D printed parts produced with this technique are mainly employed in the dental field (dental restoration, dental orthodontic, or dental surgery) as temporary replacement materials, or simply as models for surgeons to simulate some complex operations.

Initial photosensitive monomer formulations employed for the fabrication of 3D printed parts mainly focused on technical aspects, including the photopolymerization **speed**, **layer exposure time,** and **polymer viscosity**. Commonly, these printed materials comprise a photoinitiator, crosslinking agents, and monomers. In the case of acrylic-based resins, the monomeric precursors use to be cytotoxic, unlike the polymeric form. After printing, cytotoxicity may be related to the unreacted double bonds from the multifunctional monomers employed, residual photoinitiator, and eventually, other components included in the formulation that could be leached from the 3D printed part.

Two types of formulations are currently being employed in photocuring 3D printing, particularly in SLA and DLP. Two representative examples are (i) **acrylate-based photosensitive resins**, which form rigid parts by radical-free photopolymerization, and (ii) **epoxy-based formulations** that polymerize by cationic photopolymerization [80, 81]. Several differences can be found between these types of formulations and, therefore, the material selection is usually made depending on the final application of the 3D printed part. Acrylate formulations **photopolymerize faster** than epoxy-based resins so that, when time is a critical factor, this type of resin allows faster fabrications. Conversely, photopolymerizable epoxy-based resins present **lower volume shrinkage and good dimensional stability** and, as a result, **better mechanical properties** than acrylates ones.

Among the available polymers that can be processed by SLA/DLP, poly(propylene fumarate) (PPF) and photocurable derivatives of PCL, poly (lactic acid) (PLA) and medical polyurethane (PU) stand out due to their high biocompatibility for TE applications [82–84]. Notwithstanding, SLA/DLP still need to be improved for their use in biomedical applications.

Elasticity and flexibility may be critical characteristics in some cases. In this sense, photocuring 3D printing and, in particular, SLA has been employed to produce flexible and elastic three-dimensional structures that, in comparison to hydrogels typically obtained by bioprinting, presented enhanced mechanical resistance. Petrochenko et al. [82] used a photosensitive resin composed of polyurethane diacrylate (UDA) resin (namely Genomer 4215), and two photosensitive diluents: 2-hydroxy-ethyl acrylate and glycol diacrylate. This resin was used to print scaffolds with SLA 3D printing technology. A honeycomb scaffold with porosity of up to 60% was obtained and mesenchymal stem cells were able to adhere better to the surface. A representative example of flexible structure fabrication was reported by Schüller-Ravoo et al. [85]. The structures were fabricated in flexible and elastic poly(trimethylene carbonate) (PTMC) using a three-armed methacrylate PTMC macromer. Moreover, chondrocytes were able to differentiate over the material after 6 weeks of culture.

3.3 Polymer Powders, Their Use in SLS and Biomedical Applications

SLS presents **important advantages** for the fabrication of 3D printed parts with interest for different biomedical purposes. On the one hand, SLS 3D printed parts exhibit good **accurate detail** and **sharp reproduction**. These two characteristics are directly dependent upon the **powder morphology** (particle size distribution, and particle shape), fluidization, flow behavior, and printer characteristics and parameters. Other key parameters include the laser employed, such as the spot size of the laser or layer thickness [86, 87]. On the other hand, SLS 3D printed parts exhibit isotropy in terms of mechanical properties [23]. Note that 3D printed parts fabricated using SLS show comparable mechanical properties to those found in components prepared by conventional injection molding and similar techniques [88]. In addition, a major advantage of SLS is related to the **great variety of thermoplastic and thermoplastic elastomeric** materials that may be employed while its capability to process ceramics and composite powders is limited [89].

Thermoplastic and thermoplastic elastomers typically employed in SLS [90], may present either semi-crystalline or amorphous characteristics. From a morphological point of view, the main difference between these two types is related to the molecular arrangement acquired after solidification. Whereas an amorphous material exhibits randomly arranged molecular chains, a semi-crystalline material presents chains and backbone interconnections with an ordered structure. The structural variation between both parts confers different thermal properties to each type of material. Besides, this is strongly influenced by the SLS printing parameters due to the inherent physicochemical properties of crystalline polymers, as stated in Sect. 3.1. In both cases, upon heating, a glass transition temperature (T_g) is observed. The majority of semi-crystalline polymers have a glass transition temperature (T_g) below or close to room temperature (between −100 and 50 °C) while the amorphous polymers typically employed have a T_g close to 100 °C. Moreover, semi-crystalline polymers present a melting temperature (T_m) above 100 °C (between 100 and 400 °C) in which the material rapidly changes from a solid to a liquid form. This transition does not exist in amorphous polymers, which present a progressively viscous behavior with no clear phase transitions. When semi-crystalline polymers are employed in SLS, the laser heats the material above the T_m. In this situation, the material remains solid until an appropriate amount of heat is absorbed transforming it into a viscous liquid. On the opposite, amorphous polymer powders are consolidated upon heating above the T_g. In comparison to semi-crystalline polymers, amorphous materials are much more viscous and therefore require larger heating to establish the appropriate particle interconnection in order to form a stable 3D printed part.

Polymeric biomaterials and their composites employed in SLS for TE scaffolds fabrication include nylon, pure PCL, poly(D,L-lactide) (PDLLA), PEEK, poly(lactic-co-glycolic acid) (PLGA), poly(vinyl alcohol) (PVA), composite of PCL/HA or PEEK/HA [86, 87]. Among them, **Nylon** is the most extensively studied by SLS. This is due to the balance of **chemical and mechanical characteristics** of polyamides (tensile strength of 30–40 MPa [91]) that offer **flexibility, good stability, shock resistance, and rigidity**. **PCL** is also an extensively employed material for SLS due to its good **biocompatibility**, and **processability**, thus being a suitable option in TE [92] or cartilage repair applications [93]. Illustrative examples of scaffolds reported in the literature using different polymers and polymer composites are listed in Table 5.

3.4 Bioinks: 3D Printing with Living Cells

The term "bioplotting" was first introduced by Landers and Mülhaupt in the early 2000's and has gradually evolved to "bioprinting" [97]. Initially, Landers et al. [97] defined bioplotting as a type of AM technology used for cell/organ printing via the deposition L-(layer-by-layer) of **living cells**, **encapsulated** within a proper **support matrix** (typically biodegradable polymers and biopolymers), with **controlled architecture** and **defined placement** in the mild condition [98, 99]. However, more recent definitions claimed that bioprinting encompasses the utilization of different AM technologies to **3D print living cells**, and is not considered a specific AM technique on its own [6].

Bioprinting requires the use of "bioinks" to create **tissue-like micro-, macro-,** and **mesostructures** which can be used in biomedical or TE applications. This procedure is fully used for **regenerative medicine** purposes, specifically in order to provide solutions to obtain viable tissue and organs for transplantation. The proper choice of support materials to fabricate **scaffolds, cell types, growth,** and **differentiation** is among other crucial aspects to define suitable bioprinting strategies to overcome challenges and limitations in this field.

Table 5 Illustrative examples of polymeric materials employed for bio-related applications using SLS

Material	Physical properties	Biological properties	Final part	References (Adapted with permission from)
PCL	Porosity: 85% Pore size: 40–100 μm	High cell density was observed after 6 days		[94][a]
	Porosity: 40–84%	A confluent cell monolayer (elongated morphology) could be observed on the scaffold		[95][b]
Calcium phosphate (CaP)/ PHBV Carbonated HA/PLLA	Porosity: PHBV 62.6% PLLA 69.5% CHA/PLLA 66.8%	All scaffolds facilitate cell proliferation and ALP expression for SaOS-2 cells. Viability assays on sintered scaffolds were performed after 3 days of culture		[96][c]

[a] Reprinted from Yeong et al. [94], with permission from Elsevier. Copyright (2010)
[b] Reprinted from Sudarmadji et al. [95], with permission from Elsevier. Copyright (2011)
[c] Reprinted from Duan et al. [96], with permission from Elsevier. Copyright (2010)

Many bioinks have been developed from **hydrogels**, which are very well established as suitable materials for 3D cell cultures [100, 101]; they have very **good biocompatibility** and are **highly tunable** regarding their physical, mechanical, and biological properties [102, 103]. Hydrogels commonly show non-Newtonian, shear-thinning behavior, which makes them suitable for extrusion bioprinting. The first step necessary for carrying out the bioprinting methodology is to design the CAD model which will be used for printing the part. This can be accomplished via several **medical imaging techniques** such as computed tomography, magnetic resonance imaging, and optical coherence tomography, among others. Once the model is obtained (i.e., the CAD model), the next step is to choose the materials for the specific application and define the most appropriate printing path according to the **clinical needs** and **cell characteristics** [99]. This methodology can also be used for printing **heterogeneous organs** which possess different kinds of cells. The definition of the printing path is crucial because, in some cases, cell differentiation can only be achieved in certain topological and chemical environments, some examples (ear cartilage, kidney, and tooth) are shown in Fig. 6. In this case, the hydrogels are working as cell-laden vehicles, which also served as support for achieving optimal cell growth and differentiation, if it is necessary. Once the printing process is finished, in some cases, subsequential post-processing is needed in order to extract the supporting part. Then, the tissue/organs can be **implanted directly** in the patient and either permit the cells to grow freely or can be cultured externally to generate artificial tissue or organs that will be further implanted. Photocrosslinked hydrogels form a suitable matrix for tissue growth. Eventually, hydrogel matrices are biodegradable in time and they are replaced by new extracellular matrices arising from growth cells [104].

The chemical composition of the cell-laden support plays a key role in the process. For instance, Pan et al. [105] studied **gelatin/alginate** hydrogels by different crosslinking methods (alginate with $CaCl_2$ and gelatin with glutaraldehyde), and their effect on scaffold architecture, porosity, interconnectivity, pore size, and pore distribution were described. These results demonstrated that the material has the potential as a scaffold with the ability to **encapsulate cells or drugs**. The use of this technique allows for depositing different materials and immobilized cell types on porous scaffolds. Note that the disadvantage of using this method is the shear stress generated from the variable sized nozzles, which may impact negatively cell viability during the printing process [106].

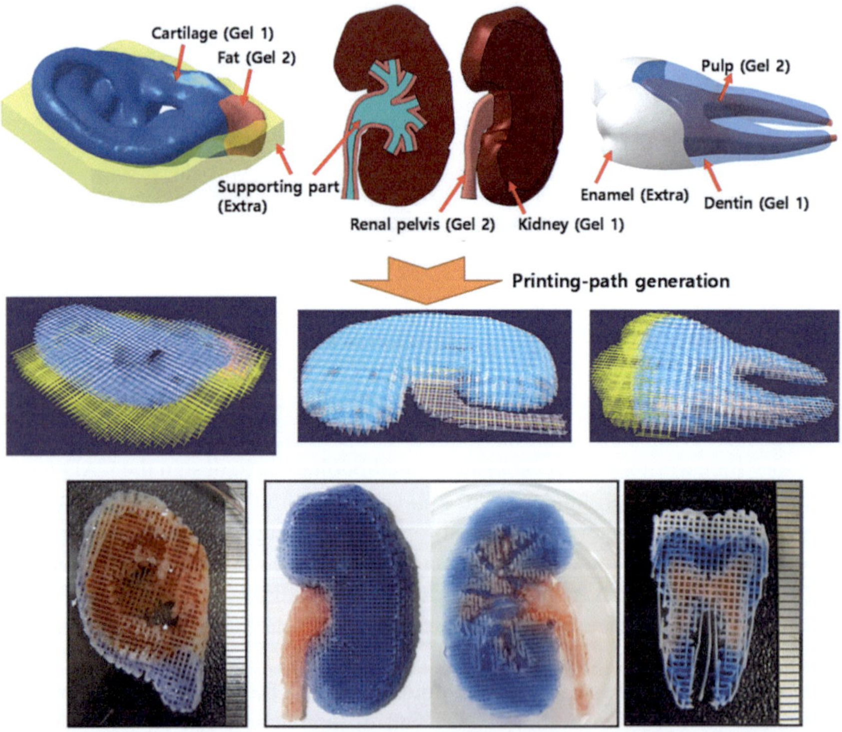

Fig. 6 CAD model acquisition, and generation of the printing path for different types of organs. Reproduced from Jung et al. [99], under the terms of the Creative Commons CC BY license. https://creativecommons.org/licenses/

Table 6 Recent works on bioinks and TE applications (ECM: extracellular matrix; hMSCs: human mesenchymal stem cells; hMPCs: human muscle progenitor cells; HUVECs: human umbilical vein endothelial cells; PAVICs: porcine aortic valve interstitial cells; SCAP: stem cells from the apical papilla; PEG: Polyethylene glycol)

Bioink	Cell type/tissue	References
Alginate	Chondrocytes	[108]
Agarose	Chondrocytes	[109]
Collagen	Hepatocytes/liver	[110]
Fibrin	hMPCs/skeletal muscle	[111]
Gelatin	HUVECs/vascular structures	[112, 113]
	hMSCs/bone	
	hMSCs/cartilage	
Gellan gum	Chondrocytes/cartilage Osteoblasts/bone	[114]
Hyaluronic acid	Fibroblasts	[115]
Silk/PEG	Fibroblasts	[116]
Tissue-derived ECM	SCAP/dentin Kidney	[117, 118]

In addition to the alginate/gelatin, other types of materials utilized using bioprinting technology are **collagen, chitosan,** or **collagen alginate-silica composites coated with HA**, among others. Guo et al. [107], studied the effect of PLGA composition on scaffold biodegradation performance, mechanical properties, and water adsorption capacity, using five different types of PLGA with variable molecular weight to repair osteochondral defects. Table 6 summarizes representative studies of different illustrative strategies to fabricate several tissues.

Despite the intensive work carried out in this area, important aspects are still under investigation including the improvement of the mechanical properties of the supports employed or the elaboration of multicellular constructs.

Although significant efforts have been made in the last decade concerning in vitro bioprinting of biological structures, standard treatments for tissue regeneration using these techniques are far to be implemented in clinical therapies.

4 Additive Manufacturing of Ceramics

Biomaterials based on ceramics (bioceramics) are widely used in orthopedics and dentistry daily to repair or replace hard tissue as they simulate the bone structure and mineral composition. Bioceramics can be considered as both **crystalline** and **amorphous inorganic** biomaterials. Their AM is of great interest to obtain patient-tailored constructs and improve their implantation success as stated in Sect. 2. In this section, the most recent developments of HA and Si-based biomaterials processed by DLS and DLP technologies will be discussed.

4.1 Calcium Phosphate-Based Ceramics

Calcium phosphate-based materials are among the most used bioceramics. Current trends focus on the study of **HA** ($Ca_{10}(PO_4)_6(OH)_2$) for bone graft applications and as a coating for dental or orthopedic implants. This is mainly due to its contribution to bone regeneration, excellent biocompatibility, and compatibility with other biomaterials. Despite the successful clinical history of conventionally processed HA-based materials, the enhancement of their mechanical properties and internal structure simulating the interconnected porosity of bone tissue can be obtained by AM.

Bioceramics also include other Ca/P-based compounds such as amorphous calcium phosphates (Ca/P: 1.2–2.2), **α-tricalcium-phosphate** (α-TCP, Ca/P: 1.5, highly resorbable, solubility 2.5 mg/L), **β-tricalcium-phosphate** (β-TCP, Ca/P: 1.5, slower resorption than the α-form, solubility 2.5 mg/L) among others [12]. β-TCP is being used as a craniofacial filling material due to its biodegradability and high wear resistance. Nevertheless, from a technological point of view, it presents the following challenge: the sintering temperature must not reach 1100 °C as it will transform into α-TCP, more reactive and soluble in physiological media [119].

In this scenario, the Ca/P ratio is the main factor that affects bioceramics outcomes such as their dissolution capacity. Calcium phosphates such as β-TCP present relatively high solubility (0.5 mg/L) and acidic nature due to the reduced Ca/P ratio (Ca/P: 1.5) compared to that of HA with a high Ca/P ratio (Ca/P: 1.67, solubility 0.3 mg/L). These properties determine the necessity of using a **polymeric precursor** for their 3D printing as composite materials [120].

Therefore, TCP- and HA-based scaffolds cannot be printed without using the combination of a biopolymer to improve their binding interaction and mechanical properties during processing [12]. As a consequence, these scaffolds are manufactured as a **composite material** when the ceramic powder is incorporated into a polymeric filament that provides **flexibility, toughness, and biodegradability**. Nevertheless, it is still a challenge to provide a homogeneous distribution of the ceramic powder within the polymer matrix [121].

State-of-the-art AM technology has addressed the combination of Ca/P-based bioceramics with both hard and soft polymeric matrices. The matrices considered as hard are focused on **load-bearing applications**. These matrices usually consist of polymers such as polylactide and PCL. By contrast, soft matrices allow the possibility to directly load and print viable cells into the composites, thus allowing a uniform cell distribution in the printed components. **Collagen** and **chitosan** have been targeted as the soft matrices most relevant to AM [122]. A recent preclinical study (Fig. 7) with an α-TCP printed as a mixture with 35 wt.% aqueous solutions of poloxamer 407 resulted in bone growth inside the porous 3D scaffolds in a sheep model (Fig. 7e) [123].

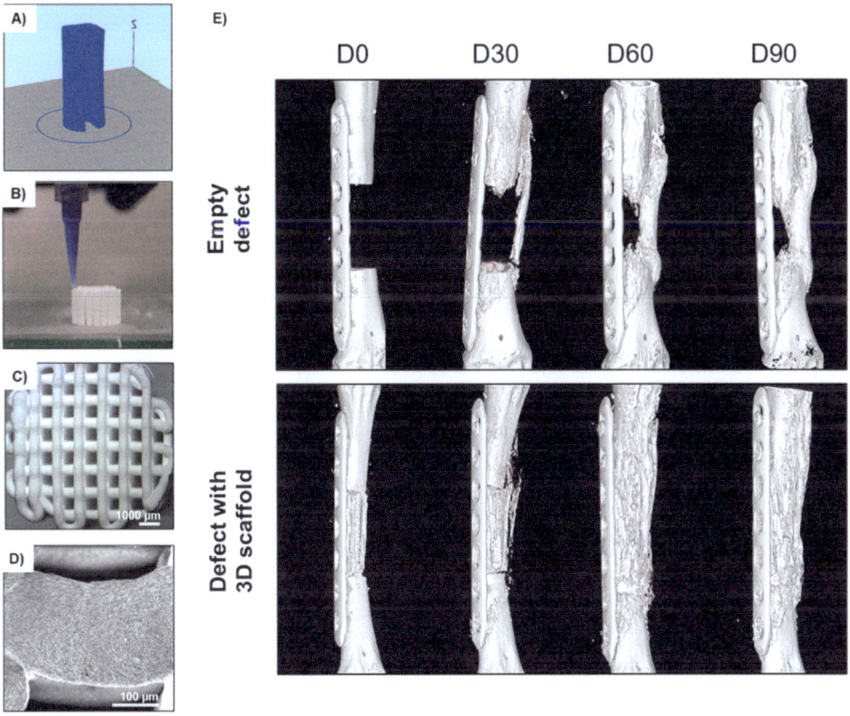

Fig. 7 Fabrication and structure of a 3D-printed customized calcium phosphate scaffold. **a**, **b** Design and printing of a calcium phosphate scaffold by 3D-microextrusion. **c**, **d** Optical and scanning electron microscope (SEM) images of the 3D-printed scaffolds. **e** 3D bone reconstructions of metatarsal defects in sheep. CT scans are shown post-surgery (D0) and after 30, 60, and 90 days (D30, D60, D90) of healing. Adapted from Vidal et al. [123], under a Creative Commons Attribution 4.0 International License. http://creativecommons.org/licenses/by/4.0/

Other examples include the use of HA with chitosan-genipin [124] for bone tissue engineering applications. Figure 8a, b shows the structure of the resulting scaffolds when the printing parameters are modified. The in vitro evaluation showed that osteoblastic cells were able to grow and proliferate over the samples (Fig. 8c, d).

3D printing of Ca/P-based ceramics presented great challenges through the years due to their **intrinsic properties**. Therefore, it is only recently that they have been evaluated for clinical use. One of the successfully evaluated strategies encompasses the use of a composite based on 90 wt.% HA and 10 wt.% PCL or PLGA. This composite material formed using extrusion-based 3D printing at room temperature has been called **hyperelastic "bone"** [125, 126]. It has already proved the **formation** and **vascularization of new bone** in preclinical models, and it is currently being under consideration to be FDA-approved.

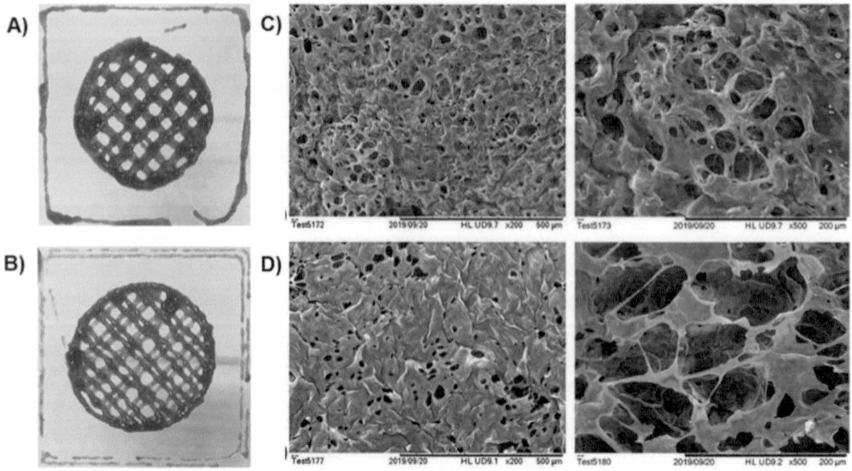

Fig. 8 **a**, **b** HA-Chitosan scaffolds fabricated by the use of different printing parameters. **c**, **d** SEM micrographs of MG63 series osteoblasts seeded on 3D printed HA-Chitosan scaffolds for 72 h of cultivation, and 7 days of cultivation. Adapted with permission from Zafeiris et al. [124]. Copyright (2021). Elsevier

4.2 Si-Based Ceramics

Within bioceramics, another worthy category of 3D printed materials employed for hard tissue replacement is Si-based biomaterials. This group encompasses **calcium silicates** (tricalcium silicates, β-calcium silicates), **bioactive glasses** (silicate-, borate-, borosilicate, phosphate-, doped-, and mesoporous bioactive glasses), and **bioactive glass–ceramics** [12]. In detail, bioglasses can **upregulate osteogenesis** and present **high tissue bonding**. Conversely, their use in load-bearing applications is challenging due to their low mechanical strength, high brittleness, and low fracture toughness [127]. To be used for these applications, new composite biomaterials based on a Biosilicate®-like matrix have been developed [128].

More recently, 45S5 bioglass® scaffolds were fabricated with biphasic calcium phosphate (BCP) by DLP technology (Fig. 9a, b). The 45S5 bioglass® and BCP bioceramic showed enhanced mechanical properties and biocompatibility compared with stand-alone BCP, thus pinpointing the DLP technology as the one enabling the fabrication of customized porous and bioactive ceramic implants [34]. A **tricalcium silicate** (Ca_3SiO_5 or C3S) has also been successfully printed at room temperature with different sizes and shapes (Fig. 9c–e) [129]. The fact that it can be printed at a physiological temperature allows the **incorporation of drugs** into the 3D bone cement and has an advantage over the use of other silicate bioceramics. The study found that the 3D-printed C3S bone cement scaffolds were bioactive in vitro and in vivo.

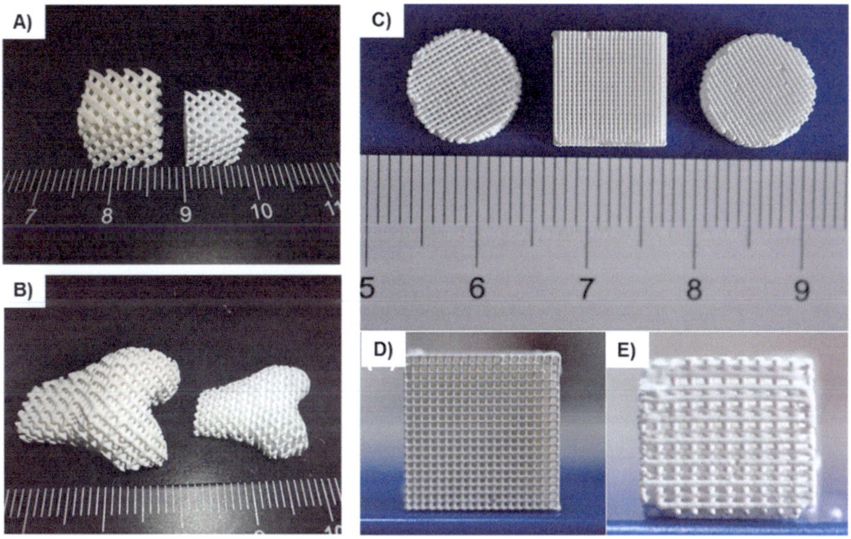

Fig. 9 **a**, **b** Diamond lattice structure BCP and 45S5 bioceramic scaffolds before and after sintering for 2 h: **a** Cube shape bioceramic scaffolds; and **b** bone shape bioceramic scaffolds. **a**, **b** Adapted with permission from Hua et al. [34]. Copyright (2021), Elsevier. **c–e** Overview of C3S bone cement scaffolds: **c** 3D-printed C3S bone cement scaffolds with different sizes and shapes; **d** top view of C3S bone cement scaffolds; **e** side view of C3S bone cement scaffolds. **c–e** Adapted with permission from Yang et al. [129]. Copyright (2017). American Chemical Society

Besides the use of polymeric materials in combination with bioceramics, other composites based on **metallic/ceramic** materials have been investigated. Nevertheless, although the possibility of printing bioglass in combination with Ti-based matrices has been recently reported, it is still essential to address systemic studies to ascertain the appropriate operating printing parameters for each type of composite and detailed biocompatibility studies [130].

In summary, ceramics printing stands as the **most challenging** of all biomaterials due to their **brittleness** and **high melting temperatures**. Besides, the dimensional accuracy ($\leq$100 μm features), reducing the multi-step processing time (including pre- and post-processing), and increasing the structural stability of large-size scaffolds are critical limitations to overcome. The complexity of the experimental procedures associated with the necessity of incorporating **living cells, growth factors,** and **biopolymers** within the printed components adds a new dimension to the challenges that AM has to face. These challenges need to be overcome to promote the use of AM bioceramic composites in a wide range of clinical practices, where to date only **craniomaxillofacial reconstruction surgery** has been implemented with the use of these materials. A complete review of the barriers to the clinical translation of AM bioceramics can be found in [131]. Among the possible solutions, new insights are focused on new **bioceramic inks** [131] that can be printed at room temperature and physiological conditions avoiding toxic components. Other studies are evaluating the use of **soft hydrogels within the bioceramics** scaffolds [132] as this strategy produces more biocompatible vascularized constructs. Also, the **incorporation of metal powders** [133] into a polymer/ceramic powder mix could be beneficial for their use in load-bearing applications providing the required mechanical strength.

5 Additive Manufacturing of Metals

5.1 3D Printing with Ti and Ti-Based Alloys

Titanium (Ti) and its alloys have been targeted as the **gold standard** metallic biomaterial for long-term implants due to their **excellent biocompatibility** and unique combination of **mechanical-anticorrosion properties**. Namely, the excellent corrosion resistance and biocompatibility of pure Ti in most physiological media are given by the formation of a stable TiO_2 passive layer. This protective film is highly adherent and favors the **proteins** and **cell bonding** with the implant surroundings [134, 135].

However, their limited osseointegration limits their long-term clinical success [136]. Besides, when compared to human bone Ti presents a **higher elastic modulus resulting in stress shielding**, thus leading to aseptic loosening and implant failure. For these reasons, there is a need for new Ti alloys design with similar mechanical performance to human bone [136, 137].

The first successful alloy was the **Ti–6Al–4V** (also known as Ti64 or Titanium Grade V), applied for artificial joints and bone fixators. This is due to its higher fatigue strength compared to pure titanium. As a limitation, Ti64 shows **poor wear resistance** and can induce **long-term neurotoxicity** due to the possible leaching of aluminum and vanadium ions. Therefore, given the toxicity of such alloying elements, advances were made in titanium alloys with no vanadium or aluminum compounds in their microstructure [138–140].

Overall, titanium alloys are divided into three categories: α-alloys, α-β alloys, and β (Fig. 10). Notwithstanding, the choice of alloying elements can reduce their biocompatibility due to the sustained release of corrosion products into body fluids. This is a critical challenge regarding obtaining a biocompatible material since most alloying elements are not even oligo-elements in the human body (e.g., rare earth elements) [137, 139].

The need for metallic implants with **complex geometries** and **improved mechanical properties** at low cost has triggered the research of new processing routes, i.e., metal additive manufacturing (MAM) [5, 141]. In the bone tissue engineering context, new materials that match **bone porosity** are needed to improve their implantation success for prosthesis applications. The patient's cell can reach the inner structure of the implant and colonize it, resulting in the formation of new bone within the biomaterial [142].

In the particular case of Ti, the most commonly used MAM technologies are EBM and SLM, as specified in Sect. 2. Since both technologies were identified as the most

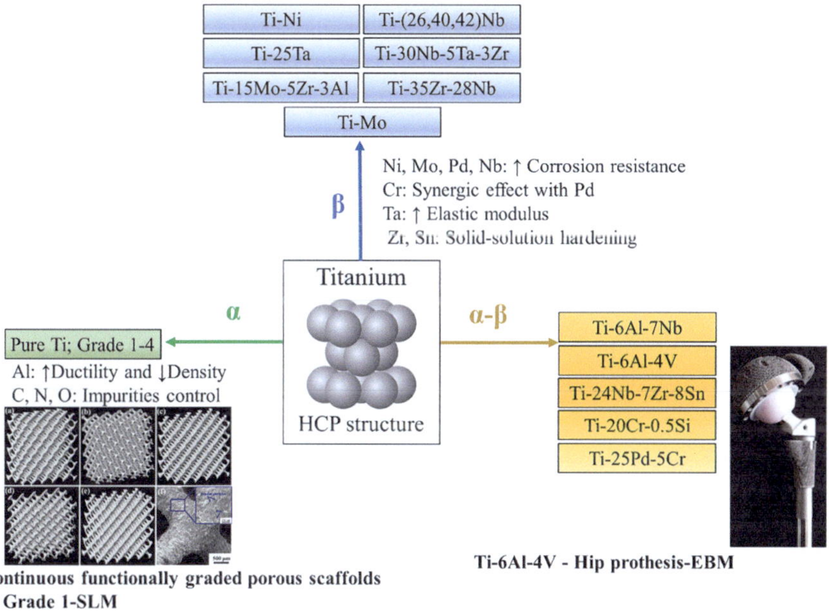

Fig. 10 Summary of the latest Ti alloys development in biomedicine. Adapted from [62, 134, 143–147]. Lower left images: Reprinted from Han et al. [143], with permission from Elsevier. Copyright (2018). Lower right image: Reprinted from Tang et al. [144], with permission from Elsevier. Copyright (2018)

suitable for Ti alloys, numerous studies aim to establish the relation between printing conditions and surface creation on new biocompatible Ti alloys. This results from the fact that MAM Ti alloys have shown an unreliable surface quality, coming from pores, roughness, or other defects and residual stresses, which negatively influence the mechanical and electrochemical properties of the implant [137]. Figure 10 shows the latest developments in Ti alloys processed by MAM technology for biomedical applications.

Several Ti alloys have been commercially used over the last several decades. Their adaptability to be processed by AM has recently been demonstrated using systems such as **Cp Ti, Ti–Ni, Ti–Mo, Ti64, Ti–6Al–7Nb, and Ti–15Mo–5Zr–3Al**. Some of them are even commercially available today, such as Ti–6Al–4V (UNS:R56400), Ti-6Al-7Nb ("TAN"; UNS:R56700), Ti-Mo (UNS:R58150), and so on. Among these alloys, much of the research effort has focused on the well-known Ti–6Al–4V alloy [138–140]. Recently, new Ti alloys have been developed using various AM techniques (Table 7).

While the alloys in Table 7 show different mechanical properties as a function of the MAM technology, it is important to notice that microstructural anisotropy (i.e., longitudinal and horizontal building orientations) is inherent in MAM technology. In turn, the mechanical and electrochemical properties are different in both building orientations [160, 161]. Besides, the electrochemical behavior of these Ti alloys

Table 7 Main features of the latest Ti alloys processed by MAM for biomedicine

Alloy	MAM method	Application	Features (microstructure, mechanical properties[a], etc.)	Refs.
Ti–Ni	SLM/EBM	Memory shape (stents)	– SLM: UTS: 880 ± 13 Mpa Recovery rate: >98% – EBM: ↑ Density of defects → Heat treatment (750 °C)	[148, 149]
Ti–Nb	SLM	Joint replacement	– Ti–26Nb: β phase is the predominant phase E(Gpa): 77.2 ± 0.4 – Ti–40Nb: β phase 85–90% β phase–α phase E(Gpa) of Ti–40Nb: 33 ± 2	[150, 151]
Ti–35Zr–28Nb	SLM	Orthopedic	Porous and solid structures were achieved with the β phase as the main constituent Porous material: E(Gpa): (1.1–1.3) ± 0.2 Solid material: E(Gpa): 57 ± 2	[152–154]
Ti–6Al–4V	SLM/EBM	Orthopedic	– SLM (UTS): 1200–1500 – EBM (UTS): 970–1100	[155, 156]
Ti–24Nb–4Zr–8Sn	SLM/EBM	Dental and Orthopedic	– SLM: Pure β phase; E(Gpa): 53 ± 1 – EBM: α + β phase and β phase after annealing (800 °C–2 h). E(Gpa): (0.7–1.5)	[157, 158]
Ti–30Nb–5Ta–3Zr	SLM	Dental and Orthopedic	β phase with minimal amount of α phase. E(Gpa): 59.5 ± 0.8 Ti–30Nb–5Ta–3Zr shows more biocompatibility than Ti–6Al–4V	[159]

[a] *Notes* Ultra tensile strength (Mpa): UTS; Elastic modulus (Gpa): E

in the physiological media remains unknown. In the particular case of Ti–6Al–4V alloy, it can be considered similar or even superior to traditionally manufactured Ti6Al4V [140, 155, 162]. Such variability in corrosion behavior depends not only on the intrinsic anisotropy in the manufacturing process itself but also depends fundamentally on surface roughness and internal porosity [62]. These parameters can be **controlled or improved** by selecting appropriate **printing parameters** and/or **surface** and **heating post-treatments**.

So far, it is still challenging to establish a benchmark as in the available studies, even for the same alloy and using the same MAM technique, different printing parameters, and post-treatments are addressed.

5.2 Fabrication of Fe-Based 3D Scaffolds

Additively manufactured **iron** (Fe) and its alloys have been identified as a promising route for designing **biodegradable scaffolds** with high structural resolution and complex geometries [54]. Pure iron is well-known for its **superior mechanical** and **anticorrosion properties** than other biodegradable materials such as Mg and Zn. Nonetheless, a higher corrosion rate of Fe may cause the accumulation of **corrosion products** in the implant surroundings, thus leading to **chronic inflammation** and **cytotoxicity** at the implantation site [54, 126, 163]. For this reason, the design of iron alloys obtained by AM has been identified as a promising route to **modulate their degradation rate**. So far, SLM of stainless steel and Fe–Mn alloys have been the most studied systems [126].

In the case of SLM Fe–Mn alloys, higher corrosion rates have been reported compared to pure conventionally processed Fe due to the formation of stable Fe–Mn galvanic pairs in simulated body fluid solution (SBF) [42, 164–166]. Although SLM created a low concentration of surface defects (e.g., porosity, cracking, delamination, etc.), there is no accepted criterion or parameters for the corrosion rate values in temporary implants [167]. In this scenario, there is a complete **lack of systematic studies** where the corrosion, mechanical and biological performance is evaluated since available studies use non-comparable test conditions (e.g., different physiological media, pH values, and so on) [165, 166].

As for stainless steel, **316L stainless steel** has been identified as the most suitable alloy due to its **availability, low cost, ease of fabrication,** and **biocompatibility**. Nevertheless, this alloy is prone to suffer pitting corrosion, corrosion fatigue, and stress corrosion cracking. Besides, the presence of nickel as alloying may be released, thereby causing cytotoxicity [42, 168]. Regarding the SLM route, the available studies reported optimal surface finishing, although there are still no studies evaluating SLM-processed 316-L steel as a possible temporary implant [168].

Apart from the alloy design, it has recently been shown that Fe biodegradation is directly proportional to its porosity (Fig. 11). Besides, increased porosity limits the elastic modulus. However, so far, there is no accepted criterion in the scientific community for the corrosion rate of absorbable implants since it is highly dependent on the on-demand application.

Although Fe-based degradable implants have not been tested in humans, the biocompatibility of SLM-manufactured Fe alloys has been recently proven [42]. However, it needs to be further improved as the natural Fe passivation results in a minimal degradation rate compared to Ti and Mg [47]. For this reason, the number

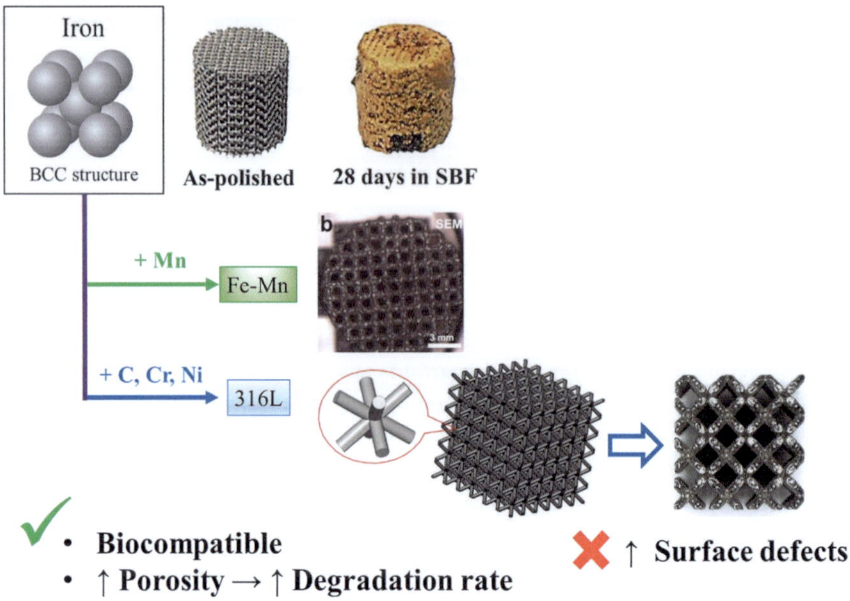

Fig. 11 Summary of the latest Fe alloys produced by SLM with possible application in biomedicine. (First two images from the right) adapted from Putra et al. [42], under the terms of the Creative Commons CC-BY license. https://creativecommons.org/licenses/by/4.0/. Fe–Mn image adapted from Shuai et al. [164], under the terms of the Creative Commons Attribution 3.0 licence. https://creativecommons.org/licenses/by/3.0/. 316L images adapted from Platek et al. [169], Creative Commons Attribution (CC BY) license (http://creativecommons.org/licenses/by/4.0/)

of scientific publications per year has been **constantly rising** from about 72 to 244 for the 2012–2021 period (used terms in the Scopus database: "iron implants" and "additive manufacturing").

The future research lines to address the use of Fe as a suitable and feasible biodegradable implant are identified as follows: (i) the **geometric** and **compositional design** of iron alloys, (ii) the **optimization of printing parameters**, and (iii) **systematic studies** to evaluate the mechanical, wear, corrosion and bio-related properties.

5.3 Mg-Based Materials for AM

Magnesium is an attractive material for low load-bearing applications such as **temporary bone fixing devices** and **cardiovascular stents** due to its biodegradability, biocompatibility, and mechanical properties ideal for avoiding the stress shielding effect [62, 170, 171]. An intensive study of magnesium alloys over the past two decades for use in temporary biomedical devices has led to the approval of Mg–Ca–Zn [172], pure Mg [173], and Mg–Y–RE–Zr [174] alloys for clinical use as bone screws; Mg–RE alloy has also been clinically approved for cardiovascular stents [175]. The biodegradation rate of bare Mg alloys can sometimes be **excessive**; in order to avoid a premature mechanical failure of the device and control the hydrogen gas evolution (by-product of Mg corrosion process) surface functionalization of Mg alloys is widely studied [62]. Both Mg–Y–RE–Zr and Mg–Ca–Zn screws showed successful **osseointegration** and **osteogenesis** in clinical trials. Complete resorption of screws is typically reported within a year [172]. Unlike screws, bone regeneration grafts need to be anatomically accurate, which is where the necessity of AM comes in. AM of Mg and its alloys is challenging due to its **low boiling point, high capacity for oxidation**, and **high vapor pressure**. Also, the porous Mg grafts tend to exhibit a higher degradation rate due to the elevated surface area, intrinsic porosity, and crevices.

Table 8 summarizes the main AM technologies that have been used for manufacturing bulk and porous Mg alloy parts. The current state of the art pinpoints SLM technology as the most promising one in terms of the resolution and the capacity to manufacture bulk parts as well as porous scaffolds.

Fabrication of AM Mg alloys is achieved either through **in situ alloying**, e.g., by mixing Mg and Zn powders in proportions required to obtain different Mg–Zn alloy compositions [176], or using pre-alloyed Mg-based powders (e.g., ZK60 and Mg–3 Zn) which can be additionally alloyed in situ alloyed with other elements (e.g., rare earth or Cu) [42]. This helps to improve the strength of alloys and achieve added-value functionality, e.g., antibacterial properties.

In terms of biodegradation behavior, SLM fabricated bulk parts of **Mg–Zn–Zr alloys** (stand-alone or in situ modified with other elements, such as Nd, Al, or Cu), can be within 1–1.6 mm/year range [190], which is considered adequate. These values are typically obtained by extrapolation of the degradation rates observed during 1–4 weeks of immersion in physiological media, such as Hank's solution, Dulbecco's Modified Eagle Medium (DMEM), or SBF at 37 °C. In biodegradation studies of AM porous scaffolds, it is more practical to express the corrosion rate as a **weight loss percentage**, although such studies are still scarce. For instance, after 28 days of in vitro degradation in r-SBF 21% weight loss has been reported for the **WE43 alloy** porous scaffold, associated with localized corrosion in the center of the scaffold [181]. Despite this degradation level, the scaffolds **maintained the mechanical properties** at a level of those for trabecular bone. Surface modification is a good strategy to adjust the biodegradation rate of porous scaffolds. For instance, the application of plasma electrolytic oxidation (PEO) coatings enabled the reduction of the weight loss of SLM printed WE43 alloy porous scaffold to 5 wt.% [183]. However, to date, only a couple of studies have been carried out on surface modification of porous Mg scaffolds, in both cases using a PEO technique [183, 184].

AM Mg alloys typically exhibit nearly 100% cell viability after 1 day of indirect cytotoxicity tests (using metal extract solutions), which can be reduced to 85–95% within 5 days of testing. Direct cytotoxicity tests reveal somewhat reduced performance, with the viability of cells seeded on porous WE43 alloy scaffolds varying

Table 8 Examples of AM processing of Mg alloy systems

Method	Energy source	Alloy system	Part architecture	Density (%)	Commentary	Refs.
PBF	Laser (SLM)	Mg–Zn–Zr (ZK60)	Bulk	94.05%	By far a method with higher accuracy and resolution, enabling a precise topological order design of porous scaffolds with strut size down to 400 μm and pore size as small as 600 μm using powders with 26–64.4 μm in diameter	[176, 177] [178, 179] [180–184] [185]
		Mg–Zn	Bulk	99.35%		
		Mg–RE (WE43)	Bulk	99.4%		
			Porous scaffolds	Porosity 67–76%		
		Mg–Sn	Bulk	Not provided		
DED	TIG[a] or MIG[b] welding torch (wire-and-arc welding)	AZ31B, AZ80M	Bulk	Fully dense	Lower accuracy than laser or EB-based PBF. Deposition and energy efficiency are ~100% and 54–88%, respectively, compared to 14% and 20–50% for laser DED	[186]
FSP	Friction heat from rotating tool	WE43	Bulk	Cavities and cracks form in hotter weld conditions	High strength and ductility of parts. Limited to bulk parts only	[187]
Binder-less jetting	Not required during 3D printing. Post-treatment at 650 °C (sintering)	Mg–Zn–Zr	Bulk	Green parts: 55.5%, after sintering: 71.1%	Compressive strength and elastic modulus of parts comparable to human bone	[43, 188, 189]

Notes [a] Tungsten inert gas; [b] Metal inert gas

within 75–97% in 3 days of culturing [190]. Notwithstanding, more research is needed to establish a correlation between **the biodegradation rate, cumulative ion release,** and **direct cytotoxicity** of AM Mg alloys, all of which depend on the immersed scaffold surface area and area-to-liquid volume ratio.

So far, still very few in vivo studies of Mg alloy porous scaffolds have been reported (e.g., for pure Mg, Mg–Al–Zn, and Mg–Li–Al–RE alloys), all of them prepared by means other than AM technology [191–193]. However, they all were found to show **good osseointegration** and **promote peri-implant bone formation**. A composite **PLGA/TCP/Mg** scaffold 3D printed using a binder-based material extrusion method was shown to promote **Osseo-** and **neoangiogenesis** (Fig. 2) [133].

5.4 Zn-Based 3D Printed Parts and Applications

Applications of Zn-based metallic biomaterials include **bone implants, cardiovascular stents, bone fixation,** and **wound closure devices** [184]. While the latter two typically have simple geometrical shapes, i.e., drilled plates, screws, and staples, easy to obtain by forging, extrusion, machining, and drawing, bone implants, and cardiovascular stents can immensely benefit from AM technologies.

Zn-based alloys were proposed as a promising alternative to Mg- and Fe-based alloys for biodegradable devices in 2013 after a successful demonstration of a considerably lower degradation rate of cast pure Zn and Zn–xMg ($x = 1, 2, 3$ wt.%) alloys in SBF, compared with Mg alloys [194]. Such behavior is expected, given that the standard electrode potential of Zn ($E^0_{SHE} = -0.76$ V) occupies an intermediate position compared to those of Mg and Fe (−2.37 V and −0.44 V, respectively). The fact that **Zn** is essential to **human osteoblastogenesis** [195], **enzyme and protein regulation**, as well as **immune and nervous systems**, makes Zn highly adaptable to tissue regeneration. It is considered particularly interesting for **stent applications** [196] due to its biocompatibility with blood vessel cells and suitable degradation rate that has been confirmed in vivo in mice and rabbit aorta models [197, 198].

In order to improve its mechanical strength (from UTS of 20–120 MPa for cast and wrought pure Zn), for stent applications in particular that require 300 MPa UTS, Zn is alloyed with various biocompatible elements that are essential to **human metabolism** (Cu, Mn) and/or act as a **nutrient** (Mg, Ca, Sr). Other alloy systems, such as **Zn–Al, Zn–Ag, and Zn–Li** have also been explored. The progress achieved in research on Zn-based biodegradable materials since 2013 has been reflected in several reviews [184, 196, 199–201]. The physical metallurgy of these alloys is summarized in [200]. The effects of alloying and processing techniques on mechanical properties and in vitro corrosion are reviewed in [184, 201].

Over the last 5 years, research on AM of Zn alloy-based porous scaffolds for bone grafting has intensified. As discussed earlier, porous scaffolds enable adjustment of the **mechanical strength of the graft**, owing to the **controlled redistribution of stress**, as well as provide an all-important permeability that is conductive to **cell ingrowth** and **tissue vascularization**. Compared with conventional manufacturing methods of porous scaffolds (i.e., casting, foaming, powder metallurgy) AM offers the possibility of **customized graft geometry** as well as precise control over the size, number, and shape of the pores [52]. The AM technique that has been most successful so far for the fabrication of porous Zn scaffolds is SLM, which belongs to the PBF family of AM methods, as stated in Table 1 and Fig. 2 (in Sect. 2).

The first effort to produce bulk Zn parts by SLM in a closed chamber undertaken in 2017 by Montani et al. showed that the low vaporization point of Zn powder particles interfered with the laser beam energy and lead to a high porosity of 12% [202]. The use of an open chamber and a jet flow of inert gas over the powder bed enabled Demir et al. to increase the part density up to 99% [60]. High porosity results either from the lack of fusion between the powder particles in case of insufficient beam energy or from gas entrapment produced by excessive metal evaporation. Following a systematic study of metal densification, Wen et al. optimized the SLM processing parameters, such as shielding gas flow and laser energy input to achieve 99.9% part density [59, 203, 204], which enabled the same group to proceed with the successful printing of cardiovascular stents [59]. Alloying Zn with other elements and optimizing the laser beam energy in order to control the liquid viscosity in the melt pool allows for greater densification of the parts. For instance, over 98% densification was for Zn–3Mg [57], Zn–2Al, and Zn–8Ag [205, 206].

Yet, the task of **printing porous Zn scaffolds** turned out to be challenging: the surface of the as-built scaffold struts was exceedingly rough due to its coverage with numerous powder particles and the processing pores outlining the strut contours [152]. A rough surface is detrimental to the fatigue and corrosion performance of metallic biomaterials; a successful interaction implant-cell interaction also depends on a favorable surface roughness range. Post-processing of as-built struts, consisting of sequential chemical and electrochemical polishing, eliminates the processing pores; although, it also reduces the diameter of the struts compared with the design value (Fig. 12).

Li et al. carried out long-term biodegradation behavior of **SLM porous Zn scaffolds** with a topologically ordered diamond structure [208]. During 4 weeks of dynamic immersion in a revised SBF they observed a 7.8% weight loss and elastic modulus values that remained within the range of those for cancellous bone. The authors also reported a good corrosion fatigue behavior of porous Zn scaffold in r-SBF, up to 80% of yield strength, which was even higher than its fatigue strength in the air due to the reinforcement of strut junctions with corrosion products layer [209]. Functionally graded design of the scaffolds appears to be a way to further improve their **corrosion fatigue behavior**.

In terms of biocompatibility in vitro, Li et al. reported that the viability of MG-63 cells after 1 day of direct contact with porous AM Zn is comparable with that of Ti–6Al–4V alloy [208]. On the other hand, hTERT-MSCs cells on Zn scaffolds exhibited low viability over 13 days [210]. Alloying of Zn with other elements appears to improve the cell viability compared with pure AM Zn, as was shown for AM Zn–3Mg alloy [57]. This is attributed to the **improved corrosion resistance** of the alloy as a result of which ion release that might provoke cytotoxicity, is reduced. Considering that the corrosion rate of porous metal scaffolds is greater than

Fig. 12 Surface quality of SLM-processed pure Zn scaffold struts: **a**, **b** as-built, **c**, **d** following chemical and electrochemical polishing; **b**, **d** are high magnification images of **a**, **c** respectively. Reproduced with permission from Wen et al. [207]. Copyright (2019), Elsevier

that of bulk metals [52], more systematic studies of the degradation behavior of AM biomedical grade Zn alloys containing essential and nutrient elements (Fe, Mg, Ca, Cu, Sr) in dynamic physiological conditions is needed to reliably correlate the combined ion release with recommended **daily allowance of these elements** [196] and **in vitro cytotoxicity** and **bioactivity** [190].

As for the biocompatibility in vivo, while conventionally processed pure Zn, Zn–Mg, Zn–Ca, Zn–Sr, Zn–Cu, Zn–Mg–Ca, Zn–Mg–Cu, Zn–Mg–Fe, and Zn–Al have been evaluated in mice, rat, rabbit, beagle dogs, and pig models [52, 184, 190, 200], the in vivo trials of Zn-alloy based AM porous scaffolds are yet to be reported.

Overall, based on the degradation behavior evolution in terms of mechanical properties and weight loss (Fig. 13), the viability of AM porous Zn scaffolds as biodegradable material for bone graft applications can be summarized as exceeding that of their Fe and Mg counterparts, due to the intermediate degradation rate and mechanical properties that increase with degradation time and are comparable with those of the human trabecular bone.

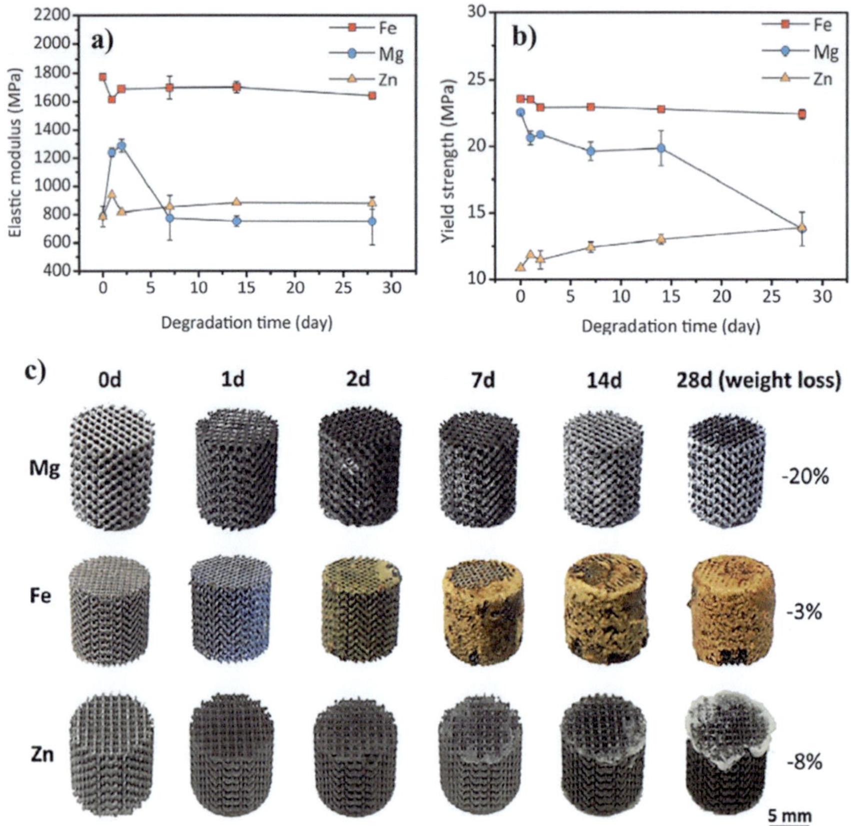

Fig. 13 Comparison of the elastic modulus (**a**), yield strength (**b**), and degradation morphology (**c**) variations with in vitro biodegradation time for Mg, Fe, and Zn porous scaffolds. Adapted from with permission Li et al. [190]. Copyright (2020), Elsevier

6 Materials Modification with Biological Cues, Drugs, or Proteins

In the last decade, 2D and 3D printing technologies have been used in combination with **drugs or biological factors** to obtain next-generation therapeutical products. These approximations have been applied both in **oral administration** (tablets) and in **clinical treatments** (medical devices, including implants), aiming to control drug delivery, improve local tissue behavior, or other systemic effects. As theoretical principles and techniques are common to both applications, constant feedback between these medical fields has been established to improve the therapeutical efficacy of their products.

6.1 3D Printed Medicines

The past decade has experienced an impressive development of 3D printing for **personalized drug delivery**. The idea of a self-sustainable method to obtain next-generation drugs has stimulated academia and the pharmaceutical industry to successfully design treatments to reach personalized medicine. These reveal that 3D printing techniques can be adapted to specific requirements related to **drug stability, administration pattern, or patient needs**. Deposing material layer by layer, AM techniques can obtain any solid oral dosage forms, called **"printlets"** [211]. Some formulations include multiple active pharmaceutical ingredients (APIs), which can be processed by obtaining different geometries and release features. Although oral dosage forms are the most used formats, **transdermal patches**, and **vaginal** and **rectal delivery systems** are other 3D printed related applications [212].

Techniques such as inkjet printing for 2D or **FDM, SLA, SSE, or SLS** for 3D printing have been extensively applied to obtain **non-toxic tablets or caplets**, usually using polymers to build drug-containing matrix. As an example, FDM has been especially successful in obtaining oral pharmaceuticals including common APIs such as **paracetamol, caffeine, or prednisolone**, and using **PVA** as an excipient [213, 214]. In 2016, US FDA approved the commercialization of a 3D-printed tablet as a pharmaceutical for the treatment of epilepsy (**SPRITAM**, Aprecia Pharmaceuticals). In this case, a highly water-soluble porous matrix printed using 3DP is applied to accelerate API (levetiracetam) administration [215].

3D printing technologies open a wide range of possibilities to **control drug delivery**, and not only by infill percentage. Changing geometry or physical dimensions directly **affects release dynamics**. Inner network design in CAD and obtained tablet density can also control future degradation [216]. **Topographical parameters** such as porosity can affect drug release dynamics [217]. Moreover, excipient composition can be in situ adapted in response to patient needs, allowing an immediate fast release [218] or a long-term administration if needed [219]. If necessary, release modifiers such as **erosive polymers** (PEG, PVA, or Hydroxypropyl methylcellulose (HPMC)) can be added to the formulation to enhance dosage administration [218]. In many situations, 3D printed tablets allowed for better control of drug release than classic tablets obtained using standard compressive techniques [220]. In addition, 3D printing can be also applied to coat conventional tablets and create a shell to tailor the release profile and improve the therapeutic efficacy [221].

6.2 3D Printed Drug-Loaded Medical Devices

As discussed before, the same perspective can be applied to the generation of 3D medical devices activated with pharmaceutical and biological factors, such as **proteins (growth factors, cytokines)** or other **active molecules** (DNA vectors). Implant devices can be designed to release cytokines or other regulatory molecules with specific spatiotemporal delivery profiles to enhance local tissue regeneration [222]. **Surgical patches** are another related approach to this kind of therapy. Providing molecular cues in the implant microenvironment can be crucial to avoiding **implant rejection**, accelerating **tissue integration,** and achieving **implantation success**. Moreover, conferring implant features such as **antimicrobial activity**, thereby extending the implant versatility in immunosuppressed situations.

In contrast to systemic treatment options, local administration through drug-eluting implants achieves sustained drug delivery **minimizing the side effects**. A target-oriented treatment could be more successful in certain pathologic scenarios, such as local infections associated with an implant.

Because of their versatility and easy handling, **ceramic** and **polymers** (synthetic and natural, including polysaccharides) are the **most used materials** to obtain drug-loaded 3D printed implants. In implant devices, most efforts have been made to **improve tissue healing** and **integration**, but other applications related to pathologies such as cancer or bacterial infections have been also explored. Thus, 3D printed drug-eluting catheters and stents, gynecological devices (IUD, vaginal rings), bone plates, surgical screws, and surgical meshes have been recently developed [223].

6.2.1 Drug-Loaded Ceramics

One advantage of 3D printing of CaP scaffolds for drug and biological factor release is the possibility of using binder jetting technique at low temperatures (below 37 °C) (Fig. 2) [224]. Using this technique homogenous drug loading into the entire piece avoids the "burst" release effect [225]. This approximation allowed for antibiotic loading and controlled release for osteomyelitis treatment [226]. Further, the possibility of using polymer-ceramic composites allows for better control of mechanical properties and a controlled release mechanism. In this scenario, TCP-based CaP cement with chitosan/dextran sulfate microparticles were used to deliver VEGF (vascular endothelial growth factor) or BSA (bovine serum albumin) proteins and affect mesenchymal stem cell activity [227].

Another approach used **C_3S bone cement scaffolds** (Fig. 14), which can be drug-loaded with rhodamine B (RHB) and calcein at low temperature and printed with a controlled nanotopography including pores, enhancing stem cells adhesion, proliferation, and differentiation. Applied in experimental surgeries, these 3D printed scaffolds in vivo have enhanced bone regeneration in comparison with conventional C_3S samples [129].

An interesting approach may consider a **hierarchical mesoporous bioceramic** composite scaffold that uses a polymer component to include anti-osteoarticular tuberculosis drugs into a 3D printable formulation. The obtained scaffolds showed improved drug-releasing behavior and an active dosage in a local environment [228].

Due to the technical challenges of printing ceramics at low temperature, mainly related to the resolution and the poor mechanical properties [229], high-temperature 3D printing and post-processing is the most common approach. Among them, surface adsorption of bone morphogenetic proteins (mainly BMP-2) using additives such as

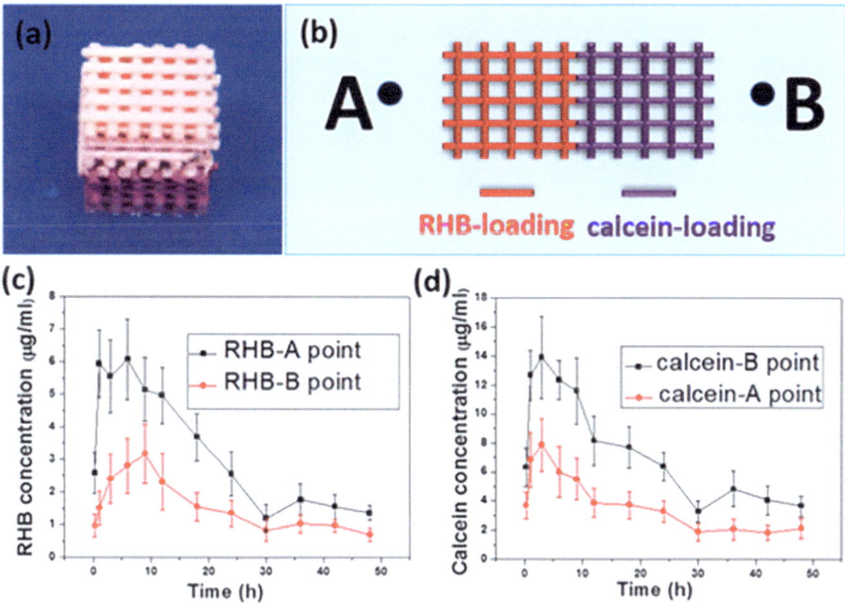

Fig. 14 Drug loading in 3D printed C3S scaffolds. **a** An image of dual-drug loaded scaffolds with RHB in one part (dark area) and calcein in another part (light area) of the scaffolds. **b** Illustration of two measuring points of drug concentrations for dual-drug release up to 48 h from codrug-loaded C3S cement scaffolds. **c** Concentration of RHB measured at point A. **d** Concentration of calcein measured at point B. Reprinted with permission from Yang et al. [129]. Copyright (2017), American Chemical Society

silicon or **dextrin** can improve drug loading and release [230]. A similar work used a coating of **collagen** and **microspheres of chitosan** and **BMP-2** over an HA scaffold to long-term controlled release and stimulate osteogenic differentiation [231].

CaP cements have also been proposed as excellent candidates for bone engineering applications [229] as stated in Sect. 4. Figure 15 summarizes three of the main strategies that have been explored with these bioceramics, including **cell delivery** after seeding cells directly into the 3D scaffolds, **drug loading** (direct mixing, drug absorption, or encapsulated drugs), and the promotion of the **vascularization** of the scaffold by (i) modulating porosity, (ii) incorporating vascular cells, angiogenic growth factors (e.g., VEGF, hypoxia-inducible factor 1α (HIF1α), placental growth factor (PLGF)), or stem cells.

6.2.2 Drug-Loaded Polymers

3D printing of polymeric structures is one the most approached and affordable techniques to create drug-loaded scaffolds aimed for TE purposes. For this goal, both synthetic and natural polymers can be used, even in combination with other

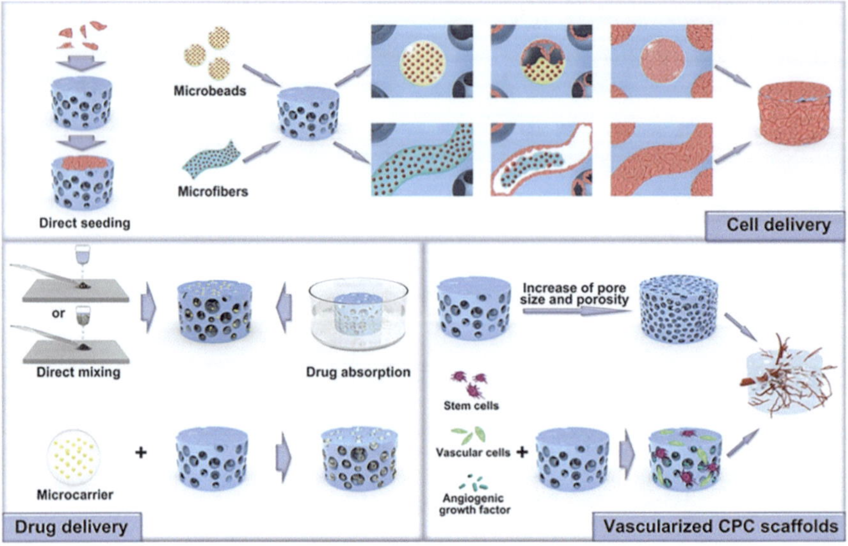

Fig. 15 Explored strategies with biological properties for CaP cements. Reproduced from Xu et al. [229], under the terms of the Creative Commons CC BY license. https://creativecommons.org/licenses/

elements (such as apatite) to reinforce mechanical properties and biological response. However, limitations concerning temperature or UV light lability of APIs are maintained, affecting some drug loading and printing strategies. As discussed above, drug release can be also controlled through physicochemical features such as **size, porosity, or degradability**. As an example, 3D printed Alg/cellulose nanofibers porous scaffolds were loaded with curcumin. In this work, a higher concentration of nanofibers provoked a lower degree of porosity of the scaffolds and a sustained release profile [232].

Polymeric **nanoparticles encapsulation** is one of the most used approaches to improve drug delivery in the 3D printing of natural polymers. Chitosan features properties (biocompatibility and antimicrobial activity) that make it an optimal candidate for this goal. An emulsion-based ink for 3D printing has been described including **chitosan-modified silica nanoparticles** and used for drug encapsulation and controlled release [233]. A similar technique using **alginate** has been used to encapsulate bacteriophage particles to confer antibacterial properties to wound dressings [234]. Antimicrobial strategies in this field also include the use of antibiotics such as cyclosporin [235] or enrofloxacin [236].

Usually, natural polymers are used in combination with synthetic polymers to synthesize scaffolds with improved properties. In this way, **PLGA drug-loaded** scaffolds have been 3D printed and sandwiched in a **gelatin-chitosan gel**. This approach allowed hemorrhage absorption after surgery and additionally showed activity against circulating cancer cells [237]. A similar approach considered the use of **antimicrobial quaternized chitosan in PLGA/HA scaffold** to treat infections in bone defect models. These scaffolds showed their efficacy, both in **anti-infection** and **bone regeneration** [238, 239]. In another example, a **PLA 3D printed tablet** with a pH-responsive polymer (Eudragit®) was used for 5-fluorouracil (5-FU) delivery in acidic environments for the treatment of colorectal cancer [240]. Once more, 5-FU was encapsulated in a CS-coated alginate bead.

Collagen is a natural polymer of peptidic nature that constitutes the ECM of organic tissues, with a contrasted performance in clinical practice. **Mineralized collagen scaffolds** can be obtained through FDM techniques, even combined with apatite to improve mechanical behavior [241]. These scaffolds resemble natural bone tissue composition, opening the possibility of obtaining mimicking constructs to stimulate bone cell infiltration.

On that basis, there is a huge clinical need to facilitate an effective loading of biological factors into 3D constructs. **Growth factors** are the natural orchestrators of the physiological healing process, in which gradient expression directs **cell migration, differentiation,** and **specialized matrix secretion**. Different strategies have been used to stimulate cell survival inside the scaffold, including guided **angiogenesis** and **vasculogenesis** [242]. In bone tissue healing, spatiotemporal signaling is crucial and is led by VEGF and BMP families. In this model, VEGF expression has a peak around the first week after an injury, whereas BMP expression increases constantly until the third week. Numerous attempts have been done to simulate this situation in 3D printed implants, mostly through polymeric encapsulation to provoke a sustained release, as direct loading commonly led to a burst effect in the first hour post-implantation [222]. This topic is attracting a great deal of attention, with evolving 3D printing technologies trying to assimilate complex signaling to mimic organic tissues. An example is a 3D printed scaffold of **PLGA with HA** that contains a lentiviral vector with the BMP-2 gene. After implantation, cell transfection is locally achieved in bone tissue, accelerating new bone formation [243].

However, there are still many more challenges to be faced. As stated above, the ability of certain drugs and proteins hinders their manufacturing using **FDM**, one of the **most affordable** and **versatile printing techniques**. Cell colonization of a drug-loaded 3D printed scaffold has real limitations, both in vitro and in vivo, especially when using common equipment and standard techniques to evaluate in situ drug release and biological response. Thus, mimicking natural tissues through AM and drug release techniques requires a signaling plan in addition to common solid design planning. Bioprinting techniques, focused on cell and matrix deposition in an organized manner to create a living tissue, are now developing new tools to include regulatory signals complexity in their constructs [244].

Finally, additional obstacles to overcome before general adoption for **patient-specific** use of **drug-loaded 3D implants** include the need for more **in vivo research** in animals and humans to prove the safety and efficacy of drug-loaded printed formulations, as well as **regulatory guidance**.

7 Smart Materials for 3D Printing: Toward 4D Printing

7.1 General Requirements for 4D Printing

In addition to the materials described above, there are some polymer materials able to react to particular **external stimuli**. These materials include **hydrogels** and **shape memory polymers** designed to respond to changes in **temperature, pH, humidity,** or **ionic strength, among others** [245]. These 3D printed materials offer unique advantages and allow the preparation of dynamic 3D printed parts instead of static and permanent components.

Despite the unclear definition for **4D printing** (4DP) due to the different existing alternatives, the main concept behind it is that printed objects can **dynamically transform,** and **their shape/function** can be actively controlled in response to certain **physical, chemical,** or **biological stimuli**. Up to date, 4D printing studies have tried to achieve this by employing various stimuli-responsive materials (SRMs) [246]. In contrast to 4D printing, 3D printed parts lack **adaptive** and **shape-morphing characteristics** [245].

In summary, the combination of these stimuli-responsive materials (also known as **smart or intelligent materials**) with AM has led to what is nowadays known as 4DP. The **fourth dimension** is the time in which the **shape is changing** and is, therefore, associated with this conformation technique to achieve the 4DP effect. More precisely, 4DP is based on the increase of stress during the one-dimensional (1D), two-dimensional (2D), or even 3D conformation of materials provoked by a particular stimulus that allows the component/part to change its shape with time [247]. This recent research area within AM has significantly evolved since the first paper on the subject with the designation of 4DP was published in 2013, where Ge and coworkers described the preparation of composite materials with shape-morphing ability over time [248].

Thus, a general prerequisite for 4DP is the use **of one or more than one material with stimuli-responsive properties**, i.e., with distinct physical properties depending on the environment which will, in turn, enable shape changes. The incorporation of stimuli-responsive materials in conjunction with the possibility to print multiple materials in a tailor-made design provides unique opportunities to fabricate complex 4D responsive objects. An illustrative example of the complexity that can be obtained is depicted in Fig. 16. Nowadays, depending on the technology employed single and multi-material objects can be directly fabricated by AM with elements placed in **particular positions** and **able to react to different stimuli**. In some cases, just one single material part can present significant morphological changes. In other cases, shape changes can be achieved using multi-stimuli arrangements based on the use of multiple stimuli-responsive materials. A concept design based on this **multi-material approach** was described by Khare et al. [249]. As is illustrated in Fig. 16 (bottom), in this work an artificial insect is employed as an example. The complex design proposed by Khare et al. involves several shape types, formed by different stimuli-responsive materials that provoke changes provided by different

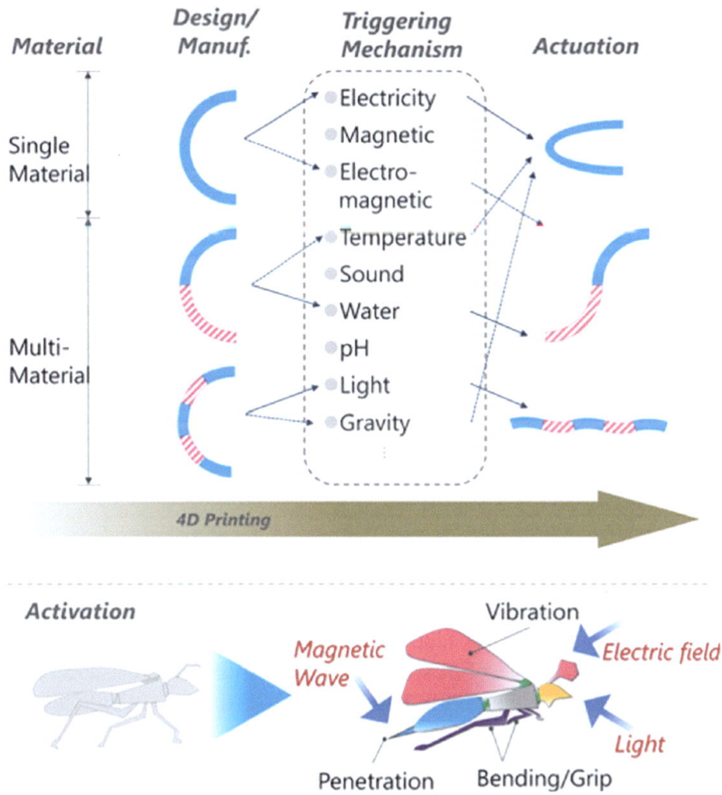

Fig. 16 **Top**: Schematic of the mechanism to trigger actuation and shape flexibility employing either single or multi-stimuli responsive materials. **Bottom**: Example of an artificial bug sample 4D printed using the above multi-stimuli activation concepts. Adapted with permission from Khare et al. [249]. Copyright (2017), Springer Nature

multi-stimuli combinations. As a result, it is possible to achieve simultaneously or separately different shape changes that include expansion, flexibility, shrinking, or morphing.

The current development phase of 4D printing technology is still in its infancy and the variety of stimuli-responsive materials and design concepts is **increasingly growing**. However, further developments to achieve complex architectures are still ongoing. Herein it will briefly describe the **most extensively employed** stimuli-responsive polymers for 4D printing and introduced the main applications of 4D printed parts in the biomedical area.

7.2 Polymers Employed in 4D Printing

The range of polymers tested in 4D printing has been growing exponentially, obliging research teams to be interdisciplinary [250–252]. It is also worth mentioning that many studies describe the combination of two or more materials to achieve 4D, by simultaneously printing these materials (as has been depicted above). Particularly, **hydrogels** and **shape memory polymers** are the most extensively employed for 4D printing.

- **4D Printed Hydrogels**: Smart hydrogels have attracted close interest in the recent literature. Particularly, those structures containing only **one** or **two layers of gels**, namely, **bi-layered structures**. In these structures, the different swelling ratio between layers is triggered by a **particular solvent, pH**, or **temperature**, which induces the structure to fold/unfold in a controlled fashion [253]. The concept of self-evolving structures has arisen from the fact that hydrophilic materials, typically hydrogels, can swell when immersed in water, thus provoking changes in the 3D printed shape (Fig. 17) [254].

Gladman et al. [255] developed a composite hydrogel ink to mimic the structure of plant cell walls (Fig. 18). This composite ink is composed of an acrylamide matrix reinforced with cellulose fibrils added to improve stiffness. Moreover, the composite includes an aqueous solution of *N*,*N*-dimethylacrylamide, Irgacure 2959 as photoinitiator, nanoclay, nanofibrillated cellulose, and glucose oxidase and glucose as oxygen

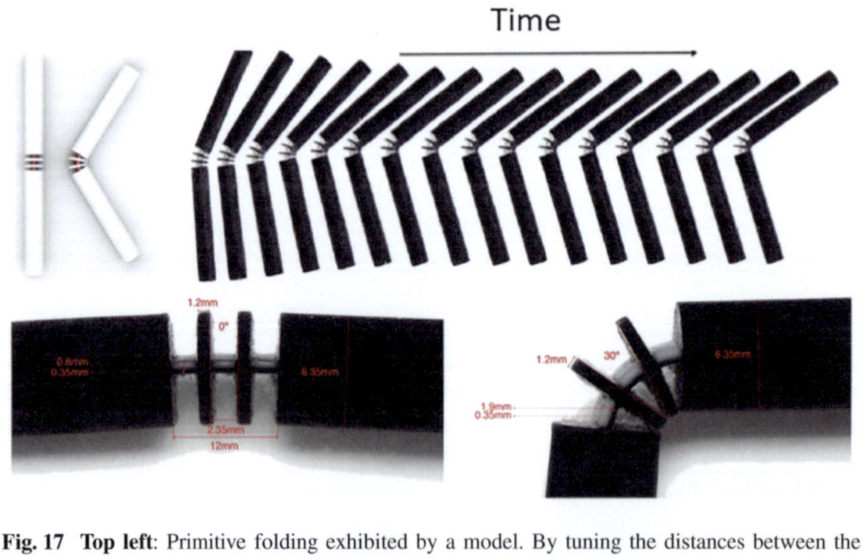

Fig. 17 **Top left**: Primitive folding exhibited by a model. By tuning the distances between the disks, it is possible to set the final folding angle. **Top right**: Sequence of the folding process in water over time. **Bottom**: Fabrication parameters of bending elements. Reproduced from Raviv et al. [254], under the Creative Commons Attribution-NonCommercial-NoDerivs 4.0 International License. http://creativecommons.org/licenses/by-nc-nd/4.0/

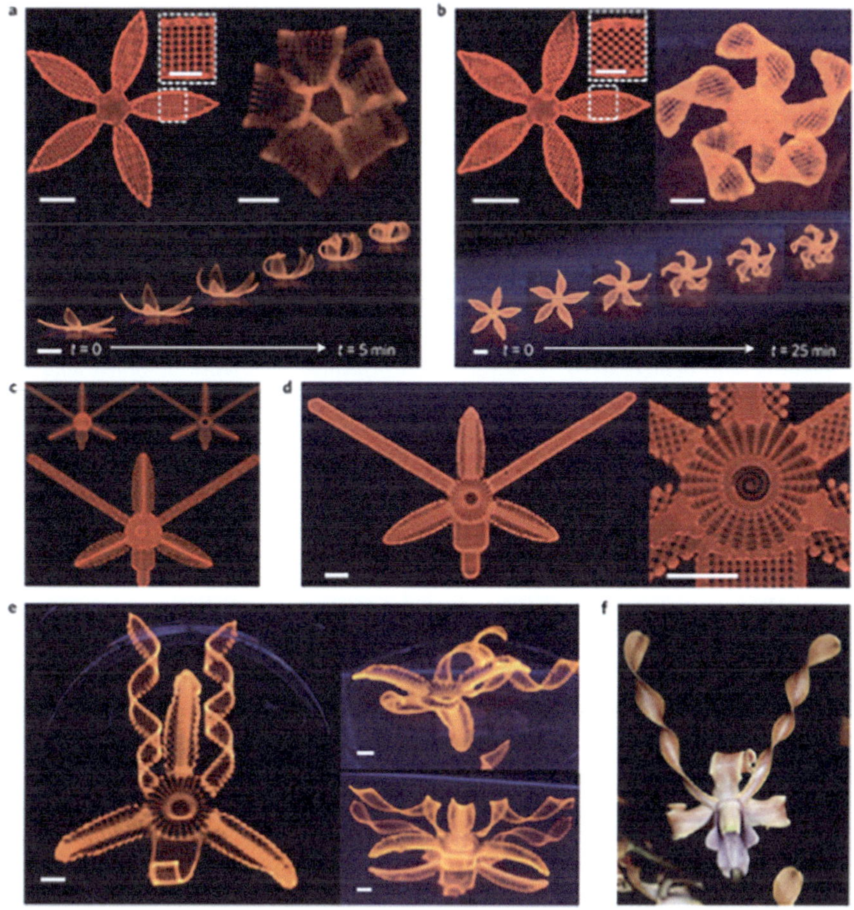

Fig. 18 Complex flower morphologies generated by biomimetic 4D printing. **a**, **b** Simple flowers composed of 90°/0° and −45°/45° bilayers oriented to the long axis of each petal, **c–f** print path (**c**), printed structure (**d**), and resulting swollen structure (**e**) of a flower demonstrating a range of morphologies inspired by a native orchid, the *Dendrobium helix* (**f**). Reproduced with permission from Sydney Gladman et al. [255]. Copyright (2016), Springer Nature

scavengers to reduce oxygen inhibition during UV curing. The nanoclay particles were used as a modifier to adjust the rheology and viscoelastic properties, necessary for obtaining a suitable ink for printing. The shape-shifting state of the 4D printed materials using this ink is irreversible. To achieve reversible shape-shifting behavior, in hot and cold water, *N*-isopropylacrylamide must be used instead of *N*,*N*-dimethylacrylamide as acrylic matrix.

– ***Shape Memory Polymers (SMP)***: SMP's is a category of stimuli-responsive materials with shape-shifting properties. These polymers have drawn increasing attention from the field of scientific research and technological development. SMP's applications may be found in several works in recent literature, owing to their capability to get transformed from shape to shape as a response to particular external stimuli, while keeping the original shape when these stimuli are discontinued. Moreover, SMP's can react to several stimuli simultaneously, namely, temperature, water, pH, pressure, light, magnetic fields, etc., which convert SMP's into multi-stimuli materials [21]. For instance, a recent work showed two models which were repeatedly cycled through different temporary shapes by changing the environmental temperature [256]. These 4D printed parts based on SMP's are fabricated using PCL. Moreover, methacrylate groups covalently linked to PCL were used as a resin to print structures able to revert to their original shape, once parts undergo a temperature above the melting point (ca. 60 °C). Note that the resin is bath heated at 90 °C before SLA printing. This work demonstrated the possibility of achieving a high final resolution in complex geometry structures. In this work not only the geometry but also the transparency could be reverted as a consequence of heat response, since transparency is a function of the crystallites melting.

7.3 Biomedical Applications of 4D Printed Structures

Over the last few years, several reviews have been published about 4D printing technology due to its **promising prospects** and **opportunities** in diverse fields. However, most of these reviews were primarily focused on specific materials, [257] the shape-changing mechanisms, [258] and the potential applications of the printed parts [252]. However, limited attention has been paid to the description of the biomedical applications of 4D printed parts. Only recently, Agarwal et al. [246] provided a detailed overview of the current state-of-the-art in biomedical applications, strictly devoted to regenerative medicine, soft robotic and medical tools, smart stents, and DDS, employing 4D printed objects/constructs. Figure 19 schematically shows the direct relationship between currently available technologies, stimuli-responsive materials employed, and the type of stimulus able to produce transformations on printed parts.

The most extensively explored application for 4D printing in the biomedical field includes **tissue reparation** and **regeneration**, the **elaboration of medical devices** and **soft robotics,** and **drug delivery**. For instance, given the preparation of devices and scaffolds for their use in TE applications, printed constructs are **static in nature** and are unable to recapitulate **native tissue dynamics** occurring due to changes in tissue conformation [259]. In contrast to this situation, 4D printing can resolve this drawback by creating **biologically active constructs** susceptible to undergoing shape change by modifying the external conditions and mimicking native tissue movements [260].

Another interesting area in which 4D printing is currently being explored is the elaboration of devices such as **stents**. Independently of the type or area that requires the implant (e.g., esophagus, airways, blood vessels, and others), a stent is either a plastic or a metallic tube-shaped device implanted aiming to maintain or even

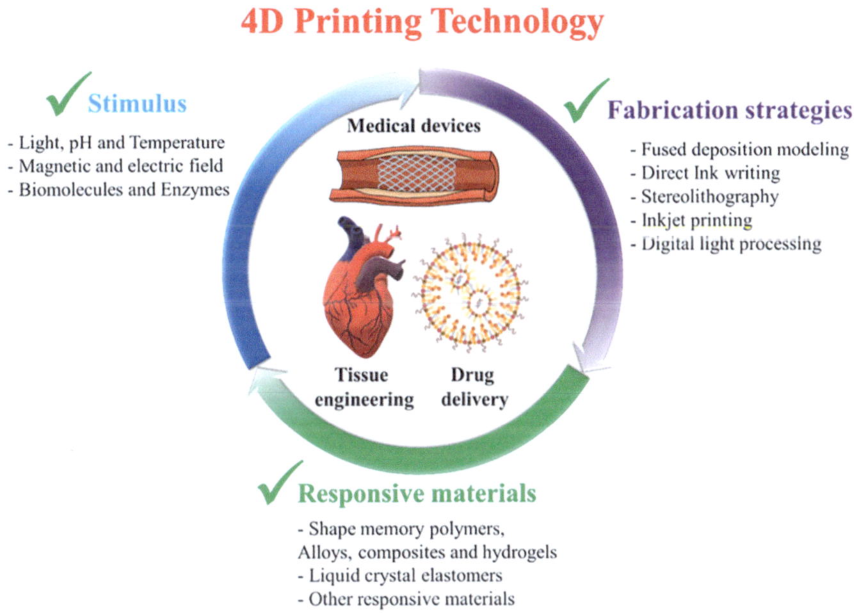

Fig. 19 Classification of different 4D printing characteristics. Fabrication strategies, different stimuli, responsive materials, and potential biomedical applications

restore lumen patency [261]. In the recent biomedical literature, there are several studies concerning 4D printing of stents. As explained by Manen et al. [262], in a recent review, the intrinsic capability of 4D printed stents to **experience shape changes** can be employed to implant parts that can be deployed at the target location in the human body to fulfill their functions by a **minimally invasive manner**.

Finally, the **encapsulation of pharmaceutical compounds** in 4D printed structures allows for modulating the absorption, distribution, and elimination to maximize the efficacy of the treatment. For instance, Wang et al. [263] described the fabrication of 4D printed **drug delivery hydrogel patches**, with shape memory behavior, combining pluronic F127 diacrylate macromere (F127DA) and sodium alginate via extrusion printing. After printing, a stable dual network structure defines the permanent shape of the patches. One part is a stable network formed by F127DA photo-crosslinking; the other one is a reversible network that is formed by Ca^{2+} cross-linked alginate. More interestingly, as evidenced by the authors, the results highlighted that the surface area and the shape (temporary/recovered) of the patches critically affected the drug release profiles. This property together with the viability of 3T3 fibroblasts that remained intact and revealed their excellent biocompatibility evidenced the potential of these systems for application in drug carriers and TE scaffolds. Therefore, although 4D printed materials still present several challenges, its unprecedented multidisciplinary perspective in the biomedicine research area is a reality.

8 Summary and Conclusions

The main conclusions may be divided into the explored sections as follows:

(i) **3D Printed Biomaterials and the Latest Drug-Loaded 3D Printed Medicines, Anatomic Biomaterials, and Their Biological Performance**: The clinical success of polymer, ceramic, and metal AM strongly depends on (i) the **selection of the material** (inherent physicochemical properties), and (ii) the **3D printing technique** according to the clinical and anatomical needs of the patient. In detail:

- **Metallic Biomaterials**: Ti has been the most widely studied as a permanent implant with a certain degree of success. In the case of Fe and Zn, numerous studies are being carried out to address their possible application as biodegradable implants with promising results. In the case of Mg, more attention is needed since current studies highlight the difficulty of being printed by the current AM technologies.

  In all cases, more systematic studies based on the AM printing processing parameters are needed to decrease the number of superficial defects.
- **Polymeric Materials**: 3D printing is a fact. Nevertheless, the use of non-toxic eluents, multi-component polymer manufacturing, and bio-inks printing with complex geometries still need to be solved for full exploitation in biomedicine.
- **Ceramic Materials**: It has been pointed out that the printing of such materials in combination with polymers can be a realistic way forward as the inherent properties of these materials hamper their stand-alone high-quality AM printing.

To overcome these limitations, specific and systematic studies are needed to establish fundamental relations between the printing parameters, pre-treatment, and post-treatment of each biomaterial for clinical use. With this approach in hand, a key factor to consider is the further combination of such biomaterials, i.e., AM of composite materials. In this way, 3D materials with good mechanical properties (metals), offering controlled drug release, low degradation rate, and controlled drug release (polymers) and, especially, highly biocompatible (ceramics) can be 3D printed.

Composite 3D printing has been targeted as a successful strategy to control biocompatibility in permanent and temporary implants, i.e., bone reconstruction/regeneration and cardiovascular stents. So far, although it seems promising, more research is needed, especially on the possibility of functionalization and in vivo studies.

(b) **4D Technology, from the Current Materials to the Future Perspective: So far,** 4D printing is experiencing rapid and significant growth in AM for biomedical applications. However, to date, researchers in this field are facing certain limitations concerning design-based and manufacturing constraints. **This is due to a gap of knowledge in** (i) **the deep understanding of the**

complexity of biological systems and (ii) **the lack of new material combinations** involving the printing of biological cells to create new materials with functional characteristics similar to those of natural biomaterials.

However, given the steady growth of this remarkable approach, it is expected that 4D printing will soon increase its full potential. For that purpose, the development of cost-effective, high-resolution printers and, more importantly, the discovery of new biocompatible smart biomaterials.

Acknowledgements The funding of the ADITIMAT-CM project (S2018/NMT-4411, Regional Government of Madrid and EU Structural and Social Funds), CSIC (PIE 202060E096, and 202060E239) and RTI2018-096391-B-C33 (MCIU/AEI/FEDER, UE) are gratefully acknowledged. J. J. Relinque acknowledges the Spanish Ministry of Universities for his Margarita Salas-Universidad de Cádiz 2021 Postdoctoral Fellowship. R. del Olmo acknowledges Marie Skłodowska-Curie Postdoctoral Fellowship (MSCA-PF), grant ID 101066947; EU651 project. A. Santos-Coquillat is grateful for financial support from Fonds de la Recherche Scientifique (F.R.S.-FNRS) as Chargée de recherches.

References

1. A.A. Zadpoor, J. Malda, Ann. Biomed. Eng. **45**(1), 1 (2017)
2. E.A. Guzzi, M.W. Tibbitt, Adv. Mater. **32**(13), 1901994 (2020)
3. E. Rezvani Ghomi, F. Khosravi, R.E. Neisiany, S. Singh, S. Ramakrishna, Curr. Opin. Biomed. Eng. **17**, 100255 (2021)
4. A.A. Zadpoor, Int. J. Mol. Sci. **18**(8), 1607 (2017)
5. S. Bose, D. Ke, H. Sahasrabudhe, A. Bandyopadhyay, Prog. Mater. Sci. **93**, 45 (2018)
6. P. Ahangar, M.E. Cooke, M.H. Weber, D.H. Rosenzweig, Appl. Sci. **9**(8), 1713 (2019)
7. D. Akilbekova, D. Mektepbayeva, in *3D Printing in Medicine*, ed. by D.M. Kalaskar (Woodhead Publishing, 2017), pp. 91–113
8. T.M. Shick, A.Z. Abdul Kadir, N.H.A. Ngadiman, A. Ma'aram, J. Bioact. Compat. Polym. **34**(6), 415 (2019)
9. G. Gao, M. Ahn, W.-W. Cho, B.-S. Kim, D.-W. Cho, Pharmaceutics **13**(9), 1373 (2021)
10. W. Jamróz, J. Szafraniec, M. Kurek, R. Jachowicz, Pharm. Res. **35**(9), 176 (2018)
11. P. Ramiah, L.C. du Toit, Y.E. Choonara, P.P.D. Kondiah, V. Pillay, Front. Mater. **7**, 76 (2020)
12. A. Kumar, S. Kargozar, F. Baino, S.S. Han, Front. Mater. **6**, 313 (2019)
13. P. Durai Murugan, S. Vijayananth, M.P. Natarajan, D. Jayabalakrishnan, K. Arul, V. Jayaseelan, J. Elanchezhian, Mater. Today Proc. **59**(Part 2), 1277 (2022)
14. M. Pérez, D. Carou, E.M. Rubio, R. Teti, Procedia CIRP **88**, 439 (2020)
15. B. Sagbas, in *Athens: ATINER'S Conference Paper Series*, No: MEC2017-2378 (2018)
16. S.A.M. Tofail, E.P. Koumoulos, A. Bandyopadhyay, S. Bose, L. O'Donoghue, C. Charitidis, Mater. Today **21**(1), 22 (2018)
17. A.K. Pal, A.K. Mohanty, M. Misra, RSC Adv. **11**(58), 36398 (2021)
18. A. Selema, M.N. Ibrahim, P. Sergeant, Energies **15**(3), 1076 (2022)
19. Fused Deposition Modeling, FDM, Manufacturing Guide. https://www.manufacturingguide.com/en/fused-deposition-modeling-fdm. Accessed 3 Mar 2023
20. A. Razavykia, E. Brusa, C. Delprete, R. Yavari, Materials **13**(17), 3895 (2020)
21. C.M. González-Henríquez, M.A. Sarabia-Vallejos, J. Rodriguez-Hernandez, Prog. Polym. Sci. **94**, 57 (2019)
22. S. Saleh Alghamdi, S. John, N. Roy Choudhury, N.K. Dutta, Polymers **13**(5), 753 (2021)

23. P. Szymczyk-Ziółkowska, M.B. Łabowska, J. Detyna, I. Michalak, P. Gruber, Biocybern. Biomed. Eng. **40**(2), 624 (2020)
24. Z.A. Starosolski, J.H. Kan, S.D. Rosenfeld, R. Krishnamurthy, A. Annapragada, Pediatr. Radiol. **44**(2), 216 (2014)
25. L. Ferrage, G. Bertrand, P. Lenormand, D. Grossin, B. Ben-Nissan, J. Aust. Ceram. Soc. **53**(1), 11 (2017)
26. J.S. Pelz, N. Ku, M.A. Meyers, L.R. Vargas-Gonzalez, J. Mater. Res. Technol. **15**, 670 (2021)
27. Y. Lakhdar, C. Tuck, J. Binner, A. Terry, R. Goodridge, Prog. Mater. Sci. **116**, 100736 (2021)
28. S. Amukarimi, S. Ramakrishna, M. Mozafari, Curr. Opin. Biomed. Eng. **19**, 100311 (2021)
29. G.P. Goldschmidt, E.G. de Moraes, A.P. Novaes de Oliveira, D. Hotza, Open Ceram. **9**, 100225 (2022)
30. D.A. Sánchez Rodríguez, A.I. Ramos-Murillo, R.D. Godoy-Silva, Bioprinting **24**, e00171 (2021)
31. A. Davoudinejad, in *Additive Manufacturing*, ed. by J. Pou, A. Riveiro, J.P. Davim (Elsevier, 2021), pp. 159–181
32. R.V. Pazhamannil, P. Govindan, Mater. Today Proc. **43**, 130 (2021)
33. D.A. Rau, M. Forgiarini, C.B. Williams, Addit. Manuf. **42**, 101996 (2021)
34. S.-B. Hua, J. Su, Z.-L. Deng, J.-M. Wu, L.-J. Cheng, X. Yuan, F. Chen, H. Zhu, D.-H. Qi, J. Xiao, Y.-S. Shi, Addit. Manuf. **45**, 102074 (2021)
35. A. Khan, K.A. Alamry, Carbohyd. Res. **506**, 108368 (2021)
36. A. Nouri, A. Rohani Shirvan, Y. Li, C. Wen, J. Mater. Sci. Technol. **94**, 196 (2021)
37. D. Grossin, A. Montón, P. Navarrete-Segado, E. Özmen, G. Urruth, F. Maury, D. Maury, C. Frances, M. Tourbin, P. Lenormand, G. Bertrand, Open Ceram. **5**, 100073 (2021)
38. International ASTM F2792-12a, Standard Terminology for Additive Manufacturing Technologies. (2012)
39. E. Davoodi, H. Montazerian, A.S. Mirhakimi, M. Zhianmanesh, O. Ibhadode, S.I. Shahabad, R. Esmaeilizadeh, E. Sarikhani, S. Toorandaz, S.A. Sarabi, R. Nasiri, Y. Zhu, J. Kadkhodapour, B. Li, A. Khademhosseini, E. Toyserkani, Bioact. Mater. **15**, 214 (2021)
40. D. Oropeza, A.J. Hart, Addit. Manuf. **48**, 102448 (2021)
41. J. Gonzalez-Gutierrez, in *Encyclopedia of Materials: Metals and Alloys*, ed. by F.G. Caballero (Elsevier, Oxford, 2022), pp. 319–329
42. N.E. Putra, M.J. Mirzaali, I. Apachitei, J. Zhou, A.A. Zadpoor, Acta Biomater. **109**, 1 (2020)
43. M. Salehi, S. Maleksaeedi, S.M.L. Nai, G.K. Meenashisundaram, M.H. Goh, M. Gupta, Acta Mater. **165**, 294 (2019)
44. L.-C. Zhang, J. Wang, Y. Liu, Z. Jia, S.-X. Liang, in *Encyclopedia of Materials: Metals and Alloys*, ed. by F.G. Caballero (Elsevier, Oxford, 2022), pp. 256–274
45. T.-S. Jang, D. Kim, G. Han, C.-B. Yoon, H.-D. Jung, Biomed. Eng. Lett. **10**(4), 505 (2020)
46. V.M.S. Muthaiah, S. Indrakumar, S. Suwas, K. Chatterjee, Bioprinting **25**, e00180 (2022)
47. M. Lowther, S. Louth, A. Davey, A. Hussain, P. Ginestra, L. Carter, N. Eisenstein, L. Grover, S. Cox, Addit. Manuf. **28**, 565 (2019)
48. L. Zhang, Y. Liu, in *Additive Manufacturing of Emerging Materials*, ed. by B. AlMangour (Springer, Cham, 2019), pp. 179–196
49. L. Dall'Ava, H. Hothi, A. Di Laura, J. Henckel, A. Hart, Metals **9**(7), 729 (2019)
50. P. Bartolo, J.-P. Kruth, J. Silva, G. Levy, A. Malshe, K. Rajurkar, M. Mitsuishi, J. Ciurana, M. Leu, CIRP Ann. **61**(2), 635 (2012)
51. G. Serin, M. Kahya, H.O. Unver, N. Durlu, O. Eroğul, Y. Gulec, Paper Presented at the The 17th International Conference on Machine Design and Production, Bursa, Turkey, 12–15 July 2016
52. Y. Qin, P. Wen, H. Guo, D. Xia, Y. Zheng, L. Jauer, R. Poprawe, M. Voshage, J.H. Schleifenbaum, Acta Biomater. **98**, 3 (2019)
53. K. Chua, I. Khan, R. Malhotra, D. Zhu, Eng. Regener. **2**, 288 (2021)
54. Y. Li, H. Jahr, P. Pavanram, F.S.L. Bobbert, U. Puggi, X.Y. Zhang, B. Pouran, M.A. Leeflang, H. Weinans, J. Zhou, A.A. Zadpoor, Acta Biomater. **96**, 646 (2019)
55. B. Song, S. Dong, S. Deng, H. Liao, C. Coddet, Opt. Laser Technol. **56**, 451 (2014)

56. B. Song, S. Dong, Q. Liu, H. Liao, C. Coddet, Mater. Des. **54**, 727 (2014)
57. Y. Yang, F. Yuan, C. Gao, P. Feng, L. Xue, S. He, C. Shuai, J. Mech. Behav. Biomed. Mater. **82**, 51 (2018)
58. C. Shuai, L. Xue, C. Gao, Y. Yang, S. Peng, Y. Zhang, Virtual Phys. Prototyp. **13**, 1 (2018)
59. P. Wen, M. Voshage, L. Jauer, Y. Chen, Y. Qin, R. Poprawe, J.H. Schleifenbaum, Mater. Des. **155**, 36 (2018)
60. A.G. Demir, L. Monguzzi, B. Previtali, Addit. Manuf. **15**, 20 (2017)
61. V. Manakari, G. Parande, M. Gupta, Metals **7**, 2 (2017)
62. A. Santos-Coquillat, E. Martínez-Campos, H. Mora Sánchez, L. Moreno, R. Arrabal, M. Mohedano, A. Gallardo, J. Rodríguez-Hernández, E. Matykina, Surf. Coat. Technol. **422**, 127508 (2021)
63. R. Karunakaran, S. Ortgies, A. Tamayol, F. Bobaru, M.P. Sealy, Bioact. Mater. **5**(1), 44 (2020)
64. N. Choudhary, V. Sharma, P. Kumar, J. Vinyl Addit. Technol. **27**(3), 612 (2021)
65. R.S. Payal, J.U. Sommer, Polymers **13**, 2078 (2021)
66. P.J. Ratri, K. Tashiro, Polym. J. **45**(11), 1107 (2013)
67. S. Wasti, S. Adhikari, Front. Chem. **8**, 315 (2020)
68. L. Yang, S. Li, Y. Li, M. Yang, Q. Yuan, J. Mater. Eng. Perform. **28**(1), 169 (2019)
69. I. Zein, D.W. Hutmacher, K.C. Tan, S.H. Teoh, Biomaterials **23**(4), 1169 (2002)
70. A. Szojka, K. Lalh, S.H.J. Andrews, N.M. Jomha, M. Osswald, A.B. Adesida, Bioprinting **8**, 1–7 (2017)
71. S. Singh, C. Prakash, S. Ramakrishna, Eur. Polym. J. **114**, 234 (2019)
72. J. Giannatsis, V. Dedoussis, J. Adv. Manuf. Technol. **40**(1), 116 (2009)
73. E. Provaggi, J.J.H. Leong, D.M. Kalaskar, Proc. Inst. Mech. Eng. H J. Eng. Med. **231**(6), 471 (2017)
74. E.O. Bachtiar, O. Erol, M. Millrod, R. Tao, D.H. Gracias, L.H. Romer, S.H. Kang, J. Mech. Behav. Biomed. Mater. **104**, 103649 (2020)
75. S. Basak, J. Macromol. Sci. A **58**(9), 579 (2021)
76. G. Cicala, G. Ognibene, S. Portuesi, I. Blanco, M. Rapisarda, E. Pergolizzi, G. Recca, Materials **11**(2), 285 (2018)
77. W. Kosorn, P. Wutticharoenmongkol, M. Fernández García, Int. J. Polym. Sci. **2021**, 6689789 (2021)
78. N.A.S. Mohd Pu'ad, R.H. Abdul Haq, H. Mohd Noh, H.Z. Abdullah, M.I. Idris, T.C. Lee, Mater. Today Proc. **29**, 228 (2020)
79. H. Quan, T. Zhang, H. Xu, S. Luo, J. Nie, X. Zhu, Bioact. Mater. **5**(1), 110 (2020)
80. A. Bagheri, J. Jin, A.C.S. Appl, Polym. Mater. **1**(4), 593 (2019)
81. C. Noè, M. Hakkarainen, Sangermano, Cationic UV-curing of epoxidized biobased resins. Polymers **13**(1), 89 (2021)
82. P.E. Petrochenko, J. Torgersen, P. Gruber, L.A. Hicks, J. Zheng, G. Kumar, R.J. Narayan, P.L. Goering, R. Liska, J. Stampfl, A. Ovsianikov, Adv. Healthc. Mater. **4**(5), 739 (2015)
83. P. Danilevicius, L. Georgiadi, C.J. Pateman, F. Claeyssens, M. Chatzinikolaidou, M. Farsari, Appl. Surf. Sci. **336**, 2 (2015)
84. J.-W. Choi, R. Wicker, S.-H. Lee, K.-H. Choi, C.-S. Ha, I. Chung, J. Mater. Process. Technol. **209**(15), 5494 (2009)
85. S. Schüller-Ravoo, S.M. Teixeira, J. Feijen, D.W. Grijpma, A.A. Poot, Macromol. Biosci. **13**(12), 1711 (2013)
86. B.K. Gu, D.J. Choi, S.J. Park, M.S. Kim, C.M. Kang, C.-H. Kim, Biomater. Res. **20**, 12 (2016)
87. L.G. Bracaglia, B.T. Smith, E. Watson, N. Arumugasaamy, A.G. Mikos, J.P. Fisher, Acta Biomater. **56**, 3 (2017)
88. H.-H. Tang, M.-L. Chiu, H.-C. Yen, J. Eur. Ceram. Soc. **31**(8), 1383 (2011)
89. S.H. Riza, S.H. Masood, R.A.R. Rashid, S. Chandra, in *Metallic Biomaterials Processing and Medical Device Manufacturing*, ed. by C.Wen (Woodhead Publishing, 2020), pp. 193–233
90. U. Ajoku, N. Hopkinson, M. Caine, Mater. Sci. Eng. A **428**(1), 211 (2006)
91. J.J. Relinque, M.G. García-Romero, J. Hernández-Saz, J. Navas, A.J. Gil-Mena, D.L. Sales, F.J. Navas, G. Morales-Cid, D. Aguilera, A. Periñán, F. Lasagni, S.I. Molina, Prog. Addit. Manuf. **3**(4), 233 (2018)

92. Y. Xia, P. Zhou, X. Cheng, Y. Xie, C. Liang, C. Li, S. Xu, Int. J. Nanomed. **8**, 4197 (2013)
93. Z. Meng, J. He, Z. Cai, F. Wang, J. Zhang, L. Wang, R. Ling, D. Li, Mater. Des. **189**, 108508 (2020)
94. W.Y. Yeong, N. Sudarmadji, H.Y. Yu, C.K. Chua, K.F. Leong, S.S. Venkatraman, Y.C. Boey, L.P. Tan, Acta Biomater. **6**(6), 2028 (2010)
95. N. Sudarmadji, J.Y. Tan, K.F. Leong, C.K. Chua, Y.T. Loh, Acta Biomater. **7**(2), 530 (2011)
96. B. Duan, M. Wang, W.Y. Zhou, W.L. Cheung, Z.Y. Li, W.W. Lu, Acta Biomater. **6**(12), 4495 (2010)
97. R. Landers, A. Pfister, U. Hübner, H. John, R. Schmelzeisen, R. Mülhaupt, J. Mater. Sci. **37**(15), 3107 (2002)
98. S.V. Murphy, A. Atala, Nat. Biotechnol. **32**(8), 773 (2014)
99. J.W. Jung, J.-S. Lee, D.-W. Cho, Sci. Rep. **6**(1), 21685 (2016)
100. O. Chaudhuri, L. Gu, D. Klumpers, M. Darnell, S.A. Bencherif, J.C. Weaver, N. Huebsch, H.-P. Lee, E. Lippens, G.N. Duda, D.J. Mooney, Nat. Mater. **15**(3), 326 (2016)
101. N.C. Hunt, L.M. Grover, Biotechnol. Lett. **32**(6), 733 (2010)
102. J.-H. Lee, H.-W. Kim, Emerging properties of hydrogels in tissue engineering. J. Tissue Eng. **9**, 1 (2018)
103. J.J. Green, J.H. Elisseeff, Nature **540**, 386 (2016)
104. T.N. Thanh, N. Laowattanatham, J. Ratanavaraporn, A. Sereemaspun, S. Yodmuang, Eur. Polym. J. **166**, 111027 (2022)
105. T. Pan, W. Song, X. Cao, Y. Wang, J. Mater. Sci. Technol. **32**(9), 889 (2016)
106. I.T. Ozbolat, H. Chen, Y. Yu, Robot. Comput.-Integr. Manuf **30**(3), 295 (2014)
107. T. Guo, T.R. Holzberg, C.G. Lim, F. Gao, A. Gargava, J.E. Trachtenberg, A.G. Mikos, J.P. Fisher, Biofabrication **9**(2), 024101 (2017)
108. F. You, X. Chen, D.M.L. Cooper, T. Chang, B.F. Eames, Biofabrication **11**(1), 015015 (2018)
109. G.R. López-Marcial, A.Y. Zeng, C. Osuna, J. Dennis, J.M. García, G.D. O'Connell, ACS Biomater. Sci. Eng. **4**(10), 3610 (2018)
110. A. Mazzocchi, M. Devarasetty, R. Huntwork, S. Soker, A. Skardal, Biofabrication **11**(1), 015003 (2018)
111. J.H. Kim, Y.-J. Seol, I.K. Ko, H.-W. Kang, Y.K. Lee, J.J. Yoo, A. Atala, S.J. Lee, Sci. Rep. **8**(1), 12307 (2018)
112. B. Byambaa, N. Annabi, K. Yue, G. Trujillo-de Santiago, M.M. Alvarez, W. Jia, M. Kazemzadeh-Narbat, S.R. Shin, A. Tamayol, A. Khademhosseini, Adv. Healthc. Mater. **6**(16), 1700015 (2017)
113. M. de Ruijter, A. Ribeiro, I. Dokter, M. Castilho, J. Malda, Adv. Healthc. Mater. **8**(7), e1800418 (2019)
114. S.R. Moxon, M.E. Cooke, S.C. Cox, M. Snow, L. Jeys, S.W. Jones, A.M. Smith, L.M. Grover, Adv. Mater. **29**(13), 1605594 (2017)
115. L. Ouyang, C.B. Highley, C.B. Rodell, W. Sun, J.A. Burdick, ACS Biomater. Sci. Eng. **2**(10), 1743 (2016)
116. Z. Zheng, J. Wu, M. Liu, H. Wang, C. Li, M.J. Rodriguez, G. Li, X. Wang, D.L. Kaplan, Adv. Healthc. Mater. **7**(6), e1701026 (2018)
117. A. Athirasala, A. Tahayeri, G. Thrivikraman, C.M. França, N. Monteiro, V. Tran, J. Ferracane, L.E. Bertassoni, Biofabrication **10**(2), 024101 (2018)
118. M. Ali, A.K. Pr, J.J. Yoo, F. Zahran, A. Atala, S.J. Lee, Adv. Healthc. Mater. **8**(7), e1800992 (2019)
119. M.C. Tronco, J.B. Cassel, L.A. dos Santos, α-TCP-based calcium phosphate cements: a critical review. Acta Biomater. **151**, 70–87 (2022). https://doi.org/10.1016/j.actbio.2022.08.040
120. J. Babilotte, V. Guduric, D. Le Nihouannen, A. Naveau, J.-C. Fricain, S. Catros, J. Biomed. Mater. Res. B Appl. Biomater. **107**(8), 2579 (2019)
121. A. Szcześ, L. Hołysz, E. Chibowski, Adv. Colloid Interface Sci. **249**, 321 (2017)
122. M. Milazzo, N. Contessi Negrini, S. Scialla, B. Marelli, S. Farè, S. Danti, M.J. Buehler, Adv. Funct. Mater. **29**(35), 1903055 (2019)

123. L. Vidal, C. Kampleitner, S. Krissian, M.Á. Brennan, O. Hoffmann, Y. Raymond, Y. Maazouz, M.-P. Ginebra, P. Rosset, P. Layrolle, Sci. Rep. **10**(1), 7068 (2020)
124. K. Zafeiris, D. Brasinika, A. Karatza, E. Koumoulos, I.K. Karoussis, K. Kyriakidou, C.A. Charitidis, Mater. Sci. Eng. C **119**, 111639 (2021)
125. A.E. Jakus, A.L. Rutz, S.W. Jordan, A. Kannan, S.M. Mitchell, C. Yun, K.D. Koube, S.C. Yoo, H.E. Whiteley, C.P. Richter, R.D. Galiano, W.K. Hsu, S.R. Stock, E.L. Hsu, R.N. Shah, Sci. Transl. Med. **8**(358), 358ra127 (2016)
126. R. Alluri, A. Jakus, S. Bougioukli, W. Pannell, O. Sugiyama, A. Tang, R. Shah, J.R. Lieberman, J. Biomed. Mater. Res. **106**(4), 1104 (2018)
127. M.J. Zafar, D. Zhu, Z. Zhang, Materials **12**(20), 3361 (2019)
128. F. Dogrul, P. Ożóg, M. Michálek, H. Elsayed, D. Galusek, L. Liverani, A.R. Boccaccini, E. Bernardo, Materials **14**(18), 5170 (2021)
129. C. Yang, X. Wang, B. Ma, H. Zhu, Z. Huan, N. Ma, C. Wu, J. Chang, A.C.S. Appl, Mater. Interfaces **9**(7), 5757 (2017)
130. N. Mani, A. Sola, A. Trinchi, K. Fox, Biointerphases **15**(6), 068501 (2020)
131. K. Lin, R. Sheikh, S. Romanazzo, I. Roohani, Materials (Basel) **12**(17), 2660 (2019)
132. N. Ashammakhi, A. Hasan, O. Kaarela, B. Byambaa, A. Sheikhi, A.K. Gaharwar, A. Khademhosseini, Adv. Healthc. Mater. **8**(7), e1801048 (2019)
133. Y. Lai, Y. Li, H. Cao, J. Long, X. Wang, L. Li, C. Li, Q. Jia, B. Teng, T. Tang, J. Peng, D. Eglin, M. Alini, D.W. Grijpma, G. Richards, L. Qin, Biomaterials **197**, 207 (2019)
134. M. Kaur, K. Singh, Mater. Sci. Eng. C **102**, 844 (2019)
135. J. Quinn, R. McFadden, C.-W. Chan, L. Carson, iScience **23**(11), 101745 (2020)
136. J. Ping, C. Zhou, Y. Dong, X. Wu, X. Huang, B. Sun, B. Zeng, F. Xu, W. Liang, Mol. Immunol. **138**, 110 (2021)
137. S.S. Sidhu, H. Singh, M.A.-H. Gepreel, Mater. Sci. Eng. C **121**, 111661 (2021)
138. A. Ataee, Y. Li, C. Wen, Acta Biomater. **97**, 587 (2019)
139. R. Balasubramanian, R. Nagumothu, E. Parfenov, R. Valiev, Mater. Today Proc. **46**, 1195 (2021)
140. A. Vance, K. Bari, A. Arjunan, Mater. Des. **160**, 1281 (2018)
141. M. Bhuvanesh Kumar, P. Sathiya, Thin-Walled Struct. **159**, 107228 (2021)
142. H.-H. Lin, D. Lonic, L.-J. Lo, J. Natl. Med. Assoc. **117**(7), 547 (2018)
143. C. Han, Y. Li, Q. Wang, S. Wen, Q. Wei, C. Yan, L. Hao, J. Liu, Y. Shi, J. Mech. Behav. **80**, 119 (2018)
144. H.P. Tang, P. Zhao, C.S. Xiang, N. Liu, L. Jia, in *Titanium in Medical and Dental Applications*, ed. by F.H. Froes, M. Qian (Woodhead Publishing, 2018), pp. 239–249
145. H. Koizumi, Y. Takeuchi, H. Imai, T. Kawai, T. Yoneyama, J. Prosthodont. Res. **63**(3), 266 (2019)
146. A. Ramezannejad, W. Xu, W.L. Xiao, K. Fox, D. Liang, M. Qian, Curr. Opin. Solid State Mater. Sci. **23**(6), 100783 (2019)
147. W. Weng, A. Biesiekierski, Y. Li, C. Wen, Materialia **6**, 100323 (2019)
148. H.Z. Lu, L.H. Liu, C. Yang, X. Luo, C.H. Song, Z. Wang, J. Wang, Y.D. Su, Y.F. Ding, L.C. Zhang, Y.Y. Li, J. Mater. Sci. Technol. **101**, 205 (2022)
149. Q. Zhou, M. Hayat, G. Chen, S. Cai, X. Qu, H. Tang, P. Cao, Mater. Sci. Eng. A **744**, 290 (2018)
150. M. Weinmann, C. Schnitter, M. Stenzel, J. Markhoff, C. Schulze, R. Bader, Int. J. Refract. Met. Hard Mater **75**, 126 (2018)
151. M. Fischer, D. Joguet, G. Robin, L. Peltier, P. Laheurte, Mater. Sci. Eng. C **62**, 852 (2016)
152. Y. Li, Y. Ding, K. Munir, J. Lin, M. Brandt, A. Atrens, Y. Xiao, J.R. Kanwar, C. Wen, Acta Biomater. **87**, 273 (2019)
153. N. Singh, P. Hameed, R. Ummethala, G. Manivasagam, K.G. Prashanth, J. Eckert, Mater. Today Adv. **8**, 100097 (2020)
154. D. Zhao, H. Liang, C. Han, J. Li, J. Liu, K. Zhou, C. Yang, Q. Wei, Addit. Manuf. **47**, 102223 (2021)

155. B. Wysocki, P. Maj, R. Sitek, J. Buhagiar, K.J. Kurzydłowski, W. Święszkowski, Appl. Sci. **7**(7), 657 (2017)
156. G. Kasperovich, J. Hausmann, J. Mater. Process. Technol. **220**, 202 (2015)
157. J. Hernandez, S.J. Li, E. Martinez, L.E. Murr, X.M. Pan, K.N. Amato, X.Y. Cheng, F. Yang, C.A. Terrazas, S.M. Gaytan, Y.L. Hao, R. Yang, F. Medina, R.B. Wicker, J. Mater. Sci. Technol. **29**(11), 1011 (2013)
158. Y.J. Liu, S.J. Li, H.L. Wang, W.T. Hou, Y.L. Hao, R. Yang, T.B. Sercombe, L.C. Zhang, Acta Mater. **113**, 56 (2016)
159. J.P. Luo, J.F. Sun, Y.J. Huang, J.H. Zhang, Y.D. Zhang, D.P. Zhao, M. Yan, Mater. Sci. Eng. C **97**, 275 (2019)
160. G. Gong, J. Ye, Y. Chi, Z. Zhao, Z. Wang, G. Xia, X. Du, H. Tian, H. Yu, C. Chen, J. Mater. Res. Technol. **15**, 855 (2021)
161. P. Moghimian, T. Poirié, M. Habibnejad-Korayem, J.A. Zavala, J. Kroeger, F. Marion, F. Larouche, Addit. Manuf. **43**, 102017 (2021)
162. Y.-L. Hao, S.-J. Li, R. Yang, Rare Met. **35**(9), 661 (2016)
163. D. Jiang, F. Ning, Y. Wang, J. Mater. Process. Technol. **289**, 116952 (2021)
164. C. Shuai, W. Yang, Y. Yang, H. Pan, C. He, F. Qi, D. Xie, H. Liang, Mater. Res. Express **7**(1), 015404 (2020)
165. I. Ferretto, D. Kim, N.M. Della Ventura, M. Shahverdi, W. Lee, C. Leinenbach, Addit. Manuf. **46**, 102071 (2021)
166. S. Thapliyal, P. Agrawal, P. Agrawal, S.S. Nene, R.S. Mishra, B.A. McWilliams, K.C. Cho, Acta Mater. **219**, 117271 (2021)
167. H. Jahr, Y. Li, J. Zhou, A.A. Zadpoor, K.-U. Schröder, Front. Mater. **8**, 628633 (2021)
168. M.I. Rahim, S. Ullah, P.P. Mueller, Metals **8**(7), 532 (2018)
169. P. Płatek, J. Sienkiewicz, J. Janiszewski, F. Jiang, Materials **13**(9), 2204 (2020)
170. A. Pandey, A. Awasthi, K.K. Saxena, Adv. Mater. Process. Technol. **6**(2), 405 (2020)
171. Y. Yang, C. He, D. E, W. Yang, F. Qi, D. Xie, L. Shen, S. Peng, C. Shuai, Mater. Des. **185**, 108259 (2020)
172. J.-W. Lee, H.-S. Han, K.-J. Han, J. Park, H. Jeon, M.-R. Ok, H.-K. Seok, J.-P. Ahn, K.E. Lee, D.-H. Lee, S.-J. Yang, S.-Y. Cho, P.-R. Cha, H. Kwon, T.-H. Nam, J.H.L. Han, H.-J. Rho, K.-S. Lee, Y.-C. Kim, D. Mantovani, PNAS **113**(3), 716 (2016)
173. D. Zhao, S. Huang, F. Lu, B. Wang, L. Yang, L. Qin, K. Yang, Y. Li, W. Li, W. Wang, S. Tian, X. Zhang, W. Gao, Z. Wang, Y. Zhang, X. Xie, J. Wang, J. Li, Biomaterials **81**, 84 (2016)
174. H. Windhagen, K. Radtke, A. Weizbauer, J. Diekmann, Y. Noll, U. Kreimeyer, R. Schavan, C. Stukenborg-Colsman, H. Waizy, Biomed. Eng. Online **12**(1), 62 (2013)
175. Y.H. Bouchi, B.D. Gogas, B Glob. Cardiol. Sci. Pract. **2015**(5), 63 (2015)
176. K. Wei, X. Zeng, Z. Wang, J. Deng, M. Liu, G. Huang, X. Yuan, Mater. Sci. Eng. A **756**, 226 (2019)
177. K. Wei, Z. Wang, X. Zeng, Mater. Lett. **156**, 187 (2015)
178. F. Baer, L. Berger, L. Jauer, G. Kurtuldu, R. Schaeublin, J.H. Schleifenbaum, J.F. Loeffler, Acta Biomater. **98**, 36 (2019)
179. S. Gangireddy, B. Gwalani, K. Liu, E.J. Faierson, R.S. Mishra, Addit. Manuf. **26**, 53 (2019)
180. Y. Li, H. Jahr, X.Y. Zhang, M.A. Leeflang, W. Li, B. Pouran, F.D. Tichelaar, H. Weinans, J. Zhou, A.A. Zadpoor, Addit. Manuf. **28**, 299 (2019)
181. Y. Li, J. Zhou, P. Pavanram, M.A. Leeflang, L.I. Fockaert, B. Pouran, N. Tümer, K.U. Schröder, J.M.C. Mol, H. Weinans, H. Jahr, A.A. Zadpoor, Acta Biomater. **67**, 378 (2018)
182. F. Witte, L. Jauer, W. Meiners, Z. Kronbach, K. Strohschein, T. Schmidt, Front. Bioeng. Biotechnol., in *Conference Abstract: 10th World Biomaterials Congress*. https://doi.org/10.3389/conf.FBIOE.2016.01.00708
183. A. Kopp, T. Derra, M. Muther, L. Jauer, J.H. Schleifenbaum, M. Voshage, O. Jung, R. Smeets, N. Kroger, Acta Biomater. **98**, 23 (2019)
184. H. Kabir, K. Munir, C. Wen, Y. Li, Bioact. Mater. **6**(3), 836 (2021)
185. Y. Zhou, P. Wu, Y. Yang, D. Gao, P. Feng, C. Gao, H. Wu, Y. Liu, H. Bian, C. Shuai, J. Alloys Compd. **687**, 109 (2016)

186. H. Takagi, H. Sasahara, T. Abe, H. Sannomiya, S. Nishiyama, S. Ohta, K. Nakamura, Addit. Manuf. **24**, 498 (2018)
187. S. Palanivel, P. Nelaturu, B. Glass, R.S. Mishra, Mater. Des. **65**, 934 (2015)
188. M. Salehi, S. Maleksaeedi, M.A. Bin Sapari, M.L.S. Nai, G.K. Meenashisundaram, M. Gupta, Mater. Des. **169**, 107683 (2019)
189. M. Salehi, S. Maleksaeedi, M.L.S. Nai, M. Gupta, Addit. Manuf. **29**, 100790 (2019)
190. Y. Li, H. Jahr, J. Zhou, A.A. Zadpoor, Acta Biomater. **115**, 29 (2020)
191. F. Witte, H. Ulrich, C. Palm, E. Willbold, J. Biomed. Mater. Res. A **81A**(3), 757 (2007)
192. M.Q. Cheng, T. Wahafu, G.F. Jiang, W. Liu, Y.Q. Qiao, X.C. Peng, T. Cheng, X.L. Zhang, G. He, X.Y. Liu, Sci. Rep. **6**, 24134 (2016)
193. N. Kleer, S. Julmi, A.K. Gartzke, J. Augustin, F. Feichtner, A.C. Waselau, C. Klose, H.J. Maier, P. Wriggers, A. Meyer-Lindenberg, Materialia **8**, 100436 (2019)
194. D. Vojtěch, J. Kubásek, J. Šerák, P. Novák, Acta Biomater. **7**(9), 3515 (2011)
195. M. Yamaguchi, M.N. Weitzmann, Mol. Cell. Biochem. **355**(1–2), 179 (2011)
196. E. Mostaed, M. Sikora-Jasinska, J.W. Drelich, M. Vedani, Acta Biomater. **71**, 1 (2018)
197. H. Yang, C. Wang, C. Liu, H. Chen, Y. Wu, J. Han, Z. Jia, W. Lin, D. Zhang, W. Li, W. Yuan, H. Guo, H. Li, G. Yang, D. Kong, D. Zhu, K. Takashima, L. Ruan, J. Nie, X. Li, Y. Zheng, Biomaterials **145**, 92 (2017)
198. P.K. Bowen, R.J. Guillory, E.R. Shearier, J.-M. Seitz, J. Drelich, M. Bocks, F. Zhao, J. Goldman, Mater. Sci. Eng. C **56**, 467 (2015)
199. G. Katarivas Levy, J. Goldman, E. Aghion, Metals **7**(10), 402 (2017)
200. D. Hernández-Escobar, S. Champagne, H. Yilmazer, B. Dikici, C.J. Boehlert, H. Hermawan, Acta Biomater. **97**, 1 (2019)
201. J. Venezuela, M.S. Dargusch, Acta Biomater. **87**, 1 (2019)
202. M. Montani, A.G. Demir, E. Mostaed, M. Vedani, B. Previtali, Rapid Prototyp. J. **23**(3), 514 (2017)
203. P. Wen, L. Jauer, M. Voshage, Y. Chen, R. Poprawe, J.H. Schleifenbaum, J. Mater. Process. Technol. **258**, 128 (2018)
204. Y. Chen, P. Wen, M. Voshage, L. Jauer, Y. Qin, J.H. Schleifenbaum, R. Poprawe, J. Laser Appl. **31**(2), 022304 (2019)
205. C. Shuai, Y. Cheng, Y. Yang, S. Peng, W. Yang, F. Qi, J. Alloys Compd. **798**, 606 (2019)
206. T. Wu, X. Chen, Y. Wang, H. Xiao, Y. Peng, L. Lin, W. Xia, M. Long, J. Tao, X. Shuai, Nanomed. NBM **14**(7), 2215 (2018)
207. P. Wen, Y. Qin, Y. Chen, M. Voshage, L. Jauer, R. Poprawe, J.H. Schleifenbaum, J. Mater. Sci. Technol. **35**(2), 368 (2019)
208. Y. Li, P. Pavanram, J. Zhou, K. Lietaert, P. Taheri, W. Li, H. San, M.A. Leeflang, J.M.C. Mol, H. Jahr, A.A. Zadpoor, Acta Biomater. **101**, 609 (2020)
209. Y. Li, W. Li, F.S.L. Bobbert, K. Lietaert, J.H. Dong, M.A. Leeflang, J. Zhou, A.A. Zadpoor, Acta Biomater. **106**, 439 (2020)
210. K. Lietaert, A.A. Zadpoor, M. Sonnaert, J. Schrooten, L. Weber, A. Mortensen, J. Vleugels, Acta Biomater. **110**, 289 (2020)
211. S.J. Trenfield, A. Awad, A. Goyanes, S. Gaisford, A.W. Basit, Trends Pharmacol. Sci. **39**(5), 440 (2018)
212. S.H. Lim, H. Kathuria, J.J.Y. Tan, L. Kang, Adv. Drug Deliv. Rev. **132**, 139 (2018)
213. J. Skowyra, K. Pietrzak, M.A. Alhnan, Eur. J. Pharm. Sci. **68**, 11 (2015)
214. A. Goyanes, M. Kobayashi, R. Martínez-Pacheco, S. Gaisford, A.W. Basit, Int. J. Pharm. **514**(1), 290 (2016)
215. A.A. Konta, M. García-Piña, D.R. Serrano, Bioengineering **4**(4), 79 (2017)
216. C. Korte, J. Quodbach, AAPS PharmSciTech **19**(8), 3333 (2018)
217. Q. Zhang, N. Willis-Fox, R. Daly, Int. J. Pharm. **599**, 120436 (2021)
218. K. Shi, J.P. Salvage, M. Maniruzzaman, A. Nokhodchi, Int. J. Pharm. **597**, 120315 (2021)
219. N. Nashed, M. Lam, A. Nokhodchi, Int. J. Pharm. **596**, 120237 (2021)
220. J. Zhang, X. Feng, H. Patil, R.V. Tiwari, M.A. Repka, Int. J. Pharm. **519**(1–2), 186 (2017)
221. M.S. Algahtani, A.A. Mohammed, J. Ahmad, E. Saleh, Polymers **12**(6), 1395 (2020)

222. F.E. Freeman, R. Burdis, D.J. Kelly, Trends Mol. Med. **27**(7), 700 (2021). Molecules **26**(13), 4066 (2021)
223. V. Domsta, A. Seidlitz, Molecules **26**(13), 4066 (2021)
224. R. Trombetta, J.A. Inzana, E.M. Schwarz, S.L. Kates, H.A. Awad, Ann. Biomed. Eng. **45**(1), 23 (2017)
225. U. Gbureck, E. Vorndran, F.A. Müller, J.E. Barralet, J. Control. Rel. **122**(2), 173 (2007)
226. J.A. Inzana, R.P. Trombetta, E.M. Schwarz, S.L. Kates, H.A. Awad, Eur. Cells Mater. **30**, 232 (2015)
227. A.R. Akkineni, Y. Luo, M. Schumacher, B. Nies, A. Lode, M. Gelinsky, Acta Biomater. **27**, 264 (2015)
228. M. Zhu, K. Li, Y. Zhu, J. Zhang, X. Ye, Acta Biomater. **16**, 145 (2015)
229. H.H.K. Xu, P. Wang, L. Wang, C. Bao, Q. Chen, M.D. Weir, L.C. Chow, L. Zhao, X. Zhou, M.A. Reynolds, Bone Res. **5**(1), 17056 (2017)
230. L.A. Strobel, S.N. Rath, A.K. Maier, J.P. Beier, A. Arkudas, P. Greil, R.E. Horch, U. Kneser, J. Tissue Eng. Regen. Med. **8**(3), 176 (2014)
231. H. Wang, G. Wu, J. Zhang, K. Zhou, B. Yin, X. Su, G. Qiu, G. Yang, X. Zhang, G. Zhou, Z. Wu, Colloids Surf. B **141**, 491 (2016)
232. R. Olmos-Juste, O. Guaresti, T. Calvo-Correas, N. Gabilondo, A. Eceiza, Int. J. Pharm. **609**, 121124 (2021)
233. M.R. Sommer, L. Alison, C. Minas, E. Tervoort, P.A. Rühs, A.R. Studart, Soft Matter **13**(9), 1794 (2017)
234. H.Y. Shen, Z.H. Liu, J.S. Hong, M.S. Wu, S.J. Shiue, H.Y. Lin, J. Control. Rel. **331**, 154 (2021)
235. T.H. Song, J. Jang, Y.J. Choi, J.H. Shim, D.W. Cho, Cell Transpl. **24**(12), 2513 (2015)
236. S. Liu, D. Huang, Y. Hu, J. Zhang, B. Chen, H. Zhang, X. Dong, R. Tong, Y. Li, W. Zhou, RSC Adv. **10**(64), 39241 (2020)
237. X. Shi, Y. Cheng, J. Wang, H. Chen, X. Wang, X. Li, W. Tan, Z. Tan, Theranostics **10**(23), 10652 (2020)
238. Y. Yang, S. Yang, Y. Wang, Z. Yu, H. Ao, H. Zhang, L. Qin, O. Guillaume, D. Eglin, R.G. Richards, T. Tang, Acta Biomater. **46**, 112 (2016)
239. Y. Yang, L. Chu, S. Yang, H. Zhang, L. Qin, O. Guillaume, D. Eglin, R.G. Richards, T. Tang, Acta Biomater. **79**, 265 (2018)
240. C.I. Gioumouxouzis, A.T. Chatzitaki, C. Karavasili, O.L. Katsamenis, D. Tzetzis, E. Mystiridou, N. Bouropoulos, D.G. Fatouros, AAPS PharmSciTech **19**(8), 3362 (2018)
241. Z. Li, T. Du, C. Ruan, X. Niu, Bioact. Mater. **6**(5), 1491 (2021)
242. A. Longoni, J. Li, G.C.J. Lindberg, J. Rnjak-Kovacina, L.M. Wise, G.J. Hooper, T.B.F. Woodfield, D.C. Kieser, K.S. Lim, Essays Biochem. **65**(3), 569 (2021)
243. H.P. Kang, H. Ihn, D.M. Robertson, X. Chen, O. Sugiyama, A. Tang, R. Hollis, T. Skorka, D. Longjohn, D. Oakes, R. Shah, D. Kohn, A.E. Jakus, J.R. Lieberman, J. Biomed. Mater. Res. A **109**(11), 2346 (2021)
244. G.L. Koons, A.G. Mikos, J. Control. Rel. **295**, 50 (2019)
245. A.C. Pinho, C.S. Buga, A.P. Piedade, Appl. Mater. Today **19**, 100611 (2020)
246. T. Agarwal, S.Y. Hann, I. Chiesa, H. Cui, N. Celikkin, S. Micalizzi, A. Barbetta, M. Costantini, T. Esworthy, L.G. Zhang, C. De Maria, T.K. Maiti, J. Mater. Chem. B **9**(37), 7608 (2021)
247. A. Lendlein, S. Kelch, Angew. Chem. Int. Ed. **41**(12), 2034 (2002)
248. Q. Ge, H.J. Qi, M.L. Dunn, Appl. Phys. Lett. **103**(13), 131901 (2013)
249. V. Khare, S. Sonkaria, G.-Y. Lee, S.-H. Ahn, W.-S. Chu, Int. J. Precis. Eng. Manuf. Green Technol. **4**(3), 291 (2017)
250. D.-G. Shin, T.-H. Kim, D.-E. Kim, Int. J. Precis. Eng. Manuf. Green Technol. **4**(3), 349 (2017)
251. J.-J. Wu, L.-M. Huang, Q. Zhao, T. Xie, Chin. J. Polym. Sci. **36**(5), 563 (2018)
252. Z. Zhang, K.G. Demir, G.X. Gu, Int. J. Smart Nano Mater. **10**(3), 205 (2019)
253. J. Kim, J.A. Hanna, R.C. Hayward, C.D. Santangelo, Soft Matter **8**(8), 2375 (2012)
254. D. Raviv, W. Zhao, C. McKnelly, A. Papadopoulou, A. Kadambi, B. Shi, S. Hirsch, D. Dikovsky, M. Zyracki, C. Olguin, R. Raskar, S. Tibbits, Sci. Rep. **4**(1), 7422 (2014)

255. A. Sydney Gladman, E.A. Matsumoto, R.G. Nuzzo, L. Mahadevan, J.A. Lewis, Nat. Mater. **15**(4), 413 (2016)
256. M. Zarek, M. Layani, I. Cooperstein, E. Sachyani, D. Cohn, S. Magdassi, Adv. Mater. **28**(22), 4449 (2016)
257. A. Subash, B. Kandasubramanian, Eur. Polym. J. **134**, 109771 (2020)
258. A.P. Piedade, J. Funct. Biomater. **10**(1), 9 (2019)
259. S.H. Kim, Y.B. Seo, Y.K. Yeon, Y.J. Lee, H.S. Park, M.T. Sultan, J.M. Lee, J.S. Lee, O.J. Lee, H. Hong, H. Lee, O. Ajiteru, Y.J. Suh, S.-H. Song, K.-H. Lee, C.H. Park, Biomaterials **260**, 120281 (2020)
260. Y.C. Li, Y.S. Zhang, A. Akpek, S.R. Shin, A. Khademhosseini, Biofabrication **9**(1), 012001 (2016)
261. Y. Kang, Biomed. Res. Int. **2019**, 9265017 (2019)
262. T.V. Manen, S. Janbaz, K.M.B. Jansen, A.A. Zadpoor, Commun. Mater. **2**, 56 (2021)
263. Y. Wang, Y. Miao, J. Zhang, J.P. Wu, T.B. Kirk, J. Xu, D. Ma, W. Xue, Mater. Sci. Eng. C **84**, 44 (2018)

Application of Biomaterials for Tissue Engineering

Abu Hashem, Ab Rahman Marlinda, M. A. Motalib Hossain, Mohammad Al Mamun, Khanom Simarani, and Mohd Rafie Johan

Abstract Cell, tissue, and organ degeneration or loss can occur as a result of a variety of injuries, including sudden trauma, congenital disability, or the wear and tear of various diseases in humans. The human body's ability to regenerate such tissues and organs is severely limited. When tissue or cells die or lose viability, they cannot be brought back to life without removal and repair. It is possible that other parts of the body will be affected if healing does not occur. Tissue or organ transplantation to treat defective parts in various forms, such as autograft, allograft, or xenograft, may have both benefits and risks. Aside from that, there aren't enough donors to meet demand. That is why articulated tissue engineering is advantageous in terms of benefit versus risk, as well as in supplementing donor insufficiency. Functional tissues can be created using appropriate biomaterials to aid in the repair, restoration, and

A. Hashem · A. R. Marlinda (✉) · M. A. Motalib Hossain · M. Al Mamun · M. R. Johan
Nanotechnology and Catalysis Research Centre (NANOCAT), Institute for Advanced Studies, Universiti Malaya, Kuala Lumpur, Malaysia
e-mail: marlinda@um.edu.my

A. Hashem
e-mail: hashemnibbd@gmail.com

M. A. Motalib Hossain
e-mail: motalib_123@yahoo.com

M. Al Mamun
e-mail: zithrox@gmail.com

M. R. Johan
e-mail: mrafiej@um.edu.my

A. Hashem
Microbial Biotechnology Division, National Institute of Biotechnology, Ganakbari, Ashulia, Savar, Dhaka, Bangladesh

M. Al Mamun
Department of Chemistry, Jagannath University, Dhaka, Bangladesh

K. Simarani
Department of Microbiology, Institute of Biological Sciences, Faculty of Sciences, Universiti Malaya, Kuala Lumpur, Malaysia
e-mail: hanom_ss@um.edu.my

S. J. Ikhmayies (ed.), *Advances in Biomaterials Research*, Advances in Material Research and Technology, https://doi.org/10.1007/978-3-031-97481-6_4

replacement of injured tissue and organs. Thus, biomaterials play an important role in tissue engineering because they can promote cell development and differentiation to aid in injury recovery. Biomaterials have a variety of physical and mechanical properties that allow cells and tissues to grow, multiply, and differentiate. Because of their valuable properties and application possibilities in tissue engineering, research on biomaterials is expanding on a daily basis. The global biomaterials market is expected to be worth 157.18 USD billion in 2024 and 169.47 USD billion in 2025 to 333.73 USD billion till 2034, with a compound annual growth rate of 7.82% over that period. The chapter focused on the most recent advances in biomaterials research and applications in the field of tissue engineering, as well as their diverse properties and characteristics, with a classification, mode of application, application scopes, purposes, and techniques. Furthermore, the scopes, prospects, bottlenecks, and likely solutions are discussed, as are strategies for future developments.

Keywords Biomaterials · Biocompatibility · Characterization · Properties · Tissue engineering

1 Introduction

The human body is an organization consisting of various tissues and organs with unique functions. The tissues consist of a highly organized structure and a unique composition, which provide mechanical, transport, and functional support to control cellular and biological functions. The hard tissues are the core parts of the body that support load bearing and movement. An adult human body has nearly 40–50% skeletal muscles. Sometimes skeletal muscles become injured by accident. That may be owing to strain, contusion, trauma, lacerations, as well as bad surgery. As the human body ages, it suffers different changes, triggering misfortune or destruction of tissues or other functions [1]. Therefore, there is a need to replace lost or severely injured tissues or organs owing to tissue injury, disease, malfunctioning, and aging [2]. Nevertheless, the human body has a low regeneration ability and inadequate self-repair abilities.

Consequently, damaged tissues could be treated by surgical medications. This surgical treatment involves local or distant autologous tissue replacement, but the deficiency of a tissue donor and donor site morbidity worsens the situation [1]. Worldwide, lots of people would benefit considerably if tissues and organs could be replaced based on demand. Conventionally, transplantation of intact tissues and organs has been the foundation for replacing diseased and injured parts of the body [3]. Therefore, the transplantation of tissue or organ to treat these defective parts in different implant forms may become a suitable solution. In this regard, tissue engineering turns out to be a suitable and improved option.

Tissue engineering was developed as an innovative approach aimed at regenerating tissues using cells, biomaterials, and growth factors. It has become one of the most dynamic and rapidly advancing fields, offering solutions for repairing or replacing damaged or diseased tissues and organs by cultivating them in vitro and subsequently transplanting them into the body [4]. This has massive importance in regenerative medicine as it is a vital element of all invasive and allied disciplines. Furthermore, it has broadened the scope of its applications in fields such as general surgery, dentistry, dermatology, pediatric surgery, plastic surgery, oncology, gynecology, ophthalmology, stomach medicine and surgery, urology, neurosurgery, thoracic and cardiovascular surgery, ENT (ear–nose–throat) surgery, and organ transplantation.

Cells serve as the fundamental building blocks of tissue, while tissues play a crucial role in bodily functions. Typically, groups of cells produce and release their own structural support system, known as the extracellular matrix. This matrix, or scaffold, not only provides structural support to cells and tissues but also acts as a conduit for various signalling molecules. Through this mechanism, cells receive information from multiple sources present in their surrounding environment. Additionally, each signal has the potential to initiate a cascade of reactions that regulate cellular activities. By understanding the mechanism of individual signals to cells, their interaction with the micro-environment, and the organization of tissues and organs, scientists have manipulated these approaches to repair injured tissues or even build new ones. Tissue engineering techniques grew from biomaterials research and development and describe the procedure of merging scaffolds, cells, and biologically active factors into functional tissues. Tissue engineering aims to build functional structures that restore, heal, maintain, or improve damaged tissues or whole organs [5].

It is crucial to choose suitable biomaterials or scaffolds for tissue engineering to support cell growth, development, and expression of other cell functions. In addition to serving as carriers for the release of growth factors or cytokines to aid in tissue repair and regeneration, the biomaterials are essential for cell adhesion, proliferation, and extracellular matrix (ECM) secretion. Thus, the tissue engineering technique is based on the careful selection of biomaterial architectures and components. A wide variety of materials have been developed and employed as scaffolding elements, including natural molecules, synthetic polymers, nanoscale materials, inorganic materials, and various composites. Since natural polymers' biochemical properties are similar to those of human extracellular matrix, they have long piqued interest in tissue engineering applications. In vivo, these naturally occurring macromolecular compounds are therefore easier to identify and accept [6]. The tissue regeneration pathways are depicted in Fig. 1.

In order to support cell proliferation and tissue regeneration, biomaterials are used in tissue engineering and are intended to provide an architectural framework akin to the native extracellular matrix [8]. Biomaterials are substances that interface with cells, tissues, or bodily fluids and interact with biological systems. It is described as any natural or synthetic material, or mixture of materials, that can be used for any length of time as a system or as an individual material that replaces, restores, or fixes any organ, tissue, or bodily function. These days, biomaterials are used extensively in everything from controlled drug delivery to the fabrication of functioning tissues to therapeutic implants [9]. The process often begins with constructing a scaffold. Cells with or without growth elements can be added to the scaffold after developing the scaffold construct. Then, the targeted tissue can be developed if the environment

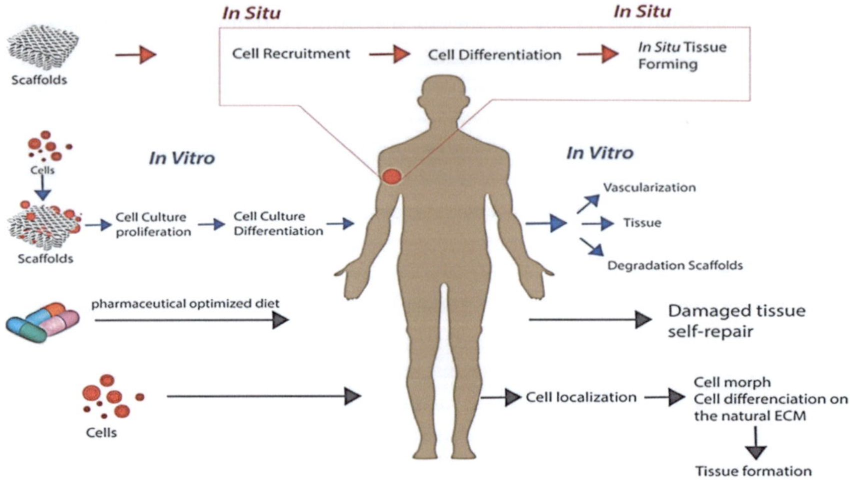

Fig. 1 Tissue regeneration pathways. Reprinted from Abdulghani and Mitchell [7], under the terms of the Creative Commons Attribution (CC BY) license. https://creativecommons.org/licenses/by/4.0/

is suitable for the construct. The tissue can self-assemble when scaffolds, cells, and growth factors are mixed together at the same time. Using an already-existing scaffold is another method of creating new tissue. This method has the potential to create customized organs that are compatible with a patient's immune system by using scaffolding made from human tissue that is removed during surgery and merging it with the patient's unique cells [5].

Thousands of biomaterial-based medical gadgets, diagnostic tools, and disposable components are already available on the market as a result of the constantly expanding practical demands of healthcare. Indeed, biomaterials have a significant impact on the practice of modern patient care in both improving quality and saving human lives. As per the market data forecast (https://www.marketdataforecast.com/market-reports/biomaterials-market), the global biomaterials market is expected to be valued at 157.18 USD billion in 2024 and 169.47 USD billion in 2025 to 333.73 USD billion till 2034, with a compound annual growth rate of 7.82% throughout that time. The demand for and usage of biomaterials in various medical applications has expanded dramatically during the last few years all around the world. Biomaterials are crucial in tissue engineering, particularly in orthopaedic procedures such as knee replacements. The need for knee replacements will skyrocket in the next few years, according to the journal of bone and joint surgery. According to a comparable source, 572,000 individuals will require total hip replacement by 2030. The biomaterials industry will soon see tremendous expansion due to the incredible surge in demand in the tissue engineering sector [10].

It is challenging to integrate engineered biomaterials owing to the limitations of onsite targets and undesirable effects such as cytotoxicity, which interferes with transporting mechanisms and immune systems. Besides these, designing and developing a method based on biological characteristics such as mechanical tension at the targeted position, nonlinear strength, complex viscoelasticity, and anisotropic mechanical properties has always been an area of consideration [2] for successful tissue engineering. Although some reviews are available on specific areas of biomaterial applications in tissue engineering [11–14], no comprehensive reviews or chapters are available covering all aspects of biomaterial applications in tissue engineering. Thus, in this chapter, we provide a comprehensive overview of biomaterials used in tissue engineering, covering different aspects such as properties required for better grafting, classification based on different aspects, materials-tissue interactions, biomaterial characteristics required for tissue engineering application, fabrication techniques, scope, limitations, and hints at future directions in this field of study.

2 Properties of Biomaterials Required for Tissue Engineering

Three factors are considered critical elements for tissue engineering: scaffold, cells, and growth factors. The scaffold gives cells the proper environment to efficiently achieve their missions, such as creating a new tissue matrix. The role of growth factors is to enable and stimulate cells to regenerate new tissue. Biomaterials can serve as a scaffolding element designed to replace, repair, grow and maintain tissue/organ structures in tissue engineering [15]. A biomaterial is a non-viable substance used in a therapeutic device; it anticipates interacting with biological systems passively. Numerous engineering ingredients, including metals and alloys, ceramics, polymers (natural or synthetic), and their composites, have been developed over the period to replace the function of biological materials. These biomaterials serve as components of the scaffold in tissue engineering [16]. The essential components of tissue engineering are depicted in Fig. 2.

The significant role of the scaffold in tissue engineering is to mimic the natural ECM. The scaffold should support cell attachment, propagation, differentiation, and routine cell function. Furthermore, a scaffold settled at the regeneration site should protect the cells from being disturbed by external factors. To fulfil the role of a scaffold in tissue engineering, it should meet several conditions: (1) it should be biocompatible; (2) it should have a porous microstructure with appropriate porosity; and (3) it should have suitable surface chemistry to permit cell adhesion, propagation, and differentiation. Scaffolds should also possess good mechanical features and controlled biodegradability [15].

Providing short-term mechanical backing and mass transfer to encourage cell adhesion, proliferation, and differentiation while regulating the shape of the regenerated tissue is the basic role of biomaterials in tissue engineering. Moreover, biomaterials in scaffolds may provide physical strength and provide chemical messages

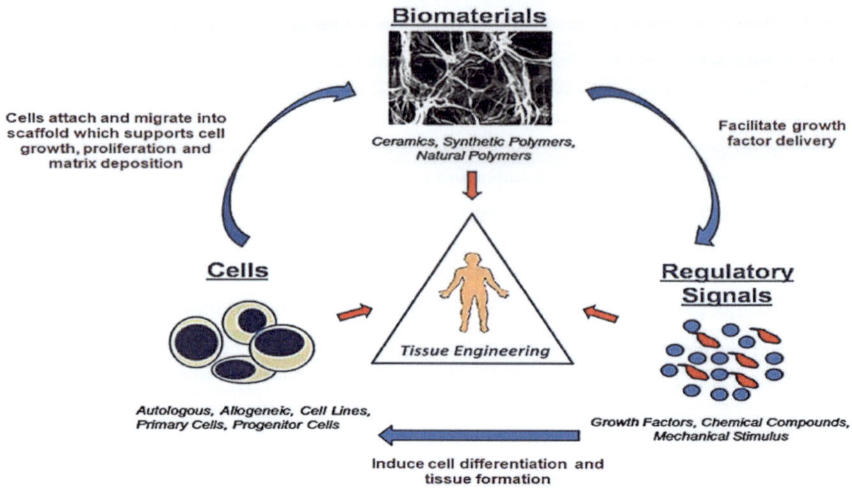

Fig. 2 The three obligatory components needed for tissue-engineering. Reprinted from Murphy et al. [17] under the Creative Commons Attribution (CC BY) license. https://creativecommons.org/licenses/by-nc-sa/4.0/

with spatiotemporal accuracy, both of which are crucial for regulating cell activity and performance and directing appropriate tissue regeneration [16].

A scaffold should ideally function as a transient construction that will gradually decompose or be reabsorbed in a controlled way in accordance with the rate at which new tissue regenerates. The freshly formed tissue of interest and naturally deposited ECM are then appropriately substituted for the biomaterial template [16].

Since the biomaterials come from external sources, they may generate an immune response. The immune response is a living network in charge of safeguarding the host from external threats and preserving homeostasis. The human immune system encompasses two arms of defensive mechanisms: the innate immune system, which stimulates a non-specific inflammatory reaction just after recognizing external elements, and the adaptive immune system, which executes precise antigen reactions and generates a long-term memory. Each part employs different cell populations in a defensive process: (1) mono-nuclear phagocyte cells (monocytes, macrophages, and dendritic cells), polymorphonuclear cells, and lymphocytes (gamma delta T-cells, natural killer cells, and innate lymphoid cells) belong to the innate immune system. On the other hand, (2) B and T lymphocytes are part of the adaptive immune system. Thus, implantation of a biomaterial-based scaffold stimulates a range of host responses toward the implant that determines the consequences of the integration and the biological functioning of the scaffold. The interaction between the biomaterial and the host immune system relies on the surrounding tissue of the implant. Therefore, the advantage and functionality of the implanted biomaterial can be reduced by developing an acute sterile foreign body inflammatory response overlaying tissue vascularization, remodelling, and ending with a fibrotic encapsulation that protects further interaction between the biomaterial and the host tissue [18]. The innate immune response toward biomaterials is described in Fig. 3.

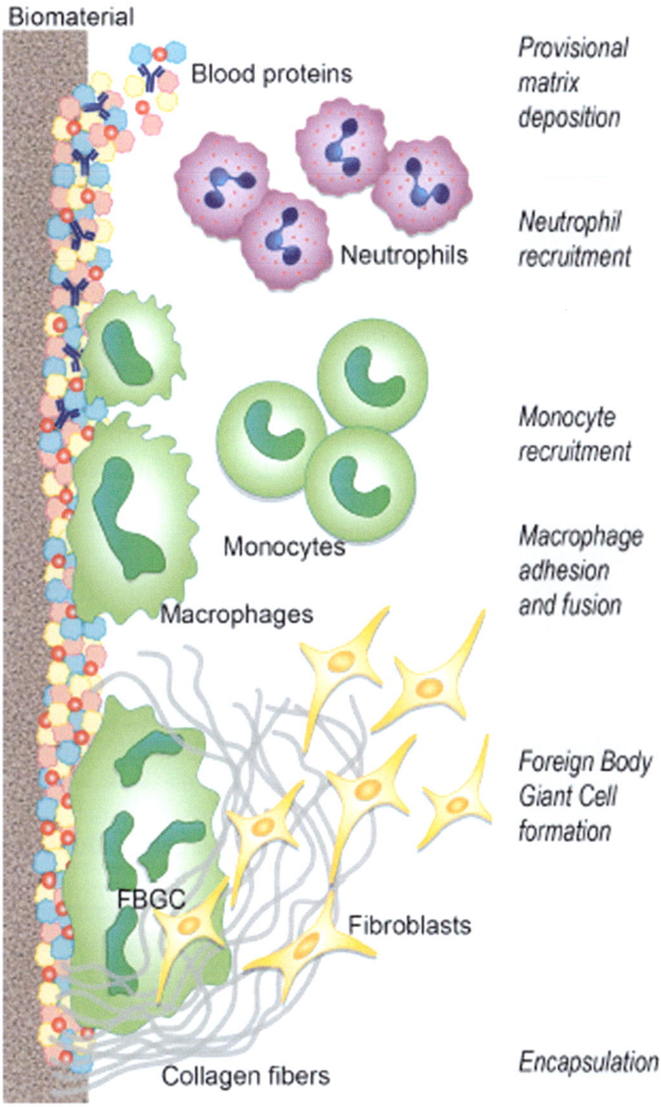

Fig. 3 Innate immune reaction toward biomaterials-based scaffolding. The steps of the foreign body reaction after implantation. The key cellular participants in the biomaterial-immune system interaction are presented. The major events, from the initial biomaterial implant to fibrous encapsulation, are schematically presented. Reprinted from Mariani et al. [18], under the Creative Commons Attribution (CC BY) license. Link: https://creativecommons.org/licenses/by/4.0/

The proteins from blood and interstitial fluids are adsorbed onto the biomaterial's surface within nanoseconds after initial contact with tissue. This coating of proteins decides the stimulation of the coagulation cascade, complement pathway, immune cells, and platelets and guides their interaction, leading to the development of a provisional matrix and the initiation of the inflammatory reaction [19].

The host response is an inescapable and inevitable outcome of any tissue or organ regeneration in tissue engineering. Reasonably, the host responds to the selected biomaterials in the scaffold to determine the engineered tissue's endurance and downstream functionality. Depending on the nature of the biomaterial chosen and the source of the cells, the immune system may induce elements of the innate and adaptive parts of the immune system. This may lead to transplant rejection, tissue damage, and foreign body reactions [20].

That is why the interactions of the biomaterials with the host immune system are crucial for their functionality. Before application of biomaterials in tissue engineering, the following matters need to be considered for the success of tissue regeneration: (a) immune cells response to biomaterials at the interface; (b) biomaterial form, shape, and physicochemical features; (c) pattern of biofilm formation on implanted ingredients; (d) tissue-specific effects of biomaterials and their consequences; and (e) immune response in a clinical context. In the arena of biomaterials, there are significant advances in using immunomodulation techniques to enhance the success rates of implantable biomaterials. To better understand such techniques, it is necessary to have full knowledge of the biomaterial-immune system interactions. This would significantly improve the development of immune-instructive biomaterials as a means of improving clinical outcomes [21].

Another crucial avenue of tissue engineering is developing biomaterials that can stimulate the regeneration process by transporting cells, growth factors, and therapeutic elements effectively and offering structural scaffolds that confer adequate mechanical strength. Moreover, the biomaterial used should be ideally degraded at an equivalent rate to that of the growth of new tissue at the implemented site [8]. As a result, the demand for high-performing implantable biomaterials that can tackle rare challenges in vascular therapy, cardiology, trauma, spine, orthopedics, dental, and wound care has been rising steadily [22].

Consequently, biodegradable polymers with distinct architectures and no immunological issues are widely used in tissue engineering. Tissue engineering makes considerable use of synthetic biodegradable polymers, such as polyesters, polyurethane, polyanhydrides, poly(glycerol sebacate), and polyphosphazenes. Newly developed polymers include amino-acid-based polymers, peptide-activated polymers, enzymatically degradable polymers, conducting polymers, and photo-responsive polymers. Synthetic polymers generally have a controlled structure, better processing flexibility, and no immunological concerns [23]. The essential factors involved in designing biomaterials for tissue engineering to persuade cells to behave like their natural counterparts are presented in Fig. 4.

It is crucial to emphasize that materials utilized as biomaterials in tissue engineering should fulfil some particular standards and should be differentiated from

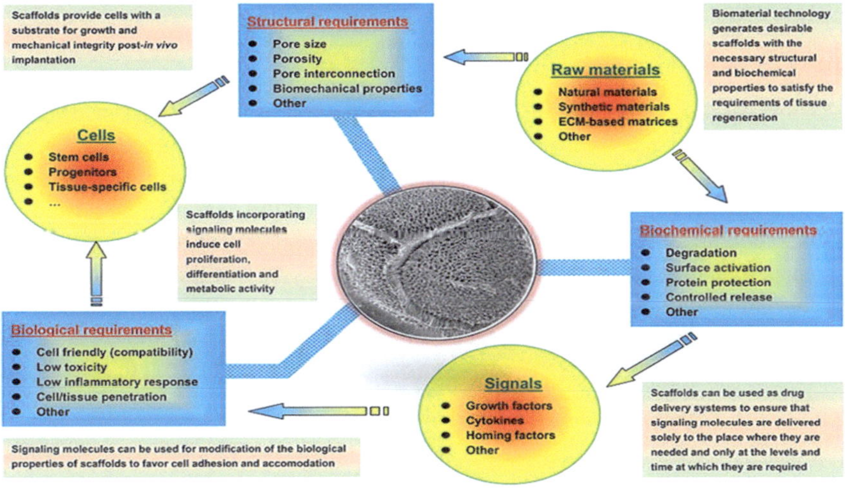

Fig. 4 Schematic interpretation of key factors (structural, biochemical, biological, and mechanical) involved in the design of biomaterials for tissue engineering that tempt cells to work in the same way as their natural in vivo counterparts. Reprinted from Chen and Liu [16] with permission Elsevier. Copyright (2016)

other classes of materials. Several biomaterials are highly desirable from an engineering perspective, but the immune system of human tissue rejects them. Therefore, care should be taken while choosing biomaterials that the patient's immune system must receive the biomaterial and must not reject it. Other characteristics that a good biomaterial should have are toughness, durability, corrosion resistance, modulus of elasticity, biocompatibility, and non-toxicity. Metals and metal alloys, glass-based materials, ceramics, and polymers are examples of artificial biological materials now employed in medical applications. They have different characteristics and therefore have different applications in human tissue engineering owing to the variation in their structures. For instance, metallic biomaterials are commonly utilized in heart valves, pacemakers, and knee and hip prostheses. Polymeric biomaterials serve various purposes, including drug delivery, orthopedic implants, and cardiovascular transplants. Meanwhile, ceramic biomaterials are frequently employed in hip replacements, dental implants, and middle ear prostheses, among other medical applications [9].

Finally, for tissue engineering to be successful, scaffolds made from biomaterials should have the following features:

- Biocompatible and must not produce any inflammatory reaction.
- The material and its degradation products must be nontoxic.
- Non-carcinogenic to the host.
- It should be serializable.

- Anticipated physical and mechanical features of the target tissue, including elasticity, load-bearing, and stress shearing capacity where they intend to be settled.
- Allow surgical manipulation, such as sewing for soft tissue implants, drilling and holding screws and hardware as needed for cartilage and bone scaffolds.
- Porosity with a suitable pore size to permit cellular migration, vascularisation, and allow passage for penetration of nutrients, cytokines, gas, and waste.
- Superconductivity for guiding and stimulating the propagation of autogenic precursor cells that transmit from neighbouring tissues into the scaffold.
- Superconductivity for induction of propagation and differentiation of autogenic precursor cells which have migrated from neighbouring tissues into the scaffold.
- Sites for cellular binding and in vivo drug delivery, such as growth factors and genes.

3 Classification of Biomaterials Used for Tissue Engineering

3.1 *Classification Based on Elements*

Typically, four individual groups of biomaterials, such as metallic, ceramics, polymers, and composites, are applied to assemble scaffolds in tissue engineering. All these biomaterial groups have some specific advantages as well as disadvantages.

3.1.1 Metallic Biomaterials

Metallic elements can be assumed to be the most valuable engineering biomaterials; they are applied as biomaterials due to their excellent thermal conductivity and mechanical characteristics [24]. These are engineered and designed to give inner support to biological tissues. They are primarily employed in joint replacements, orthopedic fixations, stents, and dental implants [22].

In general, metallic biomaterials are used for load-bearing applications and must have sufficient fatigue strength to withstand the rigors of regular activity. Metallic materials may be pure metals such as titanium and alloys, consisting of two or more components, with at least one being a metal. They have vast quantities of non-localized electrons. Many characteristics of metals are directly related to these electrons. At present, the most commonly used metallic elements for biomedical applications are cobaltchromium alloys (CoCrMo), 316L stainless steel, and titanium-based alloys (Ti–6Al–4V). Besides these, other miscellaneous elements, including tantalum, gold, dental amalgams, and specialty metals, have found application.

Metals are technologically attractive because the properties of metallic biomaterials can be modified in a more ample and versatile way during the manufacturing processes compared to ceramics and other polymeric materials [24].

Surface alteration, such as surface structuring or covering with bioactive ceramic and polymer thin films, is typically used to give bio-inert metals, such titanium (Ti) and cobalt (Co)-based alloys, multifunctionality. Bio-inert materials, often made of steel, titanium, and cobalt, are essential for many load-bearing applications. Their corrosion resistance gives outstanding long-term stability and consistent mechanical strength with minimal enduring toxicity to the host, locally or at a systemic level. These materials also develop fatigue stress, excellent tensile strength, and fracture toughness [25]. Over the years, these metal elements have found applications in artificial plates, screws, joints, braces, and dental implants in orthopedics, neurosurgical, and cardiovascular devices such as components of artificial hearts, stents, wires, and staples. Among bio-inert materials, titanium is often the material of choice due to a good combination of biocompatibility, strength, corrosion resistance, modulus of elasticity, and relatively low density and weight compared to typical steel cobaltchromium (Co–Cr) alloys [22]. Among bio-inert metals, Co–Cr alloys, surgical stainless steel, and titanium alloys are the most frequently used metals for fracture fixation, bone remodelling, and angioplasty [22, 26].

Despite the benefit of bimetallic, higher biomaterial usage may be associated with a more significant occurrence of implant-associated complications due to inadequate implant incorporation, mechanical instability, necrosis, inflammation, infections, and pain, long-lasting patient care, and sometimes loss of function. That is why care should be taken so that the metal applied is bio-compatible and does not elicit an adverse response when put into service [22].

3.1.2 Ceramic Biomaterials

Bio-ceramics are a category of ceramic materials specially applied to repair and restore the body's injured tissues [7]. Due to the helpful interactions of bio-ceramics with the human body, they can be used in various biomedical applications. Numerous dentals, medicine, tissue engineering, and surgical treatments are achieving good applications of bio-ceramics to launch more effective treatment options to enhance conventional dental and medical procedures [27]. These bio-ceramics are designed to stimulate a strong bond with bone and have emerged as an alternative to metallic implants. Bio-ceramics are currently used to repair and reconstruct both hard and soft tissue engineering [28].

According to composition, origin, structure, crystallinity, and sort of tissue response, bio-ceramics are classified into the following clusters: nearly bio-inert bio-ceramics such as alumina, zirconia, active bioglass (45% SiO_2, 24.5% CaO, 24.5% Na_2O, and 6% P_2O_5), glass-ceramics, and calcium phosphate-based absorbable ceramics [α- and β-tricalcium phosphate ($Ca_3(PO_4)_2$] [7, 28, 29]. The most employed bio-ceramics for in situ tissue engineering are those employed for bone tissue regeneration, such as calcium phosphates (CP), tri-calcium phosphate (TCP), and hydroxyapatite (HAp) [$Ca_{10}(PO_4)_6(OH)_2$] [30] with different ratio of calcium to phosphate and crystallographic structure, which in turn influences their solubility state. Such bio-ceramics show good bioactivity, osteoconductivity, biocompatibility, and osteoinductivity, which are vital for the beginning of in situ bone restoration even without the inclusion of inductive factors. In addition, it has been reported that TCP and Ha show no toxic side effects or immunogenicity [7]. The primary physical characteristics of ceramics are their high toughness, brittleness, electrical insulation, thermal insulation, and corrosion resistance. As the human body is an antagonistic system for any material, chemical inertness is the first criterion for selecting appropriate ceramic substances in biological applications [31].

Bio-ceramics are used to repair, and regenerate injured and diseased parts of the periodontal and musculoskeletal anomalies. These ceramic biomaterials have been developed for orthopedic load-carrying coatings, cement and bone implants, and dental transplants [28]. Bio-ceramics play a crucial role in the regeneration of tissue and bone and thus add value to human life's well-being. Many categories of bio-ceramics exist for tissue regeneration applications, including orthopedic applications. The selection of the bio-ceramics relies on the nature of the tissue applied to and the defects to be mended. The bio-ceramic industry progresses swiftly to optimize their synthesis with the desired characteristics required for a particular application. The mechanical properties of bio-ceramics are designed to provide a variety of applications fitness, such as supporting material, cement, and implants. Bio-ceramics with absorbable properties can transform tissue engineering, which serves as a scaffold and allows the typical tissue repair and regeneration mechanism. Thus, no surgery is needed to remove absorbable bio-ceramics, and the degraded products are also non-hazardous [32].

The fabrication of implants with nanoscale bio-ceramics has transformed their application in orthopedic treatment along with tissue engineering. The efficiency of nanomaterials in osseointegration and osseoconduction was substantially better than that of the corresponding bulk elements. Nanomaterials offer an appropriate porous network that regulates the rate of restoration of the tissue. The challenging feature in orthopedic applications is optimizing mechanical characteristics to suit load-carrying implants [32]. Bio-ceramics typically have excellent biocompatibility, osseointegrability, osteoconductivity, hard-brittle surface, and corrosion resistance. The limitations of bioceramics comprise brittleness, deficient elasticity, very high rigidity, and poor fracture toughness. Therefore, their medical applications for tissue engineering have not found applications up to the expected level [28]. Besides these, the durability of bio-ceramics is up to a maximum of 15 years within the human body. Research may enhance the life span of bio-ceramic-based implants within the living system [32].

3.1.3 Polymeric Biomaterials

The term polymer means "many parts," referring to a large molecule made up of numerous repeating subunits. These smaller repeating units are known as monomers or mers. An organic molecule composed of more than 200 monomeric units is called a polymer. The characteristics of polymers, such as their physical, mechanical, and thermal properties, rely on the bonding and structure of monomeric units, the nature of side chains, the length of the chain, and the grade of cross-linking. Polymers are generally cost-effective materials because they are primarily derived from oil [9].

Biopolymer molecules consist of monomeric units covalently bonded to form larger molecules and have found application scope in biology such as tissue engineering. The key feature differentiating biopolymers from fossil fuel-based polymers is their sustainability, particularly when combined with biological systems and requiring biodegradability. A broad range of materials that commonly originate from bioresources such as animals, plants, and microorganisms can be designated as biopolymers. Materials manufactured by synthesizing biological resources such as sugars, amino acids, proteins, vegetable oils, fats, and resins are known as biopolymers [33].

Polymeric biomaterials can be used and survive within physiological environments for long periods without generating any adverse effects (local or systemic). The biopolymers can be either natural or synthetic in origin. Their physiological and mechanical features are similar to tissues that permit their usage to replace numerous tissues. Biopolymers can contain diverse polymeric groups that offer a broad range of mechanical [34] and physical properties.

Polymeric biomaterials such as natural polymers (collagen, gelatine, fibronectin, chitin, chitosan, etc.) and synthetic polymers (polyesters, polypropylene, polytetrafluoroethylene, silicone, polydimethylsiloxane) having structures like fibres, rods, and films have a broad range of application scopes in implantation, owing to their exclusive properties and ease of fabrication. These are also nonallergic, nontoxic, and noncancerous. In addition, biopolymers have good tensile strength, elasticity, and durability. Due to their well-suited properties, biomaterials have been extensively employed as dental elements, prosthetic materials, drug delivery devices, and tissue-engineered stuffs [9, 35].

Synthetic biopolymers have some benefits over their natural counterparts, such as the reproducibility of properties owing to their capacity to be synthesized on a massive scale. Furthermore, their features, including morphological properties and mechanical stability such as pore size and shape, and crystallization pattern, can be adapted during processing. Furthermore, the same biopolymer can offer different characteristics by altering the distribution of different sized polymeric chains. For instance, commercial low-density polyethylene (LDPE) is generally used to make packaging bags that have a little mechanical strength and a low melting temperature. On the other hand, ultrahigh molecular weight polyethylene (UHMWPE) is a very threatening material that is almost fusible and is used in hip joint prostheses and mining scrapers. The major variation between these two forms of polyethylene is the length of the polymeric strings, which are about ten folds longer in UHMWPE than in LDPE. Furthermore, synthetic biopolymers are not usually rejected by the immune system as they have no sites for cell adhesion, recognition and many synthetic biopolymers are hydrophobic [34].

Natural polymers can be rejected owing to antigenicity. Their properties can vary from batch to batch and cannot be modified easily during processing. There is a risk of disease transmission, even with sufficient processing. However, the main advantages of natural biopolymers are that they allow cell adhesion, migration, recognition, propagation, and differentiation, which permits them to restore the structure and function of tissues and organs. The main major natural biopolymers include collagen, silk, elastin, alginate, polyhydroxylalkanoates, chitosan, and cellulose. They have great potential for biomedical applications to substitute synthetic counterparts [34].

3.1.4 Hybrid/Composite Biomaterials

Biomaterials perform a vital function in the success of tissue engineering processes, as they guide the structure and shape of evolving tissues, furnish mechanical stability, and give opportunities to deliver inductive factors to transplanted or migrating cells. Consequently, the choice of a suitable biomaterial may significantly affect the quality of newly developed tissue. The scaffolds fabricated from natural ECM derivatives consist of favourable structures for adhesion, multiplication, and differentiation of cells. However, in general, they do not have the appropriate mechanical features and load-carrying capability. On the other hand, scaffolds developed from synthetic biomaterials show proper mechanical stability and suitable biological characteristics. With the combined mechanical strength with biological features, the ECM-derived elements can be coupled with synthetic biomaterials. Therefore, hybrid scaffolds consisting of both the benefits of ECM derivatives and synthetic biopolymers can be considered as a dynamic device for tissue engineering applications [36]. Furthermore, the coupling of synthetic polymers with natural ones and other biomaterials has shown the ability to augment cellular contact, stimulate integration into host tissue, and give tuneable material characteristics and degradation kinetics.

Biocomposites are a combination of numerous ingredients where every element acts to complement the performance of the other elements. Each of these biomaterial elements can take various forms such as particles and fibres enveloped by the other element, which can be described as the matrix element [37]. Composite biomaterial mixtures show the necessary biocompatibility for numerous biomedical applications such as tissue engineering. Moreover, they facilitate combining the advantages of materials from different categories. It is possible to combine the advantages of the embedded and surrounding matrix from any group, such as metal, ceramic, or polymer, at the same time. Except for a few rare instances, the materials incorporated in the matrix are typically robust and have low densities, making them ductile or difficult to work with. The various biomaterials in composites, on the other hand, cooperate in a complementary manner to allow load transfer in the most efficient manner [16, 37–39].

The most widely used 3D porous scaffolds are constructed from two main categories of biomaterials. One category includes synthetic biodegradable polymers such as poly(lactic acid) (PLA), poly(glycolic acid) (PGA), aliphatic polyesters, and their copolymers, such as poly(DL-lactic-co-glycolic acid) (PLGA). The other category includes natural-origin-derived polymers such as collagen. PLA, PGA, and PLGA are biocompatible. The Food and Drug Administration of the USA has endorsed a few synthetic polymers for use in specific biomedical applications. PLA, PGA, and PLGA are commonly applied in tissue engineering scaffolds [36]. They have the advantage of being easily transformed into scaffolds having the same form as the tissue to be replaced. If designed with enough mechanical strength, they can maintain this structure until the new tissue develops. Their rate of biodegradation can also be customized to match the rate of regeneration of the new tissue. The drawbacks of these synthetic biopolymers are the deficiency of cell-recognition signals, which may result in inadequate cell adhesion and hydrophobicity [40].

Scaffolds based on cross-linked sponges of gelatine and chitosan are broadly utilized for tissue engineering such as wound dressings. The gelatine-chitosan scaffolds are successful for three reasons: (a) Gelatine contains arginine-glycine-aspartic acid-like sequences that encourage cell attachment and migration; (b) Chitosan is a bacteriostatic biopolymer that prevents infection, and (c) Gelatine can be combined with chitosan to form a polyelectrolyte complex that improves their physical characteristics. Furthermore, the gelatine-chitosan scaffolds allow the formation of composites that improve physical and biological characteristics. For example, the integration of hyaluronic acid (Ha) into the scaffold improves the development of skin cells seeded on gelatine-chitosan scaffolds. However, the efficacy of the scaffolds relies on the sort of biopolymers applied and the process for modifying them [41].

3.2 *Classification Based on Biomaterial-Tissue Interactions*

The biomaterial-to-tissue interface is directly correlated to the functional interface between the biomaterial and the host tissue [38]. When a biomaterial is placed within the human body, the host tissue responds to the implant in various ways based on the biomaterial category. The interaction mechanism between tissue and implant relies on the host tissue's response toward the implant surface. The interaction and response of biomaterials with host tissue can be classified into three general categories: bioinert, bioresorbable, and bioactive.

3.2.1 Bio-inert Biomaterials

Biologically inert or bio-inert materials are those elements that do not cause any adverse reactions when applied to a biological tissue for regeneration purposes. Therefore, introducing bio-inert materials into the body will not negatively react with the host or have little interaction with adjacent tissue, which can be mitigated.

Some examples are alumina, titanium, stainless steel, ultra-high molecular weight polyethylene, and partially stabilized zirconia. In general, a fibrous coating may form surrounding bio-inert implants; hence its bio functionality depends on tissue integration through the implant [42, 43]. These biomaterials have stable physiochemical characteristics and acceptable biocompatibility with the host tissues. When the inert biomaterials are implanted within the target tissue of the body, there will be no reaction and immunologic rejection. They are capable of retaining their biomechanical and physiochemical properties within the host. They are also corrosion, wear, and tear-resistant due to their inertness and suitable strength to resist fracture. Bio-inert materials are used as physical support in implants such as bone screws and bone plates [1].

The materials initiating a response may adversely affect the host. Initially, these inert materials were applied for vascular surgery, which does not cause blood clotting. For this reason, inert biomaterials may sometimes be called hemocompatible. In reality, there is no material that is completely bio-inert since all materials generate specific reactions after implantation within the body. Zirconia and alumina are conventional bio-inert materials used for dental and orthopedics applications [1, 16]. All biomaterials face-specific responses from the host tissue in direct connection [38].

3.2.2 Bioactive Biomaterials

The term bioactive refers to a material that interacts with the surrounding bone when implanted into the human body and, in some cases, with soft tissue. This interaction occurs due to surface modifications initiated by the implant within the living tissue. An ion-exchange reaction between the bioactive implant and adjacent bodily fluids leads to the formation of a bioactive carbonate apatite coating on the implant. This coating closely resembles the mineral phase of natural tissue in both biochemical composition and crystallographic structure. Notable examples of bioactive materials include synthetic hydroxyapatite [$Ca_{10}(PO_4)_6(OH)_2$], bioglass, and glass-ceramic [42, 43].

Bioactive materials have a constructive effect on existing tissues and can induce a response that aids in the repair, renewal, and reconstruction of the tissues. When an active biomaterial is inserted within the body, it generates a precise biological response that causes a chemical linkage between the biomaterial interface and the tissue. Bioactive materials are mainly applied to repairing craniofacial, orthopedic, dental, and chronic osteomyelitis in tissue engineering. Hydroxyapatite and calcium phosphate are typical examples of bioactive materials. Bioglass dental grafts have been a substantial success in the last decade and are a promising biomaterial. Bioactive materials form a hydroxyl apatite coating when submerged in the replicated body fluid for a few days. The creation of this artificial apatite coating is like the inorganic phase of normal human bones. Because of its similarity, the coating layer can bond with the bone via collagen fibrils. Bioactivity depends on the surface reaction of the biomaterials, which in turn relies on the material composition, and even minor changes in structure can destroy the bioactive behaviour [1].

Bioactive materials have the ability to trigger a specific tissue response, promoting better integration between the biomaterial and the surrounding biological system. In bone tissue engineering, bioactive materials are generally classified into two main categories: (a) osteoconductive and (b) osteoproductive [44]. Osteoconductive biomaterials primarily bond with hard tissues, facilitating tissue interaction on their surface. Examples of such materials include synthetic hydroxyapatite and tricalcium phosphate ceramics, both of which support bone regeneration and integration [45]. Conversely, osteoproductive biomaterials have the ability to stimulate new tissue growth beyond the initial boundary of the implant. An example of such a material is bioactive glass, which can form bonds with soft tissues like cartilage. The interaction between tissue and bioactive biomaterials is enabled by the formation of hydroxyapatite on the biomaterial's surface upon contact with bodily fluids, facilitating strong integration with the surrounding tissue [46, 47]. Hydroxyapatite film development is faster on osteoproductive biomaterials compared to those of osteoconductive biomaterials [38].

3.2.3 Bioresorbable Biomaterials

Bioresorbable refers to an element that begins to dissolve upon placement within the physique and is gradually substituted by the developing tissue of interest. Examples of resorbable biomaterials are polylactic-polyglycolic acid copolymers and tricalcium phosphate [$Ca_3(PO_4)_2$]. In addition, calcium carbonate, gypsum, and calcium oxide have been used as the common biomaterials in tissue engineering during the last four decades [42, 43]. These biomaterials can respond and break down promptly when they come into contact with the tissue fluid of the body. Bioresorbable materials entirely disintegrate into the tissue fluid because the tissue fluid can break down the chemical assembly of the biomaterials. Thus, these biomaterials do not need a second operation for the elimination of grafts from the body. They are entirely absorbed by the host tissues and become a component of the tissue. The standard metabolic pathways treat the chemicals generated by the resorption of implant biomaterials in the body without generating any toxic products or side effects. Several metallic, ceramic, and polymeric bioresorbable implants were synthesized and examined clinically. The controlled resorption of these biomaterials can be further beneficial for patients [1].

Bioresorbable materials can show a time-dependent and meticulous degradation rate, which consequently permits them to be substituted by host tissue [48, 49]. Since tissues will process a substantial amount of biomaterial, a resorbable material must be metabolically compatible. Consequently, an essential requirement for bioresorbable materials is that their degradation rate must align with the healing rate of the host tissue. Additionally, when using bioresorbable materials in clinical applications, it is crucial to ensure that their mechanical properties are suitable for effective tissue restoration. Ultimately, for a bioresorbable material to be viable, its disintegration rate must be carefully regulated [38].

3.3 Classification Based on Source Origin

Classically, three individual groups of biomaterials, such as metals, ceramics, natural and synthetic polymers are applied to fabricate implants for tissue engineering. All these categories of biomaterial have some advantages and disadvantages. As a result, the use of composite scaffolds made of various biomaterials is becoming more common [50].

3.3.1 Natural Biomaterials

Synthetic materials have adverse effects on the environment. This consciousness of the harm triggered by synthetic biomaterials on the ecosystem has led to the expansion of eco-friendly biomaterials. Scientists have demonstrated much interest in developing such biomaterials which can substitute synthetic biomaterials. Consequently, there is an upsurge in demand for commercial applications of natural fibre-based composite biomaterials for numerous manufacturing sectors. Natural fibres are sustainable biomaterials available from natural sources and have advantages such as being affordable, inexpensive, renewable, lightweight, biodegradable, and highly specific. The higher sustainability of the natural fibre-based biocomposite has increased its application scope in various industrial sectors [51].

The major natural polymers currently in application for tissue engineering are proteins and polysaccharides. They play an essential role in tissue engineering and regenerative medicine owing to their diverse chemical and physical characteristics. Amino acids and proteins are the critical elements in normal tissues, and they are well known for their regulated degradation capacity. These protein-based biomaterials are particularly valuable in suturing applications for scaffold biomaterials, as drug delivery devices, and many more. Collagen is one of the ECM proteins of the body, abundant in musculoskeletal tissues, and has been widely employed as a scaffold biomaterial for tissue engineering. Gelatine is another well-known biomaterial primarily used in numerous applications, starting from foodstuffs to tissue engineering. Elastin is a flexible insoluble protein in connective tissue, having cross-linked tropoelastin as a key element. Elastin is responsible for the contraction of vascular tissues, lungs, and skin within the body. However, natural elastin's high immune response and insolubility reduced their application in biomedical and tissue engineering.

On the other hand, the capacity of elastin to sustain the lowest platelet interaction makes it a suitable element for coatings for synthetic vascular implants. Natural silk includes 2 groups of proteins, such as serine and fibrin, produced and secreted by specific insect larvae to develop cocoons. Owing to their precise molecular structure, silk fibres have considerable biocompatibility, tensile strength, degradability, and durability compared to other biomaterials. Albumin is the abundant protein found in the plasma of human blood. This water-soluble protein functions to control the blood pH and transport fatty acids in the bloodstream. Fibrin is a natural polymeric protein like collagen, having 3 polypeptide chains joined by a fibrinopeptide bond and functions in the blood coagulation process in the human body. Fibrin has good degradability and biocompatibility and has been investigated as a probable biomaterial for some tissue engineering applications. Keratin is a big family of structural proteins that produce the components of hair, nails, and the external layer of human skin [52].

Collagen, silk, fibrin, and gelatine form the class of natural-based polymeric proteins. Collagen is the most abundant protein in humans. The FDA has permitted collagen for biomedical application owing to its easiness of preparation and nominal immune and inflammatory responses. Therefore, collagen is often used in the biomedical sector, such as artificial skin and wound dressings. Research proved that the conjugation of collagenic scaffolds with stem cell-specific antibodies could hold stem cells at a lesion location and stimulate cardiomyocyte revival in a mouse prototype [7]. Additionally, natural polymeric proteins are utilized as biomaterials in tissue engineering due to their superior biocompatibility, bioactivity, and biodegradability [53].

Natural polysaccharides are a type of polymeric biomaterials having outstanding biocompatible and biodegradable characteristics. Polysaccharides and proteins have properties that mimic the ECM, making them ideal candidates for tissue engineering applications [7]. Polysaccharides are long-chain carbohydrate molecules that consist of monosaccharide units joined by a glycoside bond. Chitosan is a polysaccharide obtained from chitin, which is present in the hard exoskeletons of crustaceans like crabs and shrimp. It is a linear polysaccharide composed of β (1–4)-glycosidic bonds linking N-acetyl-d-glucosamine and d-glucosamine units. Chitin is one of the most plentiful and cheapest polysaccharides available in nature. Chitosan is obtained by the deacetylation of chitin. Chitin is a cationic amino polysaccharide that is biodegradable in nature and can be degraded into diverse products such as chondroitin sulphate and glycosylated collagen. It is attractive for some biomedical applications owing to the availability of reactive functional groups, ease of processability, nominal foreign body rejections, and controllable mechanical characteristics. The main drawbacks of chitosan are that it is insoluble in water and has low mechanical strength, making it brittle. Alginate is a linear polysaccharide consisting of a-L-guluronic acid and b-D-mannuronic acid found in marine brown algae or seaweed. Hyaluronan is a naturally available polysaccharide consisting of repeating units of D-glucuronic acid and N-acetyl-D-glucosamine. Cellulose is the amplest polysaccharide on the planet, with thousands of D-glucose units held together in a polymeric chain. Chondroitin sulphate is a sulphated glycosaminoglycan that contains alternating chains of glucuronic acid and N-acetylgalactosamine [52].

Popular polysaccharide-based natural biopolymers with extensive applications for in situ tissue revival comprise alginate, chitosan, hyaluronic acid and fibrin. Because chitosan is cationic, it is a perfect medium for delivering growth factors, cytokines, genes, and anionic glycosaminoglycans. Hyaluronic acid is originated in the ECM of various tissues. It is commonly used in in vitro studies of cartilage restoration owing to its positive immune and biological responses, such as wound repair, morphogenesis, and proliferation. In addition, hyaluronic acid can be altered to form procedure hydrogels, promoting cartilage regeneration compared to hydrogels based on polyethylene glycol and utilized to deliver growth factors and bone structural protein [7].

Wu and colleagues designed a scaffold made of electrospun silk fibroin. They employed polydopamine to improve the surfaces of the scaffolds, followed by modification using peptide E7 with a precise likeness for bone marrow mesenchymal stem cells in a rodent calvarial bone flaw model [54]. Their findings revealed that the improved electrospun scaffolds exhibited enhanced hydrophilicity, cell adhesion, cell proliferation, and osteogenic differentiation of bone marrow mesenchymal stem cells. This was achieved by creating an optimal osteoinductive environment through the synergistic effects of polydopamine and peptide E7, both in vitro and in vivo [7, 54].

The use of natural biopolymers has limitations, such as relatively weak mechanical properties, which limit their use at structural regions that require high mechanical loading, like hard tissue regeneration [55]. In order to enhance these characteristics, natural biopolymers are commonly mixed with synthetic polymers to create hybrid biomaterials that benefit from both types of polymers without offsetting their respective impacts [7].

3.3.2 Synthetic Biomaterials

Synthetic biopolymers are higher molecular weight molecules that consist of a string of monomeric units. Based on their structural pattern, they can be divided into linear, branched, or cross-linked. Synthetic polymeric biomaterials can be made in the form of fibres, bars, films, or viscous liquids, and offer the significant benefit of modifying their mechanical characteristics and biodegradability by changing reactants and synthesis methods [56]. Synthetic biodegradable polymers are extensively employed in biology and medicine. They are vital elements of tissue engineering scaffolds, biomedical devices, and drug delivery. Therefore, there is a considerable interest in developing novel synthetic biodegradable polymers for biomedical applications. The main characteristic that regulates the function and efficiency of synthetic biomedical polymers is their degradation rate [57].

Common synthetic biomaterials include ceramics, polyesters, polyorthoesters, polyketals, and polyanhydrides. Among these, bioactive glasses and calcium phosphates are widely utilized in tissue engineering due to their strong osteoinductive and osteoconductive properties [58]. However, synthetic biopolymers may show low mechanical strength, low biocompatibility, and in vivo toxicity owing to the discharge of ions and other residual polymerization particles. Bio-erodible is the most appropriate synthetic biopolymer for scaffold production among the different synthetic biopolymers. They undergo surface degradation with the formation of non-detrimental low molecular weight elements [56].

Various synthetic polymers such as polystyrene, polyglycolic acid (PGA), poly-l-lactic acid (PLLA), and poly-dl-lactic-co-glycolic acid (PLGA) have been applied in an effort to make scaffolds [50, 56]. These biomaterials have demonstrated better results as they can be fabricated with customized architecture, and their degradation rate can be precisely regulated by changing the composition of individual polymeric units. However, they have a lower risk of rejection owing to less bioactivity. Moreover, PGA and PLLA are degraded by hydrolysis and produce carbon dioxide. Thus, reduction of local pH can lead to cell and tissue necrosis [50].

Polyesters (PGA) are primarily applied in sutures and biomedical implants owing to their high melting point, controllable solubility, crystalline nature, and tensile modulus. The weakness of PGA can be ameliorated with the addition of β-tricalcium phosphate to control polymer degradation and assist regeneration of complex tissues. PLA can be adapted to other biomaterials by utilizing copolymerizers, modifiers, physical treatments, and blending. In combination with bioactive glasses, PLA can enhance biological properties such as osteoblast cell growth and differentiation. PLGA is a very good source of biomaterials that can be adapted into diverse shapes, such as fibres, hydrogels, injectable microspheres, or scaffolds. PLGA, in combination with inorganic ingredients, is applied to promote bioactivity as well as osteoconductivity. Poly (ε-caprolactone) (PCL) is a semicrystalline hydrophobic biopolymer with a low melting point that has drawn scientists' attention owing to its capacity to promote osteoblast activity. Poly (ethylene glycol) (PEG), also known as polyoxyethylene (POE) or polyethylene oxide (PEO), is a water-soluble, non-toxic synthetic polymer. PEG polymer is recognized as a good agent for biomedical applications due to its low immunogenicity, good biocompatibility, and biodegradability [2].

Synthetic biopolymers can be tailored to include functional groups by combining them with biomolecules. As a result, they have reduced biosafety concerns, particularly concerning host tissue reactions and disease transmission compared to natural polymers. As a group of biodegradable synthetic polymers, aliphatic polyesters such as PGA, PLA, PCL, and PLGA are used in tissue engineering. When the ester groups in the backbone chains of these synthetic biopolymers hydrolyze, nonhazardous compounds are produced [7].

The characteristics of this category of biopolymers can be customized by choosing a suitable molecular weight, degradation rate, and the nature of the porous architecture for tissue regeneration by adjusting the composition of the monomer of the copolymers. These combinations of copolymers will determine their mechanical properties. Two central polymer chemistry tools are used to obtain tailored degradation rates of biopolymers: copolymerization and blending. The US FDA permits these synthetic polymers for medical applications, such as operating stitches and several implantable tools. PLGA is one of the most general artificial biopolymers and is highly applied in cartilage and bone tissue regeneration. However, traditional synthetic biopolymers have drawbacks owing to the absence of functional groups lead to in a lower ability to join with bioactive parts to reinforce their cell attraction. Artificial biopolymers with functional groups, have attained acceptance owing to their facile design and adjustment, which facilitates them in delivering bioactive signals to improve cell-biomaterial interaction [7].

4 Characteristics of Biomaterials for Tissue Engineering

4.1 Biological Characteristics

Tissue engineering integrates principles from cell biology, engineering, and materials science to create biological substitutes that replicate the structural and functional characteristics of native tissues, aiming to restore and sustain the normal function of damaged or diseased tissue [8]. Hence, scaffold biocompatibility and biomimetics are essential for performing vital cellular actions, including molecular and mechanical signalling networks [59]. Biocompatibility is defined as the feature of being biologically well-suited by not making a noxious, injurious, or immunological reaction in living tissue. The human body has a remarkable ability to determine whether an entity is foreign or self. This is part of the body's defensive mechanism against foreign invasion. If a substance is placed within the body and the immune system can detect it as foreign, then an immune response will be generated. Figure 5 depicts the details of the body's immune response to a foreign body. When an object is incorporated into the body without any immune reactions, it is biocompatible. In order for a scaffold to be biocompatible, it must abide by a very stringent set of criteria defined by the body's immune system [60].

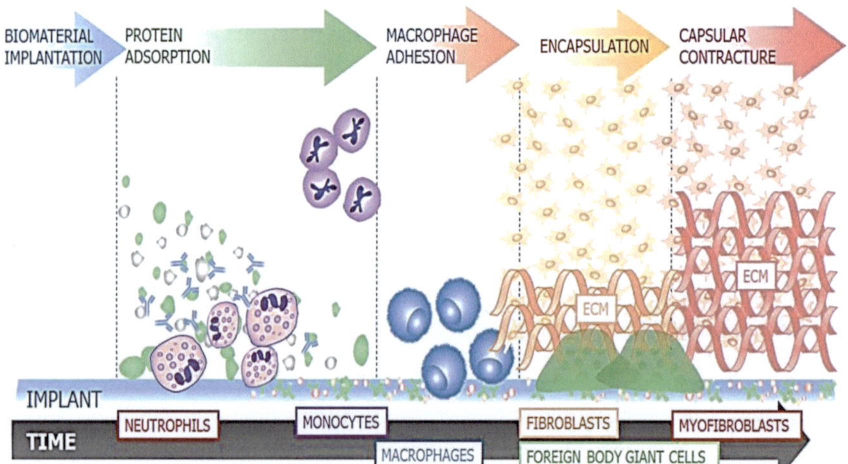

Fig. 5 The stages of the foreign body reaction upon implementation of the implant. On grafting, the biomaterial is coated in a film of protein from the adjacent wound fluid, and neutrophils move the wound site. Monocytes differentiate into macrophages, which transform into foreign body giant cells and induce the enrolment of fibroblasts, which commence to wall the implant off from the adjacent tissue by depositing collagen and the fibrous capsule. It is this capsule that contracts and hardens in some women, causing the condition known as capsular contracture. Reprinted from Barr et al. [61] under the creative common CC BY 4.0 license. https://creativecommons.org/licenses/by/4.0/

Another requirement for the suitability of biomaterials used in tissue engineering is a bioactive feature. The integration of bioactive properties such as chemical stimuli, physiochemical material properties, and mechanical forces into the selected biomaterial plays a significant role in influencing cell behaviour by making a situation that motivates a cellular response such as the host tissue [62]. Similarly, the biodegradability of a biomaterial without any cytotoxic by-products, nominal fibrosis and necrosis, and foreign body response is also crucial for effective tissue restoration. Herein, the biomaterial can be tailored to give dynamic control over bioactivity, mechanical integrity, and the scaffold's porosity [63]. It has been reported that the biodegradation rate can affect the mechanical performance of the scaffold. A quick degradation rate could lead to premature mechanical failure; for example, in bone tissue engineering, the scaffold typically needs to bear mechanical load through the regeneration process. In contrast, inflammation-related complications may occur if the degradation process is too slow [7, 64].

The biological demands of scaffolds are their nontoxicity and biocompatibility. Cells developed within scaffolds must be able to replicate and differentiate freely without interfering with new matrix generation [65]. Thus, a scaffold is considered ideal for tissue engineering applications when it mimics the features of natural ECM for perfect and complete tissue regeneration. However, the role of the supportive cell is influenced by factors such as the selected cell line, the base biomaterial, surface properties, and scaffold architecture. Additionally, biocompatibility enables the simultaneous formation of new tissue while the supporting matrix gradually degrades.

Additionally, the matrix must be non-toxic to eliminate the need for its disposal by the biological system. The biological properties of a scaffold play a crucial role in addressing this challenge, as they influence communication within scaffolds and tissues or organs. Given the limited capability of biomaterials to interact with the physiological environment, efforts have been made to incorporate bioactive scaffolds that promote proper cellular communication, tissue formation, migration, and differentiation. Furthermore, the scaffold must be designed to avoid triggering host immune responses [66, 67].

4.2 Structural Features

Biological tissue is a highly complex three-dimensional (3D) structure with intricate mechanical activities associated with weight transport characteristics. Consequently, the crucial objective of tissue engineering is to bridge this structural complexity with functional properties using biomaterial-based scaffolds that offer proteins, cells, and genes for tissue renovation and expansion. Unfortunately, biomaterials with unfavourable structures cannot replicate in multipart tissue environments, including various types of cells that interact with a variety of cytokines to yield extracellular matrices [66].

Scaffold macro and micro architecture critically influence surface adhesion and cell survival. Scaffold architecture also influences cell multiplication, differentiation, vascularisation, and precise gene expression. A scaffold should support the physical weight of the body and permit surgical treatment during embedding. It is vital to get a porous structure to circumvent cellular colonization. Therefore, equilibrium within mechanical strength and suitable porosity is a substantial challenge in scaffold construction. The porosity, pore size, and pore interconnectivity are very significant criteria for scaffold generation. Sufficient balance among these features allows cell permeability, adequate diffusion of oxygen and nutrients inside cells, the construct, vascularisation, and neo-formed ECM, confirming cell viability [56].

Pore size is a fundamental component of scaffold efficacy. The pores must have enough space to permit cell penetration and movement within the scaffold assembly but simultaneously be tiny enough to permit the binding of an adequate quantity of cells. Pores can be categorized into macropores (>50 nm), mesopores (2–50 nm), and micropores (0.1–2 nm), conferring to the dimension. Maximum scaffolds employed in tissue engineering may have a macroporous network with a precise pore size depending on the activity of the host tissue. A 20-micron pore size is recommended for fibroblast and hepatocyte growth, while the recommendation for soft tissue healing is 20–150 micron. On the other hand, researchers suggest a pore size from 200 to 400 microns for bone tissue engineering. The most commonly used techniques to get a porous network are phase separation, gas foaming, salt leaching, freeze-drying, and sintering [56].

The development of vascularized engineering scaffolds remains a significant challenge due to vascular insufficiency, which hampers the successful integration of osseous tissue, emphasizing that biomaterial selection directly influences the scaffold's final structural properties [67, 68]. To support cell growth, scaffold porosity must be optimized to enhance its structural integrity. Additionally, scaffolds with appropriately sized pores facilitate water absorption, promote cell migration [69], and improve the mass transfer of gases and nutrients throughout the structure [66].

4.3 Chemical Composition

For tissue engineering, the scaffolds must have the proper surface chemistry and structure to promote cell attachment, seeding, and proliferation [70]. It has been reported that surface chemical properties play a vital function in the amount of cell adhesion, add-on, and cellular distribution on the surface of biomaterial scaffolds [71]. Moreover, the sort of biomaterial employed to assemble the scaffold, its surface treatment, or the addition of nanoparticles may influence the surface area and enhance the scaffold bioactivity [72].

Bioactivity is a significant parameter for tissue engineering, and this indicates the capability of a biomaterial to interact with adjacent tissue to ensure cell adhesion, propagation, and differentiation [73]. Usually, biomaterials having a chemical composition similar to that of the host tissue have better bioactivity and can influence cellular recognition, inducing a particular cellular response to boost tissue development. It is possible to modify the biomaterial's surface by combining ECM supermolecules, such as fibronectin, collagen, and laminin, to make a biomimetic environment comparable to the native tissue and capable of regulating cellular response and behaviour [56].

The majority of scaffolds are traditionally composed of bioceramics, polymers, and a combination of materials, both natural and man-made [69]. Depending on the source of the biomaterials used to produce the scaffold, issues with composition, biocompatibility, and the disintegration of matrices' by-products arise, and some biomaterials are not appropriate for supporting cell growth and development within the scaffolds; for instance, natural biopolymers have uncontrolled degradation rates and little mechanical constancy, whereas synthetic polymers exhibit actual degradation rates. Synthetic polymers can be engineered into intricate shapes and exhibit enhanced cell adhesion by integrating charged chemical groups and facilitating the transport of soluble molecules [66].

Certain in vitro studies have indicated that synthetic biomaterials may hinder ex vivo tissue development when compared to natural tissue matrices. Under in vivo conditions, tissue formation can be significantly disrupted by unpredictable degradation rates, biomaterial immunogenicity, and adverse effects caused by degradation byproducts. The matrices' chemical composition, similar to natural ECM, is the most favourable in tissue engineering based on these reflections. Consequently, the recently developed strategies in tissue engineering aim to avoid non-degradable scaffolds that are resorbed at a different rate from that of skeletal tissue regeneration [66]. The combination of calcium phosphate layers with hydroxyapatite particles, leads to better bone development in porous biomaterials. By altering the surface roughness and chemistry of the scaffold, surface treatment of bone scaffolds promotes enhanced bone mineralization. Vascular tissue engineering is another fascinating example. The immobilized vascular implant must have a surface treatment applied since it is thought to be essential for preventing thrombus formation [7].

4.4 Scaffold Architecture

Concerning the use of biomaterials in tissue engineering, the present goal of science is to search for the appropriate synthetic or natural material or a mixture of both with the maximum suitable performance in vivo, such as being able to boost cell propagation and differentiation in a tissue wound in order to refurbish the natural architecture of ECM [56]. The scaffolds must be well-architected to facilitate the flow of nutrients and capillary blood vessels, hence regulating cell enrolment, proliferation, and penetration to the lesion site. Additionally, the scaffold's architecture may be able to capture bioactive elements including genes, cytokines, and growth chemicals [74]. By promoting cell-specific enrolment and proliferation, these are thought to be essential for controlling the in vivo milieu and enhancing the in-situ tissue regeneration process [75, 76]. Furthermore, the integration of biomolecules into biomaterials has improved cell recruitment. For example, the PLGA scaffolds integrated with stromal cell-originated factor 1 (SDF-1α) were observed to recruit more stem cells, enhance fresh blood vessel creation, reduce fibrotic responses and control the inflammatory cell reactions to pure PLGA scaffolds [77].

Cell seeding, response, and behaviour are also substantially affected by the architectural features of the biomaterial scaffold. It has been reported that the scaffolds applied in bone tissue reconstruction must retain optimum porosity, pore size, pore shape, proper fibre orientation, and inter-connectivity with one another [72]. A review by Bobbert and Zadpoor [72] has reported that scaffolds with minor pores within 200–300 μm represent a suitable environment for cell seeding, though they reduce cell sustainability, propagation, and differentiation. Furthermore, it has been reported that scaffolds with rectangular pores, ladder-like shapes, and large spherical pores are more prone to tissue regeneration failure than scaffolds with small, uniform, and round pore architecture [78]. Bobbert and Zadpoor [72] concluded that designing a 'tortuous void network' would help regulate the channel of the cell suspension across the scaffold, thus expanding the surface area and available time for cell adhesion will enhance the tissue regeneration process.

4.5 Mechanical Properties

The scaffold for tissue engineering must have sufficient mechanical load-bearing capacity and integrity to provide support for the implantation process. The mechanical force relies on the bonding strength of the atoms or molecules that hold them together in scaffold architecture. It is a significant factor to circumvent the solid structure distortion owing to cellular load on the scaffold or caused by scaffold management [56]. The scaffold ought to have mechanical characteristics similar to the anatomical location where it is anticipated to be embedded. From a realistic standpoint, it should be adequately strong to permit surgical procedures during implantation. Although this is crucial for all tissues, it presents some extra challenges for orthopedic and cardiovascular applications. Fabricating scaffolds with appropriate mechanical strength are one of the biggest challenges in cartilage or bone tissue engineering. The embedded scaffold must have adequate mechanical stability to function from the period of implantation to the end of the modification process, especially for these types of tissues [50].

For the scaffold to actively promote the regeneration process, its mechanical properties must be comparable to those of the natural tissue. The modulus of rigidity of a biomaterial plays a crucial role in enhancing cell adhesion, proliferation, migration, and differentiation. Compared to the rigid poly(lactic acid) (PLA) and poly(lactic-co-glycolic acid) (PLGA), studies have shown that the highly elastic poly-L-lactic acid (PLLA) scaffold significantly improves chondrogenic differentiation. By incorporating hydroxyapatite into silk sponges, Bhumiratana and Grayson [79] created a composite scaffold that encourages mesenchymal stem cells to produce mineralized bone tissue. The outcomes showed that two techniques improved mechanical strength. One resulted from the hydroxyapatite's osteoconductive properties, and the other from the mineral phase's regeneration sites.

Young's modulus determines the scaffold architecture's stiffness, another crucial feature. Through a variety of mechanisms, including protein unfolding and ion channel opening, cells react to scaffold rigidity, and as a result, rigidity influences cell division and proliferation. For instance, human cutaneous fibroblasts multiplied more rapidly when the free-floating collagen matrix stiffness increased [56]. Additionally, it is claimed that in the case of bone abnormalities, the rigidity of the scaffold affects its integration into the local natural tissue, primarily when the rigidity changes significantly. For example, scaffolds made of polymeric biomaterials would be less rigid than those made of metal or bioceramics. As a result of this mechanical change, the scaffold's tension, stress, and distortion properties would differ [7].

5 Use of Biomaterials in Tissue Engineering

5.1 *Brief Background*

The artificial regeneration of tissues, organs, or even complex living organisms was a matter of fiction and dream all over the history of humankind. Among them, regeneration of damaged tissue or organ was the priority. This dream has come to be achievable during the last decades and has recently been launched in clinical medicine [80]. Even though it began in the 1980s, tissue engineering is still a relatively new field of medicine. To establish a research center focused on living tissues, Yuan-Cheng Fung, an American bioengineering researcher, presented a proposal to the National Science Foundation (NSF). Fung introduced the concept of human tissue and expanded its use to include any living thing found in cells and organs. The NSF created the phrase "tissue engineering" in order to establish a new scientific research subject based on his idea. As a result, the Tissue Engineering Society (TES) was founded, and it subsequently changed its name to the Tissue Engineering and Regenerative Medicine International Society (TERMIS). In tissue engineering and regenerative medicine, the Tissue Engineering and Regenerative Medicine International Society (TERMIS) fosters research, development, and knowledge exchange [81].

The terminology' tissue engineering' officially came into action at an NSF workshop in 1988 to mean "the application of principles and approaches of engineering and life sciences toward the fundamental understanding of structure–function relationships in normal and pathological mammalian tissues and the development of biological substitutes to restore, maintain, or improve tissue function" [50].

Biomaterials play a vital role in tissue engineering to restore function and enable healing after disease or injury. Biomaterials are used in tissue engineering applications to support, augment, or substitute injured tissue or biological activity. When ancient Egyptians employed sutures made from animal sinew, the first historical usage of biomaterials dates to antiquity. In addition to more recent influences from materials science and tissue engineering, the modern area of biomaterials integrates biology, chemistry, physics, and medicine. Thanks to advancements in tissue engineering, the industry has expanded significantly during the past ten years.

5.2 Purpose of Tissue Engineering

Complete regeneration of injured tissues or organs and regaining their normal functions has been a vision of human beings. The discovery of tissue engineering seems to make it feasible. Tissue engineering amalgamates scaffolds, cells, and growth to restore tissues or replace injured or unhealthy tissues [82]. Tissues are the basic units of bodily function, while cells are the building blocks of tissue. Extracellular matrix is the term for the supporting structures that are typically produced and secreted by a collection of specialized cells. This scaffold serves as a structural framework that supports cell growth while acting as a hub for various signalling molecules. As a result, cells receive information from multiple sources within their surrounding environment. Each signal can trigger a cascade of reactions that influence cellular activity. By understanding how specific cells respond to signals, interact with their environment, and develop into tissues and organs, scientists have leveraged these mechanisms to regenerate damaged tissues or even create new tissues and organs [5].

The main functions of tissue engineering are to help tissue or organ repair or restoration, including calcified tissue like bone, cartilage, pancreas, cardiac, and vascular tissue. The field also researches stem cell behaviour and its application in tissue engineering. Stem cells can be developed into distinct categories of cells and can aid in the restoration of tissue or organs in the body. Stem cell tissue engineering permits scientists to produce prototypes to study numerous diseases such as heart disease and cancer. The three-dimensional feature of the tissue engineering scaffold allows tumour architecture to be examined in a more precise environment [81].

Generating a scaffold from a variety of potential materials, such as proteins or polymers, is frequently the first step in the tissue engineering process. Cells may be added to scaffolds with or without a cocktail of growth factors. The tissue grows if the conditions are right. The tissue can self-assemble in a number of situations where the scaffolds, cells, and growth ingredients are combined. Using an existing scaffold is an alternate method for creating new tissue. The residual collagen scaffold is then used to create new tissue once the donor organ cells are removed. This method has been used to develop bioengineered lung, heart, kidney, and liver tissue. This method holds the potential to use scaffold from human tissue discarded during surgical treatment and assemble this discarded tissue with a patient's cells to produce tailored tissues or organs that would not be rejected by the immune system [5].

5.3 *Fabrication Techniques*

Tissue engineering procedures usually involve using a porous scaffold, which functions as a 3D template for preliminary cell attachment and consequent tissue development both in vitro and in vivo. The scaffold offers the essential support for cell attachment, proliferation, and differentiation [83]. Scaffold fabrication is a critical and central step in tissue engineering, and it is a complex process. It entails creating a 3D functional tissue to aid in the replacement, restoration, and regeneration of a tissue or organ within the body. Scaffolds are produced for this purpose by combining cells with biomolecules like growth factors. Scaffolds are man-made or natural structures that resemble actual organs like the kidney or liver. On the scaffolds, tissue grows to resemble the biological structure that needs to be replaced [81].

Scaffolds for tissue engineering applications must be constructed to fulfil several criteria that are specific to their targeted application. Several of these criteria are well established. Others are continually changing, as our knowledge of developmental and regenerative cellular processes improves [84]. In tissue engineering, the scaffold is employed to give provisional support to the structure, assist cell growth and regeneration of tissue in a 3D matrix. Therefore, taking into account the scaffold design characteristics based on the intended tissue engineering is crucial. The structural position of regeneration, the mechanical demand on that region, and the expected assimilation rate all influence the scaffold layout choice. The essential parameters to be considered for a suitable scaffold are: first, high porosity and appropriate pore size with complete connectivity for cell passage and nutrient transfer. Second, structural features of the matrix ought to be designed to guide new tissue formation. Third, the rate of biomaterial degradation must be harmonized with the restoring rate, i.e., neither too slow nor too fast. With the knowledge of these factors, numerous fabrication techniques have been developed. They are broadly categorized as conventional and advanced techniques [85]. Each part has been subdivided into the different processes described graphically in Fig. 6.

5.3.1 Conventional Scaffold Fabrication Techniques

Conventional scaffold fabrication approaches are defined as procedures by which scaffolds with a continuous and uninterrupted porous structure can be produced but lack any long-range channelling microarchitecture. This includes porous polymer architectures, such cell attachment substrates, however obtaining complicated structures with adjustable microscale and macroscale structures is challenging [66].

The solvent casting method involves dissolving the polymer-ceramic composite in an organic solvent and then forming the solution into a pre-planned 3D mould. The porogen solvent later evaporates, producing a porous scaffold. However, in particulate leaching, various porogen elements control the scaffold's sponginess and interior pore dimension. Freeze-drying also requires water or organic solvent, removing the need for any porogenic elements. A synthetic polymer is dispersed in a proper solvent

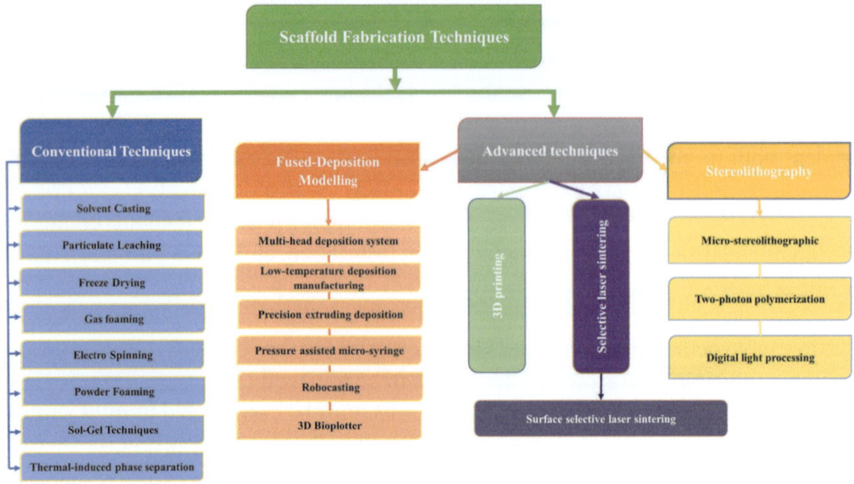

Fig. 6 Summary of the current scaffold fabrication techniques

in freeze-drying and then poured into a predesigned mould, followed by drying in a lyophilizer. Thermal-induced phase separation involves liquefying polymers with or without ceramic using an organic solvent. After rapidly cooling this polymer solution, sublimation is used to identify the porous scaffold.

A biopolymer is set in the locked chamber and saturated with high-pressure carbon dioxide in the gas foaming technique. When the pressure is promptly released, the thermodynamic unsteadiness results in nucleation and the creation of porous architecture. To achieve the definitive interconnected network combination, gas foaming and particulate leaching methods were established. In electrospinning, the polymeric non-woven scaffold is constructed with the use of an electric charge. It benefits scaffold fabrication with a broad range of fibre diameters from micrometers to nanometres with good inter-connectivity. The powder foaming process is usually employed to fabricate glass and ceramic scaffolds. In this process, a slurry of glass and ceramic particles is used to formulate the green body. Fillers such as gelatine, sucrose, spherical Poly (methyl methacrylate) microbeads, and soaking agents, which are surface-active agents, are added to the suspension to produce porosity burning during frittage. Sol–gel is an adjustable procedure involving condensation and gelation processes for making a sol by adding surfactant. The chemical reaction of metal alkoxides' inorganic polymerization forms the basis of this method [85].

5.3.2 Advanced Scaffold Fabrication Techniques

The advanced scaffold fabrication technique is also known as the rapid prototyping method. The rapid prototyping scaffold fabrication technique offers many potential opportunities for tissue engineering. First of all, multicellular structures that are desirable for intricate tissue activities can be constructed thanks to the independent control of microscale and macroscale structures. Secondly, the development of 3D vascular beds would facilitate the support of large-scale tissue growth, which would otherwise be unattainable. Thirdly, integrating medical imaging data with 3D fabrication techniques opens the door for the creation of customized scaffolds and the large-scale production of scaffold designs [66].

The conventional fabrication methods were being customized to fabricate scaffolds for defined tissue engineering applications and were not generic. In the conventional technique, each biomaterial or composite requires various processing abilities to achieve the requirements of scaffolds or tissue engineering constructs. Processibility, repeatability, and stability are still challenging among the essential requirements for scaffold construction, especially for engineering bone tissue. Various solid free-form (SFF) techniques, also known as rapid prototyping and additive manufacturing techniques, have come into reality to ameliorate these challenges. Still, this term is equivocal in two aspects. Although there is freedom concerning the geometry of the scaffold, the process is slow compared with conventional techniques.

Furthermore, the technique suggests a pilot-scale, or prototyping, use that is insufficient. This method is based on contemporary computer-aided design and manufacturing techniques in theory. The multifaceted structure is fabricated on a layer-by-layer basis, i.e., adding one layer over the other to build a digital cross-sectional 3D image via administering liquid, powder, and solid sheets according to pre-design [85]. The different rapid prototyping scaffold fabrication techniques are summarized in Fig. 6.

5.4 Medical Application Advances

In the literature, many scaffolds with numerous macro and micro architectures using different biomaterials have been reported over the period. The design of a scaffold should include adequate mechanical, physicochemical, and biological characteristics as well as considerations about sterility and commercial practicability. The performance and biological activity of 3D scaffolds, which serve as synthetic matrices, are enhanced by carefully controlling their size, shape, porosity, strength, and degradation rate. To facilitate effective tissue repair, scaffolds should be precisely tailored to mimic the anatomical structure, biomechanics, and functional characteristics of the native tissue. Furthermore, they must be resilient enough to withstand mechanical stresses and external loads generated during new tissue development while maintaining the structural integrity of the surrounding tissue [86].

Bone, the most abundant of the hard tissues, is composed of water (9%), inorganic components (69%) and organic components (22%). Both dense cortical bone and porous trabecular bone, which are both reinforced by collagen fibers, may be prominent in its hierarchical formation. Given that bones naturally regenerate and may heal themselves, the majority of scaffolds designed to promote repair and regeneration seek to provide regenerative signals to osteogenic cells [87]. Adapting to all biomaterial conditions that are contending in nature at the same time is the biggest challenge when designing a scaffold for a load-carrying application. Complex tissue scaffolds need to be easily created, blend in well with native tissue, and have a high rate of renewal while maintaining mechanical strength and a highly porous structure. Biomimetic ceramics, metals, and composites are frequently used alloplastic synthetic scaffolds as bone-implanting nanomaterials that replicate bone's native tissue and provide mechanical and structural support [86].

Researchers have developed new methods for creating complex architectures with preferred characteristics for soft tissue engineering applications through judicious selection of biomaterials and main scaffold properties. The scaffolds should be assembled so that they mimic and regenerate the anatomical structure and biological function of the targeted soft tissue to be restored. Biopolymers are generally used as materials to fabricate soft matrices broadly used to construct most transplanted organs, such as the liver and kidneys, and are effectively employed for tendon, muscle, artery, heart valve, pancreas, and bladder regeneration [86].

Despite numerous advancements in tissue engineering research, its application in treatment and patient care is not as widespread as one might anticipate. Tissue engineering has been applied in a few cases to patients' bladders, tiny arteries, skin transplants, and cartilage repair. Despite being created in the lab, tissue-engineering applications for major organs such as the liver, heart, and lungs have not yet been used in people. In addition to the risks involved with inpatient tissue engineering care, the procedures are expensive. When used in conjunction with therapeutic research, tissue engineering is advantageous, particularly when evaluating new medication compositions. Scientists can make progress in personalized medicine by using functional live tissue in a setting different from the human body. This personalized medicine testing reduces the cost of developing and testing medications on animal models and helps determine whether certain medications are effective for specific individuals depending on their genetic makeup [81].

A recent advancement in tissue engineering is the development of human liver tissue by the National Institute of Biomedical Imaging and Bioengineering, which was successfully transplanted into a rat. The liver is used by the mouse for metabolism. In order to simulate how the human liver would react to specific medications within the mouse, the implanted human liver tissue metabolizes additional substances. This helps researchers determine which medications are likely to interact with a certain medication [81].

To create engineered tissue with an integrated vascular network, researchers are experimenting with a printer capable of forming vascular-like structures using a sugar-based solution. This solution solidifies within the engineered tissue until blood circulation begins, flowing through the artificial channels. Additionally, the regeneration of a kidney using a patient's own cells is another ongoing study. Scientists have repurposed cells from donor organs, combining them with biological molecules and a collagen scaffold derived from the donor organ to cultivate new kidney tissue. This tissue has been examined for its functionality, including nutrient and oxygen absorption and urine production, both in vitro and in rats. Advancements in this field of tissue engineering have the potential to address donor shortages and mitigate diseases related to immunosuppression in organ transplant patients [81].

6 Limitations and Challenges

'Tissue engineering' was first introduced about 30 years ago. Neotissue regeneration was the main goal of this tissue engineering project. Tissue engineering techniques have become therapeutic agents to treat many of the difficult medical problems with the use of growth factors and biomaterials. This interdisciplinary engineering approach has garnered significant attention as a promising therapeutic tool that could overcome the limitations of artificial organ transplantation. These have also been used to replace organs or tissues that have been lost or seriously damaged. However, only cartilage, bone, skin, capillary, and periodontal tissues are significantly restored to patients through tissue engineering [4].

Although there is huge potential, presently, tissue engineering has a relatively minor role in inpatient care. Supplementary bladders, cartilage, skin grafts, small arteries, and even complete tracheas have been implanted in patients; however, these procedures remain experimental and costly. Conversely, more complex organ tissues, including heart, liver, and lung tissue, have been successfully cultivated in laboratory settings. However, they are not yet wholly reproducible and prepared to be inserted into a patient. These tissues can be precious in research, specifically in drug development investigation. This can hasten drug development and offer important tools for enabling tailored medicine through functioning human tissue to aid screening medication. As a result, it has the potential to reduce costs and minimize the reliance on animal testing [5].

The precision and efficacy of bioengineered scaffolds are very challenging as they need deep knowledge regarding regeneration mechanisms. New treatment strategies encourage us to design new nanostructures and constructs that can give more accurate, and target-specific scaffolds with variable functionalities. The biomimicked constructs have advanced surface topography and architecture like natural tissues or organs to recover inherent features of injured tissues and organs. Cell behaviour can be altered by using nanoenabling platforms, customised based on their unique properties. Different techniques are used, such as electrospinning, nanoenabled modelling, nanolithography, and electrochemical, to improve protein adsorption, cellular add-on, propagation, and differentiation without influencing immune response and signalling pathways. Thus, designing scaffolds to realize physiological and chemical communications with living tissues can substantially advance the fields of tissue engineering and regenerative medicine [88].

Interface tissue engineering has the potential to regenerate anatomical interfaces between different kinds of tissue. The major challenges in developing interfaces are how borders are demarcated with the tissue scaffold, re-established post-injury, and sustaining these mimicked scaffolds in the body. This will give a better understanding of the structure–function association of biomolecules with receptors at the position of the attachment and the mechanism behind interface regeneration. Engineering numerous types of tissue and related communications will play a significant role in exploring obstacles in interface design. New scaffolds are important to comprehend the theory behind the proliferation, migrations of cells, and differentiation [88].

The tissue engineering field is progressing without hard and fast firm rules for specifications for biomaterials. For suitable tissue engineering, biomaterials and scaffold templates should meet many criteria. The template ought to recapitulate the structure of the target cells and be proficient at adapting to the continually changing micro-environment. The components should be biodegradable with proper degradation kinetics, chemical, and morphological degradation profiles. The biomaterial should synchronize molecular signalling to target cells through endogenic molecules or supplement exogenic molecules. The biomaterial should possess suitable elastic or viscoelastic features favouring mechanical support and signalling to the target cells. They should be injectable when dreamed of, with proper rheological qualities and transformation mechanisms and kinetics. The biomaterial's architecture allows it to optimize cell, oxygen, nutrient, and molecule transport while also allowing blood vessel formation and nerve expansion. They need to be well-matched with the preparation techniques that concurrently pattern both the biomaterial and the cells, especially in the situation of 3D bio-printing. Finally, the material should be inherently non-immunogenic, non-cytotoxic, as well as minimally pro-inflammatory.

7 Prospects and Possible Solutions

With advancements in scaffold fabrication techniques, 3D printing, tissue engineering, and the application of animal models for in vivo studies, tissue engineering is rapidly evolving to overcome its key challenges and improve its clinical applicability [89, 90].

In addition to using biomaterials for tissue engineering, stem cells might be a different choice. Stem cell plasticity in tissue culture has generated a lot of interest worldwide since autologous stem cells are scarce for tissue culture proliferation. Mesenchymal stem cells (MSC), interstitial or circulating cells, and other regions of the body were found to contain multipotent stem cells [91]. Mesenchymal stem cells (MSCs) play a vital role in cartilage regeneration, as microencapsulation enhances their longevity and effectiveness [92]. Additionally, human embryonic stem cells (hESCs) derived from the inner cell mass of blastocysts in in vitro fertilization can differentiate into a diverse range of cell lineages in tissue culture [93]. Despite all of this, research conducted in vitro has demonstrated that the production of induced pluripotent stem cells (iPSCs) represents a major advancement [94].

There have been numerous improvements made to scaffold preparation, material uniformity, vascularization, and other areas. A three-dimensional extracellular matrix secreted by live tissues serves as a scaffold during regeneration, supporting tissue creation through a variety of molecular cues that have the necessary molecular makeup, physiological functions, and mechanical strength [95]. The native extracellular matrix (ECM) can be obtained for optimal tissue growth and development by carefully digesting living cells while retaining the highest possible amount of ECM from the targeted tissue site. This kind of successful human acellular scaffold transplantation has been performed on a number of different tissues in clinical settings. However, because natural ECM generated from non-human mammalian sources poses a risk of infection and immunogenicity, synthetic biomaterials for 3D scaffold assembly have gained popularity [96]. Collagen, fibronectin, and elastin-based artificial scaffolds are often combined with various sources of generated stem cells, such as human-induced pluripotent stem cells (hiPSCs). Scaffold fabrication has involved biodegradable materials like silk fibroin protein, alginate (a seaweed-derived polysaccharide), and apple cellulose. Additionally, synthetic polymers such as hydrogels and polyacrylamide are widely used in the production of 3D scaffolds [97]. To enhance vascular supply, extensive research has focused on developing thick, implanted viable tissue structures [98]. For instance, vascularized heart muscles have been successfully engineered using a combination of endothelial cell sheets and muscle precursors [99]. However, before these approaches can be clinically implemented, they must undergo rigorous validation.

The advent of 3D scaffold printing with targeted cells for tissue construction has fundamentally transformed tissue engineering by accelerating the creation of biomimetic living tissues [100]. These innovations are expected to drive the development of vascularized 3D soft organs, offering a potential solution to the ongoing organ transplant shortage. More recently, utilizing a layer-by-layer method, 3D printing has shown significant promise in tissue synthesis with structural control from micro- to macro-scale. The standard for 3D printed tissue engineering structures, whether scaffold-based or scaffold-free, is to offer a biomimetic structural environment that encourages tissue development and promotes host tissue integration (e.g., cellular penetration, vascularization, and active remodelling).

7.1 3D Bioprinting for Tissue Engineering

3D bioprinting is a type of additive manufacturing that prints live things layer by layer, mimicking the behaviour of real living systems by using cells and other biocompatible materials as "inks," also known as bioinks. In tissue engineering and regenerative medicine, bioprinting has emerged as a potential method. Multiple 3D printing strategies have been developed to enable bioprinting of diverse biopolymers and hydrogels. The necessity for bioinks containing autologous or patient-specific biological components for tissue engineering applications is becoming more apparent as the value of personalized treatment becomes clearer [101]. The discipline of 3D bioprinting arose from the idea of integrating additive manufacturing (AM) or 3D printing, which employs a layer-by-layer fabrication approach to produce complex tissues in vitro, with living organisms and biomaterials [102].

Using an automated dispensing device, cells and other biological components are printed and precisely placed onto a substrate or tissue culture dish. When incorporated into biocompatible materials to create functional 3D structures, this innovative technology ensures the cohesion of individual cells and cell types. The biomaterials used in this process, known as bioinks, consist of cells, fundamental structural materials, and other essential elements. To construct complex 3D tissue-mimetic structures, cell-based bioinks are shaped into the desired forms and sizes, accommodating intricate geometrical features.

This advancement presents a promising pathway for organ and tissue printing, enabling the fabrication of fully functional 3D tissues from cellular sources [103]. With scaffold-based or scaffold-free tissue and organ structures, as well as mini-tissues and organ-on-a-chip model systems, 3D printing is set to revolutionize biomedical engineering. Soon, this technology may allow for the large-scale production of functional human organs, including hearts, kidneys, skin, and livers [104–110]. To build tissue engineering constructions, a 3D bioprinter provides for the appropriate distribution and positioning of biomaterials, signalling components, and heterogeneous cells in high densities. The biocompatibility, biodegradability, cellular response, and tissue-microenvironment in which the final bioprinted scaffold is exposed all influence the quality of the final bioprinted scaffold [111]. Furthermore, 3D bioprinted structures featuring interconnected pores and large surface areas significantly improve cell adhesion, proliferation, intercellular communication, and the efficient exchange of gases and nutrients. This marks a major advancement over traditional fabrication techniques such as solvent casting, phase separation, and melt molding [112].

The potential of 3D bioprinters to produce acellular and cell-laden scaffolds with predesigned patterns, architecture, and cell and biological component distribution has sparked a lot of study in tissue engineering and regenerative medicine [113, 114]. As a result, much effort has gone into creating ways to enable high-resolution 3D printing of a range of hydrogels and biopolymers [115, 116]. Moreover, the generation of biologically relevant bioinks has been the subject of extensive investigation [116]. Biomaterials containing cells that are printed into three-dimensional scaffolds or tissue-like structures are called bioinks, and they are a crucial component of any advancement in bioprinting [116, 117]. Hydrogels have been extensively used in the creation of scaffolds for tissue engineering due to their resemblance to native extracellular matrix (ECM), as well as their adaptable physical characteristics and degradation profile [118]. Alginate is one of the most widely utilized hydrogels in fiber-based technologies, owing to its quick and reversible crosslinking into hydrogels with good mechanical properties in the presence of calcium ions [119]. Alginate has also been employed in a number of clinical trials and is FDA-approved for various biomedical purposes (Table 1) [120–125].

3D printing has come into the market as the most fruitful technique in designing and synthesising hybrid biomaterials. 3D-printed engineering tissues and organs have attracted attention as the most innovative approach. It has acceptance over the conventional engineering scaffold systems, but needs to be further developed to get better outcomes than the conventional approaches. Hence, further investigation

Table 1 Biomaterial application advancements in 3D printing for tissue engineering

No.	Biomaterials/bioink	Application employing 3D printing technology	Advancements	References
1	Synthetic PEGDA (polyethylene glycol diacrylate) into GelMA (gelatin methacrylate) hydrogel, nanospheres delivery system	Cartilage tissue engineering, tabletop stereolithography-based 3D bioprinter	Greatly improves the printing resolution, a longer retention of growth factors advantageous for cell growth and new tissue formation	[126]
2	Natural hydrogel-based alginate bioink	Bone tissue regeneration, fibroblast 3D cell-printing	Good printability, suitable environment for cell growth, proliferation, and distribution	[120, 121]
3	Natural hydrogel-based agarose bioink: cell-laden poly (D,L-lactic-co-glycolic acid) (PLGA) porous microspheres with thin encapsulation of agarose-collagen composite hydrogel (AC hydrogel)	Bone tissue regeneration, hybrid 3D bioprinting (porous micro scaffolds and extrusion-based printing)	A smooth delivery of cell-laden microspheres, immediate gelation, growth at body temperature, improved cell affinity, high cell viability, very high compressive strength	[127]
4	Natural hydrogel-based collagen bioink	Cartilage tissue engineering, commercially available 3D printer	Distinct mechanical properties, able to support and maintain cell growth	[128]
5	Polyelectrolyte gelatin-chitosan hydrogel-based bioink	Skin tissue engineering, extrusion-based 3D printer	Antimicrobial properties with ability to trigger hemostasis, good biocompatibility with fibroblast skin cells	[129]
6	Gelatin methacryloyl (GelMA), sodium alginate, and 4-arm poly (ethylene glycol)-tetra-acrylate (PEGTA) composites-based bio-ink	Large-scale vascularized tissue constructs towards applications in organ transplantation and repair using advanced multi-layered coaxial nozzle bioprinter	Satisfactorily functional tissue constructs mimicking their natural counterparts, promote the transport of oxygen, nutrients, and waste products, capable to form biologically relevant, highly organized, perfusable vessels	[122]
7	A fibrous bioink based on alginate hydrogel and polylactic acid nanofibers	Human adipose-derived stem cells engineering, 3D-bioplotted strands	Higher levels of cell proliferation within bio-printed strands, better survivability, good cell metabolic activity	[123]

(continued)

Table 1 (continued)

No.	Biomaterials/bioink	Application employing 3D printing technology	Advancements	References
8	CNT-incorporated alginate framework and cell-laden MeCol (methacrylated collagen) based hybrid hydrogel bioink	Myocardial tissue engineering, CNT-reinforced 3D-printing system	Improves electrical, mechanical, and biological behaviour, significantly improves viscoelastic behaviour and electrical conductivity of photo-cross-linked MeCol, attractive cellular proliferation, migration, and differentiation	[124]
9	Silk fibroin, Collagen I and matrigel based 3D hydrogels	Skeletal muscle tissue engineering, human primary skeletal myoblasts (hSKMs), 3D freeform bio-printing/gel casting	Able to repair defective skeletal muscles, mature cell-specific phenotypic expression, precise motoneuron integration into preformed myotubes, useful in neurite outgrowth and myogenic differentiation	[130, 131]
10	Biodegradable poly (ethylene glycol)/poly (d,l-lactide) hydrogel	Human mesenchymal stem cells engineering, stereolithography 3D bio-printing	Rapid-fabrication, high-precision, and customized-production, excellent pore interconnectivity with good mechanical properties, no need for support materials	[132, 133]
11	Polycaprolactone (PCL), poly (lactic-co-glycolic acid) (PLGA), and β-tricalcium phosphate (β-TCP) and mineralized ECM (extracellular matrix) based bio-composite ink	Bone regeneration, human nasal inferior turbinate tissue-derived mesenchymal stromal cells (hTMSCs) engineering, ECM-ornamented 3D printed scaffolds	Improves the biological functionality, tailorable physical properties with physiological presentation of biological signals	[134]
12	Chitosan-coated PCL (polycaprolactone) scaffolds	Bone-tissue regeneration., 3D photopolymerization bio-printing method,	Easy to use, good mechanical properties, solvent not required, tensile strength and complete water uptake ability, significantly increased mechanical properties	[135]

(continued)

Table 1 (continued)

No.	Biomaterials/bioink	Application employing 3D printing technology	Advancements	References
13	Microspheres of polycaprolactone (PCL) and the composite microspheres of PCL and hydroxyapatite (HA)	A feasible and effective protocol for fabricating new biomimetic bone biomaterials via SLS technique (Selective laser sintering), 3D SLS bio-printing method	No need for support materials, uniform multi-scaled porosity, moderate mechanical properties as well as good biocompatibility, excellent histocompatibility, promoted cell adhesion, supported cell proliferation and inducing cell differentiation	[136]
14	Poly (ethylene oxide) (PEO) based hydrogel bioink	Human umbilical vein endothelial Cells (HUVECs) engineering, 3D Inkjet bio-printing	Allow cell spreading and proliferation, dispensable by robotic precision, suitable stiffness to endure post gelation handling	[137]
15	Alkaline phosphates-based biomaterials	Dental pulp stem cells, deciduous tooth stem cells, and bone-marrow-derived mesenchymal stem cells regeneration, acoustic droplet ejection 3D bio-printing	A high-throughput, precise control, high resolution, good levels of bone-implant contact	[102, 138]
16	High molecular weight polyamides nylon membranes	Neuronal cells regeneration, 3D bioprinting following electrohydrodynamic jetting	A high-throughput, precise control, high resolution, sustained for at least 1 month	[102, 139]
17	Metal (Au, Ti, Ag), metal oxide (TiO_2), or photo-decomposing volatile polymer (triazene)	Regenerative medicine, laser-assisted 3D bioprinting	A high throughput and controlled printing resolution, able to reproduce the local cell micro-environment	[104]
18	Glucose, Penicillin, L-glutamine and trypsin, Polyester urethane urea	Cardiac regeneration, laser-induced-forward-transfer (LIFT) cell printing technique	Significant functional improvement of infarcted hearts, enhanced capillary density, rapid integration of human cells with functionality, could improve wound healing and functional preservation	[140]
19	Poly-ethylene glycol diacrylate and poly-caprolactone	Hybrid tissue engineering, vat polymerization bio-printing	Significantly high mechanical stiffness, material diffusion with high cell viability	[141]

(continued)

Table 1 (continued)

No.	Biomaterials/bioink	Application employing 3D printing technology	Advancements	References
20	Norepinephrine and phenylephrine as vasoactive drugs	Vasoactivity measurement, magnetic 3D bio-printing	A high throughput in vitro alternative, robustly models vasoactivity,	[142]
21	Alginate, gelatin methacryloyl	For engineering endothelialized myocardium and heart-on-a-chip, 3D microfibrous scaffolds bio-printing	A novel strategy to construct endothelialized-myocardial-tissues, also applicable for regenerative medicine, drug screening, and potentially disease modelling	[125]

needs to be performed into this aspect to use its full advantage. 3D bioprinting techniques facilitate better protein adsorption, cell adhesion, proliferation, and differentiation. Furthermore, knowledge of surface interaction, mechanisms, and their influence on bonding with living tissues/organs can help advance tissue regeneration and regenerative medicine.

8 Conclusion

Tissue engineering shows a lot of potential for addressing a variety of difficulties with new novel discoveries, and it will soon bring a lot of advances that can be successfully applied to humans. The integration of human stem cells with diverse bioscaffold fabrication techniques in tissue engineering presents new opportunities for fundamental research across multiple fields. Numerous successful clinical applications of acellular bioscaffolds have demonstrated host cell proliferation, and their potential applications are expected to expand rapidly in the near future for the benefit of humanity.

The field of tissue engineering continues to evolve, overcoming numerous challenges with innovative advancements that will soon lead to significant progress in human tissue regeneration. One of the primary obstacles in in vivo applications is the low survival rate of donor or autologous cells that are initially cultured in laboratory conditions, as well as the insufficient vascularization of implanted tissues or organs. Many isolated cells, engineered tissues, and artificially constructed organs encounter substantial clinical difficulties when transitioning from controlled culture environments to the human body. However, the future holds great promise. Ongoing efforts are focused on developing clinically viable tissue-engineered scaffolds using advanced biomaterials, paving the way for more effective and functional regenerative solutions. Advancing our understanding of the interaction between cells and biomaterials-based scaffolds has enhanced the progression towards clinical applications of tissue engineering. The advancement of 3D printing technology for scaffolds embedded with essential cells has revolutionized the approach to tissue engineering by significantly enhancing the fabrication of biomimetic living tissues. These biomedical implants, along with 3D bioprinting, have great potential to improve and enhance human life expectancy, and millions of people have experienced the advantages of incorporating implants into the human body. Biomaterial applications have included tissue cultures, engineered tissues, smart delivery systems for drugs, and hybrid organs. Although these technologies hold immense potential, their development necessitates a balanced and critical assessment, accompanied by thoughtful ethical discussions.

References

1. P. Kumar, B.S. Dehiya, A. Sindhu, Int. J. Appl. Eng. Res. **13**(5), 2744 (2018)
2. K. Sharma, M.A. Mujawar, A. Kaushik, Front. Mater. **6**, 172 (2019)
3. K. Dzobo, N.E. Thomford, D.A. Senthebane, H. Shipanga, A. Rowe, C. Dandara, M. Pillay, K.S.C.M. Motaung, Stem Cells Int. **2018**, 2495848 (2018)
4. P. Sharma, P. Kumar, R. Sharma, V.D. Bhatt, P. Dhot, J. Med. Life **12**(3), 225 (2019)
5. Tissue Engineering and Regenerative Medicine (National Institute of Health, 2021), https://www.nibib.nih.gov/science-education/science-topics/tissue-engineering-and-regenerative-medicine. Accessed 26 Jan 2021
6. M. Jin, J. Shi, W. Zhu, H. Yao, D.-A. Wang, Tissue Eng. Part B Rev. **27**(6), 604 (2021)
7. S. Abdulghani, G.R. Mitchell, Biomolecules **9**(11), 750 (2019)
8. S.J. Lee, J.J. Yoo, A. Atala, in *Clinical Regenerative Medicine in Urology*, ed. by B. Kim (Springer, Singapore, 2018), p. 17
9. Ž. Marjanović-Balaban, D. Jelić, in *Biomaterials in Clinical Practice*, ed by F. Zivic, S. Affatato, M. Trajanovic, M. Schnabelrauch, N. Grujovic, K. Choy (Springer, Cham, 2018), p. 101
10. M.D. Forecast, in *Global Biomaterials Market Size, Share, Trends, COVID-19 Impact & Growth Analysis Report—Segmented by Material Type, Application and Region—Industry Forecast (2021 to 2026)*, vol. 2021 (Market Data Forecast, 2021)
11. T.J. Keane, S.F. Badylak, Semin. Pediat. Surg. **23**(3), 112 (2014)
12. J.L. Ifkovits, J.A. Burdick, Tissue Eng. **13**(10), 2369 (2007)
13. O.F. Khan, M.V. Sefton, Trends Biotechnol. **29**(8), 379 (2011)
14. R. Parenteau-Bareil, R. Gauvin, F. Berthod, Materials **3**(3), 1863 (2010)
15. M. Pogorielov, O. Oleshko, A. Hapchenko, Adv. Tissue Eng. Regen. Med. **3**(2), 00058 (2017)
16. F.-M. Chen, X. Liu, Prog. Polym. Sci. **53**, 86 (2016)
17. C.M. Murphy, F.J. O'Brien, D.G. Little, A. Schindeler, Eur. Cell. Mater. **26**, 120 (2013)
18. E. Mariani, G. Lisignoli, R.M. Borzì, Pulsatelli, Int. J. Mol. Sci. **20**(3), 636 (2019)
19. S. Franz, S. Rammelt, D. Scharnweber, J.C. Simon, Biomaterials **32**(28), 6692 (2011)
20. B.N. Brown, S.F. Badylak, in *Regenerative Medicine Applications in Organ Transplantation* (Academic Press, Elsevier, Amsterdam, 2014), p. 151
21. N.E. Vrana, *Biomaterials and Immune Response: Complications, Mechanisms and Immunomodulation*, ed. by N.E. Vrana (CRC Press, Florida, 2018), pp. 1–14
22. K. Prasad, O. Bazaka, M. Chua, M. Rochford, L. Fedrick, J. Spoor, R. Symes, M. Tieppo, C. Collins, A. Cao, Materials **10**(8), 884 (2017)
23. B. Guo, P.X. Ma, Sci. China Chem. **57**(4), 490 (2014)
24. G. Santos, Adv. Tissue Eng. Regen. Med. **3**(1), 00054 (2017)
25. L. Pacifici, F. De Angelis, A. Orefici, A. Cielo, ORAL Implantol. **9**(Suppl 1/2016 to N 4/2016), 107 (2016)
26. M. Saini, Y. Singh, P. Arora, V. Arora, K. Jain, World J. Clin. Cases: WJCC **3**(1), 52 (2015)
27. H. Gul, M. Khan, A.S. Khan, in *Handbook of Ionic Substituted Hydroxyapatites* (Woodhead Publishing, Elsevier, Sawston, 2020), p. 53
28. S.C.A. Pina, R.L. Reis, J.M. Oliveira, in *Fundamental Biomaterials: Ceramics, Woodhead Publishing Series in Biomaterials*, ed. by S. Thomas, P. Balakrishnan, M.S. Sreekala (Woodhead Publishing, 2018), pp. 95–116
29. M. Frasnelli, V.M. Sglavo, Acta Biomater. **33**, 283 (2016)
30. D.G. Castner, B.D. Ratner, in *Principles of Regenerative Medicine*, 3rd edn., ed. by A. Atala, R. Lanza, A.G. Mikos, R. Nerem, (Academic Press, Amsterdam, 2019), pp. 523–534
31. J. Huang, S. Best, in *Tissue Engineering Using Ceramics and Polymers*, 2nd edn., ed. by A.R. Boccaccini, P.X. Ma (Woodhead Publishing, 2014), pp. 3–24
32. R. Mala, A.S.R. Celsia, *Fundamental Biomaterials: Ceramics. Woodhead Publishing Series in Biomaterials*, ed. by (s): S. Thomas, P. Balakrishnan, M.S. Sreekala (Woodhead Publishing, 2018), pp. 195–221

33. S. Mohan, O.S. Oluwafemi, N. Kalarikkal, S. Thomas, S.P. Songca, *Recent Advances in Biopolymers*, ed. by F.K. Perveen (IntechOpen, 2016), pp. 47–72
34. L. L. dos Santos, *Natural Polymeric Biomaterials: Processing and Properties*, ed. By S. Winkler, D.L. Kaplan (Elsevier, 2017), pp. 609–615
35. N. Asadi, A.R. Del Bakhshayesh, S. Davaran, A. Akbarzadeh, Mater. Chem. Phys. **242**, 122528 (2020)
36. M. Setayeshmehr, E. Esfandiari, M. Rafieinia, B. Hashemibeni, A. Taheri-Kafrani, A. Samadikuchaksaraei, D.L. Kaplan, L. Moroni, M.T. Joghataei, Tissue Eng. Part B Rev. **25**(3), 202 (2019)
37. L. Ambrosio, *Biomedical Composites. Woodhead Publishing Series in Biomaterials*, 2nd edn., ed. by L. Ambrosio (Woodhead Publishing, 2017), pp. xv–xvi
38. A. Arjunan, A. Baroutaji, A.S. Praveen, J. Robinson, C. Wang, in *Encyclopedia of Smart Materials*, ed. by (s): A.-G. Olabi, (Elsevier, 2022), pp. 86–102
39. H.R. Bakhsheshi-Rad, E. Dayaghi, A.F. Ismail, A. Madzlan, A. Akhavan-Farid, C. Xiongbiao, T. Nonferr, Metal. Soc. **29**(5), 984 (2019)
40. G. Chen, T. Ushida, T. Tateishi, Adv. Mater. **12**(6), 455 (2000)
41. C.R. Weinstein-Oppenheimer, D.I. Brown, R. Coloma, P. Morales, M. Reyna-Jeldes, M.J. Díaz, E. Sánchez, C.A. Acevedo, Mater. Sci. Eng. C **79**, 821 (2017)
42. K. Tur, Turk. J. Rheumatol. **24**(4), 206 (2009)
43. G. Heness, B. Ben-Nissan, Mater. Forum **27**, 104 (2004)
44. M. Alksne, M. Kalvaityte, E. Simoliunas, I. Rinkunaite, I. Gendviliene, J. Locs, V. Rutkunas, V. Bukelskiene, J. Mech. Behav. Biomed. **104**, 103641 (2020)
45. F. Baino, G. Novajra, C. Vitale-Brovarone, Front. Bioeng. Biotechnol. **3**, 202 (2015)
46. F. Baino, S. Hamzehlou, S. Kargozar, J. Funct. Biomater. **9**(1), 25 (2018)
47. F. Westhauser, F. Hohenbild, M. Arango-Ospina, S.I. Schmitz, S. Wilkesmann, L. Hupa, A. Moghaddam, A.R. Boccaccini, Int. J. Mol. Sci. **21**(5), 1639 (2020)
48. B. Prabhu, A. Karau, A. Wood, M. Dadsetan, H. Liedtke, T. DeWitt, in *Orthopedic Biomaterials*, ed. by B. Li, T. Webster (Springer, Cham, 2018), pp. 287–344
49. M. Prakasam, J. Locs, K. Salma-Ancane, D. Loca, A. Largeteau, L. Berzina-Cimdina, J. Funct. Biomater. **8**(4), 44 (2017)
50. F.J. O'brien, Mater. Today **14**(3), 88 (2011)
51. Y.G. Thyavihalli Girijappa, S. Mavinkere Rangappa, J. Parameswaranpillai, S. Siengchin, Front. Mater. **6**, 226 (2019)
52. X. Tang, S.K. Thankappan, P. Lee, S.E. Fard, M.D. Harmon, K. Tran, X. Yu, in *Natural and Synthetic Biomedical Polymers*, ed. by S.G. Kumbar, C.T. Laurencin, M. Deng (Elsevier, Amdterdam, 2014), pp. 351–371
53. R. Detsch, J. Will, J. Hum, J.A. Roether, A.R. Boccaccini, in *Cell Culture Technology*, ed. by C. Kasper, V. Charwat, A. Lavrentieva (Springer, Berlin, 2018), pp. 91–105
54. J. Wu, K. Zheng, X. Huang, J. Liu, H. Liu, A.R. Boccaccini, Y. Wan, X. Guo, Z. Shao, Acta Biomater. **91**, 60 (2019)
55. H.J. Haugen, S.P. Lyngstadaas, F. Rossi, G. Perale, J. Clin. Periodontol. **46**, 92 (2019)
56. A. Dolcimascolo, G. Calabrese, S. Conoci, R. Parenti, in *Biomaterial-Supported Tissue Reconstruction or Regeneration*, ed. by M. Barbeck, O. Jung, R. Smeets, T. Koržinskas (IntechOpen, London, 2019). https://doi.org/10.5772/intechopen.83839
57. N. Murthy, S. Wilson, J. Sy, *Polymer Science: A Comprehensive Reference*, ed. by K. Matyjaszewski, M. Möller (Elsevier, Amsterdam, 2012), pp. 547–560
58. W. Habraken, P. Habibovic, M. Epple, M. Bohner, Mater. Today **19**(2), 69 (2016)
59. T.M. Shick, A.Z. Abdul Kadir, N.H.A. Ngadiman, A. Ma'aram, J. Bioact. Compat. Polym. **34**(6), 415 (2019)
60. P. Suvaneeth, N. Nair, World J. Pharm. Res. **7**(10), 161 (2018)
61. S. Barr, E. Hill, A. Bayat, J. Mech. Behav. Biomed. **75**, 75 (2017)
62. S.W. Crowder, V. Leonardo, T. Whittaker, P. Papathanasiou, M.M. Stevens, Cell Stem Cell **18**(1), 39 (2016)

63. S. Lee, J. Yoo, A. Atala, in *In Situ Tissue Regeneration*, ed. by S.J. Lee, J.J. Yoo, A. Atala (Elsevier, Amsterdam, 2016), pp. 3–17
64. G. Turnbull, J. Clarke, F. Picard, P. Riches, L. Jia, F. Han, B. Li, W. Shu, Bioactive Mater. **3**(3), 278 (2018)
65. Z. Li, M.-B. Xie, Y. Li, Y. Ma, J.-S. Li, F.-Y. Dai, J. Biomater. Tiss. Eng. **6**(10), 755 (2016)
66. A. Eltom, G. Zhong, A. Muhammad, Adv. Mater. Sci. Eng. **2019**, 3429527 (2019)
67. L. Roseti, V. Parisi, M. Petretta, C. Cavallo, G. Desando, I. Bartolotti, B. Grigolo, Mater. Sci. Eng. C **78**, 1246 (2017)
68. B. Nasiri, S. Mashayekhan, Biologicals **48**, 39 (2017)
69. Y. Yang, A.C. Ritchie, N.M. Everitt, Mater. Sci. Eng. C **80**, 263 (2017)
70. Q. Li, L. Ma, C. Gao, J. Mater. Chem. B **3**(46), 8921 (2015)
71. F. Guilak, D.M. Cohen, B.T. Estes, J.M. Gimble, W. Liedtke, C.S. Chen, Cell Stem Cell **5**(1), 17 (2009)
72. F. Bobbert, A. Zadpoor, J. Mater. Chem. B **5**(31), 6175 (2017)
73. X. Yu, X. Tang, S.V. Gohil, C.T. Laurencin, Adv. Healthc. Mater. **4**(9), 1268 (2015)
74. Y.M. Ju, A. Atala, J.J. Yoo, S.J. Lee, Acta Biomater. **10**(10), 4332 (2014)
75. P.T. Thevenot, A.M. Nair, J. Shen, P. Lotfi, C.-Y. Ko, L. Tang, Biomaterials **31**(14), 3997 (2010)
76. X. Bai, M. Gao, S. Syed, J. Zhuang, X. Xu, X.-Q. Zhang, Bioactive Mater. **3**(4), 401 (2018)
77. R. Agarwal, A.J. García, Adv. Drug Deliv. Rev. **94**, 53 (2015)
78. H.-Y. Mi, X. Jing, M.R. Salick, T.M. Cordie, X.-F. Peng, L.-S. Turng, J. Mater. Sci. **49**(5), 2324 (2014)
79. S. Bhumiratana, W.L. Grayson, A. Castaneda, D.N. Rockwood, E.S. Gil, D.L. Kaplan, G. Vunjak-Novakovic, Biomaterials **32**(11), 2812 (2011)
80. U. Meyer, in *Fundamentals of Tissue Engineering and Regenerative Medicine*, ed. by U. Meyer, J. Handschel, H.P. Wiesmann, T. Meyer (Springer, Berlin, 2009), pp. 5–12
81. C. Travers, in *Health Technology*, ed. by M. Kashif, J. Piracha (Verywellhealth, 2020). https://www.verywellhealth.com/tissue-engineering-4580368. Accessed 25 Mar 2021
82. F. Han, J. Wang, L. Ding, Y. Hu, W. Li, Z. Yuan, Q. Guo, C. Zhu, L. Yu, H. Wang, Z. Zhao, L. Jia, J. Li, Y. Yu, W. Zhang, G. Chu, S. Chen, B. Li, Front. Bioeng. Biotechnol. **8**, 83 (2020)
83. D.W. Hutmacher, J. Biomat. Sci. Polym. E **12**(1), 107 (2001)
84. D.F.V. Rey, J.-P. St-Pierre, in *Handbook of Tissue Engineering Scaffolds: Volume One. Woodhead Publishing Series in Biomaterials*, ed. by M. Mozafari, F. Sefat, A. Atala (Woodhead Publishing, 2019), pp. 109–125
85. A. Bagde, A. Kuthe, S. Quazi, V. Gupta, S. Jaiswal, S. Jyothilal, N. Lande, S. Nagdeve, Irbm **40**(3), 133 (2019)
86. M.P. Nikolova, M.S. Chavali, Bioactive Mater. **4**, 271 (2019)
87. P. Chocholata, V. Kulda, V. Babuska, Materials **12**(4), 568 (2019)
88. A. Kaushik, M.A. Mujawar, K. Sharma, Front. Mater. **6**, 172 (2019)
89. A.J. Salgado, J.M. Oliveira, A. Martins, F.G. Teixeira, N.A. Silva, N.M. Neves, N. Sousa, R.L. Reis, Int. Rev. Neurobiol. **108**, 1 (2013)
90. B.T. O'Donnell, C.J. Ives, O.A. Mohiuddin, B.A. Bunnell, Front. Bioeng. Biotechnol. **7**, 95 (2019)
91. M.D. Grounds, J.D. White, N. Rosenthal, M.A. Bogoyevitch, J. Histochem. Cytochem. **50**(5), 589 (2002)
92. F. Li, V.X. Truong, H. Thissen, J.E. Frith, J.S. Forsythe, ACS Appl. Mater. Interfaces **9**(10), 8589 (2017)
93. K. Takahashi, K. Tanabe, M. Ohnuki, M. Narita, T. Ichisaka, K. Tomoda, S. Yamanaka, Cell **131**(5), 861 (2007)
94. M. Chakravarthy, E. Spangenburg, F. Booth, Cell. Mol. Life Sci. CMLS **58**(8), 1150 (2001)
95. Y. Shen, S.L. Redmond, J.M. Papadimitriou, B.M. Teh, S. Yan, Y. Wang, M.D. Atlas, R.J. Marano, M. Zheng, R.J. Dilley, Biomed. Mater. **9**(1), 015015 (2014)
96. D.J. Modulevsky, C. Lefebvre, K. Haase, Z. Al-Rekabi, A.E. Pelling, PLoS One **9**(5), e97835 (2014)

97. S. Naahidi, M. Jafari, M. Logan, Y. Wang, Y. Yuan, H. Bae, B. Dixon, P. Chen, Biotechnol. Adv. **35**(5), 530 (2017)
98. L. Perry, M.Y. Flugelman, S. Levenberg, Mol. Ther. **25**(4), 935 (2017)
99. J. Yang, M. Yamato, K. Nishida, T. Ohki, M. Kanzaki, H. Sekine, T. Shimizu, T. Okano, J. Control. Rel. **116**(2), 193 (2006)
100. M.S. Kim, B. Lee, H.N. Kim, S. Bang, H.S. Yang, S.M. Kang, K.-Y. Suh, S.-H. Park, N.L. Jeon, Biofabrication **9**(1), 015029 (2017)
101. N. Faramarzi, I.K. Yazdi, M. Nabavinia, A. Gemma, A. Fanelli, A. Caizzone, L.M. Ptaszek, I. Sinha, A. Khademhosseini, J.N. Ruskin, Adv. Healthc. Mater. **7**(11), 1701347 (2018)
102. J.M. Lee, S.L. Sing, M. Zhou, W.Y. Yeong, Int. J. Bioprinting **4**(2), 151 (2018)
103. I. Matai, G. Kaur, A. Seyedsalehi, A. McClinton, C.T. Laurencin, Biomaterials **226**, 119536 (2020)
104. F. Guillemot, B. Guillotin, A. Fontaine, M. Ali, S. Catros, V. Kériquel, J.-C. Fricain, M. Rémy, R. Bareille, J. Amédée-Vilamitjana, MRS Bull. **36**(12), 1015 (2011)
105. S. Khalil, W. Sun, J. Biomech. Eng. **131**(11), 111002 (2009)
106. A.B. Dababneh, I.T. Ozbolat, J. Manuf. Sci. Eng. **136**(6), 061016 (2014)
107. A. Atala, F.K. Kasper, A.G. Mikos, Sci. Transl. Med. **4**(160), 160rv112 (2012)
108. F.P. Melchels, W.J. Dhert, D.W. Hutmacher, J. Malda, J. Mater. Chem. B **2**(16), 2282 (2014)
109. I.T. Ozbolat, Y. Yu, IEEE Trans. Bio-Med. Eng. **60**(3), 691 (2013)
110. L.G. Griffith, M.A. Swartz, Nat. Rev. Mol. Cell Biol. **7**(3), 211 (2006)
111. J. Li, M. Chen, X. Fan, H. Zhou, J. Transl. Med. **14**(1), 1 (2016)
112. S. Knowlton, S. Anand, T. Shah, S. Tasoglu, Trends Neurosci. **41**(1), 31 (2018)
113. S.V. Murphy, A. Atala, Nat. Biotechnol. **32**(8), 773 (2014)
114. C. Mandrycky, Z. Wang, K. Kim, D.-H. Kim, Biotechnol. Adv. **34**(4), 422 (2016)
115. T. Billiet, M. Vandenhaute, J. Schelfhout, S. Van Vlierberghe, P. Dubruel, Biomaterials **33**(26), 6020 (2012)
116. K. Zhu, S.R. Shin, T. van Kempen, Y.C. Li, V. Ponraj, A. Nasajpour, S. Mandla, N. Hu, X. Liu, J. Leijten, Adv. Funct. Mater. **27**(12), 1605352 (2017)
117. C. Ferris, K. Gilmore, S. Beirne, D. McCallum, G. Wallace, M. Panhuis, Science **1**, 224 (2013)
118. A. Fallahi, A. Khademhosseini, A. Tamayol, Trends Biotechnol. **34**(9), 683 (2016)
119. J.A. Rowley, G. Madlambayan, D.J. Mooney, Biomaterials **20**(1), 45 (1999)
120. S. Adepu, N. Dhiman, A. Laha, C.S. Sharma, S. Ramakrishna, M. Khandelwal, Curr. Opin. Biomed. Eng. **2**, 22 (2017)
121. J. Park, S.J. Lee, S. Chung, J.H. Lee, W.D. Kim, J.Y. Lee, S.A. Park, Mater. Sci. Eng. C **71**, 678 (2017)
122. W. Jia, P.S. Gungor-Ozkerim, Y.S. Zhang, K. Yue, K. Zhu, W. Liu, Q. Pi, B. Byambaa, M.R. Dokmeci, S.R. Shin, Biomaterials **106**, 58 (2016)
123. L.K. Narayanan, P. Huebner, M.B. Fisher, J.T. Spang, B. Starly, R.A. Shirwaiker, ACS Biomater. Sci. Eng. **2**(10), 1732 (2016)
124. M. Izadifar, D. Chapman, P. Babyn, X. Chen, M.E. Kelly, Tissue Eng. Part C Methods **24**(2), 74 (2018)
125. Y.S. Zhang, A. Arneri, S. Bersini, S.-R. Shin, K. Zhu, Z. Goli-Malekabadi, J. Aleman, C. Colosi, F. Busignani, V. Dell'Erba, Biomaterials **110**, 45 (2016)
126. W. Zhu, H. Cui, B. Boualam, F. Masood, E. Flynn, R.D. Rao, Z.-Y. Zhang, L.G. Zhang, Nanotechnology **29**(18), 185101 (2018)
127. Y.J. Tan, X. Tan, W.Y. Yeong, S.B. Tor, Sci. Rep. UK **6**(1), 1 (2016)
128. S. Rhee, J.L. Puetzer, B.N. Mason, C.A. Reinhart-King, L.J. Bonassar, ACS Biomater. Sci. Eng. **2**(10), 1800 (2016)
129. W.L. Ng, W.Y. Yeong, M.W. Naing, Int. J. Bioprinting **2**(1), 53 (2016)
130. S. Ostrovidov, S. Salehi, M. Costantini, K. Suthiwanich, M. Ebrahimi, R.B. Sadeghian, T. Fujie, X. Shi, S. Cannata, C. Gargioli, Small **15**(24), 1805530 (2019)
131. T.A. Dixon, E. Cohen, D.M. Cairns, M. Rodriguez, J. Mathews, R.R. Jose, D.L. Kaplan, Tissue Eng. Part C Methods **24**(6), 346 (2018)

132. B.K. Gu, D.J. Choi, S.J. Park, M.S. Kim, C.M. Kang, C.-H. Kim, Biomater. Res. **20**(1), 1 (2016)
133. T.M. Seck, F.P. Melchels, J. Feijen, D.W. Grijpma, J. Control. Rel. **148**(1), 34 (2010)
134. F. Pati, T.-H. Song, G. Rijal, J. Jang, S.W. Kim, D.-W. Cho, Biomaterials **37**, 230 (2015)
135. K. Lee, G. Jin, C.H. Jang, W.-K. Jung, G. Kim, J. Mater. Chem. B **1**(42), 5831 (2013)
136. Y. Du, H. Liu, J. Shuang, J. Wang, J. Ma, S. Zhang, Colloids Surf. B. **135**, 81 (2015)
137. S.A. Irvine, A. Agrawal, B.H. Lee, H.Y. Chua, K.Y. Low, B.C. Lau, M. Machluf, S. Venkatraman, Biomed. Microdev. **17**(1), 1 (2015)
138. Y. Yamada, S. Nakamura, K. Ito, T. Sugito, R. Yoshimi, T. Nagasaka, M. Ueda, Tissue Eng. Part A **16**(6), 1891 (2010)
139. P.A. Eagles, A.N. Qureshi, S.N. Jayasinghe, Biochem. J. **394**(2), 375 (2006)
140. R. Gaebel, N. Ma, J. Liu, J. Guan, L. Koch, C. Klopsch, M. Gruene, A. Toelk, W. Wang, P. Mark, Biomaterials **32**(35), 9218 (2011)
141. Y. Shanjani, C. Pan, L. Elomaa, Y. Yang, Biofabrication **7**(4), 045008 (2015)
142. H. Tseng, J.A. Gage, W.L. Haisler, S.K. Neeley, T. Shen, C. Hebel, H.G. Barthlow, M. Wagoner, G.R. Souza, Sci. Rep. UK **6**(1), 1 (2016)
143. S.D. Anker, A.J. Coats, G, Cristian, D. Dragomir, E. Pusineri, M. Piredda, L. Bettari, R. Dowling, M. Volterrani, B.-A. Kirwan, Eur. Heart J. **36**(34), 2297 (2015)

Progress in Multifunctional and Multimodal Biomaterials for Therapeutic Applications

Sérgio R. S. Veloso, Valéria Gomes, Raquel G. D. Andrade, Beatriz D. Cardoso, Ana Rita O. Rodrigues, and Elisabete M. S. Castanheira

Abstract Humankind is witnessing a rapid progress toward multifunctional materials endowed with multimodal control of their properties. In particular, biomaterials are increasingly providing solutions to complex challenges in the nanomedicine field, from the compartmentalization and delivery of unstable molecules or even the in situ synthesis of biological active molecules to the synergistic theranostic strategies that join therapy and diagnosis in a single system. Hence, considering the complementary of different materials, we intend to provide in this chapter a discussion on the most recent and relevant progress in biomaterials for therapeutic applications, starting from hydrogels and liposomes to the combination of these materials with composites toward smart multifunctional materials.

Keywords Drug delivery · Magnetic hyperthermia · Hydrogels · Liposomes · Theranostics · Cancer therapy · Nanomedicine

1 Introduction to Nanomaterials in Nanomedicine

The progress of nanomaterials from simple structural substituents to complex multifunctional materials opened up new opportunities and strategies in several biomedical fields. The need for solutions that tackle health problems has contributed to the research and development of biocompatible materials, which led to the emergence of new fields such as nanotechnology, and nanomedicine.

Cancer remains as a disease of major concern with the mortality rising worldwide, both in high- and low-income countries, and a high incidence ratio on lungs, colorectal, breast, and prostate [1, 2]. According to the World Health Organization report on cancer, there were 18.1 million new cancer cases in 2018, expected to raise to 29.4 million by 2040 and 1/6 of annual total deaths is caused by cancer

S. R. S. Veloso · V. Gomes · R. G. D. Andrade · B. D. Cardoso · A. R. O. Rodrigues ·
E. M. S. Castanheira (✉)
Physics Centre of Minho and Porto Universities (CF-UM-UP), University of Minho, 4710-057 Braga, Portugal
e-mail: ecoutinho@fisica.uminho.pt

S. J. Ikhmayies (ed.), *Advances in Biomaterials Research*, Advances in Material Research and Technology, https://doi.org/10.1007/978-3-031-97481-6_5

[3]. Besides, the carried efforts to improve the survival rates through conventional chemotherapy have only been successful in patients with early stages of disease in high-income countries [4, 5]. This is accounted to several limitations of chemotherapeutic drug administration, which include the lack of specificity toward the neoplastic tissues, the lack of selectivity in the mechanism of action, and the high concentration required [1]. Together, these problems lead to a worsening of the patient's quality of life and high therapeutic cost, which makes it barely accessible and affordable to low-income countries. Here, nanomedicine can provide an improvement over conventional chemotherapy through the incorporation of drugs into nanosystems. These nanosystems increase drug stability and solubility, reduce systemic clearance, improve the specificity by active (e.g., surface functionalization) or passive (e.g., via enhanced permeability and retention effect, EPR) targeting, and enable synergistic therapeutic strategies [6–8].

Hereby, considering the wide interest and advances undertaken in both soft matter, and inorganic nanomaterials for therapeutic and diagnostic applications, this chapter focuses on providing a discussion of emerging and current advances of materials in nanomedicine and biomedical fields, such as hydrogels and liposomes. Each section briefly introduces the reader to the system's properties and development strategies, as well as the advancements of therapeutic strategies in various biomedical fields. In the final section, conclusions and some future perspectives are pointed out to further advance the smart multifunctional nanomaterials.

2 Trends with Hydrogels in Therapeutic Applications

2.1 Polymeric Hydrogels for Cancer Therapy

Over the past decade, many solutions have emerged to overcome some major shortcomings associated with cancer therapy, including poor water solubility of chemotherapeutic drugs, toxicity of systemic chemotherapy, short circulation half-life, the need for repeated injections and low specificity of treatment [9]. Besides the widely investigated nanoformulations such as liposomes and lipid nanoparticles [10, 11], polymeric nanoparticles [12, 13], micelles [14], magnetic liposomes [15, 16], nucleic acid and peptide assemblies [17, 18] have been developed. Also, there is a plethora of injectable hydrogel-based drug delivery systems being developed as therapy platforms for several types of cancer, such as thermosensitive hydrogels, pH-sensitive hydrogels, photosensitive hydrogels, dual-sensitive hydrogels and active-targeting hydrogels [19]. The great advantages of hydrogels are their biocompatibility, biodegradability, their reversible behavior upon external stimuli (temperature, pH, light, magnetic field) and the sustained localized release of drugs regardless of the morphology of the tumor microenvironment [20]. In general, the preparation is achieved through chemical or physical crosslinking. The former method consists in chemical reactions to form covalent bonds, such as Diels–Alder, Michael addition, enzyme-mediated and photopolymerization, while the latter is a result of non-covalent interactions such as electrostatic, hydrophobic, hydrogen bonding and host–guest interactions [21, 22]. Polymeric hydrogels used in cancer treatment studies can be composed of many materials, including the synthetic ones such as poloxamers [23, 24], polyesters [25, 26], polyacrylates [27, 28] and dendrimers [29, 30], and the natural such as chitosan [31, 32], hyaluronic acid [33, 34], alginate [35, 36] and dextran [37, 38].

2.1.1 Injectable Thermosensitive Hydrogels

Thermosensitive hydrogels are hydrogels with a Lower Critical Solution Temperature (LCST) above which they undergo a sol–gel phase transition. In this way, it is possible to prepare a polymer solution with chemotherapeutic agents at room temperature, inject it, and form a gel in situ at physiological temperature [39].

The co-delivery of chemotherapeutic drugs, such as doxorubicin (DOX), cisplatin, methotrexate (MTX) [40–43], or the development of hybrid hydrogels [44] using thermosensitive hydrogels, is not new and has demonstrated how these strategies promote the synergistic effect of the loaded components. For example, Wei et al. [45] used the difference in water solubility of DOX (hydrophilic) and the vascular disrupting drug combretastatin A4 (hydrophobic) to develop a thermosensitive hydrogel with sequential delivery of these drugs for the treatment of cervical tumor. More recently, a thermosensitive hydrogel composed of two types of poloxamer and alginate (P407/P188/alginate 20/2/1% w/v) loaded with 0.5% w/v of 5-Fluorouracil was developed by Sabbagh et al. [46]. The efficacy was studied in the context of neoadjuvant therapy and adjuvant therapy (after tumor removal) in colorectal cancer. The researchers could achieve, with negligible systemic toxicity, a flowable hydrogel at room temperature capable of intratumoral administration with tumor growth delay of 15 days.

Chung et al. [47] also obtained hopeful results with a P407 poloxamer hydrogel loaded with doxorubicin. The cell proliferation assays showed that the combination of doxorubicin with the poloxamer hydrogel enhanced the killing effect of MC-38 colon cancer cells. A non-invasive molecular imaging was also performed with indocyanine green (ICG) as a drug model for doxorubicin (they have comparable water solubility and molar weights) and the fluorescence images showed that ICG-loaded hydrogel sustained the molecule for at least 7 days at the site of injection. Salinomycin, a potential chemotherapeutic agent, was loaded in two types of thermosensitive copolymers, one composed of Pluronic F127 and the other of PLGA-PEG-PLGA (where PLGA is poly lactic acid-co-glycolic acid, and PEG is polyethylene glycol), to investigate their potential as local delivery systems for brain tumors. Both hydrogels were effective in inducing apoptosis and generate intracellular Reactive Oxygen Species (ROS); however, the Pluronic hydrogel exhibited better biocompatibility, drug release profile, and anticancer synergistic effect [48].

It is now clear how important it is to do research into this type of hydrogels in the development of a panoply of different delivery systems for local and sustained delivery of chemotherapeutic drugs at the tumor site, making use of the physiological temperature for in situ gelation. Besides temperature, there are other parameters that can be used to tune hydrogels behavior, such as pH, photosensitivity, magnetic, and photodynamic response. In the following sections, pH-sensitive and photosensitive and multimodal hydrogels will be also discussed.

2.1.2 pH-Sensitive Hydrogels

pH-sensitive hydrogels are polymeric gels that change volume in response to pH and ionic strength changes in the environment around them. This occurs because the polymer chains contain ionizable groups such as carboxyl, and amine groups, that donate or accept protons leading to conformational changes in the gel structure [49, 50]. The extracellular matrix of tumor tissues has a lower pH than normal tissues due to the acidification promoted by tumor glycolysis, while the intracellular environment has an increased alkalization [51], so the pH-dependent swelling is an important advantage for the controlled release of drugs in the tumor microenvironment [52]. Figure 1 represents the conformational changes of a pH-sensitive hydrogel in a more acidic or more basic external environment. Cationic hydrogels swell at low pH while anionic hydrogels swell at high pH due to protonation or ionization, respectively. In this way, two different delivery strategies can be performed with these hydrogels, namely the use of cationic hydrogels for controlled drug delivery at the extracellular matrix or anionic hydrogels in intracellular delivery [53].

The composition of the hydrogels can be tuned to match the desired properties. For instance, Zhang et al. [54] produced a thermosensitive hydrogel with pH-dependent drug release composed of chitosan, hyaluronic acid, and β-sodium glycerophosphate. They demonstrated that varying the percentage of hyaluronic acid (HA) can modify the gelation temperature, facilitating gel formation. Moreover, HA presence avoids the initial burst release of DOX, and its increase allows a sustained drug release. An acid-degradable nanogel was produced through copolymerization of alginate and an acid-labile monomer and then functionalized with collagenase (Co@ALG NG) with the purpose of enhancing its penetration in solid tumors, considering the high content of collagen in their extracellular matrix. These nanoformulations presented pH-triggered release of DOX with higher release rates from pH = 7.4 to pH = 5, reaching 49% at pH = 6.0 and 71% at pH = 5 after 120 h. Comparing to nanogels without collagenase and DOX alone, Co@ALG nanogels promoted the lowest tumor growth after the treatment, with a tumor growth inhibition of 77.6% [55].

Besides the pH sensitivity, the mucoadhesive properties were explored by Liang and co-workers [56]. The group developed a hydrogel based on chitosan-grafted-dihydrocaffeic acid (CS-DA) and oxidized pullulan. The authors have tested mechanical properties, such as rheological properties and mechanical strength, and DOX release from the hydrogel under different pH conditions, as well as the antitumor effect on colon tumor cells (HCT116 cells). The researchers demonstrated that at

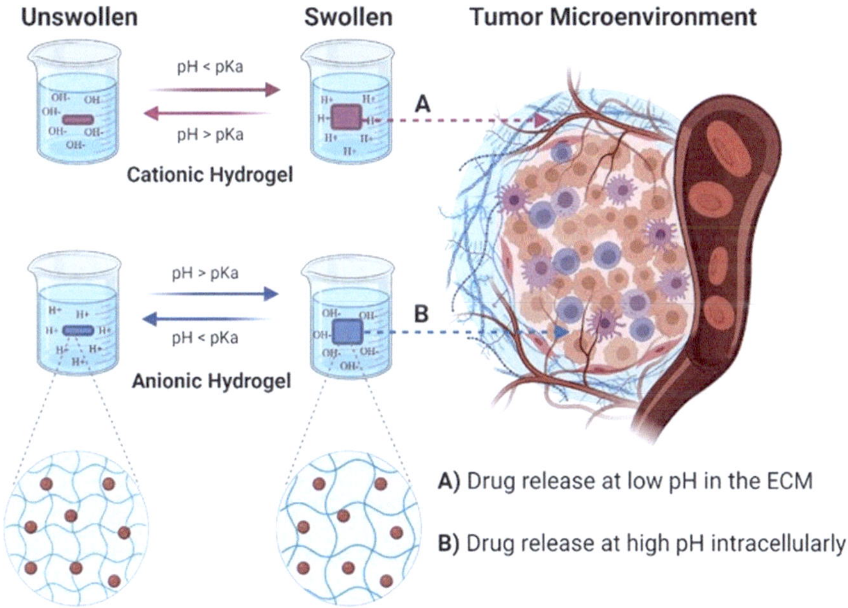

Fig. 1 Schematic representation of pH-responsiveness of cationic and anionic hydrogels at low pH in the extracellular matrix (ECM) of tumors and at high pH inside tumor cells, respectively. Reproduced from Andrade et al. [53], under the terms and conditions of the Creative Commons Attribution (CC BY) license. http://creativecommons.org/licenses/by/4.0/

pH = 5.5, the amount of DOX (87%) released from the hydrogel was much larger than the one released at pH = 7.4 (52%), indicating the pH-responsive drug release. On the other hand, the hydrogel showed quick in vivo gelation time and a strong adhesion to rat skin mucosa due to the addition of dihydrocaffeic acid, which constitutes another step further for the ideal localized delivery system. These studies show that changing the composition of a hydrogel can produce different properties which, together with the pH-responsiveness, lead to multifunctionality and more efficient hydrogels.

2.1.3 Photosensitive and Multimodal Hydrogels

Photosensitive hydrogels change their chemical or physical state when exposed to light (UV, visible or near-infrared (NIR)). The three main strategies for developing these hydrogels are photochemical cleavage at the crosslink site (irreversible process), photoisomeration (reversible process) and incorporation of photothermal agents that induce sol-to-gel or gel-to-sol transitions in thermosensitive hydrogels. In this section, some examples of these techniques are described.

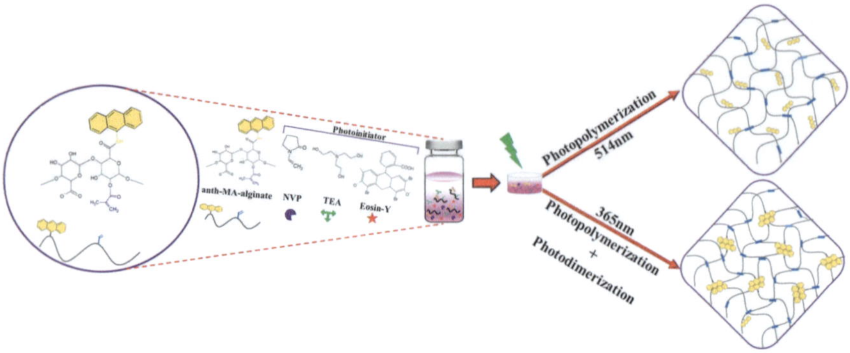

Fig. 2 Synthesis method of dual crosslinking of anthracene-methacrylated-alginate (anth-MA-alginate) through photopolymerization at 514 nm or simultaneous photopolymerization and photodimerization at 365 nm. Reprinted from Batool et al. [57], with permission from Elsevier. Copyright (2020)

For instance, Batool et al. [57] demonstrated that a new function can be added by simply modifying the structural composition of the initial polymers. They synthesized a pH-responsive alginate-based hydrogel conjugated with anthracene to give it a reversible crosslinking function. The reported method consists of a simultaneous dual crosslinking of methacrylated-alginate through free-radical photopolymerization of vinyl groups and photodimerization of anthracene molecules (Fig. 2). The incorporation of anthracene led to reversible control on the gel-sol states transition due to dimerization or dedimerization of anthracene groups at 365 nm and 254 nm, respectively, as well as better mechanical properties, increased drug release at pH = 5.0 (when compared to physiological pH), and compromised cancer cell viability without affecting healthy fibroblasts.

The insertion of photothermal agents such as gold nanoparticles is another type of strategy to provide photo-responsive hydrogels and it was used in the work of Xu and co-workers [58]. The researchers developed a NIR/pH-responsive hydrogel prepared through copolymerization of *N*-isopropylacrylamide and methacrylated β-cyclodextrin-based macromer (poly(NIPAM-co-MPCD)) for chemo-photothermal cancer therapy. The process is based on the conversion of NIR light into heat, by the gold nanorods (GNRs) incorporated in the hydrogel. The generated heat not only provides cancer hyperthermia therapy, but also promotes structural changes on the thermosensitive hydrogel, triggering the release of chemotherapeutic drugs in response to an incident NIR light (Fig. 3). Moreover, this hydrogel demonstrated pH-responsive release of DOX through cleavage of acid-labile hydrazone bond between DOX and adamantyl group (AD-DOX) in acidic environment. The performed in vivo studies showed on-demand release of drug by NIR irradiation and chemo-photothermal synergistic therapy.

The same strategy was used by Yang and co-workers [59] for local therapy of breast cancer. The developed hydrogel was composed of crosslinked hyaluronic acid and dopamine conjugated with selenium and loaded with indocyanine green (ICG)

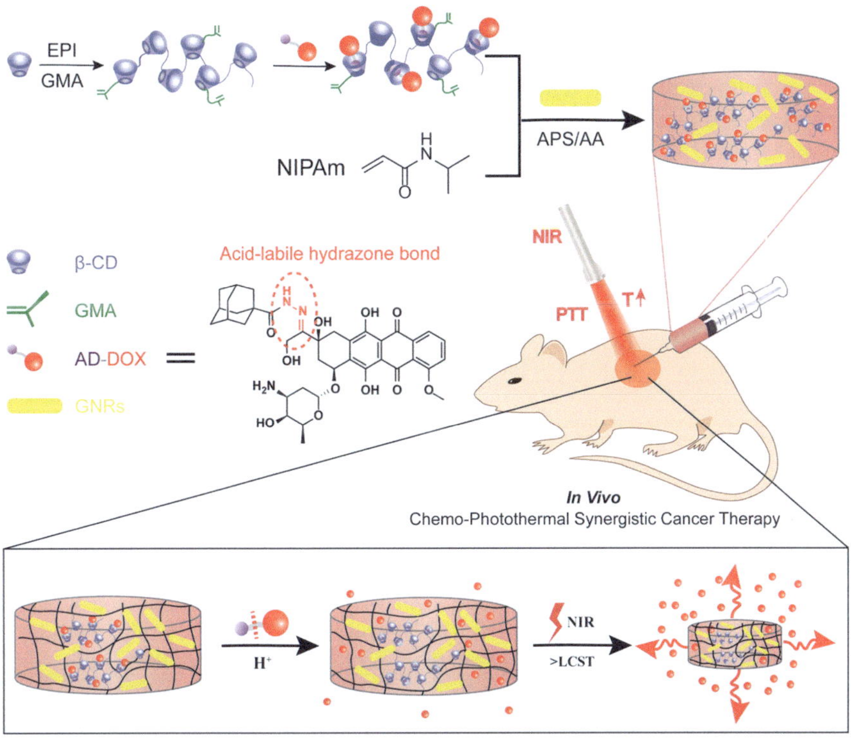

Fig. 3 Schematic representation of the hydrogel formation and its chemo-photothermal effect under NIR irradiation, as well as its pH-responsiveness. Reprinted from Xu et al. [58], with permission from American Chemical Society. Copyright (2017)

(HD/Se/ICG hydrogel). The Se moiety acts as pH modulator for polymerization agent of dopamine and as anticancer agent by pro-oxidant effect, while ICG is the photosensitizer that gives the photothermal function to the hydrogel. The performed in vivo antitumoral studies indicated constant tumor volume after 16 days with HD/Se/ICG hydrogel treatment.

Later on, the chemo-photothermal therapy was complemented with medical imaging, resulting in a novel imageable temperature-, pH- and NIR-responsive hydrogel developed by Chen et al. [60]. This hydrogel was composed of cellulose nanofibrils with embedded citric-stabilized Prussian blue nanoparticles, which acted not only as photothermal agents, but also as contrast agents for magnetic resonance imaging (MRI) and photoacoustic imaging (PAI). The biocompatibility assays in this study included acute systemic toxicity test, in vitro cytotoxicity test, skin sensitization, and skin irritation tests and could demonstrate the clinical applicability of the hydrogel. Moreover, the MRI and PAI measurements conducted on Michigan Cancer Foundation-7 (MCF-7) tumors allowed to conclude that the used nanoparticles acted as T1-weighted contrast agents and that stronger signals of PAI in the tumor site are obtained due to hydrogel's penetration in the tumor vessels. This, along with the significant tumor volume decrease promoted by the NIR irradiation and DOX triggered release, gives this hydrogel the possibility for being used in the clinical diagnosis and treatment of tumors.

It is also important to highlight the combination of localized and sustained chemotherapy with other modalities such as radiotherapy and photodynamic therapy. The work performed by Zheng et al. [61] is an example of combined drug delivery and radiotherapy in a single nanosystem. In this work, the drug sorafenib (SOR) was loaded in a thermosensitive hydrogel composed of PEG-PLGA-PEG with Lipiodol® Ultra fluid (LUF) and selenium nanoparticles (Gel-SOR-LUF-SeNPs) for hepatocellular carcinoma. The therapy was tested in HepG2 bearing-tumor mice, and the group treated with both Gel-SOR-LUF-SeNPs and X-rays had a decrease in tumor weight from 2.22 to 0.39 g in 21 days. The in vivo toxicity assays showed the biosafety of this treatment.

On the other hand, the use of photosensitizing agents can potentiate the development of new and bold strategies, such as the one reported by Chang and co-workers [62]. They developed a novel hybrid hydrogel based on polyethyleneglycol diacrylate (PEGDA) with methylene blue sensitized mesoporous titania nanocrystals for photodynamic therapy. The study consists of investigating the in situ formation of the hydrogel through photopolymerization around HeLa cells and its potential for photodynamic therapy. The obtained survival rate of HeLa cells treated with both the hydrogel and under laser irradiation of wavelength 660 nm was 14%, for a concentration of 15 μg/mL hydrogel precursor. Along with the optical and fluorescence microscopy assays, this study showed that the hydrogel precursor could effectively form a shell around the cancer cells, increasing the photoactivity effects of the trapped photosensitizers, generating highly reactive $^1O_2$ that kills cancer cells. These combined approaches in a single hydrogel platform open a variety of different strategies that can be used for local cancer treatment, and it is worth of continuous investigation.

2.2 *Peptide-Based Hydrogels*

Besides polymers, peptides are also used in nanomedicine, either for therapeutic or diagnostic purposes. They can be natural or synthetic, presenting several advantages, such as remarkable biocompatibility, low-cost and possibility of amino acid sequence tuning and functionalization [63, 64]. Peptide-based hydrogels are formed by peptide self-assembly, which involves supramolecular interactions—hydrophobic and van der Waals forces, electrostatic interactions, π–π stacking and hydrogen bonding—promoting the successive formation of hierarchical structures [65] upon triggering (Fig. 4). Considering the several triggering methods that can be used to induce gelation, which include temperature, pH, ionic force, enzymes, redox reactions, light and organic solvents, the reader is referred to references [65–68] for more detailed information on the preparation of supramolecular peptide-based gels. These materials can comprise different secondary structural motifs, forming β-sheets, β-hairpins,

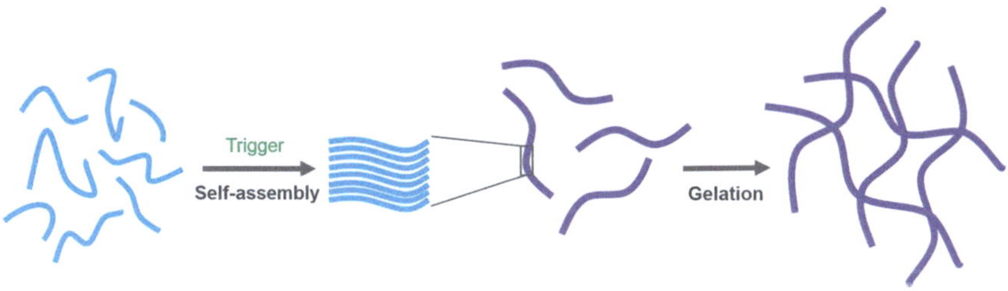

Fig. 4 Schematics of peptide self-assembly to give supramolecular hydrogels

and α-helices, which, naturally, impact their mechanical properties [69, 70] and, consequently, their possible applications [71].

2.2.1 Commonly Used Hydrogelators

Ultra-short and Short Peptides

Hydrogel-forming short peptides were first described by Gazit's group by observing that diphenylalanine (FF) self-assembled on nanotubes and that its functionalisation resulted in a hydrogel [71]. This type of peptides has less than 21 amino acids and is therefore more advantageous in terms of cost and production time compared to long sequences [72, 73]. Within this class, ultra-short peptides (less than 8 amino acids) are also included, sharing similar characteristics: they are amphiphilic, yield hydrogels based essentially on β-structures, and the gelation occurs by self-assembly [74]. Commonly, these peptides are N-capped with bulky aromatic groups to adjust their physicochemical properties, such as hydrophobicity and the ability to interact by π–π stacking, as Draper and Adams [75], and Draper et al. [76] reported. Interestingly, these authors state that even the simplest aspect in the preparation of this type of gels can affect the properties of the final system, so with the same short peptide they could obtain distinct fiber networks [75]. Also, an increasing number of studies demonstrate the suitability of short or ultra-short peptides to participate in multicomponent systems in order to reinforce them [75, 77]. This was demonstrated by Li et al. [77] who proved that mixing two ultra-short peptides with different capping groups results in a hydrogel with enhanced mechanical properties, which may constitute a strategy to adjust the characteristics and complexity of the fiber network.

Arginine-Glycine-Aspartate (RGD)-Containing Peptides

Peptides including the arginine-glycine-aspartate (RGD) motif in their sequence are among the most commonly used hydrogelators. That sequence is recognized by the $\alpha_v\beta_3$ integrin receptors, which are overexpressed in several tumors [63, 78, 79]. The corresponding hydrogels are very useful in selective drug delivery and tissue regeneration [63, 80]. Indeed, Mei and co-workers developed a series of peptides with the general form GFFYGRGDH$_n$ (with n between 0 and 2) that, in synergy with an anticancer drug, self-assemble to give different hydrogels. In addition to the rheologic improvement, there was an increase in the efficiency of the system in terms of releasing chemotherapeutic drugs and in inhibiting the growth of several tumor cell lines thanks to the integrin receptor-mediated drug endocytosis [81]. In a different work, a phenylalanyldehydrophenylalanine-based hydrogel was functionalized with a pentapeptide containing the RGD bioepitope [80]. This architecture significantly increased the adhesion and growth of mammalian cells compared to the non-functionalized hydrogel; furthermore, 3D cell growth was observed through scaffold cell invasion.

Arginine-Alanine-Aspartate-Alanine (RADA)-Containing Peptides

Arginine-alanine-aspartate-alanine (RADA) sequence, although not a bioepitope, promotes the formation and arrangement of β-strands into highly ordered fibers, mimicking the cellular microenvironment. Typically, this peptide appears as ac-(RADA)$_n$-CONH$_2$ and alternately comprises hydrophobic and hydrophilic residues, where the former is arranged on the opposite face regarding the latter (Fig. 5) [63, 82]. Due to their peculiar characteristics, RADA-containing peptides are frequently used in the production of hydrogels for therapeutic applications.

For example, Zhou et al. investigated the controlled release of the transforming growth factor β_1 from RADA$_{16}$-based hydrogels concomitantly with cell proliferation stimulation in rat bone mesenchymal stem cells, revealing a promising system for sustained release of biomolecules [84]. In another work, the ability of the hydrogel formed by the assembly of the Ac-(RADA)$_4$-CONH$_2$ peptide to encapsulate hydrophobic and hydrophilic drugs, DOX, and curcumin, was reported [85]. It is clear that this type of hydrogel-forming peptides is quite promising in enhancing the efficiency and local delivery of anticancer drugs.

Peptide-Polymer Conjugates

The polymers discussed above can also be combined with peptides to obtain hydrogels with specific properties that none of the components exhibits individually, for example, biocompatibility, elastic reversibility, or responsiveness to certain stimuli [69, 86]. In this context, a PEG hydrogel functionalized with laminin mimicking peptides was recently developed for cell encapsulation and delivery into the degenerative disc space [87]. Differently, Maslovskis et al. doped the FEFEFKFK-based hydrogel (where (F) is the hydrophobic phenylalanine, (E) is the hydrophilic and negatively charged glutamic acid, and (K) is the hydrophilic and positively charged lysine) with the poly(N-isopropylacrylamide) (PNIPAAm)-FEFEFKFK conjugate, which significantly increased the elastic modulus of the gel by additional hydrogen bonding [86]. In addition, the hydrogel acquired temperature sensitivity because of

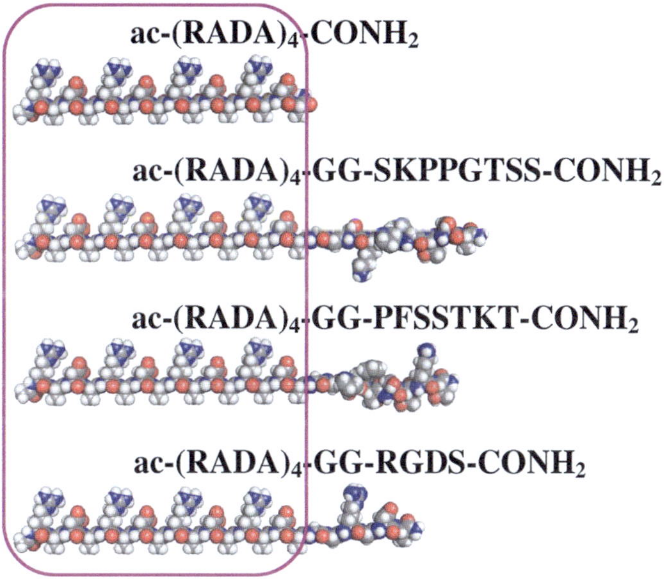

Fig. 5 Molecular models of Ac-(RADA)$_4$-CONH$_2$ hydrogel-forming peptides with and without functional motifs linked by a di-glycine spacer. Color code: gray for carbon, red for oxygen, blue for nitrogen, and white for hydrogen. Reprinted from Koutsopoulos and Zhang [83], with permission from Elsevier. Copyright (2013)

the thermoresponsive behavior of the polymer, which can be used as a trigger for drug release. Although these composites have not yet been widely explored for cell culture, they are very useful in the development of extracellular matrix mimetic materials, controlled drug/cell delivery systems, and platforms that promote biological mechanisms, such as angiogenesis [69, 88].

2.2.2 Stimuli-Responsiveness

Certain environmental circumstances, including pH, temperature, the presence of metal ions, enzymes, redox species, and light can trigger self-assembly or structural changes of peptides (Fig. 6), forming sophisticated supramolecular hydrogels as controlled therapeutic systems, for example, for drug delivery [65]. Indeed, related applications are enhanced when these materials take advantage of biological processes and in vivo conditions, since the encapsulation and release of therapeutic agents become controlled at a spatial and temporal level [65, 70, 89].

One of the most effective methods to induce gelation is the pH change, because protonation and deprotonation of acidic (such as –COOH) and basic (e.g., $–NH_2$) groups plays a huge role on their water solubility, as well as on the strength of intermolecular hydrogen bonds. Ultimately, this leads to the gelation of peptides if

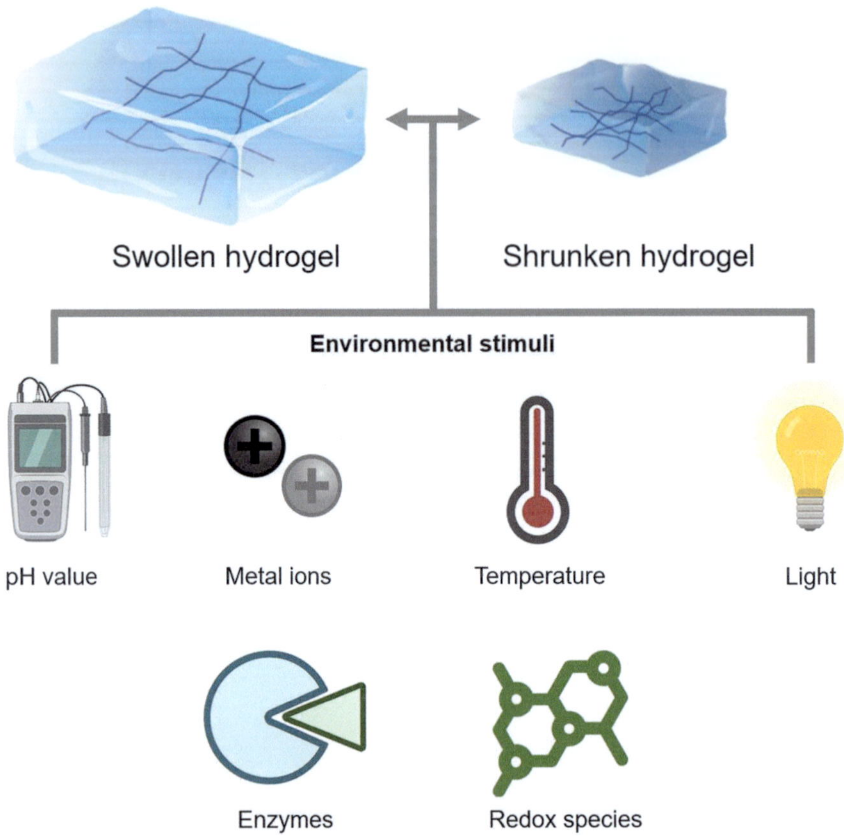

Fig. 6 Structural changes of hydrogels as a function of environmental conditions. Adapted from Shi et al. [90], under a Creative Commons Attribution 4.0 International License. http://creativecommons.org/licenses/by/4.0/

they are found at a concentration above the critical gelation concentration [89, 91]. It should be noted that complementary ionic peptides, such as RADA, are also pH responsive, considering that they have negative charge when $pH > pK_a$, where K_a is the acid dissociation constant, and $pK_a = -\log_{10}K_a$, and positive charge when $pH < pK_a$ [91]. In cancer therapy, this type of peptides is of great interest as they possess the ability to selectively deliver drugs into tumor cells (characterized by an acidic pH). Taking this in consideration, Raza et al. developed a hydrogelator composed of basic, acidic, and hydrophobic amino acids (FEFERFK). At physiological pH, the resulting gel was shown to be stable due to the absence of charge [92]. Interestingly, at lower pH, gel degradation rate increased due to electrostatic repulsion of glutamic acid residues. By incorporating the anticancer drug paclitaxel into this system, the accumulation of the drug and its retention on the target was verified, in addition to the increased activity [92].

Apart from this, metal ions are fundamental supports for the stability of several proteins, so some amino acids also interact with the presence of these entities, stabilizing and self-assembling [70]. As ions are an essential component of the body, developing supramolecular hydrogels triggered by certain metal ions is an excellent strategy to build platforms for controlled therapy [65]. Considering this, Tao and collaborators designed the EEEFFF peptide whose gelation occurs in the presence of zinc ions in the consequence of a chelation effect [93]. These forces act in synergy with the exposed –COOH groups, which are available to coordinate zinc. As organs, e.g., the prostate, are rich in these ions [94], such system is highly promising in sustained and localized drug delivery as a consequence of in situ encapsulation and gelation [93].

Differently, thermoresponsive peptide-based hydrogels undergo changes in swelling and sol–gel transition as a function of subtle changes in room temperature [89]. These changes deeply impact hydrophobic and hydrophilic interactions. In an elegant approach, De Leon-Rodriguez et al. described the synthesis of hydrogels consisting of an HβP peptide (PEELKLELKLEP or PEELELELKELKLEP, where P represents proline, E glutamic acid, L leucine, and K lysine), the substrate peptide sequence for metalloproteinase 2 (IPESLRAG), where I corresponds to isoleucine, S serine, R arginine, A alanine, while G corresponds to glycine), and the cell adhesion epitope RGD (GRGDSP-NH_2) with D corresponding to aspartic acid, and S represents serine [95]. The two lysine residue-HβP peptide showed an increased β-sheet content with increasing temperature, which points to an energy-dependent self-assembly. Thus, it is concluded that the sol–gel transition results from conformational changes (random coils to β-sheets). This hydrogel, as well as the one containing an HβP peptide with an extra glutamic acid residue, in addition to having their state of association governed by temperature, are degraded in a metalloproteinase 2 controlled manner, which makes them interesting candidates for drug delivery [95].

Light is a highly sophisticated factor in modulating the state of a hydrogel because it acts as a remote control with high spatial and temporal resolution [70, 89]. Here, peptides can be photodegradable—containing photolabile parts—or thermosensitive—including NIR-absorbing segments [91, 96]. Gels formed by the former allow drug release through fiber dissociation, and with the latter, the phenomenon occurs by dimerization/polymerization-induced volume variations. Taking this into account, Roth-Konforti envisioned that N-capping the phenylalanylphenylalanine peptide with the photolabile nitroveratryloxycarbonyl group would originate a stable hydrogel [97]. This hypothesis was verified, and irradiation with UV light caused degradation of the hydrogel matrix. The author has also tested encapsulating insulin and recorded a high efficiency of encapsulation. This system allows the release of large molecules that cannot be released by diffusion and the control of the amount of therapeutic agent released, since fiber breakage rate was found to be directly proportional to the period of gel exposure to light [97].

Finally, enzymatic reactions were pioneered in peptide gelation by Xu et al. [58]. Indeed, life at nanoscale, more specifically the formation of hierarchical structures, results from the action of enzymes and self-assembly processes [65, 70]. The main enzymes used as triggers are esterase, β-lactamase, β-galactose, glucose oxidase, peroxidase, chymotrypsin, thermolysin, thrombin, and metalloproteinase [89], so a wide range of supramolecular networks can be developed using this method. This can be useful when a hydrophilic segment is coupled to the hydrogelator and is enzymatically cleaved, self-assembling into a hydrogel [91]. Xu et al. [98] has just described that fact: the reaction of phosphatase to dephosphorylate fluorenylmethyloxycarbonyl (Fmoc)-tyrosine phosphate under basic conditions, allowing the adjustment of its solubility. In the absence of enzymes and in the presence of small redox species, such as glutathione, a similar effect can be produced in peptide hydrogelators containing redox-responsive groups, such as disulfide bonds. This type of hydrogels is excellent for the controlled release of anticancer drugs, since tumor cells have a markedly reducing environment. In this context, Wu et al. produced the anticancer drug-peptide conjugate HCPT-SA-FFE-ss-EE, in which HCPT corresponds to 10-hydroxy camptothecin, SA corresponds to succinic acid, FFE corresponds to phenylalanine-phenylalanine-glutamic acid, and "ss" represents the disulfide bond between two glutamic acid (E) residues [99]. The reduction of the latter by glutathione, which occurs abundantly in tumor tissues, induces self-assembly of the monomers due to decreased hydrophilicity, and culminates in a hydrogel with thixotropic behavior. With this, there was a slow and sustained release of HCPT resulting from two processes: diffusion of the drug from the gel matrix and cleavage of the ester bond between HCPT and the peptide chain [99].

It is clear that there are plenty of peptide gelation methods offering the possibility of several applications. Having a good understanding of the intermolecular interactions during self-assembly and gelation processes helps the rational design of hydrogelators and the use of the most physiologically convenient stimuli. In this sense, further studies are needed from a fundamental point of view on how these materials work, to enable the use of supramolecular hydrogels in the improvement of the life quality.

2.2.3 Therapeutic Applications

Peptide-based smart hydrogels have attracted a lot of attention in the last two decades due to the properties addressed so far. Bringing together so many advantages in a single platform makes them promising for several applications, either in therapy—e.g., drug delivery, wound healing, antimicrobial action, tissue engineering and 3D bioprinting—or in diagnosis—as bioimaging (Fig. 7). Indirect therapy (tissue engineering and 3D bioprinting) is widely explored. In this case, long-chain peptides, for example $RADA_{16}$ [64], are the most used, since they are the suitable media for cell growth, division, and differentiation, in addition to being able to work as functionalizable bioink in nanofabrication [64, 73]. Regarding diagnosis, these materials are excellent for conjugating or loading imaging agents, owing to their biocompatibility and ability to carry different materials [100]. For instance, hydrogel matrix can stabilize quantum dots, which can provide a platform for bioimaging, biosensing, and a means to modulate the materials optoelectronic properties [101–104]. As gelation can occur at the site of interest depending on a stimulus, peptides enable the rational design of hydrogelators for accurate and sensitive disease bioimaging in vivo [100]. Their utility indeed extends across several fields, but as this chapter focuses on therapeutic applications, direct therapeutic applications will be addressed in more detail below.

Local Drug/Bioactives Delivery

Often, peptide hydrogels, being formed by weak non-covalent interactions, are injectable due to the ability to recover their structure after breakage. This makes them ideal vehicles for the local administration of drugs and biomolecules (proteins, hormones, or genes) [65]. This application is intended to decrease the side effects of

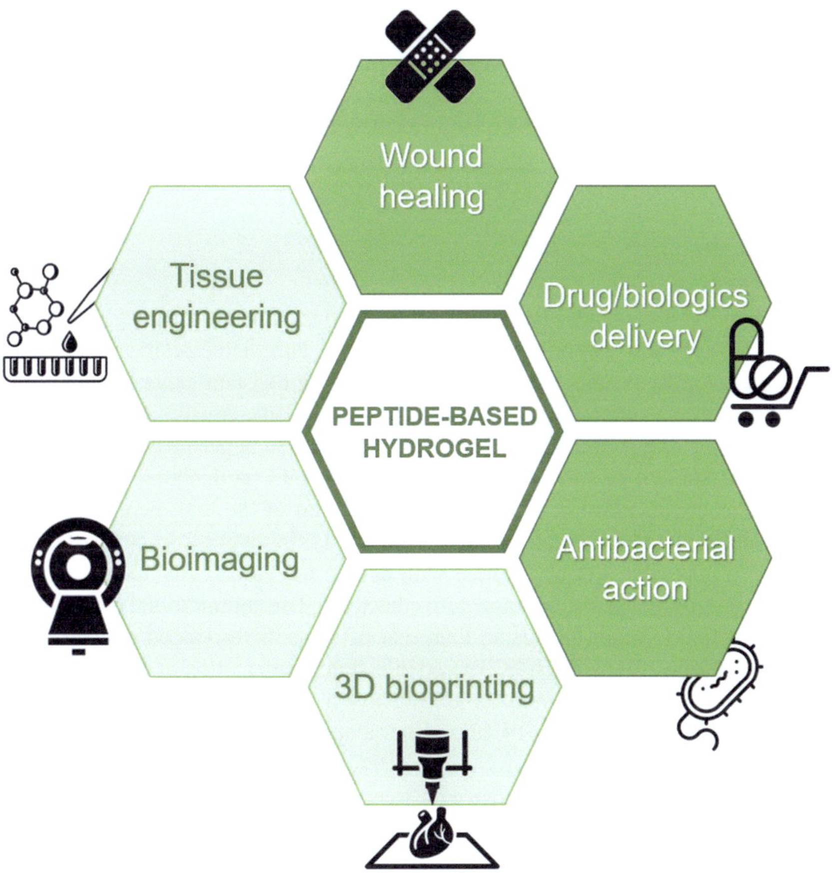

Fig. 7 Schematic of the main biomedical applications of supramolecular peptide hydrogels

systemic administration of therapeutic agents, but also to achieve sustained administration [64]. Thus, to avoid rapid degradation of the network by endogenous peptidases, D-amino acids or dehydroamino acids can be used [105, 106]. For example, Vilaça and co-workers synthesized three naproxen N-capped peptide hydrogelators whose C-terminal residue consisted of a ΔPhe, ΔAbu, or ΔAla residue [106]. The most hydrophobic one (ΔPhe) exhibited the highest elasticity and was tested for its ability to carry an antitumor thieno[3,2-*b*]pyridine derivative as drug model. The results showed that the payload settles in the hydrophobic cavities of the gel, at a distance of 2.3 nm from the fibers [106]. In addition to cancer therapy, the treatment of a number of diseases, e.g., diabetes, is envisaged [63]. Here, the therapeutic agent is often the well-known insulin protein, so a glucose responsive system is ideal for self-regulated macromolecule release [107]. Therefore, in an elegant approach, Li et al. developed a pH-responsive hydrogel loaded with the enzyme glucose oxidase (GOx) and insulin. This material swelled and shrunk as a function of the pH set by the enzyme according to the glucose concentration in the medium. The system showed to be injectable, biocompatible, and effective in regulating blood glucose levels over a long period of time [40]. However, despite the innovation, the transposition of these platforms to clinical practice requires their deep characterization in vivo and at a cellular scale [65].

Antibacterial Action

Peptide hydrogels can also act themselves as direct therapeutic agents [70]. Several natural and synthetic peptides present a fantastic antimicrobial mechanism that makes them highly promising for treating multi-drug resistant bacterial infections, suppressing, for example, the need for difficult and time-consuming sterilization procedures [72, 108]. One of those mechanisms is physical permeation and disruption of bacteria cell membrane, rather than binding to particular receptors [109, 110]. For example, Baral et al. describe the gelation of the long peptide Boc-11-aminoundecanoic acid-F at physiological pH [111]. In addition to being stable to proteolysis, injectable and compatible with human red blood cells, the gel demonstrated high activity against Gram-negative bacteria. The antimicrobial performance of this structure is more promising than that of some silver-loaded gels and does not entail its side effects, such as irreversible skin pigmentation and inflammation [111, 112]. In an approach against Gram-positive bacteria, another group developed a peptide construct consisting of central tetrapeptide connected to two antibacterial peptide sequences $–NH_2–(KIGAKI)_3–TPPG–(KIGAKI)_3–NH_2$, where T corresponds to threonine [113]. The balance between electrostatic and hydrophobic interactions and hydrogen bonds resulted in the formation of a supramolecular hydrogel when the precursor solution was exposed to changes in pH, ionic strength and temperature. The authors suggest that the ability to kill bacteria by the designed system is due to the cationic nature of the network which can disrupt the negatively charged outer membrane of bacteria by electrostatic interactions [113].

Wound Healing

Supramolecular peptide hydrogels for wound healing may or may not possess antimicrobial properties. Indeed, wound healing is a complex process, which involves tissue replacement, restoring skin function and maintaining homeostasis [114]. For this application, injectability is not such a critical factor considering the essentially topical administration. In order to develop primary coatings for burn injuries, Loo and co-workers used a moiety composed of 3–7 aliphatic amino acids to produce hydrogels that promote cell regeneration by encouraging cell migration and proliferation [115]. The Ac-LIVAGK-NH_2 (where V represents valine) hexamer allowed re-epithelialization and rapid wound closure, accelerating tissue regeneration, and reducing the risk of infection. It was also found to be much more effective than the frequently used Mepitel wound dressing [115]. For the efficient treatment of diabetic wounds, Carrejo et al. have developed a hydrogel based on the $K_2(SL)_6K_2$ peptide sequence [116]; the properties of this material, such as biocompatibility, possibility of cell infiltration, angiogenesis, neurogenesis, and hair follicle regeneration, make it an excellent candidate to replace other hydrogels—less effective—used in the clinic [116].

In fact, a great effort has been made in supramolecular chemistry to bring peptide hydrogels to clinical practice. However, despite the progress, there is still a long way to go, mainly through the theoretical understanding of peptide self-assembly and the development of biological simulation systems that allow the rational design of hydrogels and ensure their reproducibility.

Tissue Engineering

Peptide-based hydrogels can also be employed in tissue engineering, owing to the good mechanical and biological properties, and the ability to provide extracellular matrix mimicking environments, which endows the hydrogels suitable 3D scaffolds for several tissues [117–120]. The commonly employed strategy includes the chain modification and/or incorporation of a peptide sequence that enables a specific function, such as the RGD sequence for improved cell attachment and spreading properties [121–123]. For instance, in order to develop a 3D scaffold for the culture and maintenance of functional primary human hepatocytes, Macpherson et al. [122] fabricated a co-assembled Fmoc-FF/Fmoc-RGD hydrogel that even outperformed Matrigel (a commercial cell culture substrate) in the induction of cytochrome P450 expression. Rosa et al. [124] screened several multicomponent hydrogels based on Fmoc-FF and cationic peptides with a lysine residue at the C-terminus, which could strongly favor adhesion, spreading, and duplication of pre-adipocyte 3T3-L1 cells. Another strategy includes the use of hybrid hydrogels, bearing both peptides and polysaccharides as reported by Halperin-Sternfield et al. [125], who developed a Fmoc-FF/hyaluronic acid hydrogel that supports osteogenic differentiation of MC3T3-E1 pre-osteoblasts and facilitates calcium deposition, the implantation of which led to approximately 93% bone restoration.

3 Liposomes for Drug Delivery

3.1 Conventional Liposomes

In the 1960s, the work of Alec D. Bangham and colleagues led to the amazing discovery of lecithin dispersions comprising "spherulites composed of concentric lamellae" later named as liposomes (a term derived from the conjugation of two Greek words, *Lipos* meaning fat, and *Soma* meaning body) [126]. Initially, liposomes were used as artificial bilayers to study membranes' biophysical properties, including cell fusion, membrane pumps and antigen presentation [127]. Following Bangham's discovery, G. Gregoriadis established liposomal technology as a promising tool for the delivery of bioactive compounds in the treatment of diseases [128]. In 1995, DOXIL® was the first liposomal product to obtain regulatory approval by United States Food and Drug Administration (U.S. FDA) for the treatment of ovarian cancer [129, 130]. Nowadays, liposomes are considered an amazing nanosystem for widespread applications, including medical and non-medical fields, being one of the fastest growing scientific topics worldwide. Among the research areas focused on liposome formulations, oncology is in the top 10 list, corroborating the importance of this nanosystem in cancer treatment.

Liposomes are spherical vesicles composed of lipids that self-assemble in water, forming bilayers that enclose an aqueous volume [131]. The primary driving force for self-assembly of lipid molecules in water sits in the amphipathic character, owing to their hydrophilic head group (consisting of a phosphate-containing group attached to a glycerol molecule) and a hydrophobic tail (containing either saturated or unsaturated fatty acid hydrocarbon chains). Additionally, other weak intermolecular interactions including hydrogen bonding, coulombic interactions, π-stacking, and van der Waals forces contribute to self-assembling dynamics [132]. Although weak alone, the collective interactions of these forces result in stable supramolecular architectures of different sizes and shapes. The ratio between the volume of the lipophilic tail (v) and the product of the effective area of the head group (a) and lipid tail length (l_c), known as the packing parameter (p), can be used to predict the morphology of the resulting nanostructure [133]. With increasing the value of p, the aggregates goes from spherical ($p \leq 1/3$) and cylindrical ($1/3 \leq p \leq 1/2$) micelles to bilayer vesicles ($1/2 \leq p \leq 1$ and lamellar structures ($p = 1$). For packing parameters larger than 1 ($p > 1$), reverse micelles are formed. Other parameters, such as the chemical structure of the lipid, the ratio between lipid and water molecules, and solvent temperature, pH and salinity, are also important for the final size and shape of the final lipid aggregates [134]. Due to their bilayer vesicle structure, liposomes can encapsulate drugs within their aqueous core (for hydrophilic compounds) and/or in lipophilic bilayers (for lipophilic compounds). Thus, the pharmacokinetics of a drug can be controlled by the liposomal delivery system rather than the drug attributes. This is of major interest in oncology, as it is possible to make use of liposomes for the transport and delivery of toxic chemotherapeutic drugs [135, 136]. Classically, liposomes are classified in terms of their structural parameters, by the number of lamellae and vesicle size. Another important parameter is the transition temperature, which refers to the temperature at which phospholipids transit from the gel to the liquid-crystalline phase, affecting the fluidity of the lipids within the bilayer.

The primordial structure of liposomes consists in a simple lipid bilayer, but this nanostructure has been enriched with molecules capable to add or improve functions of administration or transmission of biochemical signals. This versatility offers the possibility to adapt the liposome properties toward specific applications. In accordance, the classification of liposomes has broadened, and nowadays it relies on a wide number of varying parameters, such as their membrane composition or surface sensitivity and encapsulated compounds [137]. In the following sections, an overview of the liposomes' preparation methods will be addressed, covering conventional and innovative techniques. The importance of post-production techniques for the effective clinical translation of liposomal formulations will also be discussed. A classification of the current liposome categories will be also presented.

3.2 Synthesis Techniques

3.2.1 Conventional Methods

Upon administration, the pharmacokinetic profile and fate of liposomes is connected to their physical properties. In cancer treatment, vesicle size is a key parameter for determining the in vivo circulation lifetime and the passive targeting to tumor tissues through the enhanced permeability and retention (EPR) effect. In fact, it is reported that nanosystems with sizes below 200 nm can accumulate in the tumor interstitium, as a result of the tumor rapid growth, that generates neovasculature with wide fenestrations and suppressed lymphatic drainage [138]. Hence, adjustment of the physical and chemical properties of liposomes is of major importance, and it can be done by controlling preparation conditions (temperature and pH) and formulation composition and concentration [139].

Different methods have been employed for liposome preparation. Conventional protocols include thin-film hydration, reversed phase evaporation, solvent-injection (Fig. 8), and detergent dialysis. In general, these methods consist of drying lipids from an organic solvent followed by dispersion in aqueous medium. Additional procedures, such as filtration or ultra-centrifugation, have been used to remove non-encapsulated drugs from the final liposomal product, while sonication or extrusion has been used to reduce their size [139, 140]. Traditional methods are associated with some technical problems, as well as scalability and production cost concerns. The several steps of manufacturing and purification processes to produce purified liposomes usually compromise the production cost. On the other hand, major technical issues such as the size control, weak reproducibility, and low drug encapsulation efficiencies can also influence the final product [139, 141]. Furthermore, many of the organic solvents and surfactants used in the vesicles preparation are hazardous and/or flammable, resulting in safety problems if not properly handled. On the other hand,

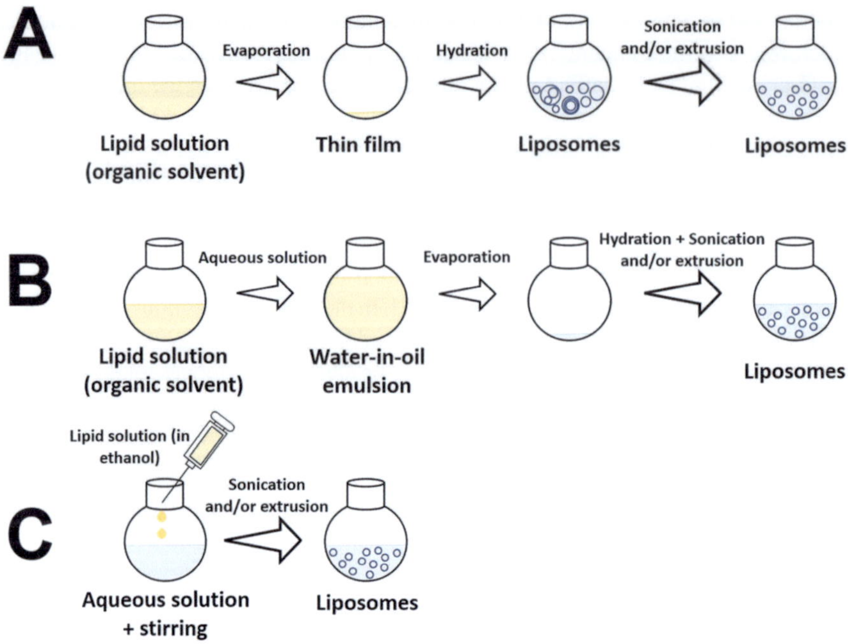

Fig. 8 Scheme of the major conventional methods for the production of liposomes. **A** Thin-film hydration; **B** reversed phase evaporation; **C** ethanolic injection

surfactants can affect membrane permeability, electrical conductance, and induce aggregation of intramembrane particles in human erythrocytes [142].

The encapsulation of drugs into liposomes can be performed by passive or active loading. The former involves co-current drug loading and liposomal formation. On opposing, in active loading, drug encapsulation occurs after liposomes synthesis by co-mixing of the vesicle suspension with a drug solution. Typically, compound internalization is driven by a transmembrane pH gradient. Hydrophobic drugs have shown to be effectively encapsulated into liposomes by passive loading, depending on the solubility of the compound in the membrane. In its turn, water-soluble drugs are more efficiently trapped inside liposomes through active loading. Ideal loading efficiencies have been reported for amphipathic weak basic ($pK_a \leq 11$) or weak acidic ($pK_a > 3$) drugs [143].

3.2.2 Progress in the Solvent-Injection Method

The solvent injection is one of the most commonly used methods because of its simple procedure. It allows to obtain a narrow distribution of small liposomes ($\approx$100 nm), simply by the injection of a lipid organic solution into a warmed aqueous phase (above lipid transition temperature) without extrusion or sonication [144]. The organic phase typically consists of ethanol or ether. The advantage of the ether injection over the ethanolic one relies on the low boiling point of ether, which vaporizes upon contacting with warm water, resulting in the removal of organic solvent from the final product [145]. Ethanol removal, due to its higher boiling point (around 78 °C), usually requires an extra step of rotary evaporation under reduced pressure.

The continuous development of methods based on the modification or improvement of conventional ones is an ongoing field of research. The goal is to overcome the traditional methods' concerns and reach an ideal methodology for large-scale production. New technologies such as the cross-flow injection, membrane contractor and microfluidic channel method are advancements of the conventional solvent-injection technique which allow the continuous and scalable fabrication of liposomes for pharmaceutical applications [146–148].

The microfluidic channel method consists in the mixture of lipid solution and water phase under highly controlled fluidic conditions at the micrometer scale [149]. A facile apparatus consists of a multichannel syringe-pump, with one of 1 mL for the lipid phase and another of 10 mL for the aqueous solution, linked through a microchannel connector. The most common devices for microfluidic production are coaxial assemblies of glass capillaries and planar microfluidic devices [150]. More advanced devices based on microfluidic chips of different types and geometries have also been developed for a more precise control over the fluids transport [151]. Among the advantages of the microfluidics method, particle size control of produced liposomes by the interfacial diffusion of the two phases and the reproducibility are the two more valued. Drawbacks include the low production rate, the need for expertise in microfabrication, as well as high manufacturing expenses that hinder its technological use on an industrial scale [149].

The cross-flow injection setup module consists of two tubes welded together forming a cross-flow connector, one for the polar phase and other for the ethanol/lipid solution which is linked to the cross-flow with a nitrogen pressure device. This setup allows the aqueous solution (buffer) to be pumped with a peristaltic pump to the cross-flow tubes, while the lipid solution is regularly pumped by the adjustable nitrogen device [152]. Experiments have demonstrated that a minimum of buffer flow rate is required to obtain batch homogeneity. Also, it has been shown that for constant injection velocities, narrow-sized vesicles are obtained even for a high lipid concentration injected [146]. Hence, one of the major advantages of the cross-flow technique is its possible use for a high lipid concentration. Another advantage associated with this method relies on the process itself, which is simple, scalable, and sterile. The main disadvantage reported for this methodology is the presence of residual organic solvents that can lead to stability problems.

Finally, the ethanol injection principle connected to a membrane contactor is a humble procedure aiming at large-scale liposomes production using a simple, rapid, and scalable process [153, 154]. In this apparatus, a lipid phase is pressed, under nitrogen flow, through a Shirasu porous glass (SPG) and the aqueous solution flows tangentially to the SPG membrane. The tubular Shirasu porous glass is made of a special type of glass prepared from volcanic ash through leaching process and is the key constituent of this device, as it can be prepared with specific dimensions and pore sizes [155]. The setup configuration allows controlling the mean size and size distribution of liposomes by varying the size of SPG pores, lipid concentration, aqueous flow rate, and organic phase pressure. Reported advantages are the accomplishment of homogenous liposomes with high encapsulation efficiency for hydrophobic drugs. On the other hand, the high-cost operation is a weakness that can compromise its use on an industrial scale.

3.2.3 Developments in Preparation Techniques

The recent advances in preparation methods mainly focus on the large-scale production of liposomes of small size with distribution control, stability, and high encapsulation of bioactive materials [156, 157]. Has and Sunthar performed an extensive and comprehensive review on preparation techniques of liposomes, with more than 20 different methodologies being discussed [141].

Supercritical fluid (SCF) method has emerged as an alternative to conventional methodologies for liposomes preparation. In this method, the lipids are dissolved in the SCF under high pressure, followed by depressurization in a low-pressure chamber and simultaneously mixing with the aqueous phase to produce the liposomes [158]. The most used SCF is the carbon dioxide, for its low critical pressure (73.8 bar) and temperature (31.1 °C), in addition to the fact that it is inexpensive and non-toxic [159]. Upon depressurization, CO_2 returns to the gas state, preventing the existence of residual solvents in the final product, typically associated with toxicity issues. This, together with the capability to control particle size, the in situ sterilization, and the possibility for large-scale production, confers serious advantages on SCF methods [160]. However, the complexity of this process is a major drawback. There are several variant types of SCF technology that can be classified in three broad categories, depending on whether the SCF is being used as a solvent, anti-solvent, or as assisted spray drying [161]. The rapid expansion of supercritical solutions (RESS), gas anti-solvent process (GAS), supercritical anti-solvent (SAS), aerosol solvent extraction systems (ASES), and supercritical reverse-phase evaporation (SCRPE) are the most common types of SCF technique.

Another innovation on liposome preparation is the dual asymmetric centrifugation (DAC) method, used for the preparation of lipid nanoparticles since 2008 [162]. In a typical procedure, liposomes are obtained by homogenizing a concentrated blend lipid solution and the aqueous solution, resulting in a viscous vesicular phospholipid gel, that is later diluted to obtain conventional liposome dispersions [163]. The key advantages come from the simple, fast, and reproducible process, which forms small liposomes with high encapsulation efficiency for hydrophilic compounds. On the other hand, the small volume limitation is a major drawback for scale-up production.

3.2.4 Post-processing Techniques

Apart from the development of new synthesis methods, post-processing techniques are also a concern to overcome instability issues of liposomes in aqueous solution. Specifically, the occurrence of self-aggregation, coalescence, and flocculation during storage can compromise liposomes composition [164]. The storage of liposomes in a dry state can be an attractive approach for long-term stability, bioactivity, and bioavailability of liposomes [165]. The process consists of the removal of water after liposomes formulation to generate dry vesicle particles. The dry powder form is convenient for storage and distribution, providing optimal flexibility and stability during sterilization.

The most used drying techniques are freeze drying, spray drying, and spray freeze drying [165]. Among all, freeze drying, also known as lyophilization, is the most employed. It consists of a first freezing step of a liposome-cryoprotectant mixture, followed by two drying steps (the first one for sublimation and the second for the removal of the residual moisture content) [166]. The major drawbacks of freeze drying are the changes in liposomes size, the membrane integrity damage, and the leakage of loaded drugs. To avoid these phenomena, the addition of a lyoprotectant agent to liposomes formulation has been reported. Examples of lyoprotectants are carbohydrates, proteins, and several alcohols [167]. Moreover, freezing rate, liposome preparation technique, bilayer composition, and residual moisture content have been shown to play an important role in the stability of the lyophilized liposomes.

Another technique is the spray drying, known for its ability to join both particle formation and drying in a continuous one-step operation [168]. This is one of the most suitable methods for scale-up of drug delivery systems. The process is based on the atomization of liquid feed into fine droplets and evaporation of the solvent using a hot drying gas. It consists of a four-stage process that begins with the liquid feedstock preparation, followed by the atomization of the feed into a spray through a nozzle and contact with the hot drying gas. Then, particles are formed by evaporative mass transfer of the liquid from droplets into the drying gas, and the final step is separation of the dried product from the gas [169]. Some strategies should be adopted to avoid liposomes aggregation that can lead to an increase in particle size and also escape of encapsulated compounds. The most common approaches are the use of charged biopolymers as protectants for electrostatic repulsion between liposomes or the coating of liposomes with chitosan. In addition, parameters such as temperature, biopolymer type, biopolymer amount, and spray gas flow can influence the characteristics and morphology of the final powder [170, 171].

An alternative approach in preserving liposomes in a solid form is by using the spray freeze drying. This technique consists of the combination of both freeze drying and spray drying, merging the advantages and compensating limitations to provide dry products with high stability. The three main stages are atomization, spray freezing and freeze drying. In the first stage, the prepared feed solution is converted into spherical droplets through an atomizer. Then, the spray freezing and freeze-drying stages consist of solidification of the droplets with a cryogenic agent and sublimation of the ice at low temperature and pressure conditions [172]. Ultimately, the spray freeze-drying process offers liposome particles with important properties, such as spherical morphology, high porosity, and high solubility, which are important characteristics for reconstitution without agglomeration after rehydration [173].

3.3 Types of Liposomes

The composition of liposomes is crucial for their safety, stability, and efficiency. Overall, the ultimate goal in liposome development for drug delivery is to increase the bioavailability and reduce the toxic side effects of the pharmaceutical drugs by releasing them at a specific site. Liposome formulations are designed considering three main components, namely the lipids (and other molecules in the bilayer), functionalization molecules, and encapsulated drugs [174]. The choice of lipids is done taking into account their charge and transition temperature. The most common lipids used are phosphatidic acid (PA, anionic), phosphatidylethanolamine (PE, zwitterionic), phosphatidylcholine (PC, zwitterionic), phosphatidylserine (PS, anionic), phosphatidylglycerol (PG, anionic), cardiolipin (CL, anionic) and phosphatidylinositol (PI, anionic) [175]. In its turn, the transition temperature is regulated by the length of the hydrophobic chain and its degree of saturation. Cholesterol is normally part of the bilayer composition for stability purposes, owing to its capacity to avoid interaction with proteins (albumin, transferrin, macroglobulin, and high-density lipoproteins) [176].

The bilayer formulation, together with the functionalization molecules and loaded compounds, determine the classification of liposomes in 5 major groups: (i) conventional liposomes; (ii) long-circulating liposomes; (iii) ligand-targeted liposomes; (iv) cationic liposomes; and (v) sensitive liposomes. Conventional liposomes were the first generation of liposomes and consist in a humble formulation composed of a lipid bilayer containing cholesterol and enclosing an aqueous volume. These liposomes are effective in reducing the toxicity of encapsulated drugs by improving their pharmacokinetics and biodistribution compared to free drug performance. Yet, the opsonisation of plasma components and uptake by reticuloendothelial system macrophages make the system to be rapidly eliminated from the bloodstream [177]. Consequently, long-circulating liposomes were introduced to enhance stability and circulation lifetime. Polyethylene glycol (PEG) was found to be effective in the design of sterically-stabilized liposomes. This steric barrier has been shown to prolong circulation time in the human body to 45 h in some cases, depending on the properties of the coating polymer [178]. In fact, the molecular weight and length of the polymer, as well as the amount grafted on the liposome surface, have a strong influence on the circulation time. PEG molecular weight and concentration determine if its conformation on liposomes surface is of mushroom or brush type. Recently, tailoring PEG into a brush conformation has been shown to best reduce the nonspecific uptake [179]. Alternative compounds have been studied to prolong the circulation of liposomes. In this regard, the modification of liposome surface with chitosan has been shown to produce a protective shell that confers long lifetime to the nanosystem and increases the stability of drug release [180].

The lack of specificity toward disease tissues led to the development of the next generation of liposomes, the ligand-targeted liposomes. The idea is to deliver drugs to target cells having specific overexpressed ligands at the disease site. There are different types of targeting moieties, for instance, antibodies, peptides/proteins, and carbohydrates [181]. Among all, antibodies or antibody fragments are the most widely used for their high specificity to target antigens. For that, a new subclass of ligand-targeted liposomes was created, the so-called immunoliposomes, an antibody-conjugated vesicle designed to selectively target antigen-expressing cells [182].

Cationic liposomes can form stable complexes with negatively charged therapeutic agents, such as nucleic acids, producing lipoplexes. The driving force for the complexation of cationic liposomes with Deoxyribonucleic acid (DNA) into well-organized structures has two main contributions. The first are Coulombic attractions between cationic liposomes and anionic nucleic acids. The second and major one is the entropy gain resulting from the release of the counter ions from both cationic lipid and DNA to partially neutralize each other [183]. Cationic liposomes were among the first nanosystems explored to deliver nucleic acids, instead of the classical viral vectors. The positively charged liposomes were found to be capable of simultaneously forming complexes with negatively charged nucleic acids and with the anionic sulfate groups of cell-surface proteoglycans on mammalian cells, triggering not only uptake by cells, but also the expression of the delivered nucleic acid [184, 185]. Since then, this class of liposomes has been of major interest in the fight against cancer, through gene therapy. Apart from plasmid DNA, other nucleic acids such as messenger Ribonucleic acid RNA (mRNA), small-interference RNA (siRNA), microRNA (miRNA) and antisense oligonucleotides (ODN) have been used for the preparation of these nanosystems. Depending on the type of nucleic acid contained in the lipoplexes, different approaches are possible [185]. For instance, DNA or mRNA lipoplexes have the ability to supplement tumor suppression genes (e.g., p53 and mda-7/IL-14) [186], while siRNA lipoplexes are ideal to repair or silence oncogenes [187]. In another approach, it is possible to stimulate the patient's immune system with mRNA lipoplexes cancer vaccines encoding tumor antigens or genes encoding immunomodulatory proteins [188]. The same approach has been recently used in some of the COVID-19 vaccines.

Sensitive liposomes have emerged as a strategy to enhance local delivery using the specific characteristics of disease microenvironment. Owing to the uncontrolled cell proliferation, tumors have a peculiar environment and metabolic network that result in specific physiognomies, such as acidic pH, overexpression of proteins and enzymes and high levels of oxidation/deoxidation. These unique features are used as stimuli to trigger controlled release of drugs in tumors, either by an endogenous trigger (pH, redox potential or enzymatic activity) or by exogenous trigger through application of an external stimulus (temperature, light, magnetic field, and ultrasound) [189, 190]. Many different liposomal formulations have been developed to explore a single stimulus and the combination of two or more stimuli, and, among all, pH and temperature-sensitive liposomes are the most studied. The controlled release of loaded drugs from pH-sensitive liposomes, at acidic pH, is promoted either by the increase in membrane destabilization or membrane fusion (intracellular activation), or by the increase of membrane permeability to release drug cargo (extracellular activation). Typically, this class of liposomes can be obtained by the use of pH-sensitive lipids or by the inclusion of pH-sensitive polymers in the formulation. pH-sensitive lipids consist of weakly acidic amphiphilic molecules with a polymorphic phase behavior as a function of its protonation state around its pK_a, owning a lamellar stable phase at $pH = 7$ and an inverted hexagonal phase (H_{II}) at $pH = 5$ [191]. The pH-sensitive polymers can be either natural or synthetic molecules that undergo structural transition/modification or own specific properties such as surface charge and bond break at acidic pH. In its turn, temperature-sensitive liposomes undergo a melting phase transition from gel to liquid-crystalline phase, inducing an increase in the bilayer permeability and consequent drug release. For this purpose, the lipid dipalmitoylphosphatidylcholine (DPPC) is one of the most widely used, owing to its clinically relevant transition temperature of 41 °C. This temperature is a few degrees above the physiological one and in the range of mild hyperthermia treatments [192]. Heat can be induced by different exogenous triggers; among them, light and alternating magnetic fields (AMF) have been the most reported [190]. For that, plasmonic and magnetic nanoparticles have been included in liposomes and used as temperature-inducing agents, under irradiation or AMF application, respectively [190, 193, 194].

Other structures derived from liposomes have been considered in the literature as potential models for drug delivery, such as vesosomes. The vesosomes are multicompartment structures consisting of heterogeneous aggregation of vesicles, in which a large liposome hosts multiple small vesicles. Owing to the possibility of encapsulating different drugs in the small vesicles and the presence of secondary barriers to prevent early leakage of encapsulated compounds, this type of structure is potentially powerful for sustained drug release [195].

As a consequence of the several types and potential formulations, liposomes have provided several solutions for biomedical applications, such as tissue engineering and bioimaging. For instance, drug-loaded liposomes have been commonly combined with scaffolds to minimize burst drug release, widen the choices of drug types, and help maintain a concentration in situ for a long period, besides other functionalities and/or delivery of growth/differentiation factors to promote regeneration in defect areas [196–198]. In addition, liposomes' ability to encapsulate inorganic particles has provided a suitable means for targeted delivery of imaging agents, such as quantum dots, whose stability and biocompatibility have also been improved [199–202]. The former advantage has been explored to further advance the theranostic field through encapsulation of magnetic nanoparticles, which is discussed in detail in the subsequent subsections.

4 Progress in Theranostics with Magnetic-Based Systems

4.1 Magnetic Liposomes

Nanomedicines and smart delivery systems are part of the second generation of drug delivery systems and include, among others, dendrimers, carbon nanotubes, liposomes, micelles, and gene nanocarriers. The first FDA-approved nanodrug (Doxil®) [203], an anticancer liposomal formulation encapsulating Doxorubicin, emerged in 1995. Since then, liposomes have been gaining recognition as biomimetic nanostructures for encapsulation and delivery of bioactive agents, with their efficiency proven by several lipid formulations FDA-approved for the treatment of diseases. In addition to biocompatibility, biodegradability, and ability to encapsulate hydrophilic and hydrophobic drugs, their clinical applicability can be manipulated by varying preparation strategies, lipid formulations, surface modifications, and conjugation with other types of organic and inorganic materials.

In recent years, superparamagnetic iron oxide nanoparticles (SPIONs) have also opened a new horizon of applicability in biomedical sciences, both for diagnosis and therapeutics. Their inherent physicochemical properties give them the potential for tissue engineering, magnetic separation, in vivo cell tracking, magnetic hyperthermia, iron detection, magnetic resonance imaging (MRI), stem cell tracking, and targeted drug or genes delivery (Fig. 9). There are already FDA-approved formulations containing magnetite nanoparticles, for instance, Feridex®/Endorem®, Nanotherm®, Feraheme™/ferumoxytol, Umirem®, and GastroMARK™ [204].

Thus, the conjugation of lipids with SPIONs can give rise to supramolecular hybrid structures that retain the properties of the materials individually, and new combined and synergistic therapeutic functions. This section will focus on the conjugation of inorganic magnetic nanoparticles (MNPs) with liposomes, which results in the formation of magnetoliposomes. Depending on the location of the nanoparticles in the lipid structure, magnetoliposomes can be categorized into: (A) aqueous magnetoliposomes (AMLs), when nanoparticles are dispersed in the aqueous pool of liposomes; (B) solid magnetoliposomes (SMLs), when a single particle or a cluster of particles is coated with a lipid membrane, without internal aqueous component; (C) liposomes with bilayer-embedded nanoparticles; and (D) liposomes with surface-bound nanoparticles (Fig. 10).

Since magnetoliposomes were first described by Margolis et al. [205] and De Cuyper and Joniau [206], the proposal has been to bring a multifunctional nanomaterial with combined properties of the two systems. The premise of using magnetoliposomes as a controlled delivery strategy is based on the cargo-controlled delivery and release from liposomes under magnetic stimuli. The magnetophoretic mobility of magnetoliposomes adds to conventional liposomes the functionality of directing the cargo to a target area when a magnetic gradient is applied. On the other hand, by applying an alternating magnetic field, thermally activated processes are induced, which are associated with the controlled release of the encapsulated cargo, either by increasing the permeability or by altering the diffusiveness of the lipid bilayer.

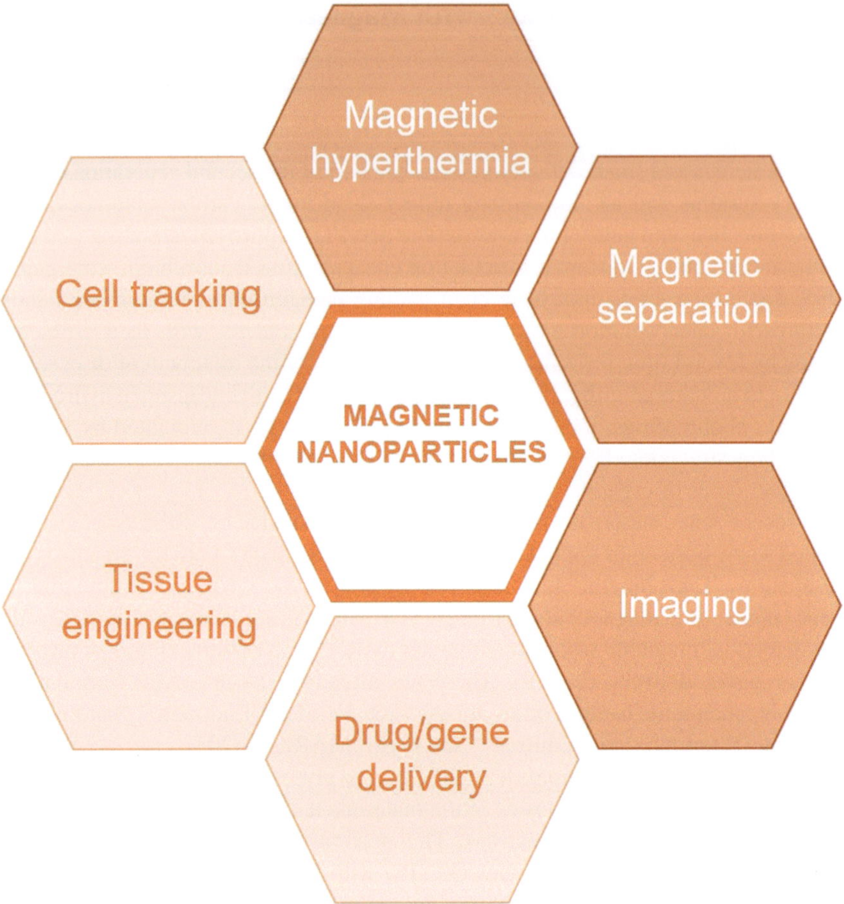

Fig. 9 Schematic of the main biomedical applications of magnetic nanoparticles

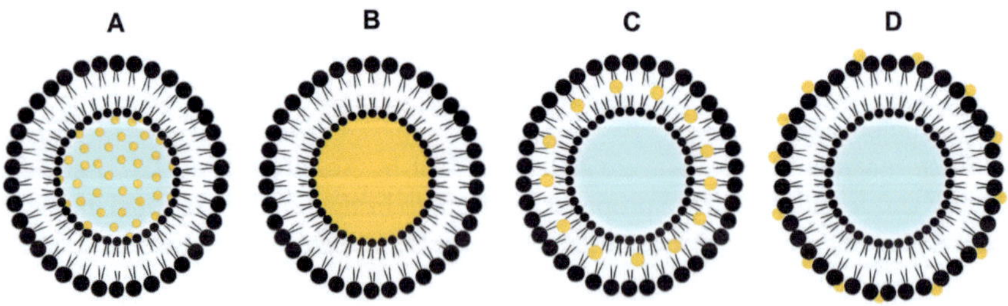

Fig. 10 Schematic representation of **A** aqueous magnetoliposomes (AMLs), **B** solid magnetoliposomes (SMLs), **C** liposomes with bilayer-embedded nanoparticles, and **D** liposomes with surface-bound nanoparticles

The possibility of choosing the lipid composition of bilayers over an extensive range of lipids with different functional characteristics allows enhancing the applicability of magnetoliposomes in several biomedical applications and offers them additional targeting and controlled delivery strategies.

At the end of this section, Table 1 summarizes the different types of magnetoliposomes designed and adjusted to specific properties that have been developed recently.

4.1.1 Fabrication Strategies of Magnetic Liposomes

Currently, the fabrication of magnetic liposomes can be achieved with several strategies that are classified as either in situ or ex-situ. In the former method, the phospholipid vesicles are used as nanoreactors for the synthesis of magnetic nanoparticles in the inner aqueous cavity [207], while in the latter method the composite is initially synthesized and then assembled with phospholipids/liposomes to afford the magnetic liposomes. For an in-depth discussion on the preparation methods, the reader is referred to the reviews [208, 209].

Regarding the ex-situ method, the commonly employed techniques, such as ethanolic injection (EI) [210–214], thin film hydration (TFH) [215, 216], and reverse-phase evaporation (RPE) [217, 218], are similar to the fabrication of liposomes, in which an aqueous suspension of nanoparticles is employed instead of an aqueous solution. Nonetheless, as the liposomes' internal volume is a small fraction of the total sample volume, several parameters can affect the loading of the nanoparticles, such as the buffer type, concentration, pH, liposome composition, components' stoichiometry, and the payload [215, 219]. Besides, after production of the liposomes, the non-encapsulated nanoparticles need to be removed, which can be achieved through size exclusion chromatography, centrifugation, and salt-induced aggregation [220]. Hydrophobic nanoparticles can also be employed to produce magnetic liposomes, which can lead to the liposomes with bilayer-embedded nanoparticles architecture [221]. The inclusion of nanoparticles has been demonstrated to improve the bilayer stability and reduce the spontaneous leakage of the loaded cargo into the inner aqueous cavity [222], but the development is limited by the thickness of the membrane, constraining the range of particle sizes that can be inserted [223, 224].

4.1.2 Advances in Biomedical Applications Using Stimuli Strategies for Enhanced Drug Release

Alternating Magnetic Fields (AMF)-Triggered Release

The exposure to a high frequency (HF) in the range (100 kHz–1 MHz) alternating magnetic field (AMF) of non-interacting magnetic nanoparticles, whose magnetization varies linearly with the applied magnetic field, leads to the heat generation due to the delay of the magnetic nanoparticles' relaxation time to the AMF [225, 226],

Table 1 Summary of magnetoliposomes types, lipid compositions, nanoparticles, encapsulated compounds, and triggers used for the aimed biomedical applications

Magnetoliposomes type	Lipid composition	Nanoparticles	Encapsulated compound	Trigger	Aimed biomedical application	Refs.
AMLs	DPPC:Chol (5:1) DPPC:DSPC:Chol (10:5:3)	Dextran-modified SPIONs (resovist)	Carboxyfluorescein	AMF	Cancer therapy (not specified)	[235]
AMLs	DOPC	Citrate-modified SPIONs	Carboxyfluorescein	AMF	Cancer therapy (not specified)	[236]
Liposomes with bilayer-embedded nanoparticles	Soybean phosphatidylcholine (PC)	SPIONs	Erythromycin	AMF	Bacterial infections	[238]
Liposomes with bilayer-embedded nanoparticles	DPPC POPC DSPC MPPC[a]	Oleic acid-coated SPIONs	Calcein	AMF	Not specified	[221]
Liposomes with surface-bound nanoparticles Liposomes with bilayer-embedded nanoparticles AMLs	DPPC	DMSA[1]-coated SPIONs Oleic acid-coated SPIONs APS[2]-coated SPIONs	Doxorubicin	AMF	Cervical cancer and triple-negative breast cancer	[239]
AMLs	PC:Chol:carboxymethyl dextran (55:40:0.5)	Citrate-coated SPIONs	Doxorubicin	LF-AMF	Human neuroblastoma	[241]
SMLs	DPPC: Chol (4:1)	Oil-soluble SPIONs	Doxorubicin	NIR	Human breast cancer	[251]
AMLs	DPPC:Chol:DSPE-PEG2000 (80:20:5)	Gold-coated SPIONs	Doxorubicin	NIR	Breast cancer	[253]
SMLs	DOPG[b]	Nickel ferrite/gold core/shell nanoparticles Nickel ferrite nanoparticles Decorated with Gold	Not specified	NIR	Cancer therapy (not specified)	[212]
AMLs	DPPC/DDAB[c]/CH	Citric-acid-coated SPIONs	Not specified	AMF NIR AMF-NIR	Glioblastoma	[255]

(continued)

Table 1 (continued)

Magnetoliposomes type	Lipid composition	Nanoparticles	Encapsulated compound	Trigger	Aimed biomedical application	Refs.
Methotrexate-functionalized Liposomes with bilayer-embedded nanoparticles	DPPC:Chol:SA:DSPE-MPEG2000:DSPE-PEG2000-MTX (67:17:13:1:2)	Oleic acid-coated SPIONs	Doxorubicin	AMF-NIR and folate receptors	Human cervical carcinoma	[258]
Immuno (anti-LIBS)-AMLs	POPC:DPPC:Chol:m-PEG_{2000}DPPE[d]	SPIONs	Not specified	GPIIb/IIIa receptor of platelets	Targeting atherothrombosis	[260]
RGD-functionalized AMLs	DMPC[e]:DMPG[f] (9:1)	SPIONs	Not specified	$\alpha_V\beta_3$ integrin	Targeting tumor cells and neovascularization	[261]
Cetuximab-functionalized	DPPC:MPPC:DSPE-mPEG-2000	Acid-coated SPIONs	Doxorubicin	NIR and EGFR	Breast cancer	[262]
Glucose-functionalized liposomes with bilayer-embedded nanoparticles	DPPC:L:Chol (4:4:1)	SPIONs	Fluorescein	AMF and glucose receptor (GLUTs)	Cancer therapy (not specified)	[263]
Transferrin-functionalized AMLs	EPC:Chol:mPEG2000-DSPE[g] (7:2:0.5)	SPION	Not specified	Magnetic force and transferrin receptor	Blood–brain barrier transmigration	[264]
Cetuximab-functionalized AMLs	DPPC:Chol:DSPE-PEG2000-NH_2[h] (90:5:5)	Citric acid-coated SPIONs	Camptosar(CPT-11)	AMF and EGFR	Brain tumor	[15]

[a] MPPC: 1-Myristoyl-2-palmitoyl-sn-glycero-3-phosphocholine
[b] DOPG: 1,2-dioleoyl-sn-glycero-3-phosphoglycerol
[c] DDAB: Didodecyldimethylammonium bromide
[d] PEG_{2000}DPPE: Methoxypolyethyleneglycol-1,2-Dipalmitoyl-sn-glycero-3-phosphorylethanolamine
[e] DMPC: 1,2-dimyristoyl-sn-glycero-3-phosphocholine
[f] DMPG: 1,2-Dimyristoyl-sn-glycero-3-phosphorylglycerol sodium salt
[g] DSPE-mPEG2000: 1,2-Distearoyl-rac-glycerol-3-Phosphoethanolamine-*N*-Polyethyleneglycol-2000
[h] DSPE-PEG2000-NH_2: 1,2-distearoyl-sn-glycero-3-phosphoethanolamine-*N*-[amino(polyethylene glycol)-2000] (ammonium salt)

which relies on both the Néel and Brownian relaxations of the nanoparticles. On the other hand, the application of a low frequency (LF) (of values < 1 kHz) AMF leads to the nanoparticles mechanical movement that can be explored to deform the hydrogel and/or stimulate cells/tissues to induce biological effects [227–229]. Regarding the use of magnetic hyperthermia, the temperature in the target tissue needs to reach high temperatures (42–45) °C, either to kill the tumor cells by apoptosis or to prompt higher susceptibility to radiation and antitumor drugs [230, 231]. In this temperature range, the cytoplasmic membrane and nuclear proteins required for the DNA synthesis are prone to denaturation. At higher temperatures, the cell death occurs through necrosis [232, 233], but can also affect the healthy cells.

The idea of using thermosensitive liposomes (TSL) for controlled drug release under a tolerable local–regional hyperthermia (41–46) °C dates to the 70 s [234]. The hypothesis of these systems is based on the lipids that undergo a solid to fluid phase transition at those temperatures, leading to enhanced permeability of membranes and controlled release of drugs. However, these systems preclude a sustained drug delivery as its cargo is released in a single burst. Magnetoliposomes are excellent candidates for temperature-induced controlled release and magnetic hyperthermia. The latter is the ability of magnetic nanoparticles to produce heat when subjected to appropriate alternating magnetic fields (AMF), whether by hysteresis losses or from Néel or Brownian relaxation processes. Aqueous magnetoliposomes were the first developed magnetoliposome type [206]. Encapsulating inorganic nanoparticles in the aqueous core of liposomes gives the advantage of non-invasively stimulating the release of therapeutic molecules by AMF trigger. In addition, by spatially restricting the presence of nanoparticles in the lumen of liposomes, the heating source occurs within the liposome itself, which is another advantage over the strategy of using hyperthermia to induce the temperature phase transition in conventional thermosensitive liposomes. Tai et al. [235] monitored the release of carboxyfluorescein (CF) from AMLs loaded with Dextran-coated iron oxide nanoparticles (Resovist) with and without the stimulation from an AMF (Fig. 11A). AMLs containing 14 mg Fe mL^{-1} (Fig. 11B-a) demonstrated significant CF release at 32 °C when subjected to an AMF, compared to the leaky temperature window (36–38) °C of non-exposed nanosystems. It was also verified that the release profile depends on the encapsulated iron content, as the nanosystems containing 7 mg Fe mL^{-1} have been shown to induce release at 34 °C instead. The controlled release of CF was also proven after the injection of lipid systems in anesthetized rats, which allowed concluding the suitability of these systems to controlled compounds release in living systems. More recently, Bi and Han [236] studied in detail the AMF thermal activation of citrate-modified magnetic nanoparticles encapsulated in dioleoylphosphatidylcholine (DOPC) AMLs, aiming their use as safe drug delivery and stimulus-induced cancer treatment. Release profiles of CF from AMLs were recorded by fluorescence microscopy. Figure 11B-b (where green core corresponds to CF solution while the red ring corresponds to DOPC membrane) shows an evident release dependence on the AMF activation (2 kHz, 20 mT). The authors also studied the extent to which the AMF parameters (magnetic field frequency, trigger time, and MNPs concentration) influence the release and concluded the proportional effect of these parameters on CF release.

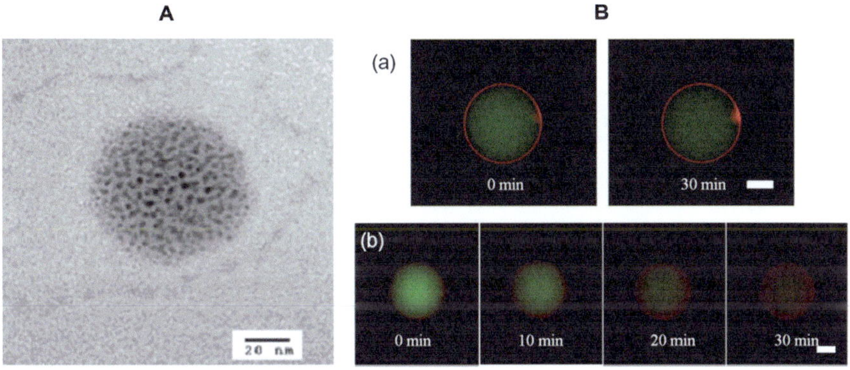

Fig. 11 **A** Transmission electron microscopy (TEM) image of a single DPPC/DSPC-based AMLs containing 14 mg Fe mL^{-1} of Resovist. Reprinted from Tai et al. [235], with permission from IOP Publishing. Copyright (2009). **B** Fluorescence images of DOPC-loaded (red) magnetoliposomes loaded with CF (green), as a function of time, (a) without and (b) with the action of an AFM (2 kHz, 20 mT). Reprinted from Bi and Han [236], with permission from Elsevier. Copyright (2018). *Note* DSPC refers to Distearoylphosphatidylcholine

Dong and coworkers [237] demonstrated that, when applying an AMF, the local temperature around the SPIONs, due to Néel relaxation, is significantly higher than the bulk temperature. This fact supports the idea of placing nanoparticles in the lipid bilayer since the lipids near each SPION manage to reach higher temperatures than the surrounding bulk. Furthermore, the enhanced membrane permeability can induce local triggered release when applying an AMF while stopping it upon removal of the AMF. Salah et al. [238] developed erythromycin-loaded liposomes based on soybean phosphatidylcholine (SoyPC) with bilayer-embedded Fe_3O_4 nanoparticles as potential candidates to address antibiotic resistance. In the erythromycin-controlled release, they studied the influence of exposing magnetoliposomes over 30 min to 5 and 15 mT AMF at 100 kHz. The results revealed a substantially larger release of erythromycin from the systems when subjected to an external magnetic field. The mechanical forces exerted by the particles on the lipid bilayer ensure a controlled release of the antibiotic without causing the rupture of the lipid structure. Shaghasemi et al. [221] studied the same type of magnetoliposomes based on different phosphatidylcholine lipids with different chain lengths. Thus, they correlated the transition temperature (T_m) of the magnetoliposomes and the susceptibility to induce thermally a local phase separation and to increase the local permeability of the magnetoliposomes. They demonstrated that dipalmitoylphosphatidylcholine (DPPC) liposomes containing 4 wt% SPION released calcein whenever an AMF pulse was applied, while that release was interrupted in the interval between pulses. However, in palmitoyloleoylphosphatidylcholine (POPC, $T_m = -2$ °C) magnetoliposomes, there was no calcein release (neither passive nor triggered). These results thus prove that the application of an AMF causes a distortion or local heating that leads to local phase separation at the T_m of the liposome membrane. Recently, Fortes Brollo et al. [239] developed a strategy for synthesizing different types of

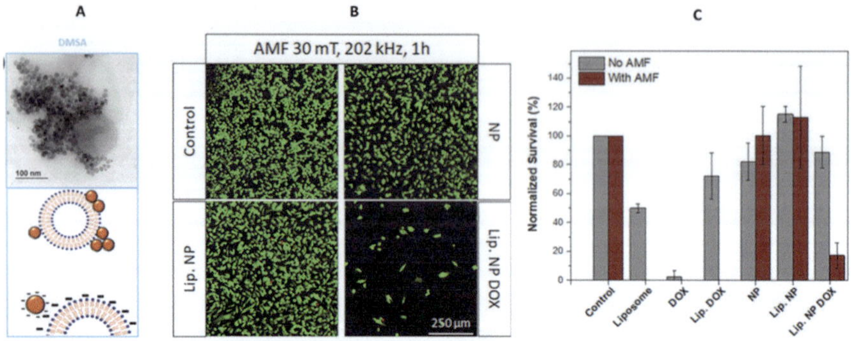

Fig. 12 **A** TEM image of liposomes with surface-bound DMSA-coated nanoparticles and its schematic representation; **B** representative confocal laser scanning microscopy images, and **C** normalized survival after 24 h of MDA-MB-231 cells after exposure to different treatments (without (gray) and with the action of an AMF of 202 kHz and 30 mT for 1 h (red)). Reprinted from Fortes Brollo et al. [239] with permission from American Chemical Society. Copyright (2020)

DPPC magnetoliposomes by controlling the nanoparticles' coating and studying the effect of nanoparticle spatial distribution on liposomes (membrane integrity, magnetic response, heating and colloidal properties). The characterization results revealed that liposomes with dimercaptosuccinic acid (DMSA)-coated nanoparticles (negatively charged) attached outside (Fig. 12A) showed better stability and heating properties compared to their counterparts. Doxorubicin was encapsulated in these systems, and the ability to induce the controlled release of DOX by AMF was studied in MDA-MB-231 cell viability assays (this cell line is used in in vitro models of triple-negative breast cancer). It was found that after applying an AMF in the configuration of 202 kHz and 30 mT for 1 h, only magnetoliposomes loaded with Doxorubicin were found to induce cell death, which is qualitatively visible in images captured by confocal laser scanning microscopy (Fig. 12B, where live cells appear in green). Figure 12C summarizes the normalized survival after 24 h of exposure to different treatments, with a cell survival of 17% after application of AMF in DOX-loaded magnetoliposomes, which corresponds to a 70% reduction compared to non-exposed cells.

AMF as an external trigger has gained value in the controlled release of therapeutic compounds from magnetoliposomes. In fact, magnetic hyperthermia-triggering is the most commonly used approach for controlled release. However, the hyperthermic effect is unsuitable when leading with highly perfused organs or brain tissues [240]. Therefore, Guo et al. [241] developed carboxymethyldextran (CMD)-coated magnetoliposomes as a non-hyperthermic DOX release and diagnostic strategy for brain cancers under low-frequency alternating magnetic field (LF-AMF). The exposure of magnetoliposomes to a 45 mT LF-AMF for 30 min led to a 74% DOX release (at pH = 5), highly contrasting to the 35% released in the absence of the LF-LMF in the same conditions. Therefore, optimizing the conditions for the use of this trigger is especially desired, because it allows a controlled release in time and space in a non-invasive manner while combining a synergistic effect of different therapies (e.g., magnetic hyperthermia with chemotherapy). Furthermore, in theory, this would lead to a reduction of the systemic effects of the therapy and respective advantages of localized therapeutics (e.g., reducing resistance to antibiotics and chemotherapeutic drugs).

Near-Infrared (NIR)-Triggered Release

Near-infrared (NIR) light (700–1000 nm) is another alternative to induce heat generation in nanoparticles with optical absorption in the NIR regions. Light-absorbing nanoparticles—photothermal agents—transform the energy provided by a laser into heat in a process known as photothermal therapy [242]. The last few years have been extensively focused on developing or exploiting nanomaterials with strong optical absorption in the NIR region to address the limited optical penetration and the possible damage caused by laser illumination to healthy tissues. Some examples include, for instance, graphene nanocomposites [243], plasmonic nanorods [244], nanoshells [245], or nanostars [246]. Beyond an optimal and selective absorption region, photothermal agents should also present low toxicity and high solubility in biocompatible solutions and allow their functionalization [247], which is often incompatible with commonly used plasmonic gold nanoparticles or carbon-based materials. Iron oxide nanoparticles, particularly magnetite (Fe_3O_4) [248, 249], have been identified as the preferred photothermal agents, due not only to their absorption in the NIR region, but also to their biodegradability, easy clearance, and biotransformation in the body [250]. Recently, Shen et al. [251] developed Fe_3O_4-loaded thermosensitive liposomes (Fe_3O_4-TSL) for NIR-laser-triggered DOX release and combined therapy. For that, oil-soluble Fe_3O_4 nanoparticles were coated with DPPC/Cholesterol (Chol) (4:1) lipid formulation, and DOX was loaded into the aqueous compartment by an ammonium sulfate gradient (Fig. 13A). The NIR laser at a power density of 2 W/cm^2 for 5 min induced a total cumulative release of 41.6% at 12 h, while for the samples irradiated twice (0 and 6 h) it was 66.7%. On the other hand, in the control sample (non-irradiated), only 10% of the drug was released at the same time point. These in vitro results were corroborated in tumor growth inhibition tests performed in rats. Only the DOX-Fe_3O_4-TSL + NIR group induced a near disappearance of the tumor after a 12 day treatment (Fig. 13B), demonstrating a dual effect of photothermal-chemotherapy in cancer treatment [251].

So far, this type of trigger is most frequently used in liposomes coupled with plasmonic nanoparticles (usually gold-based nanoparticles), so they are no longer classified as magnetoliposomes. However, efforts have been made to develop nanoparticles that combine the plasmonic and magnetic modalities to associate magnetic hyperthermia and photothermal therapy in "all in one" nanoparticles [252]. This type of combined therapy may address some of the difficulties in magnetocaloric therapy (such as low heating efficiency), potentially facilitating its clinical translation. Combining this type of nanoparticles with liposomes, obtaining plasmonic

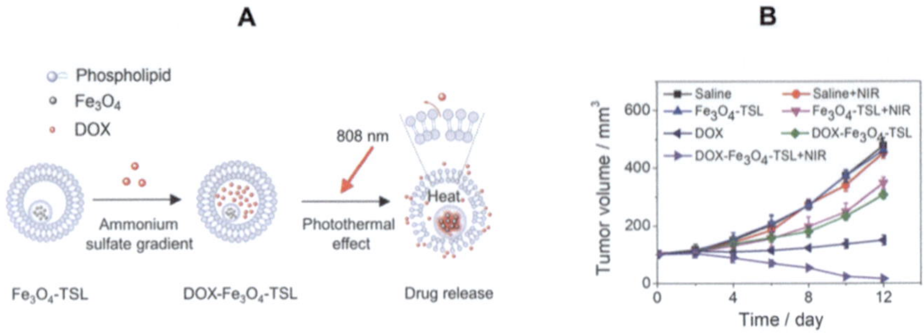

Fig. 13 **A** Schematic representation of DOX-Fe_3O_4-TSL synthesis and respective NIR-triggered DOX release; and **B** in vivo antitumor efficacy, as tumor volume changes, in different treatment groups. Reprinted from Shen et al. [251], with permission from Royal Society of Chemistry. Copyright (2019)

magnetoliposomes or magnetoplasmonic liposomes, is relatively new in the magnetoliposomes state-of-the-art. Khosroshahi et al. [253] described doxorubicin-loaded magnetoplasmonic thermosensitive liposomes (MPTL-DOX) and demonstrated the combined effect of magnetic drug targeting and laser hyperthermia-triggered drug release. The proposed system encloses poly(vinyl pyrrolidone) (PVP)-stabilized magnetite/gold nanoshells in DPPC:cholesterol:DSPE-PEG2000 (80:20:5) liposomes, where DSPE-PEG2000 corresponds to the 1,2-distearoyl-*sn*-glycero-3-phosphoethanolamine-*N*-[methoxy(polyethylene glycol)-2000] (ammonium salt). The results revealed an enhanced cell apoptosis of MCF-7 breast cancer cells induced by the synergistic effect of chemotherapy and laser-mediated hyperthermia (at 350 mW for 5 min). Rodrigues et al. [254] developed manganese ferrite nanoparticles coated with gold ($MnFe_2O_4$/Au core/shell nanoparticles) used to synthesize AMLs and SMLs, validating the potential use of both nanosystems for combined cancer therapy. This work was followed by Rio et al. [212], which developed nickel ferrite nanoparticles decorated with plasmonic gold nanoparticles and core/shell nanostructures (with a nickel ferrite core and a gold shell). The bimodal magneto-photothermal effect has proven to amplify heat generation compared to the single-mode magnetic hyperthermia or photothermia modalities. Cationic magnetoliposomes enclosing citric-acid-coated iron oxide magnetic nanoparticles demonstrated a synergy for dual therapy of human glioblastoma [255]. The action of AMF + NIR (at 4.5 mg/mL of magnetoliposomes) induced a reduction in viability of human glioblastoma U87 cells by 82%, compared to the reduction by 32–35% of single-mode treatments.

Magnetoliposomes for Ligand-Mediated Active Targeted and Combined Therapy

Magnetoliposomes, as multifunctional nanosystems, increase the bioavailability of drugs, enhance their local accumulation (either by magnetic action or by passive targeting) and induce a controlled release through external stimuli. Drug delivery strategies have evolved toward precision nanomedicine, which seeks to control the release of the drug in quantity, time, and space. Surface modification of liposomes with various targeting ligands (e.g., aptamers, antibodies, peptides) has been validated to achieve more efficient targeting, to improve systems endocytosis and the general therapeutic response [256]. These advances have been applied to develop functional magnetoliposomes with enhanced precise targeting.

The use of functionalized magnetoliposomes as a strategy for precise targeting of certain types of cells is predominant in research and development in cancer therapy. This fact is due not only to the ambition in this area, but also to the functional application of the knowledge generated over the years in molecular genetics. Folate-mediated receptor targeting is one of the most used targeting strategies for drug delivery due to the overexpression of folate receptors on the surface of several types of cancer cells. Methotrexate (MTX) is an antifolate therapeutic agent that targets folate receptors and acts as an anticancer drug [257]. As a result, Guo and coworkers [258] proposed thermosensitive magnetoliposomes (MagTSLs) functionalized with MTX (MTX-MagTSLs) for light/magnetic hyperthermia and MTX triggered release in precise cervical cancer therapy. The targeting capabilities of the magnetoliposomes were evaluated in HeLa tumor-bearing nude-mice by using NIR optical imaging technique. The authors demonstrated a much higher accumulation of MTX-MagTSLs in tumors when applying a constant magnetic field than biodistribution in other organs. These results were corroborated by in vivo antitumor activity studies. Furthermore, these studies also revealed the importance of the dual effect of AMF and NIR laser irradiation, evidencing larger tumor inhibition efficiency when both stimuli were applied simultaneously [258]. The functionalization of magnetoliposomes with Alpha-Fetoprotein (AFP) antibodies also demonstrated the importance and validity of dual-targeting (magnetic targeting and antibody targeting) for high specificity for early liver cancer diagnosis [259]. Also, the potential of immunomagnetoliposomes is being studied to detect platelets involved in atherosclerotic plaque rupture and thrombosis [260].

Integrins are a molecular family involved in tumor angiogenesis, and blockade of their signaling has been demonstrated to be efficient in inhibiting tumor growth, angiogenesis, and metastasis. Due to the overexpression of $\alpha_v\beta_3$ integrin on tumor neovascularization, the functionalization of anionic magnetoliposomes with the cyclic Arginine-Glycine-Aspartic acid (cRGD) was proposed to target the $\alpha_v\beta_3$ integrin [261]. Dorjsuren et al. [262] developed DOX-loaded thermosensitive magnetoliposomes functionalized with cetuximab (CET), a monoclonal antibody against the epidermal growth factor receptor (EGFR). In this chapter, the authors corroborated the selective binding of functionalized magnetoliposomes in cells with overexpression of EGFR. They also demonstrated the synergistic effect between chemotherapy and photothermia by reducing the cell viability of breast cancer when the treatment was combined with hyperthermia-induced by NIR laser irradiation. Recently, glucose-modified liposomes loaded with Fe_3O_4 nanoparticles also demonstrated potential for cancer cell-targeting and controlled release under the action of an AMF [263].

The efficient transport of drugs to the brain is an application that can highly benefit from the action of an external magnetic field in combination with functionalized magnetoliposomes. Ding et al. [264] demonstrated the synergistic effect of transferrin receptor-mediated transportation with magnetic action on the transmigration of magnetoliposomes across the blood brain barrier (BBB), which is resistant to the passage of several drugs. In turn, Lu et al. [15] demonstrated the ability of surface-conjugated aqueous magnetoliposomes with cetuximab (CET) in the enhanced recognition of EGFR receptors and intracellular uptake in human primary glioblastoma cells (U87). The citric acid-coated magnetic Fe_3O_4 encapsulated in the aqueous lumen of thermosensitive nanosystems induced the controlled release of the drug Camptosar (CPT-11) and enhanced cytotoxicity in U87 cells [15].

4.2 Magnetic Hydrogels

Among the classes of magnetic composite materials, magnetic hydrogels are one of the most promising materials owing to several unique properties [265]. These materials consist of magnetic nanoparticles embedded in an elastic matrix, which can endow the hydrogel matrix a magnetic response and surpass various limitations of conventional hydrogels. For instance, non-stimuli-responsive (conventional) hydrogels only display passive drug release through diffusion and matrix degradation, strongly limiting a controllable, safe and efficient delivery of drugs, in addition to displaying limited actuation properties [228]. Through the combination with magnetic nanoparticles, a wide array of micro and macroscopic properties (e.g., mechanical, electrical, optical, and thermal) can be modulated with real-time remote control [266–268]. Besides, the use of magnetic nanoparticles to implement a stimulus is a suitable alternative to the cumbersome task of modifying the hydrogel network with stimulus-responsive properties. For example, Jervis et al. [269] recently developed a photo-responsive supramolecular hydrogel based on dehydropeptides and the photo-cleavable carboxy-2-nitrobenzyl (CNB) group. However, despite the useful trigger strategy, the gel-sol transition was irreversible after UV irradiation, which is insufficient if multiple cycles of on-demand release are required in drug delivery applications. In this sense, the use of magnetic gels has been widespread to different applications, including controlled cell growth [270], drug/cell delivery [271, 272], hyperthermia [273], tissue engineering [274], biosensors [275], contrast agents [276], photothermia [277], and actuation [278].

Like the neat hydrogel, magnetic gels can be shaped as microgels and nanogels [279], but can also be fabricated into isotropic or anisotropic (obtained through gelation under an applied magnetic field). Through the tunability of the gel's shape, the formulations can be adapted for the required application. For instance, in biomedical applications, macroscopic hydrogels are suitable for surgical implants, transepithelial and transdermal drug delivery [280]. However, considering the risks associated with surgical implantation, macroscopic magnetic gels can be alternatively made injectable (e.g., in situ gelation) or formulated as microgels or nanogels. These small size particles ease the clearance and enhance penetration through the physiological barriers, which widens the administration routes to oral, pulmonary, intrabony, and systemic drug administration pathways. In addition, the endowing of the gel matrix with anisotropic microstructure affects both the mechanical and content diffusion properties, and enables the material with anisotropic magnetic-mechanical response [267, 281].

Similar to other composites, the magnetic nanoparticles can modify the hydrogel properties through several parameters, including: (1) concentration, composition, size and shape [282, 283]; (2) the density of nucleation sites induced by the nanoparticles [284, 285]; (3) the distribution of the nanoparticles within the hydrogel network; (4) the type of interactions between the nanoparticles and the hydrogel fibers (physical or chemical); and (5) the method of preparation of the magnetic gel [267, 286, 287]. Other parameters that influence the final magnetic gel properties depend on the matrix itself, such as length, chemical composition, and concentration of the hydrogel network chains, crosslinking and swelling degree of the network.

Hereby, considering the range of applications and advantages achieved through the combination of hydrogels and magnetic nanoparticles, several reviews have been published to which the reader is referred for more information about magnetic gels [228, 265, 266, 279, 281, 283, 288–290]. In this section, the reader is initially introduced to advances in the methods of preparation, followed by a critical discussion on recent biomedical applications. The novel magnetolipogels and supramolecular magnetic gel advances are also presented. Some of the discussed gels are further summarized in Table 2 at the end of the section.

4.2.1 Methods of Preparation

Currently, magnetic gels can be fabricated through different methods, which include the blending method, grafting method, in situ precipitation, and swelling method. A schematic representation of the main preparation methods is displayed in Fig. 14. For discussion on other methods, the reader is referred to the reviews [228, 266, 291].

Blending Method

The blending method is one of the most common strategies to fabricate magnetic hydrogels, which simply consists in the mixing of separately synthesized nanoparticles with the hydrogel precursor's solution, and consequent gelation induced through crosslinking/polymerization methods [292]. This method opens the possibility of synthesizing uniform nanoparticles with specific size, shape, and composition as means to endow the gel with the required properties. For example, magnetic nanoparticles can be coated with a gold shell [277] or coupled with gold nanoparticles [293] to synergistically combine the effects of magnetic hyperthermia and photothermia. Consequently, the simplicity of the strategy has also been explored in the fabrication of supramolecular magnetic gels [277, 294, 295]. However, it can suffer from three

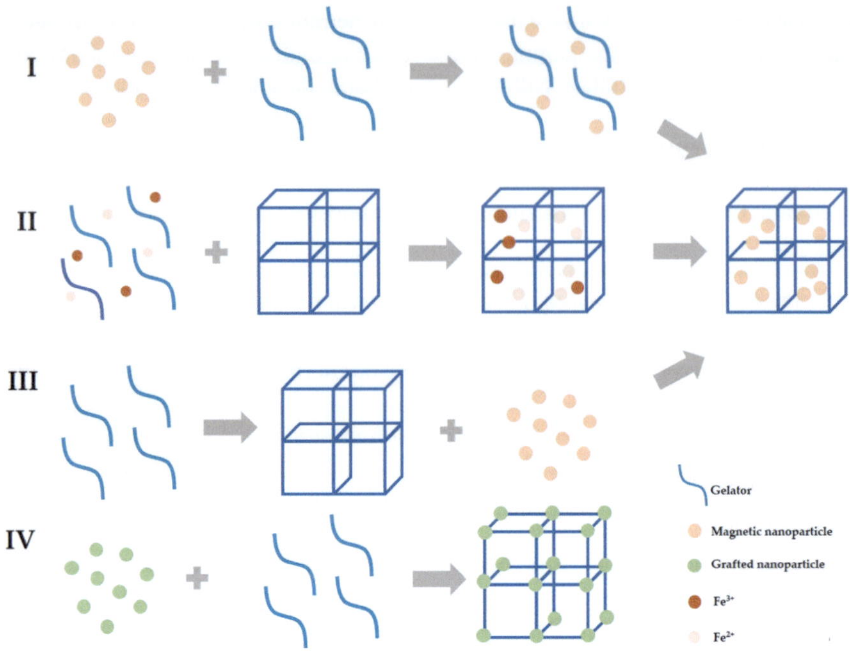

Fig. 14 Schematic representation of the preparation methods of magnetic gels: (I) blending method; (II) in situ precipitation method; (III) swelling method and (IV) grafting method

major disadvantages: (1) the nanoparticles can interfere with the network formation and final gel structure; (2) inhomogeneous distribution of nanoparticles; (3) leakage of nanoparticles upon exposure to liquid environment [265, 292]. To overcome the potential aggregation, the nanoparticles in the polymer solution need to be well dispersed through sonication, and well-stabilized with an (in)organic coating. Carvalho et al. [295] achieved a homogenous supramolecular gels with particle content up to 2.8 mg/mL owing to the steric and electrostatic stabilization provided by the polyacrylic acid (PAA) coating. Similarly, Haider et al. [296] reported the homogeneous distribution of alginate-coated nanoparticles in acrylamide-based gels owing to the electrostatic repulsion. Lie et al. [297] also reported the use of electrostatic repulsion to achieve homogeneous gels through the loading of the nanoparticles in nanofibrillated cellulose.

Grafting Method

Despite being less used, the grafting method (grafting-onto, grafting-from, and grafting through) surpasses the major disadvantages of the blending method. Like the blending method, the nanoparticles are previously synthesized, and further grafted with functional groups to work as crosslinkers with the monomers upon the polymerization [289, 292]. In this way, the method ensures that the nanoparticles remain stable and well dispersed in the polymer matrix [265]. However, it presents limitations to its use in biomedical applications, mainly the long, high-cost, and complicated fabrication processes [298].

This strategy has been commonly employed for the development of poly(acrylamide) based hydrogels [299–302]. Hu et al. [300] functionalized nanoparticles with 3-(trimethoxysilyl)propyl methacrylate to provide silane coupling agents, which are copolymerized into the polymer network through condensation with silanol or hydroxyl groups present in the matrix.

Although supramolecular gels make use of physical interactions, the grafting of groups that participate in the fiber formation has also been adapted. For instance, Yang et al. [303] fabricated magnetic nanoparticles functionalized with Dopamine-L-Phe-L-Phe-OH, which co-assembled with the gelator 2-Naph-L-Phe-L-Phe-OH in the fiber formation process. A simplification of the grafting strategy was recently achieved by Nowak et al. [304] through the addition of aspartic acid to the C-terminal of the hydrogelator, which enabled the coordination of nanoparticles' metal ions. Hence, the possibility of using the hydrogelator or simple peptides as the grafting group potentiates the use of supramolecular gels to address the current limitations of the grafting method and its use in biomedical applications.

In Situ Precipitation

The in situ precipitation method consists of the use of convenient and facile syntheses of nanoparticles through co-precipitation inside the hydrogel network, which works as a chemical reactor. The general procedure requires the initial preparation of the gel and its immersion in a concentrated aqueous solution of the required ions for the magnetic nanoparticles until reaching the swelling equilibrium. Then, the swollen gel is immersed into an alkali solution for precipitation of the magnetic nanoparticles [292]. This method has been extensively used in various polymeric gels to achieve enhanced mechanical properties [305, 306]. Besides, the nanoparticles size and shape can be tuned by changing crosslinking density of the hydrogel, iron and NaOH (precipitant) concentration, which provides a suitable pathway for the green synthesis of nanoneedles, nanorods and nano-octahedrons [307–309]. However, the translation of the method to biomedical applications is limited for the possibility of nanoparticle leakage and for only being achieved in hydrogels that possess stable networks in the alkali solution.

Swelling Method

The swelling method has preparation similarities to both the blending and in situ precipitation methods. Here, the hydrogel and nanoparticles have to be initially prepared separately. Then, the hydrogel is incubated with a concentrated nanoparticle solution, which has been commonly explored for the development of microgels. For example, Witt and coworkers [310] developed poly(*N*-isopropylacrylamide-coallylamine) (P(NIPAM-co-AA)) microgels comprising citrate-coated cobalt ferrite nanoparticles and evaluated how different microgel preparation strategies could influence the microgel homogeneity. However, the method has limitations associated to both the hydrogel and nanoparticles, as the loading can be hampered by both the hydrogel mesh size and the interactions stabilized with the nanoparticles.

4.2.2 Advances in Biomedical Applications

The wide range of properties and advantages provided by magnetic gels, mainly the magnetic response, has been of interest for various fields. Regarding the biomedical applications, they have been commonly applied in controlled drug delivery [311, 312], magnetic hyperthermia [313, 314], and tissue engineering [221, 315, 316]. In drug delivery, the magnetic response is a valuable tool to achieve controlled and on-demand release, magnetic targeting, tracking of the therapy through improved contrast in magnetic resonance imaging, and achieve the synergistic effect of chemotherapy and hyperthermia. Hence, in this section two of the major applications are discussed: magnetic hyperthermia and drug delivery.

Magnetic Hyperthermia

The repeated application of the high-frequency AMF is required to maintain the temperature [317]. However, the low retention time of nanoparticles requires additional injections to ensure the concentration for the treatment. Such problem is overcome in magnetic gels, as the matrix ensures a long-term retention of the nanoparticles in the target site [231]. The heat generation in magnetic hydrogels has been demonstrated to be enough to induce tumor cell death [265, 318, 319]. Besides, the heat generation can be modulated by changing the nanoparticles content or the AMF strength [318]. Considering the nanoparticle properties, a means to improve the heating efficiency, and thus reduce the required amount of nanoparticles to avert toxic effects is to make use of shape anisotropic nanoparticles [320]. For example, Gao et al. [321] developed hydrogels containing ferromagnetic vortex-domain iron oxide (FVIOs) nanoparticles, which could be incorporated at a 17 times lower concentration (0.6 mg mL^{-1}) than the spherical counterparts and achieve sufficient heating capacity to prevent breast cancer recurrence. Another strategy consists of the use of anisotropic magnetic hydrogels. For instance, Hu et al. [322] demonstrated that the anisotropic alignment of the magnetic nanoparticles inside the gel provided more thermogenesis than isotropic magnetic hydrogels, which could be regulated by changing the angle between the assemblies and the alternating magnetic field. The authors obtained a greater heating when the magnetic field was aligned parallel to the magnetic nanoparticle chains. The effect was associated to the magnetic coupling between the nanoparticles containing the reversal of homologous magnetic moments and thus, requires the AMF to consume more energy to induce the reversal of the magnetic moments. This effect was also explored by Wu et al. [323] to optimize the thermogenesis at a low nanoparticle dosage, which together with heat accumulation caused by the crosslinking among nanoemulsion droplets, lead to an enhancement of the hyperthermia efficiency in cancer therapy.

The magnetic gels also provide a framework for various therapeutic strategies. Qian et al. [324] developed an injectable silk fibroin-based magnetic gel, which achieved efficient in vivo heating performance and anticancer properties, besides inducing the occlusion of blood vessels. In this sense, the gel can occlude the blood supply of deep located tumors. Satarkar et al. [325] developed temperature responsive hydrogels in which the AMF application led to a uniform heating of the composite material inducing the collapse and squeezing of the loaded cargo. Campbell et al. [326] developed injectable gels bearing thermoresponsive microgels which de-swell upon heating, resulting in an increase of the free volume that leads to an enhanced drug release from the hydrogel scaffold. Once the AMF is removed, the microgel swells, and drug release rate from the hydrogel scaffold decreases.

Drug Delivery

Hydrogels offer various advantages as drug delivery systems for the good biocompatibility and tunable drug diffusion properties, which through the combination with magnetic nanoparticles, offer remote control over its release [228, 265, 281, 327–329]. One of the potential strategies includes the irradiation of the nanoparticles with radiation close to the Ferromagnetic Resonance (FMR) frequency. Savchak et al. [330] demonstrated that the energy dissipation within the particles led to heating of the gel, which induced a volume phase transition in thermoresponsive gels. Besides, the response could be tuned through changes in the particle concentration, chemical nature, frequency of the electromagnetic radiation, and time of exposure.

Under the application of a magnetic field, the drug release can be enhanced/controlled through different mechanisms. When a high-frequency AMF is used, the nanoparticles' heating generation can lead to an expanded and loose network that eases the cargo release [329]. Besides, the polymer chain relaxation can induce extended mechanical deformation followed by tensile stresses that further enhance the drug release. The heating of the gel can also lead to a gel-sol transition, which accelerates the drug leakage from the magnetic hydrogel [273]. For instance, gelatin undergoes a gel-sol transition when temperature exceeds 40 °C, which involves the transition from coil conformation to triple helical sequences [331]. Hu et al. [312] demonstrated that drug release from gelatin-based magnetic gels could be switched on and off through application of the external AMF, which decreased after each cycle due to drug depletion. Besides, the magnetic gel displayed a slower release profile than the hydrogel, potentially due to the nanoparticles working as physical barriers to drug diffusion. This observation was also reported in supramolecular peptide-based gels for curcumin [277], and doxorubicin release [272]. The shrinking of thermosensitive gels due to heating can also be used as a means to induce a burst release [326]. Although less explored, a low-frequency AMF can also be used to enhance drug release. For instance, Wang et al. [306] fabricated chitosan-based magnetic gels that achieved pulsatile release under a low frequency AMF. The system displayed a pulsatile release with quick response and a zigzag shape fraction release through the application of "15 min ON–15 min OFF" cycles.

The control can also be achieved through application of a DC magnetic field, which leads to the assembly of nanoparticles in an end-to-end configuration due to attractive forces [311, 332, 333]. This arrangement induces a reduction of the pore size and confines the cargo in the network, thus reducing its diffusion. Liu et al. [311] developed poly(vinyl alcohol) (PVA)-based magnetic gels that displayed a burst drug release after switching off the magnetic field due to reopening of the pores, which recovered the normal drug diffusion after some time.

4.2.3 Magnetic Liposome-Hydrogels: The Future of Theranostics?

Currently, a major challenge in the clinical translation of nanomedicines is the control of drug release, which can potentially be tackled through the combination of liposomes and hydrogels [334, 335]. The combination with magnetic nanoparticles is expected to further improve the control, and therapeutic efficiency and efficacy. Regarding the composites, magnetic liposomes can provide a suitable lipid-based system to combine with hydrogels as a means to improve the control. These systems can achieve both a low passive permeability of the loaded drug and high release efficiency through the heat generation of the magnetic nanoparticles under an AMF, either high- or low-frequency [209, 336, 337]. A major advantage of combining (magnetic) liposomes with hydrogels is the possibility of using the lipid-based systems as stimuli-responsive reservoirs [338–340]. Cogan et al. [341] developed magnetic lipogels bearing dipalmitoylphosphatidylcholine (DPPC) vesicle assemblies loaded with ascorbic acid-2-phosphate (AAP), a precursor that is required to induce the production of extracellular collagen. The authors achieved a smart cell culture scaffold, in which the exposure of the gel to an AMF induced release of AAP, thus enabling the production of the extracellular matrix protein collagen that was added to the alginate scaffold. Later, the same group demonstrated that magnetic lipogels also provide spatial control through the magnetic positioning in addition to the temporal control [342]. The authors developed alginate-based magnetic lipogels bearing vesicle assemblies loaded with nickel (II), which could support the cell growth through the addition of fibronectin. The exposure to an AMF led to nickel (II) being released from the vesicles, whose interaction with the alginate gel matrix slowed its diffusion out of the gel block. Through the combined effect of the slowed diffusion in the gel matrix and the magnetic patterning of the vesicles, the authors achieved spatial control over the cell apoptosis.

Besides the macroscopic formulation, magnetic lipogels can also be fabricated as microgels [343, 344]. Hanuš et al. [343] fabricated alginate-based magnetic lipogel microparticles (40–80) μm bearing DPPC/Chol (2:1) liposomes through a drop-on-demand inkjet method. The authors demonstrated that a radio-frequency (RF) alternating magnetic field could be used to control the local temperature and content diffusion from the liposomes. Pittermannová et al. [344] implemented an extractive gelation process in a microfluidic device to fabricate alginate-based magnetic lipogel microparticles (5–10) μm bearing DPPC liposomes. The authors also demonstrated that the system could provide on-demand drug release through exposure of the particles to 45 °C. Despite less explored, Kang et al. [345] achieved a fabrication method of Janus magnetic lipogel microparticles through a double syringe method. The syringe comprised a compartment loaded with magnetic nanoparticles, and the other with dipalmitoylgalloylglycerol (DPGG) and polydiacetylene (PDA) for the detection of lead (II).

Another major advantage of (magnetic) lipogels is the possibility of providing real-time monitorization of the therapy by working as contrast agents. For instance, Van Elk et al. [346] fabricated alginate-based lipogels crosslinked with holmium ions (T_2^* contrast agent, i.e., shortens the apparent transverse relaxation time in magnetic resonance imaging) and bearing thermosensitive liposomes loaded with the T_1 MRI contrast agent [Gd(HPDO$_3$A)(H$_2$O)] (ProHance®). Hence, the holmium ions enabled the visualization of the microgels by MRI, while the loaded contrast agent release from the liposomes could be explored as an indicator of the in vivo release of a therapeutic drug upon hyperthermia therapy.

In addition to the drug delivery and theranostic applications, the magnetic lipogels can also be used as bioreactors with spatiotemporal control. Ullrich et al. [347] developed alginate-based magnetic lipogels loaded with laccase and bearing liposomes loaded with the substrate 2,2′-azinobis(3-ethylbenzthiazoline-6-sulfonate) (ABTS). In this sense, the authors achieved a system that displays (I) structural function provided by the gel matrix; (II) catalytic function due to the immobilized enzymes; (III) storage function, provided by the liposomes that act as reservoirs; (IV) and communication function through translation of radio frequency signal into a biochemical reaction.

Therefore, magnetic lipogels are highly advantageous for biomedical applications compared to their separate components. Besides, the stoichiometric compartmentalization of unstable and highly reactive biologically active compounds' precursors provides suitable opportunity to attain high therapeutic efficiency and efficacy without affording enough time for a pathogen to develop resistance [347].

Table 2 Summary of magnetic gels components (hydrogel, liposomes, and nanoparticles), encapsulated compounds, and triggers used for the aimed biomedical applications

Hydrogel	Liposomes	Nanoparticles	Encapsulated compound	Trigger	Aimed biomedical application	Refs.
Chitosan	–	FVIOs	Doxorubicin	HF-AMF	Cancer therapy	[321]
Polyacrylamide	–	Fe_3O_4	Doxorubicin	HF-AMF	Hyperthermia + drug delivery	[322]
Nanoemulsion with PEGDA	–	Oleic acid-coated Zn ferrite	–	–	Hyperthermia	[323]
Silk fibroin	–	Fe_3O_4 nanocubes	–	–	Hyperthermia	[324]
Poly(*N*-isopropylacrylamide)	–	Fe_3O_4	Vitamin B_{12} Methylene blue	HF-AMF	Hyperthermia + drug delivery	[325]
Poly(NIPAM-NIPMAM) microgel	–	Fe_3O_4	4 kDa fluorescein-labeled dextran	HF-AMF	Hyperthermia + drug delivery	[326]
Starch grafted with itaconic acid	–	Fe_3O_4	Guaifenesin	pH + AMF	Wound healing	[329]
Poly(*N*-isopropylacrylamide)	–	Co nanoparticles or Fe_3O_4	–	Ferromagnetic resonance	Hyperthermia	[330]
Poly(organophosphazene)	–	Fe_3O_4	Tumor necrosis factor-related apoptosis-inducing ligand	HF-AMF	Cancer therapy	[273]
Gelatin	–	Fe_3O_4	Vitamin B_{12}	HF-AMF	Drug delivery	[312]

(continued)

Table 2 (continued)

Hydrogel	Liposomes	Nanoparticles	Encapsulated compound	Trigger	Aimed biomedical application	Refs.
Npx-L-Met-Z-ΔPhe-OH	–	$MnFe_2O_4$@Au core@shell $MnFe_2O_4$@Au-decorated	Curcumin	Photothermia	Drug delivery	[277]
Cbz-L-Met-Z-ΔPhe-OH	–	$MnFe_2O_4$	Doxorubicin	LF-AMF	Drug delivery	[272]
Chitosan	–	Fe_3O_4	Adriamycin Rifampicin	LF-AMF	Drug delivery	[306]
Poly(vinyl alcohol)	–	Fe_3O_4	Vitamin B12	DC-MF	Drug delivery	[311]
Calcium alginate	DPPC:biotin-DHPE[a]	Fe_3O_4@*N*-biotinoyl dopamine + avidin	Ascorbic acid-2-phosphate	HF-AMF	Tissue engineering + cell culture	[341]
Calcium alginate	DPPC:biotin-DHPE	Fe_3O_4@*N*-biotinoyl dopamine + avidin	Nickel(II)	HF-AMF	Cell culture	[342]
Calcium alginate microgel	DPPC:Ch (2:1)	Fe_3O_4	5(6)-Carboxyfluorescein	HF-AMF	Drug delivery	[343]
Calcium alginate microgel	DPPC:Ch (2:1)	Fe_3O_4	5(6)-Carboxyfluorescein	Radiofrequency	Drug delivery	[344]
Calcium alginate Janus microgel	DPGG:polydiacetylene	Fe_3O_4	–	–	Lead(II) detection and removal	[345]
Calcium alginate microgel	DPPC:Ch (2:1)	Fe_3O_4	ABTS	Radiofrequency	Drug delivery	[347]

[a] Biotin-DHPE: *N*-(Biotinoyl)-1,2-dihexadecanoyl-sn-glycero-3-phosphoethanolamine, triethylammonium salt

5 Conclusion and Future Perspectives

Several advances and strategies have been taken with both soft matter and inorganic nanomaterials for biological applications. From the hydrogels and liposomes to the magnetic and plasmonic nanoparticles, novel formulations are leading to promising smart multifunctional materials capable of tackling current nanomedicine and biomedical problems.

For instance, the uniqueness and tunability of hydrogels make them excellent candidates as treatment systems for cancer. As discussed in Sect. 2, several types of studies have been conducted with temperature-, pH-, and photosensitive hydrogels with the aim of developing a therapy platform the most efficient as possible. That path has been traced and hopeful results have been obtained, namely, good biocompatibility, and outstanding antitumor effects. The structural changes in response to external stimuli allow for the application of both chemotherapy and phototherapy in a single hydrogel. The rational design of these hydrogels is the key procedure for the combination of different therapy strategies in one hydrogel and consequently to obtain better performance. So, the next steps should be focused on finding different ways of maximizing the drug effect with minimal therapeutic dose.

Besides, in Sect. 2, it was also discussed how much in recent years, research on supramolecular peptide hydrogels has remarkably advanced, both on understanding the self-assembly process and on their application in therapeutics. The possibility of tuning the peptide sequence at the amino acid level makes these materials highly advantageous as they provide hydrogels with specific properties for a given application. However, there is a consensus on the need for further studies at several levels. For the moment, despite current efforts, the understanding of the relationship "amino acid sequence—fiber structure—hydrogel function" is crucial. Furthermore, gelation itself has to be thoroughly studied, considering its close dependence on numerous environmental circumstances, in order to obtain uniform and reproducible hydrogels. In this context, the effect of counterions—often resulting from peptide synthesis—on fiber quality and biological systems has to be assessed along with the development of cost-effective synthesis strategies that allow large-scale production of peptide hydrogelators. Indeed, therapeutic applications are extremely demanding, so reporting on failures would also be of great value to the scientific community. We conclude with this same challenge, in order to make research profitable and to get closer to the implementation of these materials in clinical practice.

In Sects. 3 and 4 the importance of liposomes was discussed, which role in current medicine is indisputable. Even so, their application potential continues on an upward curve due to accumulated knowledge, the development of new materials, and the combined functionalities of those materials. Therefore, the combination of inorganic magnetic nanoparticles and liposomes—magnetoliposomes—could be the next extension of application for liposomes in numerous biomedical applications. Table 1 summarizes the different types of magnetoliposomes designed and adjusted to specific features that have been developed recently. Undoubtedly, magnetoliposomes have proven to be good controlled drug delivery systems and add dual and even synergistic therapies to their conventional counterparts.

The promising results obtained in in vitro and even in vivo assays have aroused the interest of more researchers from different specialized areas. The multidisciplinarity involved in the development of magnetoliposomes (such as physics, biology, materials engineering, chemistry, and pharmacology) makes it demanding to reach a fully integrated knowledge, but on the other hand, it allows the scientific community to overlap areas of knowledge and even create new ones (e.g., biophysics). In this sense, the miscibility of scientific concepts could be fundamental in the efficient clinical translation of magnetoliposomes as a new smart material for controlled and precise drug release.

However, it should be pointed out that despite the advancements, both hydrogels and liposomes present their own limitations. For instance, some hydrogels may not be biodegradable or biocompatible, display a slow responsiveness to a trigger, and even suffer a burst and uncontrollable drug release. The latter is also a shortcoming in liposomes, which also suffer from instability and fast degradation.

To address these problems, novel systems such as magnetic gels have been provided a suitable platform to advance the development of smart multifunctional systems that integrate different composites, including magnetic liposomes and plasmonic nanoparticles. Particularly, the combination with the liposome-based systems into magnetic lipogels provides several functions, which includes structural, catalytic, storage, and communication functions with spatiotemporal control, besides minimizing the instability, and burst release of the liposomes. In addition, the stoichiometric control over different reservoirs in the magnetic lipogel and the real-time monitoring through MRI, make these systems ideal for the development of multimodal therapeutic strategies. For example, the system can comprise liposomes with different triggers or phase transition temperatures, to enable a sequential in situ administration of different therapeutic and/or diagnostic agents, which can be useful in the controlled drug delivery of unstable or easily metabolized molecules, and on the fabrication of smart cell culture scaffolds. Therefore, the magnetic lipogels represent a novel generation of materials that combines magnetic nanoparticles, hydrogel, and (magnetic) liposomes advancements, offering the opportunity to develop theranostic strategies which can potentially translate to clinical studies, and solve several challenges in nanomedicine and biomedical applications.

References

1. P. Kumari, B. Ghosh, S. Biswas, J. Drug Target. **24**, 179 (2016)
2. F. Bray, J. Ferlay, I. Soerjomataram, R.L. Siegel, L.A. Torre, A. Jemal, CA. Cancer J. Clin. **68**, 394 (2018)
3. H. Nagai, Y.H. Kim, J. Thorac. Dis. **9**, 448 (2017)
4. C.-Y. Huang, D.-T. Ju, C.-F. Chang, P. Muralidhar Reddy, B.K. Velmurugan, BioMedicine **7**, 23 (2017)

5. E. Ozdemir-Kaynak, A.A. Qutub, O. Yesil-Celiktas, Front. Physiol. **9**, 170 (2018)
6. X. Dong, Theranostics **8**, 1481 (2018)
7. R. De Souza, P. Zahedi, C.J. Allen, M. Piquette-Miller, Drug Deliv. **17**, 365 (2010)
8. WHO, *WHO Report on Cancer: Setting Priorities, Investing Wisely and Providing Care for All* (2020), p. 13
9. C. Zhang, L. Yan, X. Wang, S. Zhu, C. Chen, Z. Gu, Y. Zhao, Nano Today **35**, 101008 (2020)
10. S. Lakkadwala, B. dos Santos Rodrigues, C. Sun, J. Singh, J. Control. Release **307**, 247 (2019)
11. M. Alavi, M. Hamidi, Drug Metab. Pers. Ther. **34**, 20180032 (2019)
12. S. Zununi Vahed, N. Fathi, M. Samiei, S. Maleki Dizaj, S. Sharifi, J. Drug Target **27**, 292 (2018)
13. W. Wu, M. Chen, T. Luo, Y. Fan, J. Zhang, Y. Zhang, Q. Zhang, A. Sapin-Minet, C. Gaucher, X. Xia, Acta Biomater. **103**, 259 (2020)
14. X. Wei, L. Liu, X. Li, Y. Wang, X. Guo, J. Zhao, S. Zhou, J. Control. Release **313**, 42 (2019)
15. Y.J. Lu, E.Y. Chuang, Y.H. Cheng, T.S. Anilkumar, H.A. Chen, J.P. Chen, Chem. Eng. J. **373**, 720 (2019)
16. E.R. Riva, E. Sinibaldi, A.F. Grillone, S. Del Turco, A. Mondini, T. Li, S. Takeoka, V. Mattoli, Nanomater **10**, 2104 (2020)
17. Y. Li, X. Zheng, Q. Chu, Nano Today **38**, 101134 (2021)
18. J. Mu, J. Lin, P. Huang, X. Chen, Chem. Soc. Rev. **47**, 5554 (2018)
19. M. Norouzi, B. Nazari, D.W. Miller, Drug Discov. Today **21**, 1835 (2016)
20. M. Sepantafar, R. Maheronnaghsh, H. Mohammadi, F. Radmanesh, M.M. Hasani-sadrabadi, M. Ebrahimi, H. Baharvand, Trends Biotechnol. **35**, 1074 (2017)
21. J.H. Lee, Biomater. Res. **22**, 27 (2018)
22. W. Kang, X. Kang, Z.A. Lashari, Z. Li, B. Zhou, H. Yang, B. Sarsenbekuly, S. Aidarova, Adv. Colloid Interface Sci. **289**, 102363 (2021)
23. L. Gao, X. Wang, J. Ma, D. Hao, P. Wei, L. Zhou, G. Liu, Colloids Surf. B Biointerfaces **140**, 307 (2016)
24. G. Xu, B. Li, T. Wang, J. Wan, Y. Zhang, J. Huang, Y. Shen, RSC Adv. **8**, 21229 (2018)
25. K. Shi, B. Xue, Y. Jia, L. Yuan, R. Han, F. Yang, J. Peng, Z. Qian, Nano Res. **126**(12), 1389 (2019)
26. Y. Liang, C. Dong, J. Zhang, L. Deng, A. Dong, Drug Dev. Ind. Pharm. **43**, 972 (2017)
27. M. Zhao, F. Danhier, C. Bastiancich, N. Joudiou, L.P. Ganipineni, N. Tsakiris, B. Gallez, A. des Rieux, A. Jankovski, J. Bianco, V. Préat, Int. J. Pharm. **548**, 522 (2018)
28. Y. Wu, H. Wang, F. Gao, Z. Xu, F. Dai, W. Liu, Adv. Funct. Mater. **28**, 1801000 (2018)
29. Y.W. Lo, M.T. Sheu, W.H. Chiang, Y.L. Chiu, C.M. Tu, W.Y. Wang, M.H. Wu, Y.C. Wang, M. Lu, H.O. Ho, Acta Biomater. **86**, 280 (2019)
30. L. Xu, R.C. Cooper, J. Wang, W.A. Yeudall, H. Yang, A.C.S. Biomater, Sci. Eng. **3**, 1641 (2017)
31. L. Saeednia, L. Yao, K. Cluff, R. Asmatulu, ACS Omega **4**, 4040 (2019)
32. F.Y.J. Huang, C.C. Hung, C.W. Chang, J.H. Chao, B.T. Hsieh, Anticancer Res **38**, 4651 (2018)
33. S. Ohta, S. Hiramoto, Y. Amano, S. Emoto, H. Yamaguchi, H. Ishigami, J. Kitayama, T. Ito, Mol. Pharm. **14**, 3105 (2017)
34. X. Chen, Z. Liu, Macromol. Rapid Commun. **37**, 1533 (2016)
35. C. Wang, X. Wang, K. Dong, J. Luo, Q. Zhang, Y. Cheng, Biomaterials **104**, 129 (2016)
36. M. Liu, X. Song, Y. Wen, J.L. Zhu, J. Li, A.C.S. Appl, Mater. Interfaces **9**, 35673 (2017)
37. X. Niu, Z. Zhang, Y. Zhong, Mater. Sci. Eng. C Mater. Biol. Appl. **77**, 888 (2017)
38. O. Vittorio, M. Brandl, G. Cirillo, K. Kimpton, E. Hinde, K. Gaus, E. Yee, N. Kumar, H. Duong, C. Fleming, M. Haber, M. Norris, C. Boyer, M. Kavallaris, Oncotarget **7**, 47479 (2016)
39. Y. Xiao, Y. Gu, L. Qin, L. Chen, X. Chen, W. Cui, F. Li, N. Xiang, X. He, Colloids Surf. B Biointerfaces **200**, 111581 (2021)
40. K.H. Bouhadir, E. Alsberg, D.J. Mooney, Biomaterials **22**, 2625 (2001)
41. B. Miao, C. Song, G. Ma, J. Appl. Polym. Sci. **122**, 2139 (2011)
42. Y.Y. Chen, H.C. Wu, J.S. Sun, G.C. Dong, T.W. Wang, Langmuir **29**, 3721 (2013)

43. H. Ma, C. He, Y. Cheng, Z. Yang, J. Zang, J. Liu, X. Chen, A.C.S. Appl, Mater. Interfaces **7**, 27040 (2015)
44. V.H.G. Phan, E. Lee, J.H. Maeng, T. Thambi, B.S. Kim, D. Lee, D.S. Lee, RSC Adv. **6**, 41644 (2016)
45. L. Wei, J. Chen, S. Zhao, J. Ding, X. Chen, Acta Biomater. **58**, 44 (2017)
46. C. Al Sabbagh, J. Seguin, E. Agapova, D. Kramerich, V. Boudy, N. Mignet, Eur. J. Pharm. Biopharm. **157**, 154 (2020)
47. C.K. Chung, J. García-Couce, Y. Campos, D. Kralisch, K. Bierau, A. Chan, F. Ossendorp, L.J. Cruz, Mol. **25**, 2219 (2020)
48. M. Norouzi, J. Firouzi, N. Sodeifi, M. Ebrahimi, D.W. Miller, Int. J. Pharm. **598**, 120316 (2021)
49. C. Cheng, X. Zhang, Y. Meng, L. Chen, Q. Zhang, J. Mater. Chem. B **5**, 8487 (2017)
50. A. Hendi, M.U. Hassan, M. Elsherif, B. Alqattan, S. Park, A.K. Yetisen, H. Butt, Int. J. Nanomedicine **15**, 3887 (2020)
51. E. Persi, M. Duran-Frigola, M. Damaghi, W.R. Roush, P. Aloy, J.L. Cleveland, R.J. Gillies, E. Ruppin, Nat. Commun. **9**, 2997 (2018)
52. D.Y. Fan, Y. Tian, Z.J. Liu, Front. Chem. **7**, 675 (2019)
53. F. Andrade, M.M. Roca-Melendres, E.F. Durán-Lara, D. Rafael, S. Schwartz Jr., Cancers **13**(5), 1164 (2021)
54. W. Zhang, X. Jin, H. Li, R. run Zhang, C. wei Wu, Carbohydr. Polym. **186**, 82 (2018)
55. X. Wang, J. Luo, L. He, X. Cheng, G. Yan, J. Wang, R. Tang, J. Colloid Interface Sci. **525**, 269 (2018)
56. Y. Liang, X. Zhao, P.X. Ma, B. Guo, Y. Du, X. Han, J. Colloid Interface Sci. **536**, 224 (2019)
57. S.R. Batool, M.A. Nazeer, D. Ekinci, A. Sahin, S. Kizilel, Int. J. Biol. Macromol. **150**, 315 (2020)
58. X. Xu, Z. Huang, Z. Huang, X. Zhang, S. He, X. Sun, Y. Shen, M. Yan, C. Zhao, A.C.S. Appl, Mater. Interfaces **9**, 20361 (2017)
59. M. Yang, S.Y. Lee, S. Kim, J.S. Koo, J.H. Seo, D.I. Jeong, C.R. Hwang, J. Lee, H.J. Cho, J. Control. Release **324**, 750 (2020)
60. Z. Chen, R. Chen, C. Zhao, Z. Quan, H. Zhu, L. Wang, Q. Bu, Y. He, H. He, Chem. Eng. J. **431**, 133255 (2022)
61. L. Zheng, C. Li, X. Huang, X. Lin, W. Lin, F. Yang, T. Chen, Biomaterials **216**, 119220 (2019)
62. G. Chang, H. Zhang, S. Li, F. Huang, Y. Shen, A. Xie, Mater. Sci. Eng. C Mater. Biol. Appl. **99**, 1392 (2019)
63. C. Diaferia, E. Rosa, A. Accardo, G. Morelli, J. Pept. Sci. **28**, e3301 (2022)
64. S. Das, D. Das, Front. Chem. **9**, 770102 (2021)
65. Z. Zhang, S. Ai, Z. Yang, X. Li, Adv. Drug Deliv. Rev. **174**, 482 (2021)
66. T. Guan, J. Li, C. Chen, Y. Liu, Adv. Sci. **9**, 2104165 (2022)
67. Z. Du, B. Fan, Q. Dai, L. Wang, J. Guo, Z. Ye, N. Cui, J. Chen, K. Tan, R. Li, W. Tang, Giant **9**, 100082 (2022)
68. C. Liu, Q. Zhang, S. Zhu, H. Liu, J. Chen, RSC Adv. **9**, 28299 (2019)
69. E. Radvar, H.S. Azevedo, Macromol. Biosci. **19**, 1800221 (2019)
70. A. Dasgupta, J.H. Mondal, D. Das, RSC Adv. **3**, 9117 (2013)
71. A. Mahler, M. Reches, M. Rechter, S. Cohen, E. Gazit, Adv. Mater. **18**, 1365 (2006)
72. W.Y. Seow, C.A.E. Hauser, Mater. Today **17**, 381 (2014)
73. N. Yadav, M.K. Chauhan, V.S. Chauhan, Biomater. Sci. **8**, 84 (2020)
74. H. Wang, Z. Yang, Nanoscale **4**, 5259 (2012)
75. E.R. Draper, D.J. Adams, Langmuir **35**, 6506 (2019)
76. E.R. Draper, T.O. McDonald, D.J. Adams, Chem. Commun. **51**, 6595 (2015)
77. D. Li, Y. Shi, L. Wang, Chin. J. Chem. **32**, 123 (2014)
78. L.B. Gallagher, E.B. Dolan, J. O'Sullivan, R. Levey, B.L. Cavanagh, L. Kovarova, M. Pravda, V. Velebny, T. Farrell, F.J. O'Brien, G.P. Duffy, Acta Biomater. **107**, 78 (2020)
79. F. Yang, C.G. Williams, D.A. Wang, H. Lee, P.N. Manson, J. Elisseeff, Biomaterials **26**, 5991 (2005)

80. J.J. Panda, R. Dua, A. Mishra, B. Mittra, V.S. Chauhan, A.C.S. Appl, Mater. Interfaces **2**, 2839 (2010)
81. L. Mei, K. Xu, Z. Zhai, S. He, T. Zhu, W. Zhong, Org. Biomol. Chem. **17**, 3853 (2019)
82. J.P. Feng, R. Zhu, F. Jiang, J. Xie, C. Gao, M. Li, H. Jin, D. Fu, Biomater. Sci. **8**, 4559 (2020)
83. S. Koutsopoulos, S. Zhang, Acta Biomater. **9**, 5162 (2013)
84. A. Zhou, S. Chen, B. He, W. Zhao, X. Chen, D. Jiang, Drug Des. Devel. Ther. **10**, 3043 (2016)
85. C. Karavasili, E. Panteris, I.S. Vizirianakis, S. Koutsopoulos, D.G. Fatouros, Pharm. Res. **35**, 166 (2018)
86. A. Maslovskis, J.B. Guilbaud, I. Grillo, N. Hodson, A.F. Miller, A. Saiani, Langmuir **30**, 10471 (2014)
87. M.N. Barcellona, J.E. Speer, L. Jing, D.S. Patil, M.C. Gupta, J.M. Buchowski, L.A. Setton, Acta Biomater. **131**, 117 (2021)
88. K. Rajangam, H.A. Behanna, M.J. Hui, X. Han, J.F. Hulvat, J.W. Lomasney, S.I. Stupp, Nano Lett. **6**, 2086 (2006)
89. J. Hoque, N. Sangaj, S. Varghese, Macromol. Biosci. **19**, 1800259 (2019)
90. Q. Shi, H. Liu, D. Tang, Y. Li, X. Li, F. Xu, NPG Asia Mater. **11**, 64 (2019)
91. X.Q. Dou, C.L. Feng, Adv. Mater. **29**, 1604062 (2017)
92. F. Raza, Y. Zhu, L. Chen, X. You, J. Zhang, A. Khan, M.W. Khan, M. Hasnat, H. Zafar, J. Wu, L. Ge, Biomater. Sci. **7**, 2023 (2019)
93. M. Tao, K. Xu, S. He, H. Li, L. Zhang, X. Luo, W. Zhong, Chem. Commun. **54**, 4673 (2018)
94. R.B. Franklin, B. Milon, P. Feng, L.C. Costello, Front. Biosci. **10**, 2230 (2005)
95. L.M. De Leon-Rodriguez, Y. Hemar, G. Mo, A.K. Mitra, J. Cornish, M.A. Brimble, Acta Biomater. **47**, 40 (2017)
96. X. Wang, C. Wang, Q. Zhang, Y. Cheng, Chem. Commun. **52**, 978 (2016)
97. M.E. Roth-Konforti, M. Comune, M. Halperin-Sternfeld, I. Grigoriants, D. Shabat, L. Adler-Abramovich, Macromol. Rapid Commun. **39**, e1800588 (2018)
98. Z. Yang, H. Gu, D. Fu, P. Gao, J.K. Lam, B. Xu, Adv. Mater. **16**, 1440 (2004)
99. C. Wu, R. Li, Y. Yin, J. Wang, L. Zhang, W. Zhong, Mater. Sci. Eng. C **76**, 196 (2017)
100. X. Liu, X. Sun, G. Liang, Biomater. Sci. **9**, 315 (2021)
101. R. Misra, A. Sharma, A. Shiras, H.N. Gopi, Langmuir **33**, 7762 (2017)
102. G. Palui, J. Nanda, S. Ray, A. Banerjee, Chem. - A Eur. J. **15**, 6902 (2009)
103. J.H. Kim, S.Y. Lim, D.H. Nam, J. Ryu, S.H. Ku, C.B. Park, Biosens. Bioelectron. **26**, 1860 (2011)
104. S. Biswas, D.B. Rasale, A.K. Das, RSC Adv. **6**, 54793 (2016)
105. P.J. Jervis, C. Amorim, T. Pereira, J.A. Martins, P.M.T. Ferreira, Int. J. Mol. Sci. **22**, 2528 (2021)
106. H. Vilaça, A.C.L. Hortelão, E.M.S. Castanheira, M.J.R.P. Queiroz, L. Hilliou, I.W. Hamley, J.A. Martins, P.M.T. Ferreira, Biomacromolecules **16**, 3562 (2015)
107. X. Li, M. Fu, J. Wu, C. Zhang, X. Deng, A. Dhinakar, W. Huang, H. Qian, L. Ge, Acta Biomater. **51**, 294 (2017)
108. J. Bai, C. Chen, J. Wang, Y. Zhang, H. Cox, J. Zhang, Y. Wang, J. Penny, T. Waigh, J.R. Lu, H. Xu, A.C.S. Appl, Mater. Interfaces **8**, 15093 (2016)
109. R. Vardanyan, V. Hruby, *Synth. Best-Seller Drugs*, ed. by R. Vardanyan, V. Hruby (Academic Press, 2016), pp. 573–643
110. A.S. Veiga, C. Sinthuvanich, D. Gaspar, H.G. Franquelim, M.A.R.B. Castanho, J.P. Schneider, Biomaterials **33**, 8907 (2012)
111. A. Baral, S. Roy, S. Ghosh, D. Hermida-Merino, I.W. Hamley, A. Banerjee, Langmuir **32**, 1836 (2016)
112. A.B.G. Lansdown, Curr. Probl. Dermatol. Basel, Karger **33**, 17 (2006)
113. Y. Liu, Y. Yang, C. Wang, X. Zhao, Nanoscale **5**, 6413 (2013)
114. A.M. Jonker, D.W.P.M. Löwik, J.C.M. van Hest, Chem. Mater. **24**, 759 (2012)
115. Y. Loo, Y.C. Wong, E.Z. Cai, C.H. Ang, A. Raju, A. Lakshmanan, A.G. Koh, H.J. Zhou, T.C. Lim, S.M. Moochhala, C.A.E. Hauser, Biomaterials **35**, 4805 (2014)

116. N.C. Carrejo, A.N. Moore, T.L. Lopez Silva, D.G. Leach, I.C. Li, D.R. Walker, J.D. Hartgerink, ACS Biomater. Sci. Eng. **4**, 1386 (2018)
117. J. Li, R. Xing, S. Bai, X. Yan, Soft Matter. **15**, 1704 (2019)
118. Z. Hao, H. Li, Y. Wang, Y. Hu, T. Chen, S. Zhang, X. Guo, L. Cai, J. Li, Adv. Sci. **9**, 2103820 (2022)
119. W. Ahn, J.H. Lee, S.R. Kim, J. Lee, E.J. Lee, J. Mater. Chem. B **9**, 1919 (2021)
120. X. Ding, H. Zhao, Y. Li, A.L. Lee, Z. Li, M. Fu, C. Li, Y.Y. Yang, P. Yuan, Adv. Drug Deliv. Rev. **160**, 78 (2020)
121. B. Cheng, Y. Yan, J. Qi, L. Deng, Z.-W. Shao, K.-Q. Zhang, B. Li, Z. Sun, X. Li, ACS Appl. Mater. Interfaces **10**, 12474 (2018)
122. D. MacPherson, Y. Bram, J. Park, R.E. Schwartz, Sci. Rep. **11**, 6772 (2021)
123. Y.-L. Wang, S.-P. Lin, S.R. Nelli, F.-K. Zhan, H. Cheng, T.-S. Lai, M.-Y. Yeh, H.-C. Lin, S.-C. Hung, Macromol. Biosci. **17**, 2147 (2017)
124. E. Rosa, C. Diaferia, E. Gianolio, T. Sibillano, E. Gallo, G. Smaldone, M. Stornaiuolo, C. Giannini, G. Morelli, A. Accardo, Macromol. Biosci. **22**, 2200128 (2022)
125. M. Halperin-Sternfeld, A. Pokhojaev, M. Ghosh, D. Rachmiel, R. Kannan, I. Grinberg, M. Asher, M. Aviv, P.X. Ma, I. Binderman, R. Sarig, L. Adler-Abramovich, J. Clin. Periodontol. **50**(2), 200 (2023)
126. A.D. Bangham, R.W. Horne, J. Mol. Biol. **8**, 660 (1964)
127. A.D. Bangham, M.M. Standish, J.C. Watkins, J. Mol. Biol. **13**, 238 (1965)
128. G. Gregoriadis, Pharmaceutics **8**, 19 (2016)
129. J. Patel, J. Oncol. Pharm. Pract. **2**, 201 (1996)
130. A. Gabizon, R. Isacson, E. Libson, B. Kaufman, B. Uziely, R. Catane, C.G. Ben-Dor, E. Rabello, Y. Cass, T. Peretz, A. Sulkes, R. Chisin, Y. Barenholz, Acta Oncol. (Madr.) **33**, 779 (1994)
131. J.S. Rudra, J.H. Collier, *Comprehensive Biomaterials*, ed. by P. Ducheyne (Elsevier, 2011), pp. 77–94
132. S.Y. Fung, Y. Hong, C. Keyes-Baig, P. Chen, *Molecular Interfacial Phenomena of Polymers and Biopolymers*, Woodhead Publishing Series in Biomaterials, ed. by P. Chen (Woodhead Publishing, 2005), pp. 421–474
133. D. Lombardo, P. Calandra, L. Pasqua, S. Magazù, Materials (Basel) **13**, 1048 (2020)
134. J.M. Seddon, Biochim. Biophys. Acta - Rev. Biomembr. **1031**, 1 (1990)
135. P.M. Glassman, V.R. Muzykantov, J. Pharmacol Exp. Ther. **370**, 570 (2019)
136. T. Olusanya, R. Haj Ahmad, D. Ibegbu, J. Smith, A. Elkordy, Molecules **23**, 907 (2018)
137. S. Hossen, M.K. Hossain, M.K. Basher, M.N.H. Mia, M.T. Rahman, M.J. Uddin, J. Adv. Res. **15**, 1 (2019)
138. G. Caracciolo, Nanoscale **10**, 4167 (2018)
139. Z. Huang, X. Li, T. Zhang, Y. Song, Z. She, J. Li, Y. Deng, Asian. J. Pharm. Sci. **9**, 176 (2014)
140. P. Trucillo, R. Campardelli, E. Reverchon, Processes **8**, 1022 (2020)
141. C. Has, P. Sunthar, J. Liposome Res. **30**, 336 (2020)
142. S. Zheng, H. Alkan-Onyuksel, R.L. Beissinger, D.T. Wasan, J. Dispers. Sci. Technol. **20**, 1189 (1999)
143. D. Zucker, D. Marcus, Y. Barenholz, A. Goldblum, J. Control. Release **139**, 73 (2009)
144. P. Stano, S. Bufali, C. Pisano, F. Bucci, M. Barbarino, M. Santaniello, P. Carminati, P.L. Luisi, J. Liposome Res. **14**, 87 (2004)
145. D.W. Deamer, Ann. N. Y. Acad. Sci. **308**, 250 (1978)
146. A. Wagner, K. Vorauer-Uhl, G. Kreismayr, H. Katinger, J. Liposome Res. **12**, 259 (2002)
147. T.T. Pham, C. Jaafar-Maalej, C. Charcosset, H. Fessi, Colloids Surf. B Biointerfaces **94**, 15 (2012)
148. A. Laouini, C. Jaafar-Maalej, I. Limayem-Blouza, S. Sfar, C. Charcosset, H. Fessi, J. Colloid Sci. Biotechnol. **1**, 147 (2012)
149. D. Carugo, E. Bottaro, J. Owen, E. Stride, C. Nastruzzi, Sci. Rep. **6**, 25876 (2016)
150. G. Vladisavljević, R. Al Nuumani, S. Nabavi, Micromachines **8**, 75 (2017)
151. N. Yandrapalli, J. Petit, O. Bäumchen, T. Robinson, Commun. Chem. **4**, 100 (2021)

152. C. Charcosset, A. Juban, J.-P. Valour, S. Urbaniak, H. Fessi, Chem. Eng. Res. Des. **94**, 508 (2015)
153. C. Jaafar-Maalej, C. Charcosset, H. Fessi, J. Liposome Res. **21**, 213 (2011)
154. Y.P. Patil, S. Jadhav, Chem. Phys. Lipids **177**, 8 (2014)
155. P. Guo, J. Huang, Y. Zhao, C.R. Martin, R.N. Zare, M.A. Moses, Small **14**, 1703493 (2018)
156. L. Maja, K. Željko, P. Mateja, J. Supercrit. Fluids **165**, 104984 (2020)
157. S. Shah, V. Dhawan, R. Holm, M.S. Nagarsenker, Y. Perrie, Adv. Drug Deliv. Rev. **154–155**, 102 (2020)
158. R.K. Kankala, Y.S. Zhang, S.-B. Wang, C.-H. Lee, A.-Z. Chen, Adv. Healthc. Mater. **6**, 1700433 (2017)
159. N. Penoy, B. Grignard, B. Evrard, G. Piel, Int. J. Pharm. **592**, 120093 (2021)
160. B. William, P. Noémie, E. Brigitte, P. Géraldine, Chem. Eng. J. **383**, 123106 (2020)
161. R. Parhi, P. Suresh, J. Adv. Pharm. Sci. Technol. **1**, 13 (2013)
162. U. Massing, S.G. Ingebrigtsen, N. Škalko-Basnet, A.M. Holsæter, *Liposomes*, ed. by A. Catala (IntechOpen, 2017), pp. 3–28
163. U. Massing, S. Cicko, V. Ziroli, J. Control. Release **125**, 16 (2008)
164. M. Frenzel, E. Krolak, A.E. Wagner, A. Steffen-Heins, LWT - Food Sci. Technol. **63**, 527 (2015)
165. J.Y. Yu, P. Chuesiang, G.H. Shin, H.J. Park, Pharmaceutics **13**, 1023 (2021)
166. S. Franzé, F. Selmin, E. Samaritani, P. Minghetti, F. Cilurzo, Pharmaceutics **10**, 139 (2018)
167. F. Susa, G. Bucca, T. Limongi, V. Cauda, R. Pisano, Cryobiology **98**, 46 (2021)
168. N. Singh, P. Kushwaha, U. Ahmad, M. Abdullah, Ars Pharm. **60**, 231 (2019)
169. M. Malamatari, A. Charisi, S. Malamataris, K. Kachrimanis, I. Nikolakakis, Processes **8**, 788 (2020)
170. I. Khatib, M.Y.T. Chow, J. Ruan, D. Cipolla, H.-K. Chan, Int. J. Pharm. **597**, 120277 (2021)
171. D. Akgün, M. Gültekin-Özgüven, A. Yücetepe, G. Altin, M. Gibis, J. Weiss, B. Özçelik, Food Hydrocoll. **101**, 105532 (2020)
172. D.A. Vishali, J. Monisha, S.K. Sivakamasundari, J.A. Moses, C. Anandharamakrishnan, J. Control. Release **300**, 93 (2019)
173. M.E. Ali, A. Lamprecht, Int. J. Pharm. **516**, 170 (2017)
174. H. Daraee, A. Etemadi, M. Kouhi, S. Alimirzalu, A. Akbarzadeh, Artif. Cells, Nanomedicine, Biotechnol. **44**, 381 (2016)
175. A. Meriem, A. Linna, M. Philip, N. Franz, Front. Plant Sci. **5**, 109 (2014)
176. P. Yingchoncharoen, D.S. Kalinowski, D.R. Richardson, Pharmacol. Rev. **68**, 701 (2016)
177. S. Hua, S.Y. Wu, Front. Pharmacol. **4**, 143 (2013)
178. O. Nag, V. Awasthi, Pharmaceutics **5**, 542 (2013)
179. M. Li, S. Jiang, J. Simon, D. Paßlick, M.-L. Frey, M. Wagner, V. Mailänder, D. Crespy, K. Landfester, Nano Lett. **21**, 1591 (2021)
180. V. Quagliariello, M. Masarone, E. Armenia, A. Giudice, M. Barbarisi, M. Caraglia, A. Barbarisi, M. Persico, Oncol. Rep. **41**, 1476 (2018)
181. M.K. Yu, J. Park, S. Jon, Theranostics **2**, 3 (2012)
182. D. Wang, Y. Sun, Y. Liu, F. Meng, R.J. Lee, Expert Opin. Drug Deliv. **15**, 893 (2018)
183. R. Gaspar, F. Coelho, B.F.B. Silva, Molecules **25**, 5006 (2020)
184. B. Sternberg, F.L. Sorgi, L. Huang, FEBS Lett. **356**, 361 (1994)
185. K.K. Ewert, P. Scodeller, L. Simón-Gracia, V.M. Steffes, E.A. Wonder, T. Teesalu, C.R. Safinya, Pharmaceutics **13**, 1365 (2021)
186. S.L. Ginn, A.K. Amaya, I.E. Alexander, M. Edelstein, M.R. Abedi, J. Gene Med. **20**, e3015 (2018)
187. J.E. Zuckerman, M.E. Davis, Nat. Rev. Drug Discov. **14**, 843 (2015)
188. U. Sahin, Ö. Türeci, Science **359**, 1355 (2018)
189. H.S. El-Sawy, A.M. Al-Abd, T.A. Ahmed, K.M. El-Say, V.P. Torchilin, ACS Nano **12**, 10636 (2018)
190. A. Raza, T. Rasheed, F. Nabeel, U. Hayat, M. Bilal, H. Iqbal, Molecules **24**, 1117 (2019)
191. Y. Lee, D.H. Thompson, WIREs Nanomed. Nanobiotechnol. **9**, e1450 (2017)

192. E. Mazzotta, L. Tavano, R. Muzzalupo, Pharmaceutics **10**, 150 (2018)
193. I.S.R. Rio, A.R.O. Rodrigues, J.M. Rodrigues, M.-J.R.P. Queiroz, R.C. Calhelha, I.C.F.R. Ferreira, B.G. Almeida, A. Pires, A.M. Pereira, J.P. Araújo, E.M.S. Castanheira, P.J.G. Coutinho, Pharmaceutics **13**, 1905 (2021)
194. B.D. Cardoso, A.R.O. Rodrigues, B.G. Almeida, C.O. Amorim, V.S. Amaral, E.M.S. Castanheira, P.J.G. Coutinho, Int. J. Mol. Sci. **21**, 3641 (2020)
195. C.B. Giuliano, N. Cvjetan, J. Ayache, P. Walde, ChemSystemsChem **3**, e2000049 (2021)
196. M. Shafiei, M.N.M. Ansari, S.I.A. Razak, M.U.A. Khan, Polymers (Basel) **13**, 2529 (2021)
197. R. Cheng, L. Liu, Y. Xiang, Y. Lu, L. Deng, H. Zhang, H.A. Santos, W. Cui, Biomaterials **232**, 119706 (2020)
198. N. Monteiro, A. Martins, R.L. Reis, N.M. Neves, J.R. Soc, Interface **11**, 20140459 (2014)
199. J. Zhou, Q. Wang, C. Zhang, J. Am. Chem. Soc. **135**, 2056 (2013)
200. N.S. Awad, M. Haider, V. Paul, N.M. AlSawaftah, J. Jagal, R. Pasricha, G.A. Husseini, Pharmaceutics **13**, 2073 (2021)
201. G. Aizik, N. Waiskopf, M. Agbaria, M. Ben-David-Naim, Y. Levi-Kalisman, A. Shahar, U. Banin, G. Golomb, Nano Lett. **19**, 5844 (2019)
202. Q. Wang, Y.-M. Chao, J. Biomed. Res. **32**, 91 (2018)
203. Y. Barenholz, J. Control. Release **160**, 117 (2012)
204. D. Bobo, K.J. Robinson, J. Islam, K.J. Thurecht, S.R. Corrie, Pharm. Res. **33**, 2373 (2016)
205. L.B. Margolis, V.A. Namiot, L.M. Kljukin, Biochim. Biophys. Acta (BBA) - Biomembranes **735**, 193 (1983)
206. M. De Cuyper, M. Joniau, Eur. Biophys. J. **15**, 311 (1998)
207. C. Sangregorio, J.K. Wiemann, C.J.O. Connor, Z. Rosenzweig, J. Appl. Phys. **85**, 5699 (2016)
208. C.A. Monnier, D. Burnand, B. Rothen-rutishauser, M. Lattuada, Eur. J. Nanomed. **6**, 201 (2014)
209. S.R.S. Veloso, R.G.D. Andrade, E.M.S. Castanheira, Expert Opin. Drug Deliv. **18**, 1323 (2021)
210. A.R.O. Rodrigues, J.M.F. Ramos, I.T. Gomes, B.G. Almeida, J.P. Araújo, M.J.R.P. Queiroz, P.J.G. Coutinho, E.M.S. Castanheira, RSC Adv. **6**, 17302 (2016)
211. B.D. Cardoso, I.S.R. Rio, A. Rita, O. Rodrigues, F.C.T. Fernandes, B.G. Almeida, A. Pires, A.M. Pereira, J.P. Araújo, E.M.S. Castanheira, P.J.G. Coutinho, R. Soc, Open Sci. **5**, 181017 (2018)
212. I.S.R. Rio, A.R.O. Rodrigues, C.P. Rodrigues, B.G. Almeida, A. Pires, A.M. Pereira, J.P. Araújo, E.M.S. Castanheira, P.J.G. Coutinho, Materials (Basel) **13**, 815 (2020)
213. A.R.O. Rodrigues, B.G. Almeida, J.M. Rodrigues, M.J.R.P. Queiroz, R.C. Calhelha, I.C.F.R. Ferreira, A. Pires, A.M. Pereira, J.P. Araújo, P.J.G. Coutinho, E.M.S. Castanheira, RSC Adv. **7**, 15352 (2017)
214. A.R.O. Rodrigues, I.T. Gomes, B.G. Almeida, J.P. Araújo, E.M.S. Castanheira, P.J.G. Coutinho, Mater. Chem. Phys. **148**, 978 (2014)
215. M.R. Ghazanfari, M.R. Jaafari, S.F. Shams, M. Kashefi, Mater. Today Commun. **13**, 102 (2017)
216. T.S. Anilkumar, K.T. Shalumon, J. Chen, Curr. Pharm. Des. **25**, 1490 (2019)
217. P. Pradhan, J. Giri, R. Banerjee, J. Bellare, D. Bahadur, J. Magn. Magn. Mater. **311**, 208 (2007)
218. M. Gogoi, S. Patra, *Multifunctional Magnetic Liposomes for Cancer Imaging and Therapeutic Applications*, ed. by A.M. Holban, A.M. Grumezescu (Elsevier, 2016), pp. 743–782
219. S. Cruz dos Santos, N. Osti Silva, J.B. dos Santos Espinelli, M.A. Germani Marinho, Z. Vieira Borges, N. Bruzamarello Caon Branco, F.L. Faita, B. Meira Soares, A.P. Horn, A.L. Parize, V. Rodrigues de Lima, Chem. Phys. Lipids **218**, 22 (2019)
220. A. Mota-Cobián, C. Velasco, J. Mateo, S. Espaa, Nanotechnology **31**, 145102 (2020)
221. B.S. Shaghasemi, M.M. Virk, E. Reimhult, Sci. Rep. **7**, 7474 (2017)
222. Y. Chen, A. Bose, G.D. Bothun, ACS Nano **4**, 3215 (2010)
223. M. Namdari, M. Cheraghi, B. Negahdari, A. Eatemadi, H. Daraee, Artif. Cells, Nanomedicine Biotechnol. **45**, 1051 (2017)
224. W. Il Choi, A. Sahu, F.R. Wurm, S.-M. Jo, RSC Adv. **9**, 15053 (2019)

225. E.C. Abenojar, S. Wickramasinghe, J. Bas-Concepcion, A.C.S. Samia, Prog. Nat. Sci. Mater. Int. **26**, 440 (2016)
226. I. Obaidat, B. Issa, Y. Haik, Nanomaterials **5**, 63 (2015)
227. Y.I. Golovin, D.Y. Golovin, K.Y. Vlasova, M.M. Veselov, A.D. Usvaliev, A.V. Kabanov, N.L. Klyachko, Nanomaterials **11**, 225 (2021)
228. Z. Li, Y. Li, C. Chen, Y. Cheng, J. Control. Release **335**, 541 (2021)
229. C. Wu, Y. Shen, M. Chen, K. Wang, Y. Li, Y. Cheng, Adv. Mater. **30**, 1707572 (2018)
230. A. Hervault, N.T.K. Thanh, Nanoscale **6**, 11553 (2014)
231. Z.Q. Zhang, S.C. Song, Biomaterials **106**, 13 (2016)
232. J. Wu, P. Ning, R. Gao, Q. Feng, Y. Shen, Y. Zhang, Y. Li, C. Xu, Y. Qin, G.R. Plaza, Q. Bai, X. Fan, Z. Li, Y. Han, M.S. Lesniak, H. Fan, Y. Cheng, Adv. Sci. **7**, 1902933 (2020)
233. Q. Zhang, J. Wu, J. Wang, X. Wang, C. Wu, M. Chen, Q. Wu, M.S. Lesniak, Y. Mi, Y. Cheng, Q. Wang, Angew. Chem. - Int. Ed. **59**, 3732 (2020)
234. M.B. Yatvin, J.N. Weinstein, W.H. Dennis, R. Blumenthal, Science **202**, 1290 (1978)
235. L.A. Tai, P.J. Tsai, Y.C. Wang, Y.J. Wang, L.W. Lo, C.S. Yang, Nanotechnology **20**, 135101 (2009)
236. H. Bi, X. Han, Chem. Phys. Lett. **706**, 455 (2018)
237. J. Dong, J.I. Zink, ACS Nano **8**, 5199 (2014)
238. B.M. Salah., M. Rady, M. Abdel-Halim, H.M. Fahmy, N.S. El-Din, M.H. Gaber, Biophysics **66**, 264 (2021)
239. M.E. Fortes Brollo, A. Domínguez-Bajo, A. Tabero, V. Domínguez-Arca, V. Gisbert, G. Prieto, C. Johansson, R. Garcia, A. Villanueva, M.C. Serrano, M.D.P. Morales, ACS Appl. Mater. Interfaces **12**, 4295 (2020)
240. A. Jordan, R. Scholz, P. Wust, H. Fähling, R. Felix, J. Magn. Magn. Mater. **201**, 413 (1999)
241. H. Guo, W. Chen, X. Sun, Y.N. Liu, J. Li, J. Wang, Carbohydr. Polym. **118**, 209 (2015)
242. R.S. Riley, E.S. Day, Wiley Interdiscip. Rev. Nanomedicine Nanobiotechnology **9**, e1449 (2017)
243. Y.W. Chen, Y.L. Su, S.H. Hu, S.Y. Chen, Adv. Drug Deliv. Rev. **105**, 190 (2016)
244. R. Robinson, W. Gerlach, H. Ghandehari, J. Control. Release **220**, 245 (2015)
245. R. Bardhan, S. Lal, A. Joshi, N.J. Halas, Acc. Chem. Res. **44**, 936 (2011)
246. P. Senthil Kumar, I. Pastoriza-Santos, B. Rodríguez-González, F. Javier García De Abajo, L.M. Liz-Marzán, Nanotechnology **19**, 015606 (2008)
247. D. Jaque, L. Martínez Maestro, B. Del Rosal, P. Haro-Gonzalez, A. Benayas, J.L. Plaza, E. Martín Rodríguez, J. García Solé, Nanoscale **6**, 9494 (2014)
248. S. Shen, S. Wang, R. Zheng, X. Zhu, X. Jiang, D. Fu, W. Yang, Biomaterials **39**, 67 (2015)
249. M. Chu, Y. Shao, J. Peng, X. Dai, H. Li, Q. Wu, D. Shi, Biomaterials **34**, 4078 (2013)
250. J. Kolosnjaj-Tabi, L. Lartigue, Y. Javed, N. Luciani, T. Pellegrino, C. Wilhelm, D. Alloyeau, F. Gazeau, Nano Today **11**, 280 (2016)
251. S. Shen, D. Huang, J. Cao, Y. Chen, X. Zhang, S. Guo, W. Ma, X. Qi, Y. Ge, L. Wu, J. Mater. Chem. B **7**, 1096 (2019)
252. H. Yan, W. Shang, X. Sun, L. Zhao, J. Wang, Z. Xiong, J. Yuan, R. Zhang, Q. Huang, K. Wang, B. Li, J. Tian, F. Kang, S.S. Feng, Adv. Funct. Mater. **28**, 1705710 (2018)
253. M.E. Khosroshahi, J. Nanomed. Nanotechnol. **06**, 2 (2015)
254. A.R.O. Rodrigues, J.O.G. Matos, A.M. Nova Dias, B.G. Almeida, A. Pires, A.M. Pereira, J.P. Araújo, M.J.R.P. Queiroz, E.M.S. Castanheira, P.J.G. Coutinho, Pharmaceutics **11**, 10 (2019)
255. T.S. Anilkumar, Y.J. Lu, J.P. Chen, Int. J. Mol. Sci. **21**, 5187 (2020)
256. P. Marqués-Gallego, A.I.P.M. De Kroon, Biomed. Res. Int. **2014**, 2014 (2014)
257. P.T. Wong, S.K. Choi, Int. J. Mol. Sci. **16**, 1772 (2015)
258. Y. Guo, Y. Zhang, J. Ma, Q. Li, Y. Li, X. Zhou, D. Zhao, H. Song, Q. Chen, X. Zhu, J. Control. Release **272**, 145 (2018)
259. Y. Chen, J. Lu, J. Yang, K. Hao, M. Li, J. Nanosci. Nanotechnol. **20**, 5147 (2020)
260. S. Meier, G. Pütz, U. Massing, C.E. Hagemeyer, D. von Elverfeldt, M. Meißner, K. Ardipradja, S. Barnert, K. Peter, C. Bode, R. Schubert, C. von zur Muhlen, Biomaterials **53**, 137 (2015)

261. R. So, G. Ribeiro, S. Belderbos, P. Danhier, J. Gallo, B.B. Manshian, B. Gallez, M. Bañobre, M. De Cuyper, U. Himmelreich, 5911 (2019)
262. B. Dorjsuren, B. Chaurasiya, Z. Ye, Y. Liu, W. Li, C. Wang, D. Shi, C.E. Evans, T.J. Webster, Y. Shen, Int. J. Nanomedicine **15**, 8201 (2020)
263. Đ Cvjetinović, Z. Milanović, M. Mirković, J. Petrović, A. Vesković, A. Popović-Bijelić, Ž Prijović, D. Janković, S. Vranješ-Đurić, J. Nanoparticle Res. **23**, 252 (2021)
264. H. Ding, V. Sagar, M. Agudelo, S. Pilakka-Kanthikeel, V.S.R. Atluri, A. Raymond, T. Samikkannu, M.P. Nair, Nanotechnology **25**, 055101 (2014)
265. J. Liao, H. Huang, Biomacromolecules **21**, 2574 (2020)
266. S. Veloso, P. Ferreira, J. Martins, P. Coutinho, E. Castanheira, Pharmaceutics **10**, 145 (2018)
267. R. Contreras-Montoya, A.B. Bonhome-Espinosa, A. Orte, D. Miguel, J.M. Delgado-López, J.D.G. Duran, J.M. Cuerva, M.T. Lopez-Lopez, L.Á. De Cienfuegos, Mater. Chem. Front. **2**, 686 (2018)
268. S. Bhattacharya, S.K. Samanta, Chem. Rev. **116**, 11967 (2016)
269. P.J. Jervis, L. Hilliou, R.B. Pereira, D.M. Pereira, J.A. Martins, P.M.T. Ferreira, Nanomaterials **11**, 704 (2021)
270. H. Kim, H. Park, J.W. Lee, K.Y. Lee, Carbohydr. Polym. **151**, 467 (2016)
271. H. Kim, A. Jo, S. Baek, D. Lim, S.Y. Park, S.K. Cho, J.W. Chung, J. Yoon, Sci. Rep. **7**, 41090 (2017)
272. S.R.S. Veloso, J.F.G. Silva, L. Hilliou, C. Moura, P.J.G. Coutinho, J.A. Martins, M. Testa-Anta, V. Salgueiriño, M.A. Correa-Duarte, P.M.T. Ferreira, E.M.S. Castanheira, Nanomaterials **11**, 16 (2021)
273. Z.Q. Zhang, S.C. Song, Biomaterials **132**, 16 (2017)
274. S. Araújo-Custódio, M. Gomez-Florit, A.R. Tomás, B.B. Mendes, P.S. Babo, S.M. Mithieux, A. Weiss, R.M.A. Domingues, R.L. Reis, M.E. Gomes, A.C.S. Biomater, Sci. Eng. **5**, 1392 (2019)
275. S. Song, Y. Liu, A. Song, Z. Zhao, H. Lu, J. Hao, J. Colloid Interface Sci. **506**, 46 (2017)
276. J.I. Kim, C.J. Chun, B. Kim, J.M. Hong, J.K. Cho, S.H. Lee, S.C. Song, Biomaterials **33**, 218 (2012)
277. S.R.S. Veloso, J.A. Martins, L. Hilliou, C.O. Amorim, V.S. Amaral, B.G. Almeida, P.J. Jervis, R. Moreira, D.M. Pereira, P.J.G. Coutinho, P.M.T. Ferreira, E.M.S. Castanheira, J. Mater. Chem. B **8**, 45 (2020)
278. T. Mitsumata, Y. Horikoshi, K. Negami, Jpn. J. Appl. Phys. **47**, 7257 (2008)
279. B. Sung, M.H. Kim, L. Abelmann, Bioeng. Transl. Med. **6**, e10190 (2021)
280. J. Li, D.J. Mooney, Nat. Rev. Mater. **1**, 16071 (2016)
281. S.R.S. Veloso, R.G.D. Andrade, E.M.S. Castanheira, Adv. Colloid Interface Sci. **288**, 102351 (2021)
282. O.A. Dinislamova, A.V. Bugayova, T.F. Shklyar, A.P. Safronov, F.A. Blyakhman, Bioengineering **8**, 140 (2021)
283. F.A. Blyakhman, S.Y. Sokolov, A.P. Safronov, O.A. Dinislamova, T.F. Shklyar, A.Y. Zubarev, G.V. Kurlyandskaya, Sensors (Switzerland) **19**, 3959 (2019)
284. H. Hamidian, T. Tavakoli, Carbohydr. Polym. **144**, 140 (2016)
285. G.R. Mahdavinia, A. Mosallanezhad, M. Soleymani, M. Sabzi, Int. J. Biol. Macromol. **97**, 209 (2017)
286. M.M. Abrougui, M.T. Lopez-Lopez, J.D.G. Duran, Philos. Trans. R. Soc. A Math. Phys. Eng. Sci. **377**, 20180218 (2019)
287. A.B. Bonhome-Espinosa, F. Campos, I.A. Rodriguez, V. Carriel, J.A. Marins, A. Zubarev, J.D.G. Duran, M.T. Lopez-Lopez, Soft Matter **13**, 2928 (2017)
288. J. Zhang, Q. Huang, J. Du, Polym. Int. **65**, 1365 (2016)
289. R. Weeber, M. Hermes, A.M. Schmidt, C. Holm, J. Phys. Condens. Matter **30**, 063002 (2018)
290. M. Häring, J. Schiller, J. Mayr, S. Grijalvo, R. Eritja, D.D. Díaz, Gels **1**, 135 (2015)
291. R. Turcu, V. Socoliuc, I. Craciunescu, A. Petran, A. Paulus, M. Franzreb, E. Vasile, L. Vekas, Soft Matter **11**, 1008 (2015)

292. Y. Li, G. Huang, X. Zhang, B. Li, Y. Chen, T. Lu, T.J. Lu, F. Xu, Adv. Funct. Mater. **23**, 660 (2013)
293. S. Rittikulsittichai, A.G. Kolhatkar, S. Sarangi, M.A. Vorontsova, P.G. Vekilov, A. Brazdeikis, T.R. Lee, Nanoscale **8**, 11851 (2016)
294. S.R.S. Veloso, C.A.B. Magalhães, A.R.O. Rodrigues, H. Vilaça, M.J.R.P. Queiroz, J.A. Martins, P.J.G. Coutinho, P.M.T. Ferreira, E.M.S. Castanheira, Phys. Chem. Chem. Phys. **21**, 10377 (2019)
295. A. Carvalho, J. Gallo, D. Pereira, P. Valentão, P. Andrade, L. Hilliou, P. Ferreira, M. Bañobre-López, J. Martins, Nanomaterials **9**, 541 (2019)
296. H. Haider, C.H. Yang, W.J. Zheng, J.H. Yang, M.X. Wang, S. Yang, M. Zrínyi, Y. Osada, Z. Suo, Q. Zhang, J. Zhou, Y.M. Chen, Soft Matter **11**, 8253 (2015)
297. K. Liu, L. Chen, L. Huang, Y. Ni, Z. Xu, S. Lin, H. Wang, Cellulose **25**, 4565 (2018)
298. Z. Liu, J. Liu, X. Cui, X. Wang, L. Zhang, P. Tang, Front. Chem. **8**, 124 (2020)
299. L. Roeder, M. Reckenthäler, L. Belkoura, S. Roitsch, R. Strey, A.M. Schmidt, Macromolecules **47**, 7200 (2014)
300. X. Hu, G. Nian, X. Liang, L. Wu, T. Yin, H. Lu, S. Qu, W. Yang, A.C.S. Appl, Mater. Interfaces **11**, 10292 (2019)
301. Y. Hong Wang, M. Zhong, L. Wang, Y. liang Liu, B. Wang, Y. Li, J. Drug Deliv. Sci. Technol. **54**, 101293 (2019)
302. R. Messing, N. Frickel, L. Belkoura, R. Strey, H. Rahn, S. Odenbach, A.M. Schmidt, Macromolecules **44**, 2990 (2011)
303. Z. Yang, H. Gu, J. Du, J. Gao, B. Zhang, X. Zhang, B. Xu, Tetrahedron **63**, 7349 (2007)
304. B.P. Nowak, M. Niehues, B.J. Ravoo, Soft Matter **17**, 2857 (2021)
305. J. Tang, Z. Tong, Y. Xia, M. Liu, Z. Lv, Y. Gao, T. Lu, S. Xie, Y. Pei, D. Fang, T.J. Wang, J. Mater. Chem. B **6**, 2713 (2018)
306. Y. Wang, B. Li, F. Xu, Z. Han, D. Wei, D. Jia, Y. Zhou, Biomacromolecules **19**, 3351 (2018)
307. Y. Gao, Z. Wei, F. Li, Z.M. Yang, Y.M. Chen, M. Zrinyi, Y. Osada, Green Chem. **16**, 1255 (2014)
308. Y. Gao, C. Hu, W.J. Zheng, S. Yang, F. Li, S.D. Sun, M. Zrínyi, Y. Osada, Z.M. Yang, Y.M. Chen, ChemPhysChem 1999 (2016)
309. T. Miyazaki, A. Iwanaga, Y. Shirosaki, M. Kawashita, Colloids Surf. B Biointerfaces **179**, 334 (2019)
310. M.U. Witt, S. Hinrichs, N. Möller, S. Backes, B. Fischer, R. Von Klitzing, J. Phys. Chem. B **123**, 2405 (2019)
311. T.-Y. Liu, S.-H. Hu, T.-Y. Liu, D.-M. Liu, S.-Y. Chen, Langmuir **22**, 5974 (2006)
312. S.-H. Hu, T.-Y. Liu, D.-M. Liu, S.-Y. Chen, Macromolecules **40**, 6786 (2007)
313. M. Babincová, D. Leszczynska, P. Sourivong, P. Čičmanec, P. Babinec, J. Magn. Magn. Mater. **225**, 109 (2001)
314. R. Ramanujan, L. Lao, J. Mater Sci. Med. **5**, 1061 (2004)
315. K. Dashnyam, R.A. Perez, R.K. Singh, E.J. Lee, H.W. Kim, RSC Adv. **4**, 40841 (2014)
316. X. Zhao, J. Kim, C.A. Cezar, N. Huebsch, K. Lee, K. Bouhadir, D.J. Mooney, Proc. Natl. Acad. Sci. U. S. A. **108**, 67 (2011)
317. X. Tang, Y. Xu, J. Chen, T. Ying, L. Wang, L. Jiang, Y. Wang, Z. Wang, Y. Ling, F. Wang, L. Yao, H. Ran, Z. Wang, B. Hu, Y. Zheng, RSC Adv. **8**, 16410 (2018)
318. S.A. Meenach, J.Z. Hilt, K.W. Anderson, Acta Biomater. **6**, 1039 (2010)
319. W. Shi, J. Huang, R. Fang, M. Liu, A.C.S. Appl, Mater. Interfaces **12**, 5177 (2020)
320. R.G.D. Andrade, S.R.S. Veloso, E.M.S. Castanheira, Int. J. Mol. Sci. **21**, 2455 (2020)
321. F. Gao, W. Xie, Y. Miao, D. Wang, Z. Guo, A. Ghosal, Y. Li, Y. Wei, S.S. Feng, L. Zhao, H.M. Fan, Adv. Healthc. Mater. **8**, e1900203 (2019)
322. K. Hu, J. Sun, Z. Guo, P. Wang, Q. Chen, M. Ma, N. Gu, Adv. Mater. **27**, 2507 (2015)
323. H. Wu, L. Song, L. Chen, Y. Huang, Y. Wu, F. Zang, Y. An, H. Lyu, M. Ma, J. Chen, N. Gu, Y. Zhang, Nanoscale **9**, 16175 (2017)
324. K.Y. Qian, Y. Song, X. Yan, L. Dong, J. Xue, Y. Xu, B. Wang, B. Cao, Q. Hou, W. Peng, J. Hu, K. Jiang, S. Chen, H. Wang, Y. Lu, Biomaterials **259**, 120299 (2020)

325. N.S. Satarkar, J.Z. Hilt, J. Control. Release **130**, 246 (2008)
326. S. Campbell, D. Maitland, T. Hoare, ACS Macro Lett. **4**, 312 (2015)
327. X. Chen, M. Fan, H. Tan, B. Ren, G. Yuan, Y. Jia, J. Li, D. Xiong, X. Xing, X. Niu, X. Hu, Mater. Sci. Eng. C **101**, 619 (2019)
328. S. Forouzandehdel, S. Forouzandehdel, M. Rezghi Rami, Carbohydr. Res. **487**, 107889 (2020)
329. S. Nezami, M. Sadeghi, H. Mohajerani, Polym. Degrad. Stab. **179**, 109255 (2020)
330. O. Savchak, T. Morrison, K.G. Kornev, O. Kuksenok, Soft Matter **14**, 8698 (2018)
331. V. Crescenzi, A. Fancescangeli, A. Taglienti, Biomacromolecules **3**, 1384 (2002)
332. L. He, R. Zheng, J. Min, F. Lu, C. Wu, Y. Zhi, S. Shan, H. Su, J. Magn. Magn. Mater. **517**, 167394 (2021)
333. T.Y. Liu, T.Y. Chan, K.S. Wang, H.M. Tsou, RSC Adv. **5**, 90098 (2015)
334. S. Grijalvo, J. Mayr, R. Eritja, D.D. Díaz, Biomater. Sci. **4**, 555 (2016)
335. S. Hua, M.B.C. de Matos, J.M. Metselaar, G. Storm, Front. Pharmacol. **9**, 790 (2018)
336. E. Reimhult, N. Biotechnol. **32**, 665 (2015)
337. M. Amiri, T. Gholami, O. Amiri, A. Pardakhti, M. Ahmadi, A. Akbari, A. Amanatfard, M. Salavati-Niasari, J. Alloys Compd. **849**, 156604 (2020)
338. M. Ullrich, J. Hanuš, J. Dohnal, F. Štěpánek, J. Colloid Interface Sci. **394**, 380 (2013)
339. R.J. Mart, K.P. Liem, S.J. Webb, Chem. Commun. 2287 (2009)
340. R.I. Robert Blumenthal, J. Nanomedicine Biother. Discov. **04**, 1000130 (2014)
341. F. de Cogan, A. Booth, J.E. Gough, S.J. Webb, Angew. Chem. **123**, 12498 (2011)
342. F. de Cogan, A. Booth, J.E. Gough, S.J. Webb, Soft Matter **9**, 2245 (2013)
343. J. Hanuš, M. Ullrich, J. Dohnal, M. Singh, F. Štěpánek, Langmuir **29**, 4381 (2013)
344. A. Pittermannová, Z. Ruberová, A. Zadražil, N. Bremond, J. Bibette, F. Štěpánek, RSC Adv. **6**, 103532 (2016)
345. D.H. Kang, H.S. Jung, N. Ahn, S.M. Yang, S. Seo, K.Y. Suh, P.S. Chang, N.L. Jeon, J. Kim, K. Kim, A.C.S. Appl, Mater. Interfaces **6**, 10631 (2014)
346. M. Van Elk, C. Lorenzato, B. Ozbakir, C. Oerlemans, G. Storm, F. Nijsen, R. Deckers, T. Vermonden, W.E. Hennink, Eur. Polym. J. **72**, 620 (2015)
347. M. Ullrich, J. Hanuš, F. Štěpánek, Chem. Eng. Sci. **125**, 191 (2015)

Inorganic Particles Incorporated into Electrospun Nanofibers: Bioactivity and Antibacterial Properties

Verônica Ribeiro dos Santos Borges, Lilian de Siqueira, Alexandre Luiz Souto Borges, and Eliandra de Sousa Trichês

Abstract Natural or synthetic polymer nanofibers present great properties that allow their use as scaffolds for tissue engineering, such as biocompatibility, good mechanical resistance, high superficial area, and porous structure, allowing cell infiltration, adhesion, and proliferation. These properties are achieved by the morphology control—a result of the parameters set used during the electrospinning process—but, no matter what parameter is studied, the lack of bioactivity intrinsic to polymers cannot be changed. Additionally, considering the scaffold's final application, the capability to prevent bacterial proliferation should also be a concern. In this scenario, the incorporation of inorganic bioactive particles into the nanofibers' porous structure has been performed to optimize its biological properties. Encapsulation, in-situ deposition, and electrospray deposition of bioglasses, glass–ceramics, calcium phosphates, and metallic nanoparticles are being widely explored to enhance both nanofibers' bioactivity and antibacterial activity. Considering the exposure, this chapter aims to promote an overview regarding the electrospinning technique and ways to overcome nanofibers' biological limitations by inorganic particle addition by incorporation techniques.

Keywords Electrospinning · Inorganic particles · Antibacterial activity · Encapsulation · Electrospray

V. R. S. Borges · E. de Sousa Trichês (✉)
Bioceramics Laboratory, Institute of Science and Technology, Sao Jose dos Campos, Federal University of Sao Paulo (Unifesp), Sao Paulo, Brazil
e-mail: eliandra.sousa@unifesp.br

V. R. S. Borges
e-mail: veronica.ribeiro@unifesp.br

L. de Siqueira · A. L. S. Borges
Department of Dental Materials and Prosthodontics, Institute of Science and Technology of Sao Jose dos Campos, Sao Paulo State University (Unesp), Sao Paulo, Brazil
e-mail: lilian.siqueira@unesp.br

A. L. S. Borges
e-mail: alexandre.borges@unesp.br

S. J. Ikhmayies (ed.), *Advances in Biomaterials Research*, Advances in Material Research and Technology, https://doi.org/10.1007/978-3-031-97481-6_6

1 Introduction

Human organs and tissues are prone to damage by traumas, infectious diseases, inflammation, acquired and congenital abnormalities, and chronic non-healing wounds. With the development of modern medicine, traditional treatment methods usually employed are been substituted by what is often called the tissue engineering triad; cells, scaffolds, and signals [1]. Assembled to generate a wide range of biomimetic structures for the replacement and regeneration of damaged and diseased tissues, these engineered alternative treatments can solve clinical problems not achieved by traditional therapies and procedures.

Scaffolds are defined as three-dimensional structures to promote cellular growth and interactions. By mimicking the structure and function of the native tissue extracellular matrix (ECM), the scaffold promotes accelerated regeneration [2]. Being biomaterial-based, scaffolds must have the main characteristic of this material category; biocompatibility, which refers to the absence of rejection between the biomedical device and the native tissue in the exposure period [3]. Furthermore, general requirements include mechanical support for cell adhesion and infiltration, capability of transportation of nutrients and wastes, and ability to degrade in the human body as the regenerated tissue is formed (biodegradability), besides being able to promote an effective interaction scaffold-tissue by strong chemical bonding (bioactivity) [4–6]. Although the requirements may vary according to the final application site, recently antimicrobial action is being explored in all fields of tissue regeneration to prevent inflammation by opportunistic pathogens [7–9]. Together with the bioactivity, the antimicrobial action may not be intrinsic to the biomaterial used as a scaffold, and strategies to promote these properties lie in the incorporation of bioactive inorganic particles such as bioglasses and glass–ceramics, calcium phosphates, and metallic nanoparticles [10].

Scaffolds production includes a variety of materials and morphologies, although polymeric nanofibers stand out due to their biocompatibility, good mechanical resistance, high surface area, and interconnected porous structure, allowing cell infiltration, adhesion, and proliferation. Nanofibers scaffolds have been studied in a wide range of applications in biomedicine [11]; bone [12], dental [13], and wound healing [14], although the lack of bioactivity and antimicrobial activity is still a concern.

Electrospinning is a simple and versatile electro-hydrodynamic method for producing nanofibers, granting the morphology control for obtaining determined porous structures by the technique parameters adjustment [15]. Additionally, this technique allows the development of nanofiber composites design with organic and inorganic components to overcome the scaffolds' needs. Encapsulation is the common method of inorganic particles incorporation, which is achieved by particles addition into the polymeric solution for electrospinning. Another strategy, however, consists in its deposition. In-situ and electrospray deposition are two techniques that allow the incorporation of inorganic nanoparticles [10, 16].

Considering the incorporation of the inorganic particles into polymeric nanofibers to promote bioactivity and antimicrobial activity, this chapter aims to perform an overview on obtaining nanofibers by the electrospinning technique, and to study the incorporation of inorganic particles such as bioactive glasses, glass–ceramics, calcium phosphates, and metallic nanoparticles into polymeric nanofibers by encapsulation and electrospray deposition, then compare the performance of these composites with recent ones founds in literature.

2 Electrospinning

Electrospinning is a classic technique, that can be considered as one of the simplest, most versatile, and economical techniques for manufacturing fibers at micro and nanometer scales [17]. It is considered a promising and relevant method for industry and academia to fabricate polymer nanofibers [18]. It was discovered in 1600 when William Gilbert has the first record of the electrostatic attraction of a liquid. In 1902, Morton [19] and Cooley [20] filed the first electrospinning patent. Further research on the electrospinning process was carried out by scientist Zeleny [21] who studied in detail the process that was years later patented by Anton Formhals [22, 23]. Between 1964 and 1969, Sir Geoffrey Taylor worked on electrically driven jets laid the theoretical foundations for electrospinning. Academic visibility occurred in 1990, initiated by the Reneker group that popularized the terminology of electrospinning [24]. From the 2000s onwards, there was an exponential growth of publications as a highly functional material was obtained [25].

The system used on a laboratory scale for this technique consists of four main components: a high voltage power supply, a syringe with pump, a metal tip needle, and a conducting collector [26]. Currently, there are two main standard setups for the electrospinning technique: vertical and horizontal configuration and different types of collectors (plate collector, rotary drum collector, grid type collector, edge type collector, mandrel collector, steel sheet collector, dual rings collector) are used by the researchers, based on the applications and characteristics of fiber produced [27]. Figure 1 illustrates the electrospinning apparatus. Of the spinnerets, the most used are the clip spinneret, tube-less spinneret, co-axial spinneret, and heating spinneret [27].

This setup is mainly used for the production of nanofibers in small volumes in a laboratory setting, for evaluating the spinnability of new materials, or for optimizing the parameters for electrospinning prior large-scale production. However, this setup cannot satisfy the requirements of industrial applications. In this way, modifications have been made to this general setup aiming to increases the capability and versatility of electrospinning [29].

Although polymeric nanofibers represent a large part of publications on electrospinning [30], the popularization of the technique and technological advances combined with new application needs has expanded the adaptation of the technique to obtain nanofibers with improved capabilities beyond the production of nanofibers from other materials [29]. Depending on the modification made, the products of electrospinning techniques can be classified into several types, based

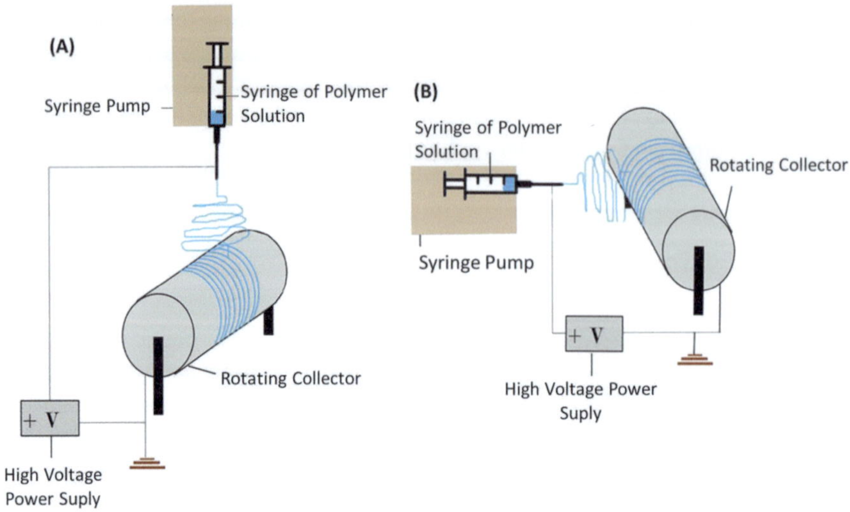

Fig. 1 Electrospinning apparatus: **a** Typical vertical setup, and **b** Horizontal setup. Redrawn after Ibrahim and Klingner [28]

on several factors: (i) chemical composition: electrospun nanofibers can be classified into inorganic, organic, carbonaceous, and inorganic–organic hybrid nanofibers. (ii) nanofiber shape: electrospinning nanofibers can be classified into round nanofibers, beaded nanofibers, porous nanofibers, side-by-side nanofibers, grooved nanofibers, nanograin nanofibers, nanobelt nanofibers, hollowed nanofibers, multi-channel nanofibers, bilayer core–shell nanofibers, trilayer core–shell nanofibers, and complex nanofibers such as porous core–shell nanofibers. (iii) orientation of nanofibers: random-distributed nanofibers, aligned nanofibers, crimped nanofibers, etc., and the (iv) difference due to the type of used collector: the products can be classified into plate, layer-by-layer stacked films, electrospun nanofibrous balloon, electrospun nanofibrous bundles, twisted nanofibrous bundles, etc. [31]. In this way, it is possible to obtain desirable characteristics of nanofibers that allow their use in various tissue engineering applications.

Thus, different types of nanofibers have been obtained successfully in the literature as shown by the works of Ge et al. [32], Duan et al. [33], and Xu et al. [34]. Recently, Atila et al. [35] studied and optimized the coaxial electrospinning of core/shell poly(methyl methacrylate) (PMMA)/silk fibroin (SF) fibers using a new collector design for obtaining homogeneously deposited mats from the collector screen to provide ECM-like microenvironment for new tissue formation. SF-shell was produced to increase cell-affinity of fiber surfaces whereas PMMA-core was designed to support the tissue mechanically during regeneration.

2.1 Technique Fundamentals

The electrospinning technique involves a strong potential difference between a polymer-based solution flowing through a capillary metal tip and a metallic collector [36]. In an experimental apparatus, one electrode is connected to a capillary tube and the other to a grounded collector. An electric field is applied to create a potential difference between the capillary tip and the collector. The polymeric fluid is held only by its surface tension, forming a droplet at the tip of the capillary tube. The applied electric field induces a charge on the surface of the fluid and, at the same time, a force acts in a direction opposite to the surface tension. As the electric field strength is increased, the drop undergoes elongation, acquiring a conical shape known as a Taylor cone [37] that extends linearly and subsequently undergoes whipping movements, due to flexural instability [37, 38]. When the applied potential reaches a critical value necessary to overcome the surface tension of the liquid, a charged jet of fluid is ejected from the tip of the Taylor cone. During the jet stretching process, the solvent evaporates, leaving behind a charged polymer fiber [39, 40].

2.2 Influence of Parameters on Nanofibers Morphology

In particular, the morphology and structure of the deposited fibers are strongly dependent on the deposition method and collector model. The production of fibers by the electrospinning process is influenced by electrostatic forces and the viscoelastic behavior of the polymer [41]. The quality of the deposition resulting from electrospinning depends on the parameters that affect the process. The main factors pointed by Mishra [42] as critical process parameters and manufacturing attributes that affect electrospun nanofibers are respectively: process parameters (voltage, flow rate, needle tip to collector distance, needle diameter, and motion), systemic parameters (polymer type, molecular weight, polymer architecture, and solvent), solution parameters (viscosity, concentration, conductivity, surface tension, charge of jet, and dielectric constant), and physical parameters (humidity, temperature, and air velocity) [42]. Among them, the most critical parameters for controlling fiber dimension are voltage, flow rate, the needle tip to collector distance, and needle diameter [43]. Table 1 presents a summary of the effects of electrospinning parameters on fiber morphology, while Fig. 2 illustrates typical outcomes (random, aligned and beaded nanofibers).

2.3 Electrospinning of Synthetic and Natural Polymers

A wide range of polymeric materials (natural and synthetic), as well as a combination of them, have been successfully electrospun to form fine nanofibers within

Table 1 Effect of parameters on produced fibers using electrospinning technique

Parameters	Variable affected	Effect	References
Processing			
Voltage	Fiber diameter	Decreasing with increasing applied tension: As the voltage rises, the charge density on the jet's surface rises, which leads to a significant effect of jet stretching	[44–46]
Flow rate	Fiber diameter Material transfer rate	Increase with flow rate Increase with flow rate	[47, 48]
Tip-collector distance	Feasibility	Low: insufficient time for evaporation High: formation of beads	[47]
Solution			
Concentration	Feasibility	Low: formation of beads High: fiber formation interrupted	[49]
Molecular weigh	Fiber diameter	The higher the molecular weight, the thicker the fiber	[50]
Conductivity	Fiber diameter	Conductivity increases within a reasonable range; fiber diameter decreases and increases again; fiber diameter is not controllable	[46]
Environmental			
Relative humidity	Fiber surface	High: formation of pores and grooves due to water droplet condensation on the fiber surface	[47, 51, 52]
Temperature	Fiber diameter	High: fiber diameter decreases. The temperature has the greatest influence on the viscosity of the solution. The viscosity of the solution reduces as the temperature rises. The intermolecular Coulomb force is lessened in this scenario	[46, 49, 51]

Fig. 2 SEM images of random, aligned and beaded poly(3-hydroxybutyrate-co-3-hydroxyvalerate) (PHBV) nanofibers

the submicron range and used for varied applications such as tissue engineering scaffolds, filtration membranes, and in various biomedical applications [53]. Over the years, more than 200 polymers with the potential to be electrospun have been reported in the literature [25].

2.3.1 Natural Polymers

Natural polymers are macromolecules that can be found in nature, generally are biodegradable, biocompatible, and have low antigenicity. Some exhibit antimicrobial and anti-inflammatory properties [54]. In biological environments, structures similar to natural ECMs can be created, which allow cells to perform vital tasks including signal transmission, proliferation, differentiation, and migration [55]. Examples of these polymers include Deoxyribonucleic acid (DNA), silk fibroin, fibrinogens, dextran, collagen, gelatin, chitosan (Ch), chitin, and poly(3-hydroxybutyrate) (P3HB) [29, 54]. Among the natural polymers, one of the most used is chitin, a precursor of chitosan. Chitin is generally extracted from crab shells and converted to chitosan after removing acetyl groups via alkaline treatment. Although somewhat expensive, it possesses a unique set of characteristics suitable for a range of biomedical applications, such as surgical sutures, dental implants, artificial skin, and rebuilding of bone [56].

Sato et al. [57] successfully obtained chitosan fibers containing nanohydroxyapatite particles (nHAp) incorporated by electrospinning. In this study, the effects of the addition of nHAp on the morphological, chemical, physical, and biological properties of the fibers were evaluated. Their results demonstrated chitosan/hydroxyapatite (ChHa) fibers with a bead-like morphology. Despite showing similar cell viability and alkaline phosphatase (ALP) activity to Ch fibers, the ChHa fibers demonstrated greater antibacterial potential against most tested bacteria (except for *P. intermedia*), and higher antibiofilm activity against *P. gingivalis* biofilm. Efforts have also been directed toward the production of nanofibrous materials containing collagen for application in tissue engineering, as shown by the work of Noh et al. [58]. Noh et al. [58] manufactured composite scaffolds composed of bacterial cellulose (BC) and collagen (Col) in different ratios (1:1, 3:1, 5:1) by electrospinning. The composite scaffolds exhibited a well-organized interconnected porous structure, significantly better physical stability than Col scaffold, and more water uptake up to 400%. They were also favorable with cell attachment and growth. BC/Col (5:1) showed more up-regulated osteogenic markers (collagen type 1, osteocalcin, bone sialoprotein) and significantly elevated proteins, and calcium deposition after osteogenic induction of umbilical cord blood-derived mesenchymal stem cells (UCB-MSCs) for 3 weeks, and when the red fluorescent dye (PKH-26) pre-labeled MSC-loaded scaffolds were subcutaneously transplanted in a mouse model, they showed many PKH-26-labeled cells and positive signals of α-smooth muscle actin, for neovascularization. The results demonstrated that BC/Col composites may be promising as a bone tissue-engineered scaffold. However, the use of collagen due to the high price, low melting point, and the melting point of nanofibers in aquatic environments is restricted.

The combination of natural polymers is a simple strategy to provide suitable substrates for cellular activities. However, its main disadvantages include rapid biodegradation, poor processability, and poor mechanical characteristics [59]. Combining natural polymers with synthetic polymers such as polycaprolactone (PCL), polyurethane, and polyaniline is another strategy for producing artificial ECMs, as shown by the works of Goh et al. [60] and Chen et al. [14]. Combined with this fact, it is important to emphasize that the preparation of nanofibers from pure natural polymers is still a challenge, due to the relatively low molecular weight of some of these materials and the limited availability of solvents that are both suitable for the electrospinning process and for dissolving them [61].

2.3.2 Synthetic Polymers

Synthetic polymers often offer many advantages over natural ones, as they can be tailored to give a wide range of properties, such as necessary mechanical properties (viscoelasticity and strength), and degradation rates [25]. Thus, the synthetic polymer is often used in admixture with the natural polymer, or the natural polymer is chemically modified to be processed using high voltage electrical forces [62]. The polymer to be used can provide a wide range of properties to the electrospun fabric, including strength, weight, porosity, and, in some cases, even a functionalized surface can be obtained.

Among them, synthetic polymers such as polystyrene (PS) and poly(vinyl chloride) (PVC) have been electrospun into nanofibers for commercial applications related to environmental protection [29]. Polyesters (e.g., Poly(DL-lactide) (PDLLA), Polylactic coglycolic acid (PLGA), and PCL) are a very common group of polymers used to construct biodegradable nanofibrous materials for tissue engineering scaffolds [63]. Among them, PLGA has received attention due to its mechanical characteristics, degradation rate, and biocompatibility when implanted in vivo [64].

Santos et al. [65] obtained membranes of poly(L-lactic acid) (PLLA) composed of PLLA sub-micro fibers with a highly porous surface by low voltage electrospinning (6 kV) that showed biocompatibility with the type of implanted cell. Singhal [66] investigated the formation of nanofibers of solutions poly (ε-caprolactone) (PCL) in the weight/volume percent (w/v) range of 10–15% at a constant voltage of 18 kV by electrospinning technique. The impacts of different parameters, including stream rate, and solution parameters, for example, concentration and solubility were likewise inspected. While with increasing the concentration, the morphology changed from beaded filaments to uniform fibers. The bioactivity of PCL nanofibers with physiological fluids was evaluated by standard assay procedures using simulated body fluids and proved the formation of hydroxyapatite on the surface of the fibers. However, in the literature, an increasing number of authors have reported better cellular response during in vitro cell culture of electrospun polymer ultrathin fibers containing HA [67–69] and bioactive glass [70–73].

2.4 Advantages and Limitations of Polymers' Nanofibers

Electrospun nanofibers are considered promising support materials widely investigated in tissue engineering, as they have unique properties [74, 75]. Their morphology, high surface area-to-volume ratio, and inter/intra fibrous porosity offer more binding sites for cell membrane receptors and adsorption of serum proteins [76]. Their applications include drug and gene delivery, artificial blood vessels, artificial organs, and medical masks, drug delivery directly to internal tissues, as the fiber diameter is much smaller than the blood cell (7.5–8.7 μm) [77]. Furthermore, electrospun fibers are comparatively lighter than other fibers produced through different methods [76].

In vitro biological assays show that electrospun nanofibers can stimulate special biological responses in cells when cells are grown on them; such as enhanced cell adhesion, stimulated cell growth, protein adsorption, and they aid in cell differentiation [78, 79].

For example, poly (lactic acid) (PLA) and polycaprolactone (PCL) have emerged prominently and are among the most promising and widely studied alternatives [80, 81], as they combine all desirable characteristics for use as biomaterials, i.e. biocompatible, bioresorbable, biodegradable, and produce promising results in clinical use. However, these polymers have some limitations, such as low bioactivity and hydrophobic surfaces, which may decrease cell adhesion [82].

On the other hand, it is known that the electrospinning technique has some disadvantages. First, in the manufacture of organic nanofibers, the types of polymers used are limited and the behavior of nanofibers is not well discussed. Second, inorganic nanofibers have been limited due to their friability after the calcination process. Third, the high-cost production of large diameter electrospun nanofibers represents a significant challenge in nanofiber fabrication [18]. Furthermore, the cost of polymers, health hazards, solvent vapor, packaging, shipping, and handling of nanofibers are still issues to be overcome.

2.5 Nanofibers Applications in Tissue Engineering

The repair or regeneration of tissues that have been lost as a result of trauma or disease involves the integration of support material (scaffolds), cells and/or biological factors. In this sense, electrospun nanofibers are materials highly attractive and widely studied for a wide number of biomedical applications due to their porous structure being suitable for use as an artificial extracellular matrix. Nanofibers have the ability to mimic the composition, length scale, and typical architecture of the ECM, and because they have a high surface/volume ratio, they can improve cell adhesion and nutrient transport [30].

Owing to the electrospinning parameters control, nanofibers morphologies can be modified to mimic the features of the extracellular matrix and its porous structure, and this mimics natural physiological conditions for cellular growth and allows their use as scaffolds. Besides the fiber diameter, porosity, and alignment, nanofibers' characteristics related to the polymer of choice composition, surface chemistry, and structural and physical properties must also be considered before defining its final application. Flexible nanofibers, for example, are more suitable for wound healing applications rather than bone regeneration. In the following sections, the use of nanofibers for biomedical applications will be addressed.

2.5.1 Dental

A major problem faced in dentistry concerns the loss of dental support tissues [83], which can occur due to various periodontal and endodontic diseases and surgical sequelae suffered by the patient [84, 85]. In addition to periodontal diseases, other reasons can lead to tooth loss, such as caries disease and dental trauma, and consequently, bone loss and atrophy of the alveolar bone crest can occur [86, 87].

Tissue engineering is one of the fields of regenerative medicine that uses methods to promote cell growth through the manipulation of various artificial or natural biomaterials that provide support for controlled growth in specific tissue types. Currently, tissue regeneration techniques are increasingly used for the treatment of periodontal defects, reconstruction of lost, damaged, and atrophied bone tissue, dental implant procedures, or for the preservation of alveolar cavities after tooth extraction [88]. Figure 3 illustrates the possible application sites of electrospun nanofibers used in dental tissue regeneration.

More than two-thirds of the global population suffers from tooth decay, which results in cavities with various levels of lesion severity [89]. Therefore, various forms of therapies aim to regenerate the roots of the teeth (apicogenesis) that suffered trauma. Although various drugs, such as calcium hydroxide, ferric sulfate, and mineral trioxide aggregate, are used as materials to regenerate pulp and root dentin, some cases still result in internal tooth resorption [90, 91]. In this way, electrospun nanofibers can be considered an excellent candidate for a variety of applications, such as tooth regeneration, guided tissue regeneration, caries prevention, drug delivery, wound, and mucosal repair, reinforcement of resin composites, implant modification cartilage regeneration [92], however, wok is still required to validate the use of electrospun nanofibrous scaffolds in the clinical stage [36].

2.5.2 Orthopedic

Bone tissue damage occurs as a consequence of aging, accidents, diseases, or congenital malformations requiring the use of grafts to repair or replace damaged bone tissue [93]. These grafts serve as structural support and act as an osteoconductive framework for the formation of new bone tissue [94]. The most used procedure is the transplant of the patient himself or a compatible donor, but this procedure presents

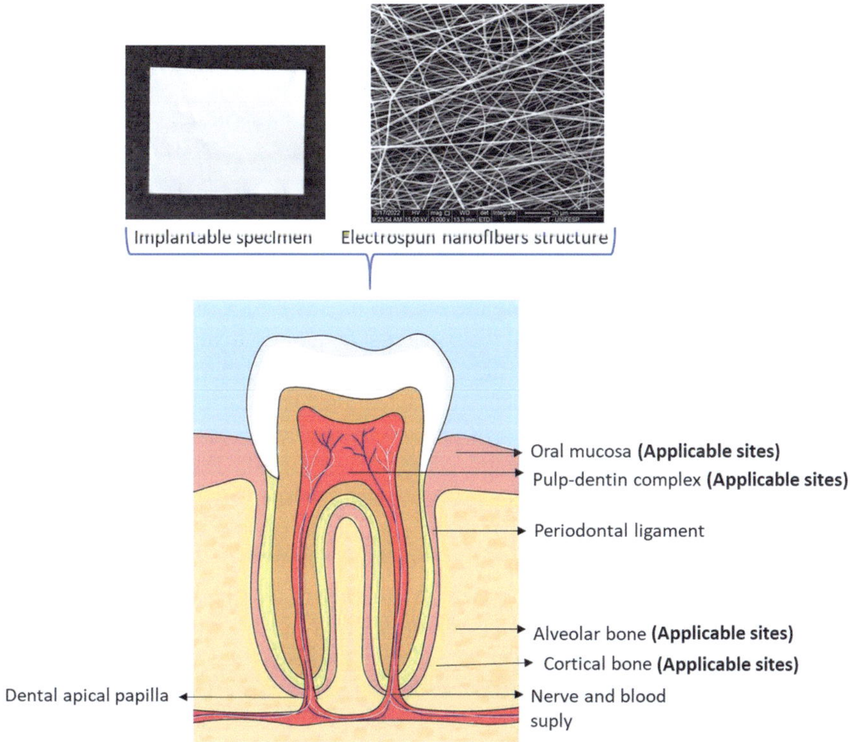

Fig. 3 Applicable sites for electrospun nanofibers used as scaffolds for dental tissue regeneration [36]

problems such as the lack of donors, the risk of disease transmission, and immunological rejection of the graft. In addition, prostheses have limitations because they are not aimed at regenerating damaged tissue, but rather at replacing certain functions [95, 96]. In addition to these factors, the high number of people who suffer from these problems and the resulting socio-economic costs demonstrate the need to develop alternative therapies to address such tissue loss or damage. In this context, tissue engineering arises, which is one of the fields of regenerative medicine and uses methods to promote cell growth through the manipulation of various biomaterials, artificial or natural, that provide support for controlled growth in specific types of tissue.

A major challenge of the current research is to produce scaffolds with physicochemical and morphological characteristics such that, when implanted in the injured regions, they stimulate osteogenesis and promote the formation of structures similar to the extracellular matrix elements, facilitating the mobilization, expansion, and integration of internal regenerative cell populations, promoting the repair of injuries or the renewal of degenerated tissues. In this context, polymeric scaffolds, specially electrospun nanofibers, are frequently explored as tissue substitutes for the regeneration of tissue physiology and specific functions because of their high surface-area-to-volume ratio and porosity [97].

2.5.3 Wound Healing

The human body's homeostasis maintenance and protection from the external environment are performed by its largest organ; the skin. Besides regulating vital functions such as the hydration state, temperature, and metabolism, the skin prevents pathological infections by isolating the interior organs [98]. Considering its importance, acute, chronic, and surgical wounds must be rapidly treated to prevent major interruption in skin normal functions and additional structural damages. Acute wounds are generated by mechanical stress or exposure to heat, chemicals, light, and electric shock, while chronic wounds are an outcome of diseases such as vascular insufficiency, prolonged inflammation, and diabetes [99]. The healing of small wounds is performed naturally for the human body, while in extended damages the wound healing process is delayed owing to opportunist pathogens exposure, inflammation, poor blood supply, excessive production of exudates, absence of nutrients, and adequate oxygenation [100]. Thus, ECM mimicking nanofibers emerge in wound healing management as dressings not only to support cell adhesion, proliferation, and maturation for skin regeneration and remodeling but also to cover the wounded area to prevent microorganisms' exposure, absorb wound secretions, and protect the site from mechanical stress [101]. Moreover, the natural ECM is not only passive structure support for cells, but also performs effective interactions with cells and growth factors [102]. Thus, nanofibers dressings for wound healing must perform these interactions as well, which can be achieved with the incorporation of inorganic particles to promote both bioactivity and antimicrobial action to further treat the skin damage.

3 Inorganic Particles for Enhancement of Bioactivity and Antimicrobial Properties

The ability of a material to undergo surface reactions to form a hydroxyapatite-like (HA) layer when in contact with body fluid is often taken as bioactivity, once this compound forms a strong chemical bond with calcified and soft tissues [103]. Thus, this property becomes a requirement regarding biomedical implants and devices owing to the high biocompatibility, especially with bone and teethhard tissues where apatite is the main component. Ceramic biomaterials are highly bioactive owing to their composition, solubility, and ion release, and with the incorporation of specific ions therapeutic functions such as antimicrobial activity may be performed, a property that is intrinsic of metallic nanoparticles. Considering microorganisms are constantly evolving to become multi-drug resistant, the development of antimicrobial materials is a necessity in tissue engineering, as well as bioactivity.

3.1 Bioceramics: Bioactive Glasses, Glass–Ceramics, and Calcium Phosphates

While ceramic biomaterials such as alumina (Al_2O_3) and zirconia (ZrO_2) are largely used as structural-support medical devices and implants, their chemical inertness results in limited tissue ingrowth, cellular non-adherence, and long-time wear [104]. To overcome these drawbacks, what is often called second-generation ceramics were developed and designed to resemble natural calcified tissue surfaces, induce biomineralization and promote osseous tissue formation at the same rate as its biodegradation [105]. Able to form apatite in its surfaces and forms a strong bond with tissues, bioactive glasses, glass–ceramics, and calcium phosphates (e.g. hydroxyapatite (HA, $Ca_{10}(PO_4)_6(OH)_2$) and tricalcium phosphate (TCP, $Ca_3(PO_4)_2$)) are relevant bioactive ceramics that effectively interact with the human body physiological environment [105–107]. Compared to the first, the second-generation bioceramics have optimized biocompatibility, bioactivity, and biodegradability, while the third-generation bioceramics applied these materials as scaffolds for living tissue regeneration by driving cells and molecules [106]. Figure 4 illustrates the generations of bioceramics used for biological applications.

3.1.1 Bioactive Glasses and Glass–Ceramics

Bioactive glasses (BG) and glass–ceramics have proven capability to form HA both in vitro and in vivo, being inorganic particles usually employed to promote bioactivity to non-bioactive biomaterials [109]. While BG is oxide-based material of an amorphous structure, glass–ceramics are formed by thermal treatment of glasses to induce its controlled crystallization into one or more phases embedded in a residual amorphous network [110].

Melt-quenching technique and sol–gel processes are two routes of BG obtention. The first is a traditional method that consists in melting a mixture of oxides in a furnace followed by quenching in water, allowing the formation of amorphous structures at the macroscale [111, 112]. The sol–gel process, however, is a chemical

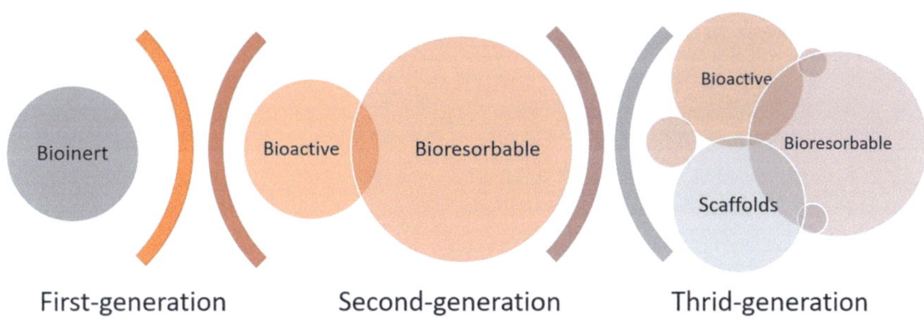

Fig. 4 First, second, and third generation bioceramics characteristics [108]

method based on the formation of sol from the precursors and its condensation into a three-dimensional molecule (gel), followed by aging, drying, and calcination steps [113]. Although both techniques allow the obtention of bioactive glasses and glass–ceramics, one of the main advantages of the sol–gel method over the melt-quenching is the resulted particle size, shape, and porosity; physical and textural properties that cause a direct impact on the surface area and bioactivity [112, 114]. Figure 5 illustrates the property of the final glasses by these two approaches of obtention. Recently, it was demonstrated that a mesoporous structure with internal connectivity results in higher conversion rates in SBF and proliferation of human periodontal ligament cells (hPDLCs), when compared with compositions with higher silica content of lower porosity [115]. On the other hand, while the obtention of amorphous phases by melt-quenching is quite direct, by the sol–gel method the organic matter and cytotoxic components removal thermal treatment temperature may overcome the crystallization one, leading to the formation of a glass–ceramic structure rather than amorphous, as demonstrated by Fiume et al. [114].

The silicate-based bioactive glasses Hench's original Bioglass® 45S5 [(46.1)SiO_2–(26.9)CaO–(24.4)Na_2O–(2.6)P_2O_5 (mol.%)] and BonAlive S53P4 (BonAlive Biomaterials, Finland) [(53)SiO_2–(20)CaO–(23)Na_2O–(4)P_2O_5 (mol.%)], together with the borate bioactive glass-based wound healing dressing MIRRAGEN™ (ETS Wound Care LLC, USA), are three formulations with regulatory approval for clinical applications. Due to their dissolution rates, silicate-based BGs are often applied in bone [116, 117] and dental applications [118, 119], while borate glasses are explored in wound healing [120, 121], although exchanges are being studied. The basic unit of silicate-based BG is silica tetrahedra (SiO_4) of high chemical stability, while bioactive borate-based glasses are composed of boron

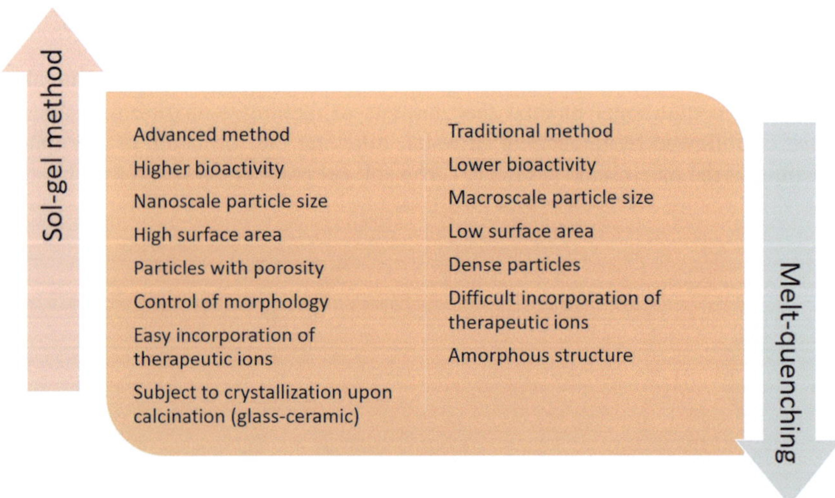

Fig. 5 Sol–gel method and melt-quenching derived glasses and their properties

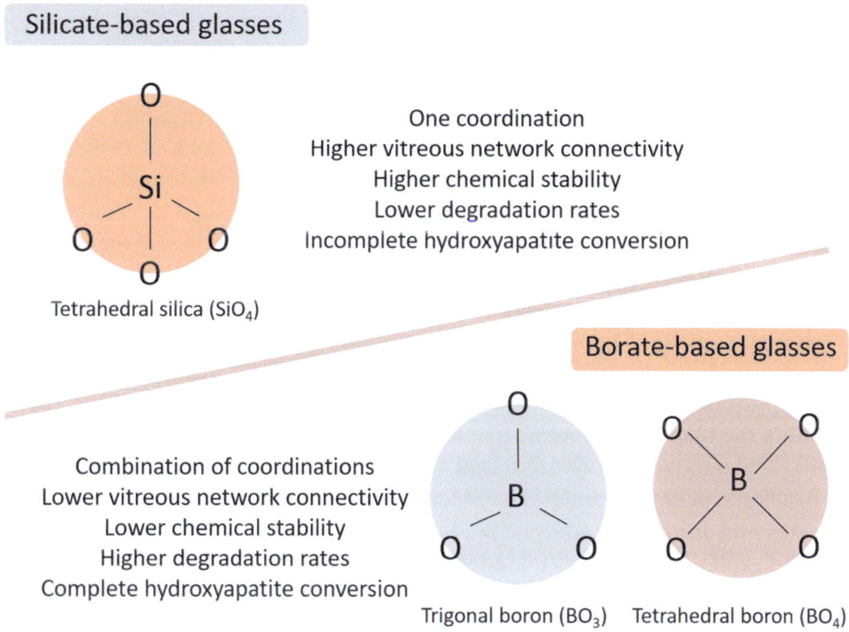

Fig. 6 Properties of silicate-based and borate-based glasses

with both trigonal (BO_3) and tetrahedra (BO_4) coordination [122]. Due to this combination, borate-based glasses have low vitreous network connectivity, which reflects in their low chemical stability, higher degradation rates, besides faster and complete hydroxyapatite conversion when compared to silicate-based BGs [122–124]. Figure 6 synthesizes this information.

Although 45S5 and S53P4 are the only silicate-based bioactive glasses approved to be used in clinical applications, other formulations are widely investigated by researchers, such as 58S [(60)SiO_2–(36)CaO–(4)P_2O_5 (mol.%)] [125] and 13–93 [(54.6)SiO_2–(22.1)CaO–(7.9)K_2O–(7.7)MgO–(6.0)Na_2O–(1.7)P_2O_5] [126]. All these bioactive glasses compositions present great conversion rates into hydroxyapatite, thus biocompatibility, whereas S53P4 has one feature that stands out: its intrinsic antimicrobial activity [127]. The S53P4 mechanisms of action against bacteria are based on its rapid component's dissolution, which increases the local pH and the osmotic pressure, thus inhibiting bacterial proliferation [128]. Owing to its bioactivity, osteostimulative and antimicrobial activity, together with its exceptional mechanical resistance [53, 129, 130], the S53P4 bioactive glass is especially explored for the treatment of an inflammatory and infectious bone disease that eventually may lead to its destruction; the osteomyelitis [127, 131]. Van Vugt et al. [132] introduced this antimicrobial bioactive glass for the treatment of cavitary long bone osteomyelitis and performed a 12 to 46-months follow up period, in which the infection, caused by *staphylococcus aureus* in 36% of the patients, was eradicated in

85% of the 78 patients treated. Another highlight of the authors' clinical report is the number of stages performed to this successful result obtention; 69 of the 78 patients were treated in a one-stage procedure (surgical debridement followed by S53P4 implantation), with no further intervention [132]. De Giglio et al. (2021) also performed a successful clinical trial aiming at the treatment of diabetic foot osteomyelitis, where the implantation of S53P4 resulted in a higher resolution rate and absence of additional antibiotic therapy during the treatment when compared to the patients who underwent the conventional procedure [116]. Thus, it is evident the therapeutic functions provided by S53P4 under clinical applications, being an effective bone growth stimulator and antimicrobial agent against osteomyelitis caused by microorganisms (e.g. *Staphylococcus aureus*, *Escherichia coli*, *Pseudomonas aeruginosa*, *Acinetobacter* spp., and *Candida albicans*) [127, 133, 134].

Compositions derived from widespread silicate-based glasses have been reproduced for the obtention of borate glasses by fully replacing SiO_2 with B_2O_3. Islam et al. [135] produced both solid and highly interconnected porous (70 $\pm$ 5%) microspheres of three borate glasses formulation namely 45B5, B53P4, and 13-93B (derived from their equivalents 45S5, S53P4, and 13-93) by a novel flame spheroidization method, and investigated their conversion into hydroxyapatite in three different media (K_2HPO_4, Na_2HPO_4, and SBF) in a 21 days study. The authors reported fully and rapid conversion in HA especially for the porous microspheres in all media, except for 13-93B; in which the conversion occurred only when in Na_2HPO_4 attributable to the presence of potassium into the glass network and in the K_2HPO_4 and SBF solutions, causing inhibition in the ion exchange reaction. Furthermore, the converted porous HA microspheres enhanced the pro-osteogenic properties when compared with their bioactive glasses counterparts [135]. Similarly, Deliormanlı [136] produced by the sol–gel method the 19-93B bioactive glass and performed an in vitro mineralization assay, where conversion into HA was observed right on the first day, confirming the rapid dissolution of sol–gel derived bioactive borate glasses and their biocompatibility in vitro.

Considering the therapeutic functions of BGs are provided by their ionic dissolution, besides the bioactivity, it is possible to incorporate into the glass network-specific dopant ions for targeting properties attribution, such as antibacterial, angiogenic, and osteogenic activities [137]. While angiogenesis refers to the biomaterial capability to stimulate the formation of blood vessels, osteogenic activity induces bone formation [138]. Figure 7 illustrates the dopant ions with the biological activity provided. As observed, the antimicrobial activity may be enhanced with the incorporation of cerium, zinc, gallium, copper, or silver into the bioactive glass network.

Aiming to provide the antimicrobial activity to a mesoporous bioactive glass from the binary system SiO_2–CaO, Kurtuldu et al. [117] incorporated 1, 3 and 5 mol.% of Ce_2O_3 or Ga_2O_3 into the vitreous network by the microemulsion assisted sol–gel method and tested the glasses against two bacteria strains. Compared to the control group, the incorporation of the dopants was effective to decrease the viability of bacteria. After 6 h of incubation and especially at 3 mol.% Ce_2O_3 and 5 mol.% Ga_2O_3, the inhibition growth performed in the gram-positive *S. aureus* was superior

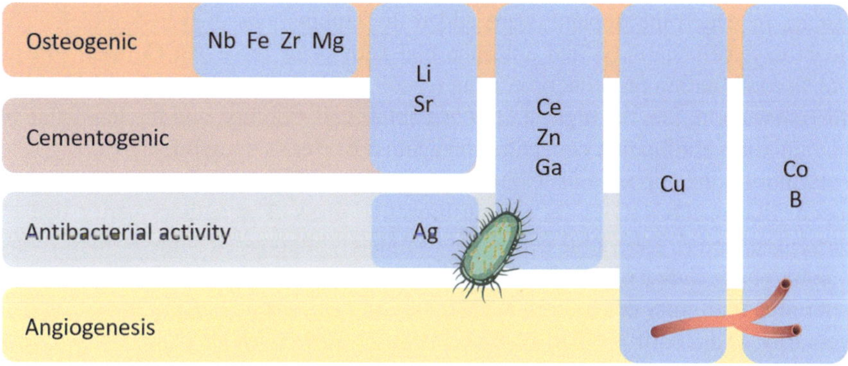

Fig. 7 Therapeutic ions and their targeting biological activities [95, 139]

to one of the gram-negative *E. coli* due to the absence of the peptidoglycan layer in the first. Furthermore, a cytotoxicity assay performed with the glasses extracts on MG-63 human osteoblast-like cell line resulted in high viability at low concentrations (1, 0.1, and 0.01% w/v); the high Ga^{3+} amount in 10% w/v extract reflected in a negative effect in cell metabolism, leading to toxicity. Overall, gallium and cerium incorporation resulted not only in the enhancement of antimicrobial properties, but also increased the cell viability of the bioactive glass [117]. The studies in continuation with the cerium doped bioactive glasses by the authors indicated an anti-inflammatory response of proinflammatory RAW 264.7 macrophage cells by the decreasing nitric oxide release [140], a cerium effect which was also demonstrated by Zheng et al. [141].

Naseri et al. [121] developed silver-doped sol–gel derived bioactive borate glass with the composition denominated B60 [(60)B_2O_3–(27)CaO–(4-X)P_2O_5–(X)Ag_2O where X = 0, 0.3, 0.5, and 1 (mol.%)], aiming the wound healing application. The release of silver ions from the glass network was effective to inhibit *E. coli* and *S. aureus* growth in a dose-dependent manner. Indeed, even at low concentrations, silver ions have a strong effect against both gram-positive and gram-negative bacteria, as demonstrated by other research groups [142–144]. In a posterior study, Naseri et al. [145] evaluated the efficacy of the ionic dissolution products of the 1 mol.% Ag_2O content B60 bioactive glass against pre-formed *Pseudomonas aeruginosa* biofilms (opportunistic pathogen in chronic wounds) and observed a 97% reduction of the colonies forming unit counts [145]. Furthermore, the viability and metabolic activity of keratinocytes were evaluated for all silver-doped B60 bioactive glasses. It was reported a reduction of metabolic activity for B60 attributable to the toxicity caused by the high concentration of boron released at a glass concentration of 1.5 mg/mL. With the incorporation of silver, however, this effect in the cell metabolisms was compensated once an enhancement of the viability was noticed for the lower silver content glasses (0.3 and 0.5 Ag_2O mol.%) [145].

Simultaneous antimicrobial activity of silver and zinc was recently evaluated by Azizabadi et al. [146] in their study of the SiO_2–P_2O_5–CaO–SrO–Ag_2O–ZnO system, in which the dopants were added in combinations of 0 to 5 mol.%. The bioactivity, cell viability, and antimicrobial activity of the 2 Ag_2O_3 and 3 Zn_2O mol.% combination presents promising results, as demonstrated by the hydroxyapatite conversion rate, the highest G29 osteoblast cell viability, and the low value of the minimum elimination concentration against *E. coli* (1.6 mg/mL) [146].

Exploring the copper ions, Chitra et al. [147] developed a 1.5 and 2.5% copper-doped SiO_2–CaO–NaO–P_2O_3 system bioactive glass. The authors verify that *P. aeruginosa* and *E. coli* viable cells were completely suppressed while *B. subtilis* and *S. aureus* have shown minimal growth when in contact with copper-doped glasses by a minimal inhibitory concentration (MIC) assay performed in agar plates. Complementarily, an agar diffusion assay performed in *C. albicans*, *E. faecalis,* and *S. aureus* revealed inhibition zones comparable to control (tetracycline), although the bactericidal activity was more significant on gram-positive bacteria than gram-negative bacteria [147]. The antimicrobial activity of copper is widely explored [148–150], as well as its angiogenic properties [151].

Overall, bioactive glasses and glass–ceramics are inorganic particles able to promote essential biological properties for the regeneration of damaged tissue, especially when associated with therapeutic ions. The incorporation of gallium, cerium, silver, copper, and zinc into the glass networks has proven efficacy in optimizing both bioactivity and antibacterial properties of bioactive glasses and, except for silver, also promotes osteogenic activity. Copper ions, however, not only influence these biological activities, but also provide angiogenesis [152, 153]. As demonstrated, these particles have great potential to optimize non-bioactive materials in the tissue engineering field.

3.1.2 Calcium Phosphates

As main constituents in bone (≈60 wt.%) and tooth enamel (90%), calcium phosphates (CaP) represent the family of minerals containing Ca^{2+} together with a phosphate group (PO_4^{3-}, PO_3^{-}, or $P_2O_7^{4-}$), which may also contain H^+ or OH^- ions [154]. Classified according to their Ca/P ratio (Fig. 8), the main relevant CaP are hydroxyapatite (Ca/P = 1.67) and tricalcium phosphates (Ca/P = 1.5) owing to their biocompatibility, bioactivity, and biodegradability.

Hydroxyapatite

The hydroxyapatite and its surface formed apatite have both structural composition and chemistry as the natural bone and teeth mineralized phase, being the widely applied bioceramic as a restorative material for orthopedics, dentistry, and coating on medical devices and implants for biocompatibility, biodegradability, osteoconductivity, and osteointegration enhancement purposes [105, 155].

Panda et al. [156], Sokolova and Epple [157], and Gomes et al. [158] recently reviewed the synthesis procedures performed for hydroxyapatite obtention, which

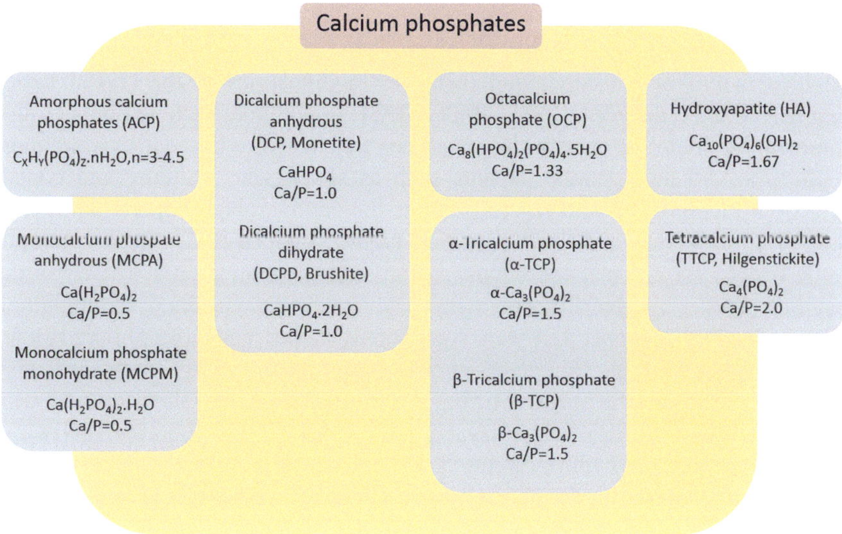

Fig. 8 Calcium phosphates and their respective Ca/P ratio

includes dry, wet, hydrothermal, biogenic, and high-temperature methodologies and processes. Although different methodologies are employed for the production of hydroxyapatite, what defined the route and parameters applied is the desired morphology of the particle; needles, rods, fibers, or sheet-like crystals, nano and microspheres, and porous particles may be achieved [159]. In the hydrothermal method, for example, crystal size is controlled by supersaturation, growth rate by pH, and crystal orientation by the selection of precursors, while additives allow complex geometries [159]. Most used techniques, however, are based on (i) solid-state reaction between $Ca(OH)_2$ or $CaCO_3$ and $CaHPO_4{\cdot}2H_2O$ or $\beta\text{-}Ca_3(PO_4)_2$ at high temperatures (950–1200 °C) [158, 160], (ii) precipitation method of dropwise addition of H_3PO_4 to a $Ca(OH)_2$ aqueous suspension under stirring and with pH control (NH_4OH) [161, 162] and (iii) sol–gel method by mixing solutions of $10Ca(NO_3)_2{\cdot}4H_2O$ and $6(NH_4)0.2HPO_4$ at controlled pH of 11 (NH_4OH) [158, 163].

While the stoichiometric hydroxyapatite has the general formula $Ca_{10}(PO_4)_6(OH)_2$, the biological apatite has the inclusion of many other ions such as Mg^{2+}, K^+, Zn^+, Na^+, Cl^-, and F^-, resulting in crystal structure partial substitutions [159, 164]. For example, the apatite structure of phosphorous either with chloride (Cl^-) or fluoride (F^-) ions produces the parenting compositions chloroapatite and fluoroapatite, respectively [163]. Tooth enamel contains the fluoroapatite of limited solubility, thanks to its high stability; the inclusion of the small-diameter F^- ion doesn't produce expressive alterations into the apatite structure [165]. The hydroxyapatite OH^- ions, however, require an atomic rearrangement that reduces the stability of the compound, which increases its solubility when compared with fluoroapatite [164]. On the other hand, the hydroxyapatite has a Ca/P of 1.67 but, when calcium deficient, the lower ratio increases its solubility [107]. As observed, any ion or groups alterations into the apatite structure change the solubility—thus the bioactivity—of the compound owing to modifications of the crystal lattice parameters and reduction of crystallinity [155]. In addition, particle porosity also optimizes the solubility once the larger surface area increases the contact with fluids.

Metallic and non-metallic dopants such as silver, zinc, gallium, and copper have been incorporated into HA to promote the antimicrobial property. Citradewi et al. [166] developed a silver nanoparticle (AgNPs)-doped HA by in situ incorporation under hydrothermal conditions, and studied the antimicrobial activity of the resulting powder by agar diffusion assay. The observed diameters of the inhibition zones of the AgNPs-doped HA were significantly higher than those of pure HA for all the four tested strains (*E. coli*, *K. pneumoniae*, *S. aureus*, and *S. pyogenes*), also revealing a bacteriostatic property owing to the reduction of the inhibition zone with increasing time [166]. Similarly, Vijayaraghavan et al. [167] incorporated AgNPs into chicken bone extracted hydroxyapatite (Black Sumatra and Fighter Cock), producing composite materials with the ability to affect biofilm formation and antimicrobial resistance against gram-negative and gram-positive bacteria (*E. coli*, *K. pneumoniae*, *E. aerogenes*, *S. aureus*, *S. pyogenes* and *B. cereus*). Other authors also tested the incorporation of silver into HA [168–170], being a field still with many opportunities.

The incorporation of zinc into HA is also generating crescent interest owing to its antimicrobial properties. Lima et al. [171] produced zinc-doped mesoporous hydroxyapatite (0.5, 1, and 2 mol.%) of larges pores and high specific area by co-precipitation method using phosphoprotein as the porous template. The authors demonstrated the expressive dependence of Zn^{2+} content in the antimicrobial action of their Zn-doped HA against *E. coli* and *S. aureus*, in which the 2 mol.% zinc-doped showed the maximum bacterial inhibitory concentration [171]. To optimize the zinc antimicrobial effect, Shokri et al. [172] performed binary doping of Zn^{2+} and Ga^{3+} into HA by the sol–gel method, obtaining nanopowders with spherical morphology and lower crystallinity compared with pure HA (85–75%). Besides promoting an antimicrobial activity against *E. coli* (68%) and *S. aureus* (84%), this binary doping was non-toxic towards human mesenchymal stem cells (hMSCs), enhanced the alkaline phosphatase activity and the mineralization after 14 days. Thus, compared with pure HA, the incorporation of Zn^{2+} and Ga^{3+} presents great antimicrobial activity and osteogenic properties [172]. Binary systems were also studied by Lv et al. [173] with their Mg, Zn, and Ce-doped CuO/HA composites, where they found that all systems exhibited the highest antimicrobial activity and complete growth inhibition at 15 wt.% CuO content against *E. coli* and *S. aureus*. As demonstrated, the mentioned authors and their respective work collaborate to illustrate the broad range of antimicrobial therapeutic ions to enhance the biological properties of hydroxyapatite.

Tricalcium Phosphate

Tricalcium phosphates (TCP, $Ca_3(PO_4)_3$) are widely used bioceramics due to their osteoconductive and osteoinductive properties, being also highly biocompatible and bioactive, besides they are more resorbable than HA owing to their Ca/P ratio of 1.5. TCP has two polymorphic forms of interest: β-TCP phase of rhombohedra structure stable at temperatures below 1125 °C, and α-TCP phase of monoclinic structure stable between 1125 and 1430 °C [174]. Once both α-TCP and β-TCP phases set under reaction with water forming calcium phosphate cement (CPC), tricalcium phosphates pastes are molded and shaped to fill complexes bone defects sites and allowed to set in situ, being explored as a self-setting bone substitute [175]. However, and compared with β-TCP, the α-TCP phase easily hydrolyzes in aqueous solutions—thus has a higher dissolution rate—and forms calcium deficient hydroxyapatite, defining this high-temperature phase as the powder component of bone cement while β-TCP is applied as a component in bioceramic and composites [174, 176].

The solid-state reaction is a simple method for TCP obtention, in which $CaCO_3$ and $CaHPO_4$ are mixed, homogenized, and then thermally treated. This approach, however, usually led to the crystallization of secondary phases containing other calcium phosphates if the reagents ratio and thermal treatment aren't well performed. Nifant'ev et al. [177] recently developed a method of β-TCP obtention without crystallization of impurities by the assumption that α-TCP can fully be converted into β-TCP by an annealing thermal treatment at lower temperatures. Indeed, the authors found that a ratio of >2:1 of $CaHPO_4$ and $CaCO_3$ allows first the obtention of α-TCP at 1350 °C and secondly pure β-TCP by subsequent annealing treatment at 850 °C [177].

The ionic substitution into β-TCP structure has been performed not only to promote antimicrobial activity, but also to improve other biological properties. It has been reported that copper-substitute TCP (Cu-TCP) in 1.4 wt.% enhanced the crystal growth and particle porosity, which positively reflects on the cytocompatibility of the bioceramic wherein no cytotoxicity was detected on mouse fibroblast NCTC L929 cell [178]. Regarding the copper's ability to inhibit bacterial growth, Cu-TCP has been demonstrated to effectively act against *E. coli* and *S. enteritidis* by agar diffusion assay, whereas pure β-TCP ceramic has shown no activity [178]. In a similar manner, Ag, Ti and Ag–Ti co-doping provided β-TCP ability to inhibit *S. aureus* growth [179]. Additionally, it was suggested the incorporation of titanium increases β-TCP solubility once the cumulative release of Ca^{2+} ions was enhanced [179].

3.2 Metallic Nanoparticles

Chronic bacterial infections are associated with high mortality rates especially owing to bacterial resistance to antibiotics. One of the mechanisms of bacterial resistance is based on increasing its intracellular life cycle [180]. Considering most drugs act in the extracellular space, medicines with the ability to penetrate multi-host cells and still reach therapeutic levels are not yet developed. Small enough to permeate intracellular organisms, metallic nanoparticles remain their therapeutic functions and can act solo or even as drug carriers helping to prevent the medicine degradation, while breaking microbial resistance [181, 182]. The antibacterial effect mechanisms of metal nanoparticles are still being investigated, although assumptions are based on two approaches occurring on the nanoparticle surface; toxicity derived from the surface metal dissolution and oxidative stress from reactive oxygen species (ROS) [183, 184]. Furthermore, electrostatic attraction between positively charged nanoparticle surfaces and negatively charged bacteria surfaces may contribute in inhibiting bacterial normal metabolism leading to its lysis [180].

While a strong antimicrobial agent according to their size, morphology, and crystallinity, metallic nanoparticles are subject to oxidation and aggregation at a level that depends on their origin metal. Additionally, their elevated surface area provides great drug-loading efficiency and chemical reactivity, whereas it contributes to their still unclear cytotoxic potential—although its intimate relationship with their shape, particle size, and concentration is well established [185]. Thus, current researches are not only focused on understanding metallic nanoparticles' antimicrobial mechanism of action, but also on their nanotoxicity. Silver, gold, copper, titanium, and zinc-containing metallic nanoparticles have a strongly antimicrobial effect and have been studied to promote their intrinsic property to medical materials, implants, and devices.

Metallic nanoparticles synthesis may be performed by top-down (bulk metal into nanoparticles using stabilizing agents) or bottom-up (transformation or nucleation followed by stabilization) approaches, while protocols and procedures consider the applied method—physical, chemical, or biological—for its definition, besides particles size and shape [186]. Physical methods are high energy-consuming and lead to a non-uniform nanoparticles size distribution [187]. Similarly, chemical methods, although being inexpensive and efficient, use environmentally toxic chemicals [187]. Biological methods, however, overcome the drawbacks by employing biological systems (e.g., plants, microorganisms, algae, proteins, and cells) in nanoparticles production. The green chemistry approach contemplated by biological methods is based on plant extracts and has been receiving special interest owing to its productivity, economy, availability, eco-friendless, and properties derived by the bioactive phytocompounds from plant extracts such as functionalization and stability [186, 188, 189].

The physical method is usually performed by the evaporation-condensation technique, in which a gas phase route coupled with a tube furnace synthesizes nanoparticles at atomic pressure [190]. A base metal source in a container is evaporated into the carrier gas in the tube furnace center, leading to the obtention of nanoparticles with size, shape, and yield according to the design of reaction facilities [191, 192]. Another physical method of production is laser ablation, where a pulsed laser is irradiated in a liquid environment containing a bulk metal source that releases nanoparticles [193]. Although there is no use of hazardous chemicals as radiation is used as a reduction agent, the NPs obtention by physical methods often results in agglomeration once capping agents aren't used. In addition, the great consumption of energy and the complex equipment needed limit the application of this route [191, 194]. Other physical methods include chemical vapor deposition, microwave irradiation, pulsed laser method, sonochemical reduction, and gamma irradiation [195].

Chemical methods are based on colloidal dispersion of water or organic solvents by reducing metal salts into nanoparticles [191]. The overall phenomenon is often described as coprecipitation and occurs in five phases: (1) reduction; (2) nucleation; (3) growth; (4) coarsening; and/or (5) agglomeration [196]. The chemical method is dependent on a set of conditions and synthesis reagents; temperature, pH, salt precursors, reduction, stabilizer and, capping agents, besides the ratio between these components. All these parameters assist the size distribution, shape, and dispersion of the resulting NPs, as well as the state of aggregation or its inhibition. Cele [195] and Karthikeyan et al. [197] listed methodologies of nanoparticles' synthesis by chemical approaches, including the polyol (seed growth), microemulsion, thermal decomposition, electrochemical synthesis, coprecipitation, sol–gel, photochemical and hydrothermal/solvothermal methods [197]. The reduction agent influence is well illustrated by these two common routes of AgNPs synthesis; the Brust–Schiffrin synthesis (BSS) and the Turkevich method, using citrate and sodium borohydride as chemical reductants, respectively [191]. The use of trisodium citrate (Na_3Ct) results in better control of shape and size distribution due to the slower reduction rate [198], while a stronger reductant such as borohydride tends to result in a large range of size distribution owing to the formation of larger-sized AgNPs [199]. The use of capping agents not only avoids agglomeration by electrostatic or steric repulsion, but also protects from oxidation [200]. Examples of capping agents include chitosan, polyethylene glycol (PEG) and, poly (amidehydroxyurethane) [195]. Additionally, the surface of the nanoparticles produced by chemical methods may be conjugated with biomolecules to induce specific cellular responses, being a great advantage over the physical method [191]. The employment of chemical methods results in high yield at low cost, but the use of toxic and hazardous materials results in low purity of NPs, whose surfaces are often sedimented with residual materials and byproducts [194].

The employment of biological systems for nanoparticle synthesis characterizes biological methods (green chemistry). This approach is considered simple, cost-effective, and environmentally friendly [194]. Living microorganisms such as plant extracts, bacteria and fungi, algae, and small biomolecules have the ability to reduce metal ions and perform toxic management, which allows the intracellular synthesis of high stability nanoparticles [201]. Plant extract and microbial metabolites compositions have terpenoids, phenols, flavonoids, and other protein compounds that help with the reduction and stabilization of metal nanoparticles [202]; dispensing the use of reducing and capping agents as they are naturally present in biological systems [196]. This ability of live organisms overcomes the major drawback of the chemical method once no hazardous and toxic materials are needed. Parmar et al. [203] reviewed the synthesis of AgNPs in bacterial, fungi, and plant-mediated routes and highlighted the main mechanisms and organisms already used.

Recently, the properties of zinc oxide nanoparticles (ZnO NPs) produced by the green synthesis (using fruit extract) and chemical precipitation method were performed and compared regarding their antimicrobial properties [204]. Besides a difference in particle size of 11–21 nm for the green synthesis and 30–40 nm for the precipitation method, an agar diffusion assay performed on *B. subtilis*, *S. aureus*, and *S. typhimurium* demonstrated higher diameter inhibition zones for green synthesized ZnO NPs [204]. As observed, the method applied for the metallic nanoparticles synthesis highly influences their final physical and antimicrobial properties. Similarly, AgNPs produced by green synthesis from *Guettarda Speciosa* leaf extract generate 30 nm-sized nanoparticles that greatly inhibit the growth of *B. subtilis, S. aureus,* and *E. coli* [205], while 63 nm-sized AuNPs produced from the medicinal plant *Kaempferia parviflora* is proven to be highly effective against *E. coli* and *P. aeruginosa* [206].

3.2.1 Silver Nanoparticles (AgNPs)

AgNPs had been showing proven efficacy in a wide range of applications in orthopedics, wound healing, and dentistry as a drug delivery system, being the most studied nanoparticles for biomedical applications owing to their low cost and cytotoxicity, great immunological response, and remarkable antimicrobial action [207, 208]. Besides therapeutics, AgNPs are explored in medical imaging, molecular diagnostics, and as an anticancer agent by enhancing tumor-killing effects in anticancer drugs [209]. Just as nanoparticles in general, the biological action of AgNPs depends on their physicochemical properties; size and distribution, particle shape and morphology, surface chemistry and agglomeration, composition, ion release efficiency, and dissolution rate, among others [210, 211]. The development of controlled structures with the set of desired physicochemical properties is imperative for biomedical applications once it defines biological distribution, targeting cells, and penetration into cellular membranes, thus defining their toxicity.

The antimicrobial mechanism of silver nanoparticles had been proposed by several researchers, which is caused not only by the release of silver ions, but also by the nanoparticles themselves, as seen in Table 2. Furthermore, the antimicrobial action of AgNPs is affected by their dissolution efficiency, the surrounding media components, and pH. The release of silver ions is highly dependent on their specific surface area and surface modification (capping agents) [208]. Regarding the media, organic and inorganic components may promote aggregation and chemical reaction, while acidic pH results in faster release than neutral media [212]. The microorganism's membrane thickness must also be considered when assessing the antimicrobial action. Penetration and uptake of Ag^+ and AgNPs occur differently in gram-positive (30 nm) and gram-negative (3–4 nm) bacteria [213, 214].

3.2.2 Zinc Oxide

Just as AgNPs, the application of Zinc oxide nanoparticles (ZnONPs) is widely investigated in biomedicine owing to their antimicrobial, and anti-inflammatory activities, ability to perform wound healing management and faster epithelialization, cytostatic activity (cell growth and multiplication inhibition) over cancer cells and also in

Table 2 AgNPs and Ag^+ antimicrobial action, and involved mechanism

Antimicrobial action	Mechanisms	References
Disrupt cell wall and membrane by Ag^+	The released Ag^+ adhere to and permeates the cell wall and cytoplasmic membrane by electrostatic attraction and affinity owing to their negative charge (presence of carboxyl, phosphate, and amino groups)	[213, 215]
Disrupt membrane by ROS	The deactivation of respiratory enzymes after cell uptake of free Ag^+ produces ROS, which disrupts the cytoplasmic membrane	[216]
Denature ribosomes	Ag^+ denature ribosomes, leading to protein synthesis inhibition	[216, 217]
Denature membrane by AgNPs	AgNPs accumulated in the cell wall pits caused membrane denaturation	[218]
Interrupt ATP production	The ROS generated at the deactivation of the respiratory chain inhibits ATP production	[219]
Interfere DNA replication	The interaction of Ag^+ and ROS with DNA's sulfur and phosphorous from deoxyribonucleic acid alters DNA replication	[213, 220]
Perforate membrane by AgNPs	Organelles are released from the cells as AgNPs move across the membrane, which eventually leads to cell lysis	[212]
Bacterial signal transduction	The peptide's substrates tyrosine residues dephosphorylation by AgNPs disrupts signal transduction, leading to apoptosis	[216]

bioimaging due to its chemiluminescent property [221, 222]. ZnO has a highly oxidative character attributed to ROS formation—a pathway for its antimicrobial action against a wide range of strains [223]. ROS generation occurs by different mechanisms: bacteria surface adsorption, generation of electron/hole pairs, the reaction between water and oxygen, and formation of intermediates [224, 225]. Besides ROS formation, the mechanisms proposed for ZnONPs antimicrobial activity also considered the release of Zn^{2+} as a dissolution product of ZnO [226], and the interaction of the NPs with the microorganisms [225].

Mendes et al. [223] determined the inhibitory concentration (IC_{100}) of a complex bead and rod-shaped ZnONPs against a set of clinically relevant bacteria (*B. subtilis*, *E. coli*, *S. aureus* and *P. aeruginosa*), and evaluated the caused effect in the membrane integrity. The IC_{100} values obtained for *E. coli* and *P. aeruginosa* (gram-negatives) were 0.6 mM, while for *B. subtilis* and *S. aureus* (gram-positives) the estimated concentrations were between 0.8 and 1.0 mM. After a 15 min ZnONPs exposure at IC_{100}, cytoplasmic membrane disruption was observed in all tested strains. The authors [223] discussed the mechanisms of action of ZnONPS: (1) Generation of ROS owing to the strong oxidizing character of ZnO as a transition metal and semiconductor leading to inhibition of protein synthesis and DNA replication; (2) Zn^{2+} from the dissolution of ZnONPs has a strong affinity to sulfur groups, thus inhibiting glycolytic enzyme through the thiol group oxidation; (3) Damage to cell membrane originated from electrostatic forces between the membrane carboxyl groups and Zn^{2+} from ZnONPs, leading to electrostatic gradient differences with eventual membrane disruption [223].

3.2.3 Platinum

Platinum nanoparticles (PtNPs) have gained increasing interest in biomedical applications owing to their outstanding biocompatibility, antimicrobial properties, high stability, and surface chemistry that allow unique catalytic properties [227]. Although conventional physical and chemical methods are used to obtain PtNPs, such as laser ablation, sol–gel, co-precipitation, and photochemical reduction [228], the green chemistry has emerged as a primary obtention route for eco-friendly and non-toxic nanoparticles with high stability and small size [229, 230]. Recently, Jan et al. [202] reviewed the PtNPs green synthesis from a diversity of biological sources (plant extracts, bacteria, fungi, and algae) and compile the resulted size and shape, besides listing the influence of synthesis parameters (precursors, temperature and pH) in the final morphology. The antimicrobial activity of PtNPs has been reported to be interconnected with adenosine triphosphate (ATP) production and mitochondrial membrane potential [202]. Furthermore, the formation of ROS results in down-regulation of DNA and oxidative stress leading to apoptosis [231].

3.2.4 Gold

Nanoparticles from the transition metal gold (AuNPs) are widely explored in biomedicine owing to their antimicrobial, anticancer, antioxidant, anti-inflammatory, and wound healing management properties [232, 233]. Just like the other metallic nanoparticles, the green chemistry approach to production stands out over the physical and chemical methods. AuNPs have been already produced from *Sambucus wightiana extract* [234], Chinese lettuce leave extract (*Brassica rapa* var. *pekinensis*) [235], and black ginger (*Kaempferia parviflora* extract) [208], proving their efficiency against *S. aureus*, *P. aeruginosa*, *E. coli*, *B. subtilis*, *S. typhimurium*, *A. niger*, *C. albicans* and *C. glabrata* [206, 234, 235]. The mechanisms of action involved are described in terms of direct contact by adhesion onto the cell wall surface, allowing penetration and ion-mediated microorganisms killing [206]. Aghamirzaei et al. [235] illustrated the mechanisms by observing the morphology of *S. aureus* after contact with AuNPs, where disrupted and collapsed surfaces were revealed.

3.2.5 Titanium Dioxide

Owing to its reliability and long-term photostability, biological and chemical inertness, besides great photocatalytic and antimicrobial activity, titanium dioxide nanoparticles (TiO_2NPs) unique properties have been evaluated in biomedicine [236, 237]. Although these nanoparticles have been already produced by physical and chemical methods such as sol–gel, microwave irradiation, and solid-state reaction, the green chemistry had been revealing the main route for metallic nanoparticles obtention, and could not be different for titanium dioxide [214]. A variety of plant materials have been already used to produce TiO_2NPs; *Morinda citrifolia* [214], *Acacia catechu* [238], *Calotropis gigantea* [239], *Trigonella foenum-graecum* extract [240], among others, resulting in nanoparticles with different shapes and small sizes. Sundrarajan et al. [214] studied the antimicrobial activity of TiO_2NPs using *M. citrifolia* against clinically relevant microorganisms (*Staphylococcus aureus*, *Escherichia coli*, *Bacillus subtilis*, *Pseudomonas aeruginosa*, *Candida albicans*, and *Aspergillus niger*). The authors suggested the antimicrobial mechanisms of action of their TiO_2NPs, which consists of their size and specific surface area; thickness and structure of bacterial cell wall; the releasing of Ti^{4+} as a result of the nanoparticles dissolution; the generation of hydrogen peroxide (H_2O_2) and ROS at the surface of TiO_2 nanoparticles [214]. Once the outer layer of gram-negative bacteria is composed of protein, lipids, and lipopolysaccharides, while the gram-positive bacteria is composed of a thick peptidoglycan layer, the TiO_2 nanoparticles can easily penetrate and disrupt the cell wall and lead the microorganisms to apoptosis. Sundrarajan et al. [214] results from their agar diffusion assay indicated great activity over the studied strains, although their green synthesized nanoparticles were more effective over the gram-positive bacteria.

4 Incorporation of Inorganic Particles in Electrospun Nanofibers

A major challenge of current research is to produce scaffolds with physicochemical and morphological characteristics that, when implanted in the injured regions, stimulate osteogenesis and promote the formation of structures similar to extracellular matrix elements, facilitating the mobilization, expansion, and integration of internal regenerative cell populations, promoting injury repair or degenerate tissue renewal. In this sense, polymeric scaffolds are often exploited as tissue substitutes for the regeneration of tissue physiology and specific functions. When incorporated with inorganic nanoparticles, however, specific properties such as bioactivity and antimicrobial action may be achieved.

Inorganic particles embedded polymer matrix for the formation of organic/inorganic composites depends on several parameters, including the proportion and nature of the components and inorganic particles distribution and size (generally

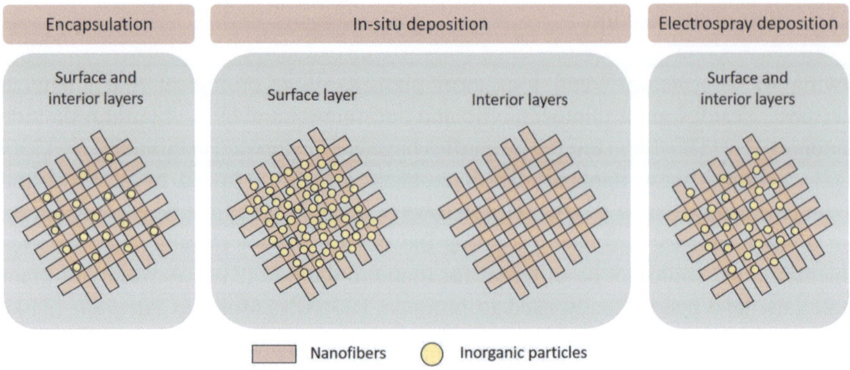

Fig. 9 Schematic illustration of methods for inorganic particles incorporation into nanofibers and the resulting morphology

in the nanometer range) [10]. A common approach for the fabrication of these composites is the physical blending, where inorganic particles are mechanically mixed with the polymer solution prior to electrospinning, encapsulating the particles within the nanofiber. Another approach is based on deposition, and may be performed in-situ or by electrospray. In the in-situ deposition method, the nanofibers are first obtained, and then serve as a substrate for the inorganic particle precipitation on its surface in a second treatment. In the electrospray method, however, polymer nanofiber obtention and inorganic particles deposition occur simultaneously by allaying electrospinning with electrospray—a parenting electrohydrodynamic methodology. The great advantages of electrospray over in-situ deposition are; (i) the absence of the second treatment of incorporation, and the fact that (ii) the inorganic particles are not only deposited on the polymer matrix surface, but are also present into and between the nanofibers and their pores. Figure 9 illustrates the final morphology obtained by encapsulation and both methods of deposition. The following sections will address these methods and recent approaches considering the composite biomedical applications.

4.1 Encapsulation

The physical blending method of embedding bioceramics particles into nanofibers results in their encapsulation. By adjusting the concentration of both components, the mechanical stirring speed of mixing, and the preparation method, this methodology is a simple way to combine different polymers and inorganic particles to produce composites for biomedical applications. Besides providing bioactivity and antimicrobial action, incorporation of inorganic particles may also cause alterations in contact angle, mechanical properties, and surface roughness, for example.

Fonseca et al. [241] produced through electrospinning polycaprolactone (PCL) membranes combined with two types of 45S5 bioactive glass (sol–gel and melt-quenched). The physical and chemical analysis results showed success in manufacturing bioactive membranes with significantly enhanced cell viability and osteoblast differentiation. The newly formed bone from the in vivo experiment was similar to that observed in the control group. Although the microstructure of PCL membranes is influenced by the bioactive glass production method both seem to be promising biomaterials for tissue engineering when incorporated into PCL [241].

Ahmed et al. [242] incorporated co-doped Bi/Sr-Hydroxyapatite (Bi/Sr-HA) into polycaprolactone (PCL) nanofibers and performed a series of characterizations. Compared to pure PCL nanofibers, the encapsulation of Bi/Sr-HA increased the surface roughness and the specific surface area, while highly decreasing the contact angle. All these physical alterations reflected in the mechanical properties—there was an increase of the tensile strength and decrease in young modulus and toughness—as well as in cell attachment and antimicrobial activity. Owing to the higher adhesion promoted by roughness and wettability in the Bi/Sr-HA doped PCL nanofibers, fibroblast cells attachment was higher. The antimicrobial agar diffusion assay indicated the great capability of the composite to inhibit *S. aureus* and *E. coli* growth, while no activity was observed for the pure nanofiber [242]. Similar results were found by El-Naggar et al. [243] in their study of Tl doped HA inorganic particles encapsulated PCL nanofibers, where they found increase in roughness, decrease in contact angle, alterations in mechanical properties, and high cell viability.

β-TCP encapsulated collagen-based nanofibers were studied by Bian et al. [244]. Before analysis, collagen cross-linking was performed to lower its water solubility, and thus the degradation rate. Considering bone repair regeneration, the obtained cross-linked scaffold mechanical properties were evaluated while wet to simulate in vivo environment. Tensile strength has been reported to increase by nearly two thousand times that of the dry scaffold. During biomineralization assay, it was found that there was a significant deposition of a hydroxyapatite mineral layer on the scaffold surface, and a tendency of HA to line up with its fiber's orientation owing to β-TCP dissolution. Cytotoxicity assay performed on bone marrow mesenchymal stem cells (BMSCs) show not only that the non-toxicity of the composite, but also its ability to induce cell proliferation. An agar diffusion assay on *E. coli* demonstrated higher inhibition zones to the β-TCP incorporated collagen scaffolds when compared with pure nanofibers. Additionally, turbidity analysis on *E. coli* reported no activity of the collagen scaffold, but when β-TCP was added, an inhibition of 66% was reported. However, when the assay was performed on *S. aureus*, poor activity was observed. The authors proposed the less negative charge of *S. aureus* cell membrane collaborated to this result, as Ca^{2+} ions released from the β-TCP induce ROS burst in bacteria, thus damaging its structures and leading to cell lysis. Due to these optimizations of properties, the bone repair field would benefit from this composite [244].

Metallic nanoparticles were also incorporated into nanofibers by encapsulation. Mostafa et al. [245] embedded silver nanoparticles into polystyrene nanofibers and evaluated their antimicrobial activity by agar diffusion assay on *E. coli* and *S. aureus*.

They found that inorganic particle incorporation into the nanofibers demonstrated a great inhibition zone, while in the pure membrane no activity was observed [245]. As already illustrated, the mechanism of action of nanoparticles is based on opposite charges between NPs surfaces and cell membrane, leading to ROS or oxidative stress formation followed by interruption of cell normal functions and its lysis.

Considering metallic nanoparticles, owing to their stability a homogeneous solution in organic solvents will hardly be obtained, leading to the formation of agglomerates and irregular ion release [46]. Furthermore, the presence of NPs affects the spinnability and morphology of the obtained nanofiber. To address these problems several approaches may be performed: (1) apply physical methods of rapid stirring to uniformly disperse the NPs in the organic solution; (2) use different solvents for each component; and (3) modify the nanoparticles using surfactants [46, 246]. Away from homogenizing NPs into the polymer solution aiming a single-fluid electrospinning process, the use of multi-fluids in core–shell structures may also provide NPs encapsulation into polymer nanofibers. Nanocomposites were already fabricated using the multi-fluid technique [247, 248], and the same fundamentals may be applied to nanoparticles.

As demonstrated, the incorporation of inorganic particles not only provides bioactivity and antimicrobial action, but the morphology alterations also reflect in other features, especially for ceramic nanoparticles. Although an easy approach to embed inorganic particles into polymer matrices, challenges related to uniform distribution within the whole scaffold, besides rapid dissolution of the nanofibers, are limitations of the technique [10].

4.2 Electrospray Deposition

Electrospraying is a method of atomizing liquids based on similar principles of the electrospinning process capable of producing micro and nanoparticles with different morphologies [16]. Figure 10 illustrates the setup for electrospinning and electrospray techniques combined. However, the difference between the two techniques consists of changing the properties of the solution, such as; concentration, solvent, and viscosity, and the process parameters such as; flow rate, distance from the tip of the needle to the collector, and voltage. Through the adjustment of solution factors and processing parameters, carriers in different structures can be obtained, such as hollow microspheres [249, 250], nanocups [251], porous microcarriers [252–254], cell-shaped microparticles [255, 256], core–shell/multilayered microspheres [257, 258], and Janus particles [259, 260].

Compared with the other conventional methods to produce particles, electrospraying has advantages, as it does not involve high temperatures and does not require additional mechanical energy, other than that of the electric field energy required for the atomization of the liquid medium, which makes the production of carriers for biomedical applications attractive [261]. Recently, this technique continued to

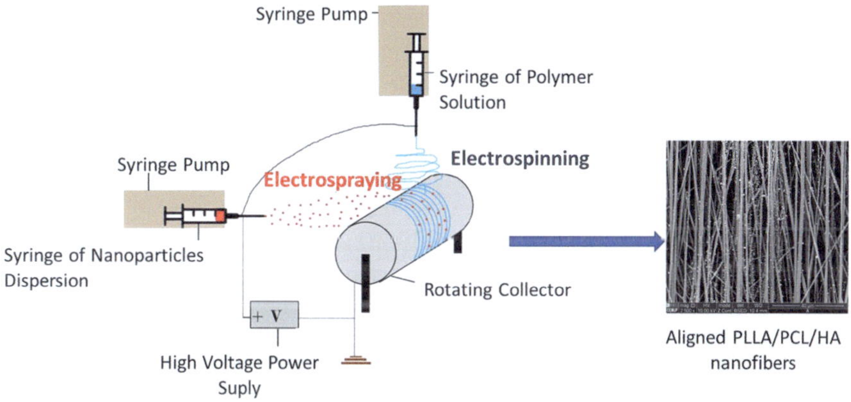

Fig. 10 Electrospinning/electrospraying apparatus: aligned PLLA/PCL/HA nanofibers production

evolve in the field of bioengineering. Bioactive compounds, as well as cells, have been electrosprayed apart from conventional drugs [261].

Currently, there is a tendency for the development of composite scaffolds by incorporating nanoparticles into a polymer solution to improve or obtain new properties of this polymer. On the other hand, an innovative approach has been employed to obtain polymer fibers partially coated with nanoparticle aggregates resulting from the simultaneous electrospinning of a polymer solution and the electrospraying of a nanoparticle aggregate dispersion [262–267]. This combination of electrodynamic techniques gives rise to a greater extent of exposed nanoparticles, thereby increasing the surface-dependent properties of fibers such as mechanical, bioactive, antimicrobial, optical, and catalytic properties, among others.

Siqueira et al. [262] obtained an aligned and random nanofiber blend of PLLA and PCL associated with electrosprayed HA and compared their mechanical and biological behaviors. The authors found a 15-fold increase in Young's Modulus and an eightfold increase in tensile strength as a result of the alignment, while the deposition of HA acted as a reinforcement, and enhanced the response to tensile stress. Osteoblast cells were able to adhere and proliferate on the scaffolds, although the cellular metabolic activity was superior on aligned nanofibers. Furthermore, an antimicrobial assay on *S. aureus* indicated a lower number of colonies forming units (CFU) for the aligned composite [262].

4.3 *Inorganic Particles Incorporation in Ready Nanofibers*

Inorganic particle incorporation may be performed in ready nanofibers, as in immersion and in-situ deposition. Due to the different hydrophilic properties of nanoparticles and nanofibers, the incorporation by immersion of ready nanofibers into a NP suspension results in significant rejection and inefficiency [46]. The modification of nanofiber surface with polar groups to overcome this limiting factor is a strategy already used [268]. In the in-situ deposition method, the nanofibers act as a foothold for the precipitation of the inorganic particle on the membrane surface forming a coating. Attachment of the nanoparticles on nanofiber surface occurs by adsorption of functional group action [46]. Pulsed laser deposition (PLD) [269] and catechol redox chemistry [270] are examples of techniques applied to perform the precipitation of inorganic particles in-situ. This methodology allows a large number of inorganic precursors and great dispersion in the matrix, however only suitable polymers can be used once the network may be damaged during the coating process. Furthermore, the stability of the precipitated nanoparticle cannot be predicted. Another methodology consists of sputtering of NPs by magnetron sputtering technique [271].

5 Final Considerations and Future Outlook

The human body tissues and their interactions with the physiological environment are highly complex and the development of a material that faithfully reproduces these conditions is still a challenge for biomedicine. The electrospinning allows the obtention of biopolymer scaffolds with a wide range of morphologies by controlling the technique parameters; diameters at micro and nanoscale, aligned or random oriented, beaded, porous, dense or hollowed structures, etc., which determined final properties in regenerative medicine. The obtention of scaffolds that not only provide an optimum environment for cell attachment, growth, and proliferation but also promote bioactivity and antimicrobial properties is imperative. The incorporation of inorganic particles such as bioactive glass, glass–ceramics, calcium phosphates, and metallic nanoparticles into the inert polymer nanofibers allows the obtention of composites that fulfill this need. The inorganic particle incorporation has been performed before and after the obtention of the nanofibers with techniques such as encapsulation, depositions and immersion. Much progress has been made in the development of these composites, although limitations of the techniques may restrict the incorporation to their surfaces or too deep inside each fiber, besides lacking homogeneity of distribution owing to incompatibility between the composite components. Current research on nanofibers incorporated with inorganic particles aims to overcome these limitations, customizing both nanofibers' scaffolds and the applied techniques. Besides developing a material that mimics tissue structures and interactions, the regulatory control of the different types of cells in the scaffold is imperative to achieve their full integration into the living biological system. Thus, not only efforts on techniques and applied methods are being performed on inorganic particles incorporated nanofibers, but also in understanding the caused interaction effects between cells and biomaterials.

Acknowledgements The authors would like to thank FAPESP—Fundação de Amparo à Pesquisa do Estado de São Paulo—(2019/10877-3), and CAPES-PRINT—Programa Institucional de Internacionalização (88887.572851/2020-00) for the financial support.

References

1. Z. Qian, D. Radke, W. Jia, M. Tahtinen, G. Wang, F. Zhao, in *Encyclopedia of Biomedical Engineering*, ed. by R. Narayan (Elsevier, 2019), pp. 444–461
2. L. Li, H. Lu, Y. Zhao, J. Luo, L. Yang, W. Liu, Q. He, Mater. Sci. Eng. C **98**, 1241 (2019)
3. S. Poovathikkal, N.D. Nair, World J. Pharm. Res. **7**, 161 (2018)
4. A. Hudecki, G. Kiryczyński, M.J. Łos, in *Stem Cells and Biomaterials for Regenerative Medicine*, ed. by M.J. Łos, A. Hudecki, E. Wiecheć (Elsevier, 2019), pp. 85–98
5. A.M. George, S.P. Reddy Peddireddy, G. Thakur, F.C. Rodrigues, in *Biopolymer-Based Formulations*, ed. by K. Pal, I. Banerjee, P. Sarkar, D. Kim, W.-P. Deng, N.K. Dubey, K. Majumder (Elsevier, 2020), pp. 717–749
6. M. Hussain, S.M. Khan, K. Al-Khaled, M. Ayadi, N. Abbas, W. Chammam, Mater. Today Commun. 103167 (2022)
7. M.E. El-Naggar, E.S. Shalaby, A.H. Abd-Al-Aleem, M.A. Abu-Saied, A.M. Youssef, J. Mater. Res. Technol. **15**, 3447 (2021)
8. Q. Yang, Z. Xie, J. Hu, Y. Liu, Mater. Sci. Eng. C **128**, 112319 (2021)
9. H.-T. Lu, G.-Y. Huang, W.-J. Chang, T.-W. Lu, T.-W. Huang, M.-H. Ho, F.-L. Mi, Carbohydr. Polym. **281**, 119035 (2022)
10. X. Wang, J. Chang, C. Wu, Appl. Mater. Today **11**, 308 (2018)
11. Z. Mohammadalizadeh, E. Bahremandi-Toloue, S. Karbasi, React. Funct. Polym. **172**, 105202 (2022)
12. C. Huang, J. Dong, Y. Zhang, S. Chai, X. Wang, S. Kang, D. Yu, P. Wang, Q. Jiang, Mater. Des. **212**, 110240 (2021)
13. D.Y. Zaki, E.M. Safwat, S.M. Nagi, H.N. Salem, T.M. Hamdy, L.M. Moharam, M.L. Hassan, E.M.A. Hamzawy, Carbohydr. Polym. Technol. Appl. **2**, 100035 (2021)
14. Q. Chen, J. Wu, Y. Liu, Y. Li, C. Zhang, W. Qi, K.W.K. Yeung, T.M. Wong, X. Zhao, H. Pan, Mater. Sci. Eng. C **105**, 110083 (2019)
15. E.T. Stace, P.-A. Mouthuy, A.J. Carr, H. (Cathy) Ye, in *Comprehensive Biotechnology*, 3rd edn., ed. by M. Moo-Young (Elsevier, 2019), pp. 424–441
16. R.M.D. Soares, N.M. Siqueira, M.P. Prabhakaram, S. Ramakrishna, Mater. Sci. Eng. C **92**, 969 (2018)
17. G. Liu, Z. Gu, Y. Hong, L. Cheng, C. Li, J. Control. Rel. **252**, 95 (2017)
18. A. Al-Enizi, M. Zagho, A. Elzatahry, Nanomaterials **8**, 259 (2018)
19. W.J. Morton, US Patent, 705691, (1902)
20. J.F. Cooley, US Patent, 692631 (1902)
21. J. Zeleny, Phys. Rev. **3**, 69 (1914)
22. A. Formhals, US Patent, 1975504 (1934)
23. A. Formhals, US Patent, 2158415 (1939)
24. N. Tucker, J.J. Stanger, M.P. Staiger, H. Razzaq, K. Hofman, J. Eng. Fiber. Fabr. **7**, 155892501200702S10 (2012)
25. A. Rossin, É. de Oliveira, F. de Moraes, R.S. da Júnior, D. Scheidt, W. Caetano, N. Hioka, D. Dragunski, Quim. Nova **43**, 613 (2020)
26. S. Subianto, D. Cornu, S. Cavaliere, in *Electrospinning for Advanced Energy and Environmental Applications*, ed. by S. Cavaliere (CRC Press, Florida, 2016), pp. 1–28
27. M.S. Islam, B.C. Ang, A. Andriyana, A.M. Afifi, S.N. Appl, Science **1**, 1248 (2019)
28. H.M. Ibrahim, A. Klingner, Polym. Test. **90**, 106647 (2020)
29. J. Xue, T. Wu, Y. Dai, Y. Xia, Chem. Rev. **119**, 5298 (2019)
30. L. Mercante, R. Andre, J. Macedo, A. Pavinatto, D. Correa, Quim. Nova **44**, 717 (2021)
31. T. Wu, M. Ding, C. Shi, Y. Qiao, P. Wang, R. Qiao, X. Wang, J. Zhong, Chin. Chem. Lett. **31**, 617 (2020)
32. Y. Ge, R. Chen, H. Li, Z. Guo, Mater. Lett. **228**, 301 (2018)
33. G. Duan, A. Greiner, Macromol. Mater. Eng. **304**, 1800669 (2019)
34. T.-C. Xu, D.-H. Han, Y.-M. Zhu, G.-G. Duan, K.-M. Liu, H.-Q. Hou, Chin. J. Polym. Sci. **39**, 174 (2021)

35. D. Atila, V. Hasirci, A. Tezcaner, J. Mech. Behav. Biomed. Mater. **128**, 105105 (2022)
36. S.-J. Seo, H.-W. Kim, J.-H. Lee, J. Nanomater. **2016**, 1 (2016)
37. G. Taylor, Proc. R. Soc. Lond. **313**, 453 (1969)
38. F.E. Ahmed, B.S. Lalia, R. Hashaikeh, Desalination **356**, 15 (2015)
39. M. Cloupeau, B. Prunet-Foch, J. Electrostat. **25**, 165 (1990)
40. J. Doshi, D.H. Reneker, J. Electrostat. **35**, 151 (1995)
41. T. Subbiah, G.S. Bhat, R.W. Tock, S. Parameswaran, S.S. Ramkumar, J. Appl. Polym. Sci. **96**, 557 (2005)
42. R.K. Mishra, P. Mishra, K. Verma, A. Mondal, R.G. Chaudhary, M.M. Abolhasani, S. Loganathan, Environ. Chem. Lett. **17**, 767 (2019)
43. P. Ginestra, E. Ceretti, A. Fiorentino, Procedia CIRP **49**, 8 (2016)
44. X. Zhang, W.J. Goux, S.K. Manohar, J. Am. Chem. Soc. **126**, 4502 (2004)
45. B. Kim, H. Park, S.-H. Lee, W.M. Sigmund, Mater. Lett. **59**, 829 (2005)
46. M. Zhang, W. Song, Y. Tang, X. Xu, Y. Huang, D. Yu, Polymers (Basel) **14**, 351 (2022)
47. C.J. Thompson, G.G. Chase, A.L. Yarin, D.H. Reneker, Polymer (Guildf) **48**, 6913 (2007)
48. T.J. Sill, H.A. von Recum, Biomaterials **29**, 1989 (2008)
49. S. Ramakrishna, K. Fujihara, W.-E. Teo, T.-C. Lim, Z. Ma, in *An Introduction to Electrospinning and Nanofibers* (World Scientific, 2005), pp. 22–90
50. C. Wang, C.-H. Hsu, J.-H. Lin, Macromolecules **39**, 7662 (2006)
51. S. De Vrieze, T. Van Camp, A. Nelvig, B. Hagström, P. Westbroek, K. De Clerck, J. Mater. Sci. **44**, 1357 (2009)
52. A. Raksa, P. Numpaisal, Y. Ruksakulpiwat, Mater. Today Proc. **47**, 3458 (2021)
53. L. Aalto-Setälä, P. Uppstu, P. Sinitsyna, N.C. Lindfors, L. Hupa, Materials (Basel) **14**, 4834 (2021)
54. A.D. Juncos Bombin, N.J. Dunne, H.O. McCarthy, Mater. Sci. Eng. C **114**, 110994 (2020)
55. M. Keshvardoostchokami, S.S. Majidi, P. Huo, R. Ramachandran, M. Chen, B. Liu, Nanomaterials **11**, 21 (2020)
56. M. Stevanović, in *Nanostructured Polymer Composites for Biomedical Applications*, ed. by S.K. Swain, M. Jawaid (Elsevier Science, 2019), pp. 1–19
57. T.P. Sato, B.V.M. Rodrigues, D.C.R. Mello, E.A. Münchow, J.S. Ribeiro, J.P.B. Machado, L.M.R. Vasconcellos, A.O. Lobo, M.C. Bottino, A.L.S. Borges, Clin. Oral Investig. **25**, 3095 (2021)
58. Y.K. Noh, A. Dos Santos Da, Y.S. Costa, P. Park, I.-H. Du, K.P. Kim, Carbohydr. Polym. **219**, 210 (2019)
59. C. Patra, S. Talukdar, T. Novoyatleva, S.R. Velagala, C. Mühlfeld, B. Kundu, S.C. Kundu, F.B. Engel, Biomaterials **33**, 2673 (2012)
60. Y. Goh, M. Akram, A. Alshemary, R. Hussain, Appl. Surf. Sci. **387**, 1 (2016)
61. N. Ribeiro, S.R. Sousa, C.A. van Blitterswijk, L. Moroni, F.J. Monteiro, Biofabrication **6**, 035015 (2014)
62. C. Vineis, A. Varesano, in *Electrofluidodynamic Technologies (EFDTs) for Biomaterials and Medical Devices*, 1st edn., ed. by V. Guarino, L. Ambrosio (Elsevier, 2018), pp. 275–294
63. L.G. Griffith, Acta Mater. **48**, 263 (2000)
64. N. Iqbal, A.S. Khan, A. Asif, M. Yar, J.W. Haycock, I.U. Rehman, Int. Mater. Rev. **64**, 91 (2019)
65. A.L. dos Santos, M.A.T. Duarte, S.H. Pezzin, L. Silva, J.A. Domingues, Mater. Res. **23**, e20190468 (2020)
66. P. Singhal, Mater. Today Proc. **37**, 2997 (2021)
67. G. Liao, S. Jiang, X. Xu, Y. Ke, Mater. Lett. **82**, 159 (2012)
68. I.P. Mendes Soares, C. Anselmi, F.A. Kitagawa, R.A. de O. Ribeiro, M.L. Leite, C.A. de Souza Costa, J. Hebling, Dent. Mater. **38**, 960 (2022)
69. S. Mohandesnezhad, Y. Pilehvar-Soltanahmadi, E. Alizadeh, A. Goodarzi, S. Davaran, M. Khatamian, N. Zarghami, M. Samiei, M. Aghazadeh, A. Akbarzadeh, Mater. Chem. Phys. **252**, 123152 (2020)
70. N. Shankhwar, M. Kumar, B.B. Mandal, A. Srinivasan, Mater. Sci. Eng. C **69**, 1167 (2016)

71. H. Lian, Z. Meng, Mater. Sci. Eng. C **84**, 123 (2018)
72. L. Hidalgo Pitaluga, M. Trevelin Souza, E. Dutra Zanotto, M. Santocildes Romero, P. Hatton, Materials (Basel) **11**, 400 (2018)
73. F. Serio, M. Miola, E. Vernè, D. Pisignano, A. Boccaccini, L. Liverani, Nanomaterials **9**, 182 (2019)
74. Q.P. Pham, U. Sharma, A.G. Mikos, Tissue Eng. **12**, 1197 (2006)
75. M. Li, M.J. Mondrinos, M.R. Gandhi, F.K. Ko, A.S. Weiss, P.I. Lelkes, Biomaterials **26**, 5999 (2005)
76. K.M. Raghvendra, L. Sravanthi, Mod. Chem. Appl. **05**, 206 (2017)
77. M. Diez-Silva, M. Dao, J. Han, C.-T. Lim, S. Suresh, MRS Bull. **35**, 382 (2010)
78. F. Anjum, N.A. Agabalyan, H.D. Sparks, N.L. Rosin, M.S. Kallos, J. Biernaskie, Sci. Rep. **7**, 10291 (2017)
79. G.-Y. Liao, L. Chen, X.-Y. Zeng, X.-P. Zhou, X.-L. Xie, E.-J. Peng, Z.-Q. Ye, Y.-W. Mai, J. Appl. Polym. Sci. **120**, 2154 (2011)
80. F. Yang, R. Murugan, S. Wang, S. Ramakrishna, Biomaterials **26**, 2603 (2005)
81. K.H. Lee, H.Y. Kim, M.S. Khil, Y.M. Ra, D.R. Lee, Polymer (Guildf) **44**, 1287 (2003)
82. P.X. Ma, Mater. Today **7**, 30 (2004)
83. Q. Shi, B. Zhang, H. Xing, S. Yang, J. Xu, H. Liu, Front. Physiol. **9**, 33 (2018)
84. D. Daubert, R.M. Black, V. Chrepa, G.A. Kotsakis, J. Endod. **46**, 444 (2020)
85. J.-W. Kim, H. Lee, H.-K. Lim, J.-W. Kim, S.-H. Byun, Y.-J. Choi, U.-L. Lee, J. Oral Rehabil. **45**, 967 (2018)
86. R.C.K. Jordan, Periodontology **2000**(34), 217 (2004)
87. B.L. Pihlstrom, B.S. Michalowicz, N.W. Johnson, Lancet **366**, 1809 (2005)
88. M.N. Sela, D. Kohavi, E. Krausz, D. Steinberg, G. Rosen, Clin. Oral Implants Res. **14**, 263 (2003)
89. D.G. Moussa, C. Aparicio, J. Tissue Eng. Regen. Med. Term **13**, 58 (2019)
90. S. Damle, H. Bhattal, A. Loomba, J. Clin. Pediatr. Dent. **36**, 263 (2012)
91. M.E. Odabas, A. Alacam, H. Sillelioglu, C. Deveci, Eur. J. Paediatr. Dent. **13**, 118 (2012)
92. M. Zafar, S. Najeeb, Z. Khurshid, M. Vazirzadeh, S. Zohaib, B. Najeeb, F. Sefat, Materials (Basel) **9**, 73 (2016)
93. L. de Siqueira, R.F. Gouveia, L. Grenho, F.J. Monteiro, M.H. Fernandes, E.S. Trichês, J. Mater. Sci. **53**, 10718 (2018)
94. S.H. Barbanti, C.A.C. Zavaglia, E.A.R. Duek, Polímeros **15**, 13 (2005)
95. A. Hoppe, N.S. Güldal, A.R. Boccaccini, Biomaterials **32**, 2757 (2011)
96. J.R. Jones, in *Bio-Glasses: An Introduction*, ed. by J.R. Jones, A.G. Clare (Wiley, 2012), pp. 179–204
97. G.-Y. Liao, X.-P. Zhou, L. Chen, X.-Y. Zeng, X.-L. Xie, Y.-W. Mai, Compos. Sci. Technol. **72**, 248 (2012)
98. D. Sousa Coelho, B. Veleirinho, T. Alberti, A. Maestri, R. Yunes, P. Fernando Dias, M. Maraschin, in *Nanomaterials—Toxicity, Human Health and Environment*, ed. by S. Clichici, G.A. Filip, G.D. do Nascimento, (IntechOpen, 2020). https://doi.org/10.5772/intechopen.81530
99. E. Mele, J. Mater. Chem. B **4**, 4801 (2016)
100. H. Mahmood, I.U. Khan, M. Asif, R.U. Khan, S. Asghar, I. Khalid, S.H. Khalid, M. Irfan, F. Rehman, Y. Shahzad, A.M. Yousaf, A. Younus, Z.R. Niazi, M. Asim, Int. J. Biol. Macromol. **166**, 483 (2021)
101. A. Gul, I. Gallus, A. Tegginamath, J. Maryska, F. Yalcinkaya, Membranes (Basel) **11**, 908 (2021)
102. P. Chhabra, K. Bhati, in *Recent Advances in Wound Healing*, ed. by S.N. Aghaei (IntechOpen, 2021). https://doi.org/10.5772/intechopen.97298
103. M. Cannio, D. Bellucci, J.A. Roether, D.N. Boccaccini, V. Cannillo, Materials (Basel) **14**, 5440 (2021)
104. P. Kumar, B.S. Dehiya, A. Sindhu, Int. J. Appl. Eng. Res. **13**, 2744 (2018)
105. G. Gautam, S. Kumar, K. Kumar, Mater. Today Proc. **50**, 2206 (2022)

106. H. Ismail, H. Mohamad, Materials (Basel) **14**, 5193 (2021)
107. J.G. Neves, D. Navarro da Rocha, C.C. Lopes, M.H. Prado da Silva, M.A.C. Sinhoreti, L. Correr-Sobrinho, M.A.A. Fraga, A.B. Correr, Cerâmica **67**, 236 (2021)
108. M. Vallet-Regí, Comptes Rendus Chim. **13**, 174 (2010)
109. F. Baino, in *Encyclopedia of Materials: Technical Ceramics and Glasses*, vol. 3, ed. by M. Pomeroy (Elsevier, 2021), pp. 614–623
110. H.R. Fernandes, A. Gaddam, A. Rebelo, D. Brazete, G.E. Stan, J.M.F. Ferreira, Materials (Basel) **11**, 2530 (2018)
111. Z. Khurshid, S. Husain, H. Alotaibi, R. Rehman, M.S. Zafar, I. Farooq, A.S. Khan, in *Biomedical, Therapeutic and Clinical Applications of Bioactive Glasses. A Volume in Woodhead Publishing Series in Biomaterials*, ed. by Gurbinder Kaur (Elsevier, 2019), pp. 497–519
112. B.A.E. Ben-Arfa, R.C. Pullar, Processes **8**, 615 (2020)
113. D. Bokov, A. Turki Jalil, S. Chupradit, W. Suksatan, M. Javed Ansari, I. H. Shewael, G.H. Valiev, E. Kianfar, Adv. Mater. Sci. Eng. **2021**, 1 (2021)
114. E. Fiume, C. Migneco, E. Verné, F. Baino, Materials (Basel) **13**, 540 (2020)
115. C. Wen, N. Bai, L. Luo, J. Ye, X. Zhan, Y. Zhang, B. Sa, Ceram. Int. **47**, 18094 (2021)
116. R. De Giglio, G. Di Vieste, T. Mondello, G. Balduzzi, B. Masserini, I. Formenti, S. Lodigiani, D. Pallavicini, B. Pintaudi, A. Mazzone, J. Foot Ankle Surg. **60**, 292 (2021)
117. F. Kurtuldu, N. Mutlu, M. Michálek, K. Zheng, M. Masar, L. Liverani, S. Chen, D. Galusek, A.R. Boccaccini, Mater. Sci. Eng. C **124**, 112050 (2021)
118. S.A.M. El-Sayed, M. Mabrouk, M.E. Khallaf, B.M. Abd El-Hady, E. El-Meliegy, M.R. Shehata, J. Drug Deliv. Sci. Technol. **57**, 101757 (2020)
119. H. Palza Cordero, R. Castro Cid, M. Diaz Dosque, R. Cabello Ibacache, P. Palma Fluxá, J. Dent. **105**, 103575 (2021)
120. R. Chen, Q. Li, Q. Zhang, S. Xu, J. Han, P. Huang, Z. Yu, D. Jia, J. Liu, H. Jia, M. Shen, B. Hu, H. Wang, H. Zhan, T. Zhang, K. Ma, J. Wang, Chem. Eng. J. **426**, 130299 (2021)
121. S. Naseri, W.C. Lepry, V.B. Maisuria, N. Tufenkji, S.N. Nazhat, J. Non Cryst. Solids **505**, 438 (2019)
122. A. Krishnamurthy, V.K. Michaelis, S. Kroeker, J. Phys. Chem. C **125**, 8815 (2021)
123. K. Schuhladen, U. Pantulap, K. Engel, P. Jeleń, Z. Olejniczak, L. Hupa, M. Sitarz, A.R. Boccaccini, Int. J. Appl. Glass Sci. **12**, 293 (2021)
124. W.C. Lepry, S.N. Nazhat, Mater. Adv. **1**, 1371 (2020)
125. B.R. Barrioni, E. Norris, S. Li, P. Naruphontjirakul, J.R. Jones, M. de M. Pereira, J. Mater. Sci. Mater. Med. **30**, 86 (2019)
126. Z.I. Dhary, S.B.H. Farid, A.A. Atiyah, Mater. Sci. Forum **1039**, 510 (2021)
127. M. Gonzalez Moreno, M.E. Butini, E.M. Maiolo, L. Sessa, A. Trampuz, Colloids Surf. B Biointerfaces **189**, 110853 (2020)
128. L. Drago, M. Toscano, M. Bottagisio, Materials (Basel) **11**, 326 (2018)
129. N.A.P. van Gestel, D.J.W. Hulsen, J. Geurts, S. Hofmann, K. Ito, J.J. Arts, B. van Rietbergen, J. Mech. Behav. Biomed. Mater. **69**, 301 (2017)
130. R. Björkenheim, E. Jämsen, E. Eriksson, P. Uppstu, L. Aalto-Setälä, L. Hupa, K. Eklund, M. Ainola, N. Lindfors, J. Pajarinen, Eur. Cells Mater. **41**, 15 (2021)
131. K.E. Kojima, F.B. de Andrade e Silva, M.C. de Leonhardt, V.C. de Carvalho, P.R.D. de Oliveira, A.L.L.M. Lima, P. Roberto dos Reis, J.S. dos Silva, Injury **52**, S23 (2021)
132. T.A.G. Van Vugt, J. Heidotting, J.J. Arts, J.J.W. Ploegmakers, P.C. Jutte, J.A.P. Geurts, J. Bone Jt. Infect. **6**, 413 (2021)
133. M.T. Cunha, M.A. Murça, S. Nigro, G.B. Klautau, M.J.C. Salles, B.M.C. Infect, Diseases **18**, 157 (2018)
134. T. Grønseth, L.K. Vestby, L.L. Nesse, M. von Unge, J.T. Silvola, Ups. J. Med. Sci. **125**, 217 (2020)
135. M.T. Islam, L. Macri-Pellizzeri, V. Sottile, I. Ahmed, Biomater. Sci. **9**, 1826 (2021)
136. A.M. Deliormanlı, Mater. Technol. **37**, 1808 (2022)
137. V. Krishnan, in *Biomedical, Therapeutic and Clinical Applications of Bioactive Glasses. A Volume in Woodhead Publishing Series in Biomaterials*, ed. by Gurbinder Kaur (Elsevier, 2019), pp. 331–353

138. J. Crush, A. Hussain, K.T.M. Seah, W.S. Khan, Front. Cell Dev. Biol. **9** (2021)
139. A.J. Salinas, M. Vallet-Regí, J. Non Cryst. Solids **432**, 9 (2016)
140. F. Kurtuldu, H. Kaňková, A.M. Beltrán, L. Liverani, D. Galusek, A.R. Boccaccini, Mater. Today Bio **12**, 100150 (2021)
141. K. Zheng, E. Torre, A. Bari, N. Taccardi, C. Cassinelli, M. Morra, S. Fiorilli, C. Vitale-Brovarone, G. Iviglia, A.R. Boccaccini, Mater. Today Bio **5**, 100041 (2020)
142. S. Akhtach, Z. Tabia, M. Bricha, K. El Mabrouk, Ceram. Int. **47**, 29036 (2021)
143. S. Mortazavi, M. Rahsepar, S. Hosseinzadeh, Ceram. Int. **48**, 8276 (2022)
144. N. Pajares-Chamorro, Y. Wagley, C.V. Maduka, D.W. Youngstrom, A. Yeger, S.F. Badylak, N.D. Hammer, K. Hankenson, X. Chatzistavrou, Mater. Sci. Eng. C **120**, 111693 (2021)
145. S. Naseri, G. Griffanti, W.C. Lepry, V.B. Maisuria, N. Tufenkji, S.N. Nazhat, J. Am. Ceram. Soc. **105**, 1711 (2022)
146. N. Azizabadi, P.A. Azar, M.S. Tehrani, P. Derakhshi, J. Non Cryst. Solids **556**, 120568 (2021)
147. S. Chitra, P. Bargavi, M. Balasubramaniam, R.R. Chandran, S. Balakumar, Mater. Sci. Eng. C **109**, 110598 (2020)
148. S. Akhtach, Z. Tabia, K. El Mabrouk, M. Bricha, R. Belkhou, Ceram. Int. **47**, 424 (2021)
149. F. Baino, Bioengineering **7**, 45 (2020)
150. H. Luo, J. Xiao, M. Peng, Q. Zhang, Z. Yang, H. Si, Y. Wan, J. Non Cryst. Solids **532**, 119856 (2020)
151. Z. Tan, B. Zhou, J. Zheng, Y. Huang, H. Zeng, L. Xue, D. Wang, Stem Cells Int. **2021**, 1 (2021)
152. S. Kargozar, M. Mozafari, S. Ghodrat, E. Fiume, F. Baino, Mater. Sci. Eng. C **121**, 111741 (2021)
153. A. Jacobs, G. Renaudin, C. Forestier, J.-M. Nedelec, S. Descamps, Acta Biomater. **117**, 21 (2020)
154. N. Eliaz, N. Metoki, Materials (Basel) **10**, 334 (2017)
155. T. Baskaran, N.F. Mohammad, S.S. Md Saleh, N.F. Mohd Nasir, F.D. Mohd Daud, J. Phys. Conf. Ser. **2071**, 012008 (2021)
156. S. Panda, C.K. Biswas, S. Paul, Ceram. Int. **47**, 28122 (2021)
157. V. Sokolova, M. Epple, Chem. A Eur. J. **27**, 7471 (2021)
158. D.S. Gomes, A.M.C. Santos, G.A. Neves, R.R. Menezes, Cerâmica **65**, 282 (2019)
159. A. Ruffini, S. Sprio, L. Preti, A. Tampieri, in *Biomaterial-Supported Tissue Reconstruction or Regeneration*, ed. by M. Barbeck, O. Jung, R. Smeets, T. Koržinskas (IntechOpen, 2019). https://doi.org/10.5772/intechopen.85091
160. M. Jamil, B. Elouatli, H. Khallok, A. Elouahli, M. E. Gourri, Ezzahmouly, F. Abida, Z. Hatim, J. Mater. Environ. Sci. **9**, 2322 (2018)
161. K. Labgairi, A. Borji, M. Kaddami, A. Jourani, Int. J. Chem. Eng. **2020**, 1 (2020)
162. M. Khalid, S.S.B. Jikan, S. Adzila, Z. Murni, N.A. Badarulzaman, R. Rosley, M.U. Hameed, Biointerface Res. Appl. Chem. **12**, 5663 (2021)
163. A. El-Ghannam, P. Ducheyne, in *Comprehensive Biomaterials II*, 2nd edn., ed. by K. Healy, P. Ducheyne (Elsevier, 2017), pp. 204–234
164. K. Pajor, L. Pajchel, J. Kolmas, Materials (Basel) **12**, 2683 (2019)
165. K. Herman, M. Wujczyk, M. Dobrzynski, D. Diakowska, K. Wiglusz, R.J. Wiglusz, Materials (Basel) **14**, 3747 (2021)
166. P.W. Citradewi, H. Hidayat, G. Purwiandono, I. Fatimah, S. Sagadevan, Chem. Phys. Lett. **769**, 138412 (2021)
167. P. Vijayaraghavan, M.A. Rathi, K.S. Almaary, F.S. Alkhattaf, Y.B. Elbadawi, S.W. Chang, B. Ravindran, J. King Saud Univ. Sci. **34**, 101749 (2022)
168. P.N. Silva-Holguín, S.Y. Reyes-López, Dose-Response **18**, 155932582095134 (2020)
169. J. Hou, Y. Liu, Z. Han, D. Song, B. Zhu, Mater. Sci. Eng. C **120**, 111655 (2021)
170. Q. Zhou, T. Wang, C. Wang, Z. Wang, Y. Yang, P. Li, R. Cai, M. Sun, H. Yuan, L. Nie, Colloids Surf. A Physicochem. Eng. Asp. **585**, 124081 (2020)
171. C.O. de Lima, A.L.M. de Oliveira, L. Chantelle, E.C. Silva Filho, M. Jaber, M.G. Fonseca, Colloids Surf. B Biointerfaces **198**, 111471 (2021)

172. M. Shokri, M. Kharaziha, H.A. Tafti, M.B. Eslaminejad, R.M. Aghdam, Mater. Sci. Eng. C 112684 (2022)
173. Y. Lv, Y. Chen, Y. Zheng, Q. Li, T. Lei, P. Yin, Colloids Surf. B Biointerfaces **209**, 112194 (2022)
174. T.V. Safronova, I.I. Selezneva, S.A. Tikhonova, A.S. Kiselev, G.A. Davydova, T.B. Shatalova, D.S. Larionov, J.V. Rau, Bioact. Mater. **5**, 423 (2020)
175. N.W. Kucko, R.-P. Herber, S.C. G. Leeuwenburgh, J.A. Jansen, in *Principles of Regenerative Medicine*, 3rd edn., ed. by A. Atala, R. Lanza, A. G. Mikos, R. Nerem (Elsevier, 2019), pp. 591–611
176. A. Kareiva, G. Inkrataitė, L. Daumantas, Vilnius Univ. Proc. **11**, 1 (2020)
177. I.E. Nifant'ev, V.K. Besprozvannykh, A.V. Shlyakhtin, A.N. Tavtorkin, M.P. Smirnova, I.S. Levin, P.V. Ivchenko, Mendeleev Commun. **31**, 379 (2021)
178. I.V. Fadeeva, B.I. Lazoryak, G.A. Davidova, F.F. Murzakhanov, B.F. Gabbasov, N.V. Petrakova, M. Fosca, S.M. Barinov, G. Vadalà, V. Uskoković, Y. Zheng, J.V. Rau, Mater. Sci. Eng. C **129**, 112410 (2021)
179. L. Stipniece, I. Skadins, M. Mosina, Ceram. Int. **48**, 10195 (2022)
180. H. Zazo, C.G. Millán, C.I. Colino, J.M. Lanao, in *Antimicrobial Nanoarchitectonics: From Synthesis to Applications*, ed. by A.M. Grumezescu (Elsevier, 2017), pp. 411–444
181. M.J. Mitchell, M.M. Billingsley, R.M. Haley, M.E. Wechsler, N.A. Peppas, R. Langer, Nat. Rev. Drug Discov. **20**, 101 (2021)
182. Neha Desai, M. Momin, T. Khan, S. Gharat, R.S. Ningthoujam, A. Omri, Expert Opin. Drug Deliv. **18**, 1261 (2021)
183. M. Thapa, S. Roy Choudhury, *Comprehensive Analytical Chemistry,* vol. 94, ed. by S.K. Verma, A.K. Das, (Elsevier, 2021), pp. 49–79
184. V.V. Kumar, S.P. Anthony, in *Surface Chemistry of Nanobiomaterials*, ed. by A. M. Grumezescu (Elsevier, 2016), pp. 265–300
185. C. Egbuna, V.K. Parmar, J. Jeevanandam, S.M. Ezzat, K.C. Patrick-Iwuanyanwu, C.O. Adetunji, J. Khan, E.N. Onyeike, C.Z. Uche, M. Akram, M.S. Ibrahim, N.M. El Mahdy, C.G. Awuchi, K. Saravanan, H. Tijjani, U.E. Odoh, M. Messaoudi, J.C. Ifemeje, M.C. Olisah, N.J. Ezeofor, C.J. Chikwendu, C.G. Ibeabuchi, J. Toxicol. **2021**, 1 (2021)
186. R. Güzel, G. Erdal, in *Silver Nanoparticles—Fabrication, Characterization and Applications*, ed. by M. Khan (InTech, 2018). https://doi.org/10.5772/intechopen.75363
187. M. Abass Sofi, S. Sunitha, M. Ashaq Sofi, S.K. Khadheer Pasha, D. Choi, J. King Saud Univ. Sci. **34**, 101791 (2022)
188. G.M. Nair, T. Sajini, B. Mathew, Talanta Open **5**, 100080 (2022)
189. O. Maťátková, J. Michailidu, A. Miškovská, I. Kolouchová, J. Masák, A. Čejková, Biotechnol. Adv. **58**, 107905 (2022)
190. D.H. Park, Y.H. Joe, J. Hwang, J.H. Byeon, Chem. Eng. J. **381**, 122639 (2020)
191. S. Lee, B.-H. Jun, Int. J. Mol. Sci. **20**, 865 (2019)
192. A.V. Vodop'yanov, D.A. Mansfeld, A.V. Samokhin, N.V. Alekseev, Y.V. Tsvetkov, Radiophys. Quant. Electron. **59**, 698 (2017)
193. M. Kim, S. Osone, T. Kim, H. Higashi, T. Seto, KONA Powder Part. J. **34**, 80 (2017)
194. X.-F. Zhang, Z.-G. Liu, W. Shen, S. Gurunathan, Int. J. Mol. Sci. **17**, 1534 (2016)
195. T. Cele, in *Engineered Nanomaterials—Health and Safety*, ed. by S. Marius Avramescu, K. Akhtar, I. Fierascu, S. Bahadar Khan, F. Ali, A.M. Asiri, (IntechOpen, 2020). https://doi.org/10.5772/intechopen.90771
196. G. Pal, P. Rai, A. Pandey, in *Green Synthesis, Characterization and Applications of Nanoparticles*, ed. by A. K. Shukla, S. Iravani (Elsevier, 2019), pp. 1–26
197. B. Karthikeyan, R. Govindhan, M. Amutheesan, in *Noble Metal-Metal Oxide Hybrid Nanoparticles: Fundamentals and Applications. A Volume in Micro and Nano Technologies*, ed. by S. Mohapatra, T.A. Nguyen, P. Nguyen-Tri (Elsevier, 2019), pp. 179–188
198. H. Tyagi, A. Kushwaha, A. Kumar, M. Aslam, Nanoscale Res. Lett. **11**, 362 (2016)
199. P. Slepička, N. Slepičková Kasálková, J. Siegel, Z. Kolská, V. Švorčík, Materials (Basel) **13**, 1 (2019)

200. R. Javed, M. Zia, S. Naz, S.O. Aisida, N. ul Ain, Q. Ao, J. Nanobiotechnol. **18**, 172 (2020)
201. J.K. Patra, K.-H. Baek, J. Nanomater. **2014**, 1 (2014)
202. H. Jan, R. Gul, A. Andleeb, S. Ullah, M. Shah, M. Khanum, I. Ullah, C. Hano, B.H. Abbasi, J. Saudi Chem. Soc. **25**, 101297 (2021)
203. S. Parmar, H. Kaur, J. Singh, A.S. Matharu, S. Ramakrishna, M. Bechelany, Nanomaterials **12**, 1115 (2022)
204. B. Bekele, A. Degefa, F. Tesgera, L.T. Jule, R. Shanmugam, L. Priyanka Dwarampudi, N. Nagaprasad, K. Ramasamy, J. Nanomater. **2021**, 1 (2021)
205. S.K. Deivanathan, J.T.J. Prakash, Chem. Data Collect. **38**, 100831 (2022)
206. B.A. Varghese, R.V.R. Nair, S. Jude, K. Varma, A. Amalraj, S. Kuttappan, J. Taiwan Inst. Chem. Eng. **126**, 166 (2021)
207. K.S. Siddiqi, A. Husen, R.A.K. Rao, J. Nanobiotechnol. **16**, 14 (2018)
208. I.X. Yin, J. Zhang, I.S. Zhao, M.L. Mei, Q. Li, C.H. Chu, Int. J. Nanomed. **15**, 2555 (2020)
209. A. Pugazhendhi, T.N.J.I. Edison, I. Karuppusamy, B. Kathirvel, Int. J. Pharm. **539**, 104 (2018)
210. A. Albanese, P.S. Tang, W.C.W. Chan, Annu. Rev. Biomed. Eng. **14**, 1 (2012)
211. L.A.P. Gontijo, E. Raphael, D.P.S. Ferrari, J.L. Ferrari, J.P. Lyon, M.A. Schiavon, Matéria (Rio Janeiro) **25**, e12845 (2020)
212. V.T. Noronha, A.J. Paula, G. Durán, A. Galembeck, K. Cogo-Müller, M. Franz-Montan, N. Durán, Dent. Mater. **33**, 1110 (2017)
213. Y. Qing, L. Cheng, R. Li, G. Liu, Y. Zhang, X. Tang, J. Wang, H. Liu, Y. Qin, Int. J. Nanomed. **13**, 3311 (2018)
214. M. Sundrarajan, K. Bama, M. Bhavani, S. Jegatheeswaran, S. Ambika, A. Sangili, P. Nithya, R. Sumathi, J. Photochem. Photobiol. B Biol. **171**, 117 (2017)
215. A. Abbaszadegan, Y. Ghahramani, A. Gholami, B. Hemmateenejad, S. Dorostkar, M. Nabavizadeh, H. Sharghi, J. Nanomater. **2015**, 1 (2015)
216. T.C. Dakal, A. Kumar, R.S. Majumdar, V. Yadav, Front. Microbiol. **7**, 2016 (1831)
217. M.K. Rai, S.D. Deshmukh, A.P. Ingle, A.K. Gade, J. Appl. Microbiol. **112**, 841 (2012)
218. C. Liao, Y. Li, S. Tjong, Int. J. Mol. Sci. **20**, 449 (2019)
219. M. Chichova, M. Shkodrova, P. Vasileva, K. Kirilova, D. Doncheva-Stoimenova, J. Nanopart. Res. **16**, 2243 (2014)
220. Y. Tao, T. Aparicio, M. Li, K.W. Leong, S. Zha, J. Gautier, Nucleic Acids Res. **49**, 5074 (2021)
221. S.V. Gudkov, D.E. Burmistrov, D.A. Serov, M.B. Rebezov, A.A. Semenova, A.B. Lisitsyn, Front. Phys. **9**, 641481 (2021)
222. S.M.I. Rayyif, H.B. Mohammed, C. Curuțiu, A.C. Bîrcă, A.M. Grumezescu, B. Ștefan Vasile, L.M. Dițu, V. Lazăr, M.C. Chifiriuc, G. Mihăescu, A.M. Holban, Materials (Basel) **14**, 3084 (2021)
223. C.R. Mendes, G. Dilarri, C.F. Forsan, V.M.R. de Sapata, P.R.M. Lopes, P.B. de Moraes, R.N. Montagnolli, H. Ferreira, E.D. Bidoia, Sci. Rep. **12**, 2658 (2022)
224. B. Abebe, H.C.A. Murthy, E. Amare, Environ. Nanotechnol. Monit. Manag. **14**, 100336 (2020)
225. B. Abebe, E.A. Zereffa, A. Tadesse, H.C.A. Murthy, Nanoscale Res. Lett. **15**, 190 (2020)
226. A. Joe, S.-H. Park, K.-D. Shim, D.-J. Kim, K.-H. Jhee, H.-W. Lee, C.-H. Heo, H.-M. Kim, E.-S. Jang, J. Ind. Eng. Chem. **45**, 430 (2017)
227. M.S. Jameel, A.A. Aziz, M.A. Dheyab, Green Process. Synth. **9**, 386 (2020)
228. S.A. Fahmy, E. Preis, U. Bakowsky, H.M.E.-S. Azzazy, Molecules **25**, 4981 (2020)
229. A.S. Eltaweil, M. Fawzy, M. Hosny, E.M. Abd El-Monaem, T.M. Tamer, A.M. Omer, Arab. J. Chem. **15**, 103517 (2022)
230. N.S. Al-Radadi, Arab. J. Chem. **12**, 330 (2019)
231. S. Gurunathan, M. Jeyaraj, M.-H. Kang, J.-H. Kim, Cells **8**, 444 (2019)
232. X. Hu, Y. Zhang, T. Ding, J. Liu, H. Zhao, Front. Bioeng. Biotechnol. **8**, 990 (2020)
233. J. Toczek, M. Sadłocha, K. Major, R. Stojko, Biomedicines **10**, 679 (2022)
234. F. Khuda, Z. Ul Haq, I. Ilahi, R. Ullah, A. Khan, H. Fouad, A. Ali Khan Khalil, Z. Ullah, M. Umar Khayam Sahibzada, Y. Shah, M. Abbas, T. Iftikhar, G. El-Saber Batiha, Arab. J. Chem. **14**, 103343 (2021)

235. M. Aghamirzaei, M.S. Khiabani, H. Hamishehkar, R.R. Mokarram, M. Amjadi, Mater. Today Commun. **29**, 102831 (2021)
236. S. Sagadevan, S. Imteyaz, B. Murugan, J. Anita Lett, N. Sridewi, G.K. Weldegebrieal, I. Fatimah, W.-C. Oh, Green Process. Synth. **11**, 44 (2022)
237. K.S. Khashan, G.M. Sulaiman, F.A. Abdulameer, S. Albukhaty, M.A. Ibrahem, T. Al-Muhimeed, A.A. AlObaid, Appl. Sci. **11**, 4623 (2021)
238. K. Chand, D. Cao, D.E. Fouad, A.H. Shah, M.N. Lakhan, A.Q. Dayo, H.J. Sagar, K. Zhu, A.M.A. Mohamed, J. Mol. Liq. **316**, 113821 (2020)
239. S. Marimuthu, A.A. Rahuman, C. Jayaseelan, A.V. Kirthi, T. Santhoshkumar, K. Velayutham, A. Bagavan, C. Kamaraj, G. Elango, M. Iyappan, C. Siva, L. Karthik, K.V.B. Rao, Asian Pac. J. Trop. Med. **6**, 682 (2013)
240. S. Subhapriya, P. Gomathipriya, Microb. Pathog. **116**, 215 (2018)
241. G.F. da Fonseca, S.O.M. de Avelino, D.C.R. de Mello, R.F. do Prado, T.M.B. Campos, L.M.R. de Vasconcellos, E. de Sousa Trichês, A.L.S. Borges, J. Mater. Sci. Mater. Med. **31**, 41 (2020)
242. M.K. Ahmed, S.F. Mansour, R. Al-Wafi, E. Abdel-Fattah, J. Mater. Res. Technol. **13**, 2297 (2021)
243. M.E. El-Naggar, O.A. Abu Ali, D.I. Saleh, M.A. Abu-Saied, M.K. Ahmed, E. Abdel-Fattah, S.F. Mansour, E.-R. Kenawy, J. Mater. Res. Technol. **15**, 2909 (2021)
244. T. Bian, N. Pang, H. Xing, Mater. Chem. Phys. **273**, 125059 (2021)
245. M. Mostafa, N.G. Kandile, M.K. Mahmoud, H.M. Ibrahim, Heliyon **8**, e08772 (2022)
246. S. Karagoz, N.B. Kiremitler, G. Sarp, S. Pekdemir, S. Salem, A.G. Goksu, M.S. Onses, I. Sozdutmaz, E. Sahmetlioglu, E.S. Ozkara, A. Ceylan, E. Yilmaz, A.C.S. Appl, Mater. Interfaces **13**, 5678 (2021)
247. Y. Ding, C. Dou, S. Chang, Z. Xie, D.-G. Yu, Y. Liu, J. Shao, Polymers (Basel) **12**, 2034 (2020)
248. S. Chang, M. Wang, F. Zhang, Y. Liu, X. Liu, D.-G. Yu, H. Shen, Mater. Des. **192**, 108782 (2020)
249. M. Jafari-Nodoushan, J. Barzin, H. Mobedi, Polym. Adv. Technol. **26**, 502 (2015)
250. F.-L. Zhou, A. Chirazi, J.E. Gough, P.L. Hubbard Cristinacce, G.J.M. Parker, Langmuir **33**, 13262 (2017)
251. A.S. Kranthi Kiran, K. Madhumathi, T.S. Sampath Kumar, Colloid Interface Sci. Commun. **12**, 17 (2016)
252. Y. Gao, Y. Bai, D. Zhao, M.-W. Chang, Z. Ahmad, J.-S. Li, Polymers (Basel) **7**, 2701 (2015)
253. M.S.A. Karimian, S. Mashayekhan, H. Baniasadi, Int. J. Biol. Macromol. **88**, 288 (2016)
254. X. Huang, J. Gao, N. Zheng, W. Li, H. Xue, R.K.Y. Li, Colloids Surf. A Physicochem. Eng. Asp. **517**, 17 (2017)
255. K.K. Khanum, S.B.S.P.C. Ramamurthy, RSC Adv. **5**, 60419 (2015)
256. X. Ju, X. Wang, Z. Liu, R. Xie, W. Wang, L. Chu, Particuology **30**, 151 (2017)
257. M. Rasekh, Z. Ahmad, R. Cross, J. Hernández-Gil, J.D.E.T. Wilton-Ely, P.W. Miller, Mol. Pharm. **14**, 2010 (2017)
258. C. Zhang, Z.-C. Yao, Q. Ding, J.J. Choi, Z. Ahmad, M.-W. Chang, J.-S. Li, A.C.S. Appl, Mater. Interfaces **9**, 21485 (2017)
259. B. Sanchez-Vazquez, A.J.R. Amaral, D.-G. Yu, G. Pasparakis, G.R. Williams, AAPS PharmSciTech **18**, 1460 (2017)
260. K. Li, P. Li, Z. Jia, B. Qi, J. Xu, D. Kang, M. Liu, Y. Fan, Sci. Rep. **8**, 17117 (2018)
261. J. Wang, J.A. Jansen, F. Yang, Front. Chem. **7**, 258 (2019)
262. L. de Siqueira, N. Ribeiro, M.B.A. Paredes, L. Grenho, C. Cunha-Reis, E.S. Trichês, M.H. Fernandes, S.R. Sousa, F.J. Monteiro, Materials (Basel) **12**, 3879 (2019)
263. Higuchi, Fortunato, Woźniak, Chodara, Domaschke, Męczyńska-Wielgosz, Kruszewski, Dommann, Łojkowski, Nanomaterials **9**, 1625 (2019)
264. H. Bae, J. Lee, J. Ind. Eng. Chem. **40**, 99 (2016)
265. W. Zhu, F. Masood, J. O'Brien, L.G. Zhang, Nanomed. Nanotechnol. Biol. Med. **11**, 693 (2015)
266. D. Braghirolli, F. Zamboni, G.A. Acasigua, P. Pranke, Int. J. Nanomed. **10**, 5159 (2015)

267. J. Ramier, T. Bouderlique, O. Stoilova, N. Manolova, I. Rashkov, V. Langlois, E. Renard, P. Albanese, D. Grande, Mater. Sci. Eng. C **38**, 161 (2014)
268. Z. Liu, L. Jia, Z. Yan, L. Bai, New J. Chem. **42**, 11185 (2018)
269. J. Li, D. Zhai, F. Lv, Q. Yu, H. Ma, J. Yin, Z. Yi, M. Liu, J. Chang, C. Wu, Acta Biomater. **36**, 254 (2016)
270. A. GhavamiNejad, A. Rajan Unnithan, A. Ramachandra Kurup Sasikala, M. Samarikhalaj, R.G. Thomas, Y.Y. Jeong, S. Nasseri, P. Murugesan, D. Wu, C. Hee Park, C.S. Kim, ACS Appl. Mater. Interfaces **7**, 12176 (2015)
271. D. Valerini, L. Tammaro, R. Vitali, G. Guillot, A. Rinaldi, Coatings **11**, 345 (2021)

Selected VI-Element Based Glasses, Glass–Ceramics and Nanoparticles: Synthesis and Antibacterial/Bactericidal Activity

Dariush Souri

Abstract Exploring the non (low)-toxic antibacterial/bactericidal materials which can prevent the growth of bacteria/kill bacteria is the subject of many recent works. Glasses and nanomaterials are novel agents in this field and in bio-imaging when employing the luminescent agents. In this chapter, some VI-element based materials containing transition metals (such as silver or sliver-free telluro-vanadate glasses and glass ceramics, bare and doped ZnSe nanoparticles) are discussed to give more comprehension into their opto-biological features and capabilities. Besides the synthesis and fabrication methods, structural and optical properties such as the formation of desired phases, precise optical band gap and optical absorption edge are presented and discussed. Furthermore, their detected antibacterial/bactericidal potential employing serial dilution and/or agar disc-diffusion bioassays are presented and discussed. The results emphasized the high antibacterial/bactericidal potential of the 30%-sliver containing glasses and glass–ceramics against the studied bacteria from both Gram-negative and Gram-positive categories. Moreover, in the cases of zinc selenide nanoparticles synthesized under various conditions, luminescence and antibacterial potential were evaluated as aspects dependent on dopant content and particle size. Such mentioned materials are promising agents in optical and biological applications. This chapter discusses these subjects, and gives sheds light on their applications in the bio-optical fields.

Keywords Glasses and glass–ceramics · Zinc selenide nanoparticles · Antibacterial and bactericidal activities · Transition metals · Optical properties · Bioassays

---

D. Souri (✉)
Department of Physics, Faculty of Science, Malayer University, Malayer, Iran
e-mail: d.souri@malayeru.ac.ir; d.souri@gmail.com

S. J. Ikhmayies (ed.), *Advances in Biomaterials Research*, Advances in Material Research and Technology, https://doi.org/10.1007/978-3-031-97481-6_7

1 Introduction

Nowadays, exploring and engineering of new and functional materials play the vital roles in advancement of science and technology, which can serve as improvement in human life. Many efforts have been done for these requests by different groups throughout the world, but it is believed that more and more theoretical and experimental works are needed. Such continuous works would be permanently needed during the human life, in which some of them may open/solve some problems. Surely, groups of researchers work on the synthesis/fabrication, characterization, and/or engineering of glasses (Gs), glass–ceramics (GCs), and nanoparticles (NPs) as fundamentally and technologically interested materials. Among them, this work aims at provide more insight on the synthesis and opto-biological aspects of some VI-elements based materials such as tellurite glasses and glass–ceramics, and Se-containing nanoparticles.

Consistently, an attempt has been made to introduce these categories and to assess a bird view to the subject and considering the hygroscopic limitation of phosphate glasses [1–7], multicomponent tellurite, tellurium-vanadate glasses. Special attention and focus has been given to glasses and glass ceramics containing transition metal oxides (TMOs) due to their fundamental and technological aspects, their exceptional ambient durability, thermal stability, electro-optical and thermo-mechanical properties [8–28], high-bandwidth infrared waveguides, amplifying features in optical communications [14, 18], and due to their luminescence [29, 30]. Also, there are some reports on the biological features of some glasses such as TeO_2-V_2O_5 [24] and Ag_2O-TeO_2-V_2O_5 [7]. On the other hand, some other research conducted on tellurium-based glasses have led to the production of the important class of beneficial engineering materials as glass–ceramics [1–3]. Glass–ceramics are prepared from thermally stimulated crystallization of glasses in a controlled procedure involving a hybrid of vitreous (glassy) state and crystallites (i.e. in a well-controlled process leading to the creation of nanoparticles) algebraically implanted in the amorphous background [3, 7, 27, 31]. In a comparative trend, GCs may possess higher structural strength and elastic moduli. The properties of GCs are controlled by the physical properties of the single crystals (SCs), and also by the mutual relationship between the grown crystals and remaining amorphous matrix [7, 27, 31]. Glasses and GCs have different benefits such as in optical fibers [32], luminescence [29, 30] antibacterial/bactericidal [7, 24] and in dental applications [33].

The fabrication method has effect on the material properties. In the case of nanomaterials a key point is to select a fabrication method that enables a more controllable particle size process, because particle size affects the properties especially the optical features such as the energy band gap. Therefore, the knowledge about the size of the nanoparticles is vital. There are several formalisms to estimate the crystallite size and lattice strain (i.e. Scherrer approach, Williamson-Hall method [34], Williamson–Smallman method [35] and Halder–Wagner approach [36]).

Taking into account the above, another material of scientific importance will be reviewed in this chapter is ZnSe in its bare, doped, co-doped, and core–shell states. Among the prominent and broadly investigated II–VI nano-structured semiconducting materials such as CdS, CdSe, ZnS and CdTe, ZnSe seems to be more beneficial and needs more works. On the other hand, it is worth mentioning that Zn-based materials such as ZnSe with no toxicity making them a good alternative to cadmium-containing agents especially in biological applications [37, 38]. ZnSe as an n-type semiconductor has a direct bulk band gap energy of about 2.72 eV, and can be used in technology of LEDs, solar cells, bio-sensors etc. [39, 40], and luminescence [41–43]. There are several reports on the different fabrication methods of ZnSe nanocrystals (NCs) including photochemical approach [44], soft chemical method [45], chemical precipitation [46], sol–gel method [47], co-precipitation [48] and microwave irradiation (MWIR) method [41, 44, 49–62]. The microwave-assisted chemical route is an interesting, fast, and non-expensive method with products of tunable to narrow size distribution [49–62].

In the field of biology and environmental science, despite the development and social life civilization, microorganisms (such as viruses, bacteria, and microbes) have a large and increasing diversity and are in close contact with our bodies and environment, which is characterized by its unsafe special effects [63]. For example, antibacterial medicines (such as penicillin as a conventional antibiotic) have been used to inhibit and kill bacteria; their long-time and sometimes excessive use has led to a decrease in the efficacy due to the high degrees of drug-resistances [64].

With the emergence of nano-science and nanotechnology, a revolution has been occurred in many fields such as environmental sciences, food industry (food protection and packaging), engineering of wound healing tissues, immunology, pharmaceutical industry and others [65, 66]. Biology has been benefited from the wonderful tactic of employing NPs or QDs in new/novel class of drugs, bio-imaging, and biopharmaceutical applications. It is surprising to known that such nano-products can kill the highest content or even all undesired microorganisms without side effects [67, 68]. There are some efforts on the antibacterial and biological properties of TeO_2-V_2O_5 glasses [24], Ag_2O-TeO_2-V_2O_5 glasses and glass–ceramics, and nanoparticles such as silver [69, 70], gold [49, 71], Cu and copper oxide [50, 72], ZnS [50], Chitosan nanoparticles [73], ZnSe [41, 44, 48–62, 74, 75], and ZnO [35, 76].

Briefly, in the present chapter, some telluro-vanadate glasses and special series of their silver-containing glasses and glass–ceramics, and also ZnSe NPs are introduced as some important VI-element based agents, are reviewed for their optical and antibacterial/bactericidal properties. In the case of biological aspects, to detect the antibacterial/bactericidal potential of the studied samples, some important Gram-negative bacteria in hospital and environment (i.e. Escherichia coli, Klebsiella pneumoniae and Pseudomonas aeruginosa) and Gram-positive bacteria (Staphylococcus aureus and Bacillus cereus) are considered as test stains using disc-diffusion and serial dilution bioassays. In this chapter, parts of our previous outcomes on the opto-physical and antibacterial features of some VI-elements based materials (i.e. multicomponent tellurium-based glasses and glass–ceramics, and zinc selenide NPs) are reviewed.

After the introductory remarks on the mentioned functional materials in this section, the experimental procedures including synthesis, characterization, and antibacterial/bactericidal bioassays of the desired agents are presented in Sect. 2. The bio-optical results of the mentioned samples are reviewed and discussed in Sect. 3, and finally, a conclusion and further remarks on the antibacterial potential and bio-optical advantages of the studied materials are given in Sect. 4.

2 Experimental Procedures

The present section includes subsections that address the experimental techniques used in the synthesis of the desired materials and their characterization with particular emphasis on the detection approaches of their antibacterial potential. Therefore, as followed, in addition to the fabrication methods used for the synthesis of telluro-vanadate glasses and the corresponding glass–ceramics, as well as for ZnSe NPs, disc-diffusion and serial dilution bioassays are presented.

2.1 Fabrication of the Desired Materials

It will be separately presented in this sub-section, fabrication details of $40TeO_2$-(60-x)V_2O_5-xA_yO_z (or A) (where A and A_yO_z are another element and another oxide as the 3th constituent, respectively; i.e. MoO_3, NiO, Sb_2O_3, Ag_2O, Bi_2O_3, Sb), and their corresponding glass–ceramics as well as doped/co-doped and core–shell ZnSe NPs are presented in Sects. 2.1.1, 2.1.2, and 2.1.3 (see Table 2), respectively.

2.1.1 Fabrication of Te-Based Glasses

Binary and ternary vitreous samples which are reviewed in the present chapter are of the general form of $40TeO_2$-(60-x)V_2O_5-xA_yO_z (or A) (i.e. the 3rd component will be as MoO_3, NiO, Sb_2O_3, Ag_2O, Bi_2O_3, Sb); x denotes the molar content of the 3rd component, which has a different interval in each compositional series depending on its glass-forming region. However, among the various fabrication methods of glasses, herein utilizing the low melting point of tellurate glasses, the simple and standard melt quenching approach was used, which reserve the good stoichiometry [4–6, 25–27]. For the sample preparation, initial powders of TeO_2, V_2O_5 and the third components were purchased from Merck High Purity Products. Depending on the mole fraction of each component of a desired sample, the required powders were accurately weighed using a balance with an accuracy of 0.1 mg. All of the components were well-mixed in a mortar for $\approx$15 min, and then the mixture was poured into a porcelain/alumina pot. To obtain a homogeneous melt of the desired sample, first the preheating/dehydration process was carried out in an electric furnace to avoid any volatilization, and then, temperature was increased step by step. Also during the melting route, the obtained fluid was stirred every 5 min to obtain a homogeneous melt. As is known, the melting point is composition-dependent, which for the mentioned samples in this work falls within the range of 600–980 °C. Different forms of each sample may be used, which is related to the type of required characterization. For the film form of the samples, blowing of the melt using hollow-silica tube was done. Such samples are suitable for spectroscopy and optical characterizations such as (ultraviolet-Visible (UV-vis) spectra). On the other hand, in the mentioned melt quenching technique, bulk glasses were obtained by pouring the melt onto a polished steel mold, which was immediately pressed by another polished mold at room temperature (RT). Such treatment ensures the production of uniform disc samples having two flat, parallel surfaces. For some of characterizations, such shape of the samples is not necessary (i.e. for Fourier transform Infrared Spectroscopy (FTIR): a few milligrams of the powdered sample are sufficient), but for electrical measurement, size of the disc and the uniform shape are important. To obtain suitable bulk glasses that would withstand (or eliminate) the least mechanical stresses raised from rapid quenching, they were directly annealed for 2 h at a temperature ($\approx$50 °C) below their glass transition temperature (T_g) [25]. Annealing at the mentioned appropriate temperature prevents the crystallization of the samples, and keeps their vitreous nature. The color of the aforementioned ternary telluro-vanadate glasses, except those containing Sb or Sb_2O_3, seems as black (for bulk samples), and for blown layers the color changed by increasing the third component from brown to yellowish brown. It should be mentioned that in the case of tellurate-vanadate glasses containing Sb or Sb_2O_3, the films changed from dark green to light green. For example, to guide the eyes, Table 1 shows the abbreviations of samples, their constituents and their bulk/film images for the glassy system with the formula $xMoO_3$-(60-x)V_2O_5-40TeO_2 (hereafter abbreviated as TVMx) [25].

2.1.2 Preparation of Glass–Ceramics (GCs)

During the previous decades, glass–ceramics have been as extensively attentive research areas for their scientific and technological requests such as uses in solid state lasers, frequency converters [77–80], and their biological features [7]. Although, there are various methods (such as pulsed laser deposition (PLD)) to produce glass–ceramics, nanocrystals (NCs) may be grown within the glassy matrix by heat treatment [7, 27, 31]. In such a process, GCs can be fabricated employing an electric furnace under a specific thermal procedure. In the thermally stimulated crystallization process used in our recent works for Yb^{3+}-doped chlorophosphate glasses of the form (60-x)P_2O_5-29$CaCl_2$-10NaCl-1Sb_2O_3-xYb_2O_3 with (x = 0.5, 1, 1.2, 1.5, and 4 mol%) [31], and xAg_2O-40TeO_2-(60-x)V_2O_5 (x = 0, 10, 20, 30, 40, and 50 mol%) glasses (abbreviated as TVAgx), a one-pot procedure was used. As is known, the temperature needed to induce phase transition/crystallization in a glass generally lies (suitably) between the glass transition temperature (T_g) and the crystallization temperature (T_{Cr}). So, for the aforementioned chlorophosphate glasses, the heat treatment was done at a temperature between their glass transition T_g and T_{Cr} obtained by differential scanning calorimetry (DSC) [31]. The DSC thermograph of the sample owing 1 mol% of Yb_2O_3 is shown in Fig. 1.

Table 1 Compositions, initial constituents, and bulk/blown film-images for TVMx systems [25]

Abbreviated name	MoO_3 mol%	V_2O_5 mol%	TeO_2 mol%	Image (bulk and film)
TVM0	0	60	40	
TVM10	10	50		
TVM20	20	40		
TVM30	30	30		
TVM40	40	20		
TVM50	50	10		
TVM60	60	0		

Table 2 Precursors needed for the synthesis of bare, doped, co-doped, and core–shell NPs of ZnSe

Precursors	Company/country
Zinc acetate ($Zn(Ac)_2$)	Merck Co./Germany
Selenium powder	
Sodium thiosulfate ($Na_2S_2O_3$)	
Sodium hydroxide (NaOH)	
Sodium borohydride ($NaBH_4$)	
Copper sulfate ($CuSO_4$)	
Thioglycolic acid (TGA) (another commercial name: mercaptoacetic acid)	
Acetone	
Deionized water (DI)	–

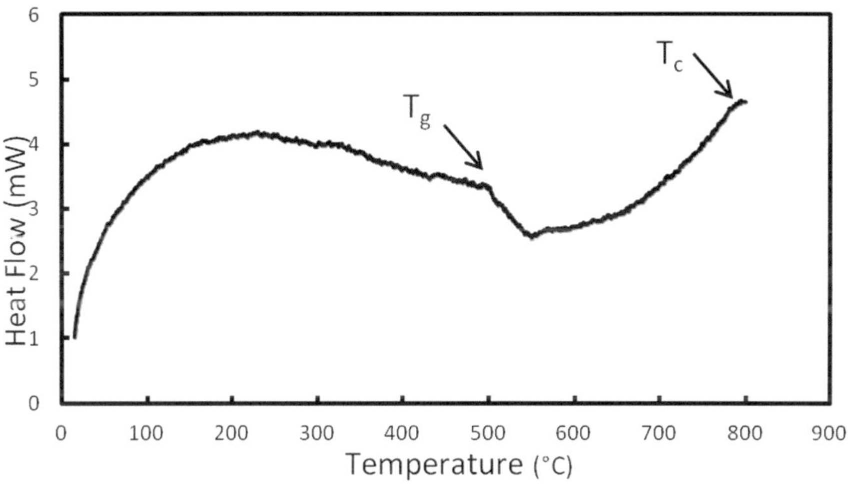

Fig. 1 DSC thermograph of $59P_2O_5$, $29CaCl_2$, $10NaCl$, $1Sb_2O_3$, $1Yb_2O_3$. T_g is about 500 °C and T_{Cr} is about 800 °C; selected annealing temperature for heat treatment has been selected as 650 °C [31]

For different glassy systems such as TVAgx: xAg_2O-$40TeO_2$-(60-x)V_2O_5, TVMx: $40TeO_2$-(60-x)V_2O_5-xMoO_3, TVNx: $40TeO_2$-(60-x)V_2O_5-xNiO and etc., calorimetric DSC investigations have been done to determine the characteristic temperatures, thermal stabilities, and so on. For example, Fig. 2 shows the DSC thermographs of TVAg20 and TVM20 at the specified heating rates. In the case of the main GCs focused on them in this work, ternary silver telluro-vanadate TVAgx glasses were heat treated in a one-step thermal procedure. The obtained GCs were denoted as ATVAgx GCs in accordance with their glass names. The glass preparation process was generally described in Sect. 2.1.1, but details of the preparation method of TVAgx have been detailed in Refs. [4–7, 27]. Then, their corresponding glass–ceramics (i.e. ATVAgx) were made-up by heat treatment (annealing) of the corresponding glasses at their T_{Cr} available in Ref. [13]. In contrast to the aforementioned chlorophosphate glasses, no nucleation of NCs occurred inside the glasses (certified by X-ray diffraction XRD during the work) at the annealing temperatures between T_g and T_{Cr}. So, annealing temperatures were chosen as T_{Cr} of each glass. Crystallization temperature has been reported as the best suitable temperature for nucleation and growth of nanocrystals [81]. Procedure of the thermal treatment was as follows: each glass was heated using an electric furnace (Azar, Frances, F11L 1250, Iran) to reach the crystallization temperature T_{Cr} during one hour under a constant heating rate. When the temperature reached T_{Cr}, the sample was kept at this temperature for 3 h to form nanocrystals in the glassy matrix. Such one-step process guarantees the simultaneous nucleation and growth of NCs. The heat treatment process should continue as long as the particle size is too large to completely cover the glass matrix and not too small

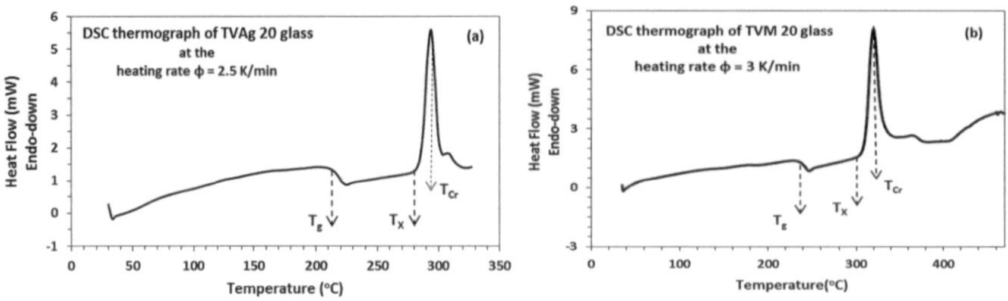

Fig. 2 DSC thermographs of **a** TVAg20 glass ($20Ag_2O$-$40TeO_2$-$40V_2O_5$) at the heating rate of 2.5 K/min [13], and **b** TVM20 glass ($40TeO_2$-$40V_2O_5$-$20MoO_3$) at the heating rate of 3 K/min [25]

to trap the active ions in them [44]. The thermal features, characteristic temperatures, and other important physical properties of specified glasses are summarized in Tables 3 and 4 in Sect. 3.1.1.

2.1.3 Synthesis of Bare and Doped ZnSe Nanoparticles by on the Microwave-Assisted Aqueous Method

Before dealing with the synthesis details of ZnSe NPs, some points are instructive to be mentioned. Usually, dopants/impurities are entered to a material in order to regulate and improve its desired properties such as optical, electronic, thermal, structural, and biological properties [82]. This is true in the case of II–VI semiconducting NPs such as ZnSe. It should be noted that the valence state and ionic radius ratio of the host and dopant ions are important in the fabrication of doped II–VI nano-metered semiconductors. Furthermore, alongside the strategy of doping, there is an emphasize on the more controllable, rapid and echo-friendly approach through the use of microwave irradiation (MWIR).

The adopted synthesis strategy of ZnSe NPs is echo-friendly/green (i.e. the solvents are environmentally/user-friendly) owing to low energy consumption, low cost, and low time consumption features. Therefore, to avoid the use of chemical catalysts and further ensure the green synthesis approach of purposeful materials, microwave (MW) heating provides the simplicity of use, speed of the reactions, controllable growth of NPs, narrow distribution and tunable size capability [44, 51–53, 56, 59, 60]. As stated in Sect. 1, there are several published works on the different synthesis methods for the preparation of the transition metal (TM)-doped ZnSe NCs [83–88], the simple MW-assisted aqueous approach is described in this chapter, as can be followed.

To synthesize a series of bare, doped, co-doped, and core–shell NPs of ZnSe, initial raw ingredients used were provided from the high purity Merck products and scheduled in Table 2.

- ZnSe NPs (abbreviated as Spure) and $Zn_{1-x-y}Cu_xMn_ySe$NPs (x = 0, 0.1, 0.75, 1.5%, y = 0, 0.1, 1, 1.5, 2 and 4%) were synthesized employing the MWIR

Table 3 Selected tellurate glasses: collected data of density (ρ), molar volume (V_m), oxygen packing density (OPD), oxygen molar volume (V_O^*), refractive index at the absorption edge (n), optical band gap (E_{gap}) and the related evaluation method, as well as the power m denoting the type of the optical transition of charge carriers. N.R.: is the abbreviation of (not reported)

Glass abbreviation	Composition	ρ (g cm^{-3})	V_m (cm^3 mol^{-1})	V_O^* (cm^3 mol^{-1})	OPD (g atom L^{-1})	n	E_g (eV)	Gap determination approach	m
10BaTe	10BaO-90TeO_2	5.582 [30]	28.46 [30]	N.R.	N.R.	2.355	3.16	Qualitative eye observation of the band edge	N.R.
15BaTe	15BaO-85TeO_2	5.561 [30]	28.53 [30]	N.R.	N.R.	N.R.	N.R.		
20BaTe	20BaO-80TeO_2	5.521 [30]	28.68 [30]	N.R.	N.R.	N.R.	N.R.		
1Eu9BaTe	1Eu_2O_3-9BaO-90TeO_2	5.654 [30]	28.46 [30]	N.R.	N.R.	2.357	3.15		
1Dy9BaTe	1Dy_2O_3-9BaO-90TeO_2	5.655 [30]	28.49 [30]	N.R.	N.R.	2.369	3.11		
1Er9BaTe	1Er_2O_3-9BaO-90TeO_2	5.668 [30]	28.45 [30]	N.R.	N.R.	2.357	3.15		
1Nd9BaTe	1Nd_2O_3-9BaO-90TeO_2	5.646 [30]	28.47 [30]	N.R.	N.R.	2.379	3.07		
$\{[(TeO_2)_{0.70}(B_2O_3)_{0.30}]_{0.7}$ $(ZnO)_{0.3}\}_{0.95}(Er_2O_3)_{0.05[green]}$		N.R.	N.R.	N.R.	N.R.	2.319	3.31	Qualitative eye observation of the band edge	
TVMx	**TVM0**: 40TeO_2-60V_2O_5	3.857 [10]	45.722 [107]	11.823	83.11	2.879	1.70 [108]	Tauc	2
							–	ASF	

(continued)

Table 3 (continued)

Glass abbreviation	Composition	ρ (g cm^{-3})	V_m (cm^3 mol^{-1})	V_O^* (cm^3 mol^{-1})	OPD (g atom L^{-1})	n	E_g (eV)	Gap determination approach	m
	TVM10: $40TeO_2$-$50V_2O_5$-$10MoO_3$	3.880 [10]	44.325 [107]	12.134	83.22	2.781	1.90 [108]	Tauc	
							–	ASF	
	TVM20: $40TeO_2$-$40V_2O_5$-$20MoO_3$	4.053 [10]	40.805 [107]	12.064	83.32	2.736 2.723	2.00 [108]	Tauc	
							2.03 [108]	ASF	
	TVM30: $40TeO_2$-$30V_2O_5$-$30MoO_3$	4.290 [10]	37.667 [107]	11.837	84.96	2.666 2.654	2.17 [108]	Tauc	
							2.20 [108]	ASF	
	TVM40: $40TeO_2$-$20V_2O_5$-$40MoO_3$	4.394 [10]	35.908 [107]	12.023	83.55	2.635 2.631	2.25 [108]	Tauc	
							2.26 [108]	ASF	
	TVM50: $40TeO_2$-$10V_2O_5$-$50MoO_3$	4.596 [10]	33.506 [107]	N.R.	83.57	2.530 2.524	2.55 [108]	Tauc	
							2.57 [108]	ASF	
	TVM60: $40TeO_2$-$60MoO_3$	4.895 [10]	30.686 [107]	N.R.	84.73	2.439 2.436	2.85 [108]	Tauc	
							2.86 [109]	ASF	

(continued)

Table 3 (continued)

Glass abbreviation	Composition	ρ (g cm^{-3})	V_m (cm^3 mol^{-1})	V_O^* (cm^3 mol^{-1})	OPD (g atom L^{-1})	n	E_g (eV)	Gap determination approach	m
TVNx	**TVN0**: $40TeO_2$-$60V_2O_5$	3.856 [98]	44.858 [110]	N.R.	N.R.	2.728	2.02 [98]	Tauc	2
	TVN5: $40TeO_2$-$55V_2O_5$-5NiO	3.872 [98]	43.289 [110]	N.R.	N.R.	2.741	1.99 [98]		
	TVN10: $40TeO_2$-$50V_2O_5$-10NiO	3.900 [98]	41.604 [110]	N.R.	N.R.	2.814	1.83 [98]		
	TVN20: $40TeO_2$-$40V_2O_5$-20NiO	4.019 [98]	37.703 [110]	N.R.	N.R.	2.951	1.57 [98]		
	TVN30: $40TeO_2$-$30V_2O_5$-30NiO	4.243 [98]	33.189 [110]	N.R.	N.R.	2.911	1.64 [98]		
TVSbx	TVSb0: $40TeO_2$-$60V_2O_5$	3.710 [113]	46.623 [113]	12.281 [115]	81.50 [116]	2.238 [22]	1.73 [113]	ASF	2
	TVSb5: $40TeO_2$-$55V_2O_5$-5Sb	3.759 [113]	46.240 [113]	12.686 [115]	76.77 [116]	2.218 [22]	2.14 [113]		
	TVSb8: $40TeO_2$-$52V_2O_5$-8Sb	3.882 [113]	44.260 [113]	12.655 [115]	76.82 [116]	2.193 [22]	1.57 [113]		
	TVSb10: $40TeO_2$-$50V_2O_5$-10Sb	3.888 [113]	43.836 [113]	12.861 [115]	75.28 [115]	2.191 [22]	1.85 [113]		
	TVSb12: $40TeO_2$-$48V_2O_5$-12Sb	4.031 [113]	41.353 [113]	12.621 [115]	77.38 [115]	2.163 [22]	N.R.		
	TVSb15: $40TeO_2$-$45V_2O_5$-15Sb	4.144 [113]	40.320 [113]	12.615 [115]	75.64 [115]	2.143 [22]	N.R.		

(continued)

Table 3 (continued)

Glass abbreviation	Composition	ρ (g cm^{-3})	V_m (cm^3 mol^{-1})	V_O^* (cm^3 mol^{-1})	OPD (g atom L^{-1})	n	E_g (eV)	Gap determination approach	m
TVSx	**TVS0**: $40TeO_2$-$60V_2O_5$	3.710 [99]	46.623 [112]	N.R.	N.R.	N.R.	1.74 [99]	ASF	2
	TVS5: $40TeO_2$-$55V_2O_5$-$5Sb_2O_3$	3.778 [99]	47.235 [112]	N.R.	N.R.	N.R.	1.98 [99]		
	TVS8: $40TeO_2$-$52V_2O_5$-$8Sb_2O_3$	3.848 [99]	47.230 [112]	N.R.	N.R.	N.R.	1.72 [99]		
	TVS10: $40TeO_2$-$50V_2O_5$-$10Sb_2O_3$	3.922 [99]	46.898 [112]	N.R.	N.R.	N.R.	1.37 [99]		
TVAgx	**TVAg0**: $40TeO_2$-$60V_2O_5$	3.587 [4]	48.214 [4]	12.70 [6]	78.82 [6, 13]	2.598 [13]	2.41 [4]	ASF	0.5 [4]
							2.35 [4]	DASF	0.623 [4]
	TVAg10: $40TeO_2$-$50V_2O_5$-$10Ag_2O$	4.038 [4]	44.066 [4]	12.99 [6]	77.16 [6, 13]	2.617 [13]	2.29 [4]	ASF	0.5 [4]
							2.30 [4]	DASF	0.510 [4]
	TVAg20: $40TeO_2$-$40V_2O_5$-$20Ag_2O$	4.756 [4]	38.457 [4]	12.86 [6]	78.01 [6, 13]	2.658 [13]	2.16 [4]	ASF	0.5 [4]
							2.19 [4]	DASF	0.483 [4]
	TVAg30: $40TeO_2$-$30V_2O_5$-$30Ag_2O$	5.126 [4]	36.657 [4]	14.21[6]	68.20 [6, 13]	2.617[13]	2.41 [4]	ASF	0.5 [4]
							2.30 [4]	DASF	0.869 [4]

(continued)

Table 3 (continued)

Glass abbreviation	Composition	ρ (g cm^{-3})	V_m (cm^3 mol^{-1})	V_O^* (cm^3 mol^{-1})	OPD (g atom L^{-1})	n	E_g (eV)	Gap determination approach	m
	TVAg40: $40TeO_2$-$20V_2O_5$-$40Ag_2O$	5.717 [4]	33.736 [4]	15.44 [6]	65.21 [6, 13]	2.477 [13]	2.72 [4]	ASF	0.5 [4]
							2.72 [4]	DASF	0.551 [4]
	TVAg50: $40TeO_2$-$10V_2O_5$-$50Ag_2O$	6.398 [4]	30.925 [4]	17.29 [6]	58.21 [6, 13]	N.R.	N.R.	N.R.	N.R.
							N.R.	N.R.	N.R.
TeO_2-Bi_2O_3-ZnO core-clad system	$72.5TeO_2$-$10Bi_2O_3$-17.5ZnO (core)	N.R.	N.R.	N.R.	N.R.	2.1583 at 630 nm [45]	1.26	N.R.	N.R.
	$70TeO_2$-$10Bi_2O_3$-20ZnO (clad)	N.R.	N.R.	N.R.	N.R.	2.1492 at 630 nm [32]	1.27	N.R.	N.R.
BTVx	**BTV0**: $60Bi_2O_3$-$40TeO_2$	5.950 [114]	57.700 [114]	N.R.	45.06 [114]	2.441 [114]	2.95 [114]	Tauc	0.5 [114]
	BTV5: $60Bi_2O_3$-$35TeO_2$-$5V_2O_5$	5.411 [114]	63.654 [114]	N.R.	43.202 [114]	2.453 [114]	2.86 [114]		
	BTV10: $60Bi_2O_3$-$30TeO_2$-$10V_2O_5$	4.670 [114]	76.618 [114]	N.R.	39.192 [114]	2.514 [114]	2.83 [114]		
	BTV15: $60Bi_2O_3$-$25TeO_2$-$15V_2O_5$	4.448 [114]	82.203 [114]	N.R.	39.134 [114]	2.536 [114]	2.81 [114]		

Table 4 Reviewed binary and ternary Te-based oxide glasses: glass transition temperature (T_g), crystallization temperature (T_{Cr}), the onset temperature of crystallization (T_x), melting temperature (T_m), two different thermal stability criteria of studied glasses (S_1 and S_2) and luminescent wave lengths. These thermal features accompanied with the optical data of Table 1, can be considered as vital information to decide about the capability of glass in optoelectronic applications

Glass abbreviation	Composition	T_g (°C)	T_{Cr} (°C)	T_x (°C)	T_m (°C)	Thermal stability criteria		Luminescent wavelengths (nm)	Region of luminescence
						$S_1 = T_{cr} - T_g$	$S_2 = T_x - T_g$		
10BaTe	$10BaO\text{-}90TeO_2$	321 [30]	461 [30]	N.R.	582 [30]	140	N.R.	N.R.	N.R.
15BaTe	$15BaO\text{-}85TeO_2$	326 [30]	450 [30]	N.R.	585 [30]	124	N.R.	N.R.	N.R.
20BaTe	$20BaO\text{-}80TeO_2$	335 [30]	435 [30]	N.R.	600 [30]	100	N.R.	N.R.	N.R.
1Eu9BaTe	$1Eu_2O_3\text{-}9BaO\text{-}90TeO_2$	319 [30]	445 [30]	N.R.	520 [30]	126	N.R.	593, 616, 702 [30]	Visible
1Dy9BaTe	$1Dy_2O_3\text{-}9BaO\text{-}90TeO_2$	322 [30]	447 [30]	N.R.	530 [30]	125	N.R.	576, 665 [30]	Visible
1Er9BaTe	$1Er_2O_3\text{-}9BaO\text{-}90TeO_2$	322 [30]	443 [30]	N.R.	534 [30]	121	N.R.	N.R.	N.R.
1Nd9BaTe	$1Nd_2O_3\text{-}9BaO\text{-}90TeO_2$	316 [30]	438 [30]	N.R.	496 [30]	122	N.R.	N.R.	N.R.
$\{[(TeO_2)_{0.70}(B_2O_3)_{0.30}]_{0.7}(ZnO)_{0.3}\}_{0.95}(Er_2O_3)_{0.05[green]}$		N.R.	N.R.	N.R.	N.R.	N.R.	N.R.	538, 559 [29]	Green [29]
TVMx	**TVM0**: $40TeO_2\text{-}60V_2O_5$	226.09 [25]	265.74 [25]	253 [25]	600 [25]	39.65 [25]	26.91 [25]	N.R.	N.R.
	TVM10: $40TeO_2\text{-}50V_2O_5\text{-}10MoO_3$	232.15 [25]	290.10 [25]	273.5 [25]	620 [25]	57.95 [25]	41.35 [25]		
	TVM20: $40TeO_2\text{-}40V_2O_5\text{-}20MoO_3$	239.24 [25]	320.01 [25]	293.5 [25]	630 [25]	80.77 [25]	54.26 [25]		
	TVM30: $40TeO_2\text{-}30V_2O_5\text{-}30MoO_3$	251.41 [25]	303.75 [25]	294.6 [25]	655 [25]	52.34 [25]	43.19 [25]		

(continued)

Table 4 (continued)

Glass abbreviation	Composition	T_g (°C)	T_{Cr} (°C)	T_x (°C)	T_m (°C)	Thermal stability criteria		Luminescent wavelengths (nm)	Region of luminescence
						$S_1 = T_{cr} - T_g$	$S_2 = T_x - T_g$		
	TVM40: $40TeO_2$-$20V_2O_5$-$40MoO_3$	265.97 [25]	309.64 [25]	289.1 [25]	685 [25]	43.67 [25]	23.13 [25]		
	TVM50: $40TeO_2$-$10V_2O_5$-$50MoO_3$	279.64 [25]	317.02 [25]	292.4 [25]	740 [25]	37.38 [25]	12.76 [25]		
	TVM60: $40TeO_2$-$60MoO_3$	300.44 [25]	330.24 [25]	311.1 [25]	780 [25]	29.8 [25]	10.66 [25]		
TVNx	**TVN0**: $40TeO_2$-$60V_2O_5$	227.70 [117]	267.16 [117]	N.R.	720 [22]	39.46 [22]	N.R.	N.R.	N.R.
	TVN5: $40TeO_2$-$55V_2O_5$-$5NiO$	N.R.	N.R.	N.R.	750 [22]	N.R.	N.R.		
	TVN10: $40TeO_2$-$50V_2O_5$-$10NiO$	246.30 [117]	355.80 [117]	N.R.	820 [22]	109.5	N.R.		
	TVN20: $40TeO_2$-$40V_2O_5$-$20NiO$	287 [117]	357.30 [117]	N.R.	900 [22]	70.3	N.R.		
	TVN30: $40TeO_2$-$30V_2O_5$-$30NiO$	N.R.	N.R.	N.R.	980 [22]	N.R.	N.R.		
TVSbx	TVSb0: $40TeO_2$-$60V_2O_5$	227.70 [12]	267.16 [12]	249.26 [115]	680 [12, 113]	39.46 [113]	21.56 [115]	N.R.	N.R.
	TVSb5: $40TeO_2$-$55V_2O_5$-5Sb	243.030 [12]	312.22 [12]	280.90 [115]	700 [12, 113]	78.25 [113]	37.60 [115]		

(continued)

Table 4 (continued)

Glass abbreviation	Composition	T_g (°C)	T_{Cr} (°C)	T_x (°C)	T_m (°C)	Thermal stability criteria		Luminescent wavelengths (nm)	Region of luminescence
						$S_1 = T_{cr} - T_g$	$S_2 = T_x - T_g$		
	TVSb8: $40TeO_2$-$52V_2O_5$-8Sb	N.R.	N.R.	N.R.	705 [12, 113]	N.R.	N.R.		
	TVSb10: $40TeO_2$-$50V_2O_5$-10Sb	249.98 [12]	340.39 [12]	317.40 [115]	710 [12, 113]	119.43 [113]	67.42 [115]		
	TVSb12: $40TeO_2$-$48V_2O_5$-12Sb	256.64 [12]	411.69 [12]	338.2 [115]	720 [12, 113]	191.29 [113]	81.56 [115]		
	TVSb15: $40TeO_2$-$45V_2O_5$-15Sb	260.69 [12]	377.36 [12]	339.30 [115]	730 [12, 113]	146.45 [113]	78.81 [115]		
TVSx	**TVS0**: $40TeO_2$-$60V_2O_5$	227.70 [111]	267.15 [111]	249.26 [111]	680 [111, 112]	39.45	21.56	N.R.	N.R.
	TVS5: $40TeO_2$-$55V_2O_5$-$5Sb_2O_3$	253.10 [111]	326.84 [111]	291.88 [111]	720 [111, 112]	73.74	38.78		
	TVS8: $40TeO_2$-$52V_2O_5$-$8Sb_2O_3$	269.60 [111]	397.60 [111]	371.82 [111]	740 [111, 112]	128.0	102.2		
	TVS10: $40TeO_2$-$50V_2O_5$-$10Sb_2O_3$	273.91 [111]	423.11 [111]	401.80 [111]	750 [111, 112]	149.2	127.89		
TVAgx	**TVAg0**: $40TeO_2$-$60V_2O_5$	235.10 [13]	277.64 [13]	N.R.	660 [13]	48.22 [13]	29.17 [13]	N.R.	N.R.

(continued)

Table 4 (continued)

Glass abbreviation	Composition	T_g (°C)	T_{Cr} (°C)	T_x (°C)	T_m (°C)	Thermal stability criteria		Luminescent wavelengths (nm)	Region of luminescence
						$S_1 = T_{cr} - T_g$	$S_2 = T_x - T_g$		
	TVAg10: $40TeO_2$-$50V_2O_5$-$10Ag_2O$	240.04 [13]	284.73 [13]	N.R.	680 [13]	51.75 [13]	33.20 [13]		
	TVAg20: $40TeO_2$-$40V_2O_5$-$20Ag_2O$	217.8 [13]	294.36 [13]	N.R.	700 [13]	88.33 [13]	63.75 [13]		
	TVAg30: $40TeO_2$-$30V_2O_5$-$30Ag_2O$	192.37 [13]	N.R.	N.R.	710 [13]	N.R.	N.R.		
	TVAg40: $40TeO_2$-$20V_2O_5$-$40Ag_2O$	166.48 [13]	N.R.	N.R.	720 [13]	N.R.	N.R.		
	TVAg50: $40TeO_2$-$10V_2O_5$-$50Ag_2O$	139.38 [13]	188.52 [13]	N.R.	730 [13]	46.17 [13]	29.12 [13]		
TeO_2-Bi_2O_3-ZnO core-clad system	$72.5TeO_2$-$10Bi_2O_3$-17.5ZnO (core)	334 [32]	N.R.	512 [32]	850 [32]	N.R.	178 [32]	N.R.	N.R.
	$70TeO_2$-$10Bi_2O_3$-20ZnO (clad)	N.R.	N.R.	N.R.	850 [32]	N.R.	N.R.	N.R.	N.R.

technique used in [41, 44, 52, 53, 59–62, 89] for the synthesis of $ZnSe:Cu^{2+}$, and pure (bare) ZnSe NPs. For an overview, the synthesis process of bare, single-doped, and co-doped ZnSe includes 7 steps (see the flowchart in Fig. 3a), where, some steps or solutions may be omitted from the process depending on the type of target NPs. For example, to fabricate ZnSe:Cu, step 4 (providing solution D) is not included. Initially, selenium source ($NaHSe^{-}$) was prepared in a 3-necked flask by stirring under the flow of argon gas (solution A: 0.024 g Se powder and 0.04 g $NaBH_4$ (as a reduction agent) were dissolved in 2 cc of deionized water). Then solution A was moved to a filtered syringe. The chemical reaction is as shown in Eq. (1) [62]:

$$Se_{powder} + NaBH_4 \rightarrow NaHSe^{-} \quad (1)$$

Then, zinc ions source (solution B) was made ready by solving 0.1314 g zinc acetate in 75 cc of deionized water. To prepare the Cu ion source (solution C), $CuSO_4$ powder was dissolved in deionized water. Likewise, Mn source was prepared (solution D) by solving the Mn powder. pH of the solution was regulated at the planned values of 8.0, 10.2, 11.2 or 12.2 using dropwise addition of 1 M NaOH prepared solution (≈3 mL) (the pH was fixed at 11.2 to guarantee the best crystallinity [53, 54]). After that, solution D, C and B were mixed, then TGA was added (under stirring) as the capping agent to inhibit the agglomeration of NCs. Finally, the resulting solution was kept for 20 min under argon flow, which resulted in obtaining solution E. In the next step, solution A was added to solution B, which contained the dopant sources of Cu and Mn dopants. At this step, the nucleation process and the production of $ZnSe:Cu^{2+}$, Mn^{2+} nanoparticles (NPs) is initiated. At the last step of the fabrication route, the solution was heated/irradiated by a microwave system with a constant power of 1000 W for MWIR exposure times of 0, 1, 2, 3, 4 and 6 min.

- In the case of ZnSe@ZnS core–shell structure, although different techniques have been reported [62], our recent approach may be considered as a new photochemical microwave-assisted colloidal method [44], where this method takes advantage of the previously mentioned advantages of MW. For the production of the ZnS shell according to this synthesis method [44, 90], an amount (0.117 g) of sodium thiosulfate is needed, which should be dissolved in deionized water (7.5 mL) and exposed to UV illumination for 80 min. To produce S ions, the UV sensitive material $Na_2S_2O_3$ is used, where, it dissociates and produces S_2^- and free electrons according to the following chemical reaction [44, 62]:

$$Na_2S_2O_3 + UV_{illumination} \rightarrow 2Na^{+} + S_2^{-} + SO_3^{-} \quad (2)$$

Next for production of Zn^{2+} ions, an amount of 0.033 g of zinc acetate is dissolved separately in 7.5 mL deionized water. Both the S and Zn sources are mixed together and added to 30 mL of the bare ZnSe:Cu core-containing solution that was prepared as stated previously. Next, pure TGA is added, and pH is set to the desired value by dropwise addition of NaOH solution. Finally, mixed final solution

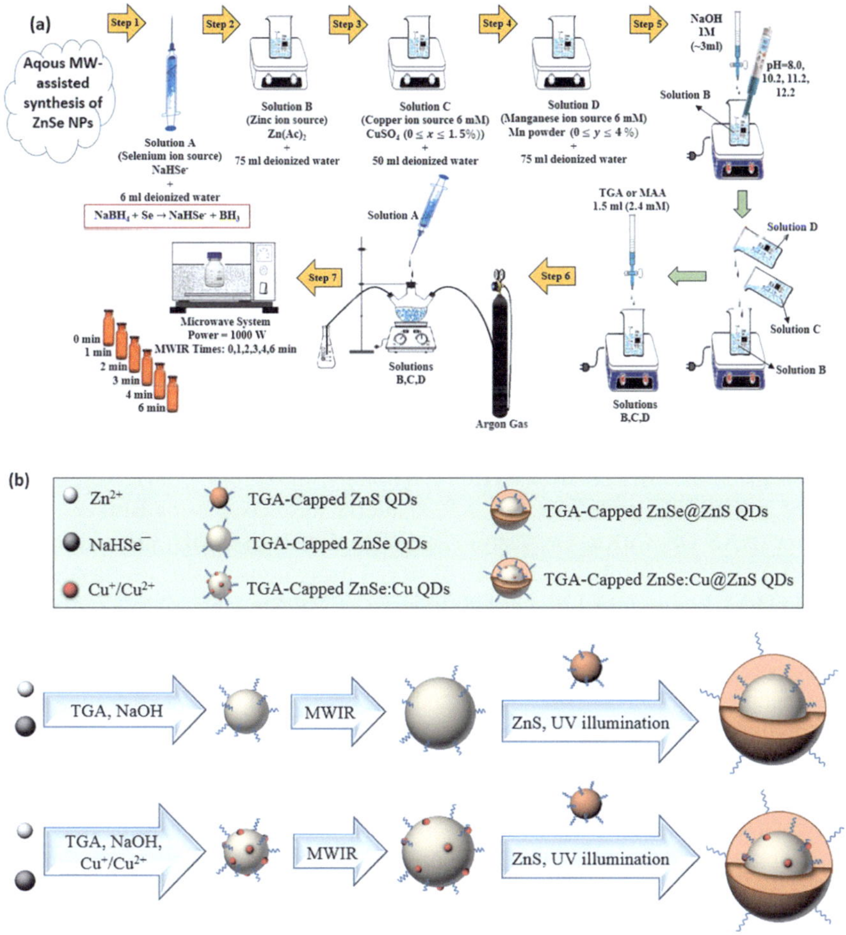

Fig. 3 A general flowchart depiction for microwave-assisted fabrication of **a** aqueous ZnSe:Cu doped and ZnSe:Cu-Mn co-doped NPs (i.e. one can omit the excess step or each of solutions C or D in consistency with the aim of the synthesis of doped or co-doped ZnSe NPS) [59]. **b** ZnSe@ ZnS core–shell NPs in an aqueous photochemical approach based on the general MW-assisted synthesis of bare and doped Cu-doped ZnSe cores

was illuminated by UV only for 90 min, and the ZnSe:Cu@ZnS core/shell NPs were obtained. The general related chemical reaction for this new photochemical MW-assisted approach is shown in Eq. (3). In the case of un-doped ZnSe@ZnS core-shells, the synthesized ZnSe NCs (see Fig. 3a) are used instead of ZnSe:Cu NCs:

$$\begin{aligned} &\text{ZnSe:Cu NCs} + \text{MWIR} + \text{Zn} + Na_2S_2O_3 \\ &\quad + \text{TGA} + \text{UV - illumination} \rightarrow \text{ZnSe:Cu@ZnS NCs} \end{aligned} \tag{3}$$

So, the obtained NPs-containing solution is ready to use for UV-Vis. photoluminescence (PL) spectroscopy, and TEM imaging. Centrifuging the final solution (after addition of acetone for deposition of NPs) results in obtaining the sediments, which need to be dried in air. After drying the sediments, a suitable powder of NPs is acquired, which can be used for some characterizations such as XRD, FTIR, DSC and etc. The schematic of ZnSe:Cu@ZnS core/shell QDs steps is depicted in Fig. 3b.

2.2 *Physical Chemical Characterizations*

Generally, various characterizations are needed to acquire different parameters or properties of the fabricated materials (i.e. glasses, GCs and NPs) directly or indirectly. The fabricated samples were characterized in powder or solution states using UV-Vis spectroscopy, PL, XRD, FTIR, morphology, and antibacterial features. In the case of glasses, GCs and ZnSe NPs, XRD patterns were taken using Bruker diffractometer (AXS D8, CuKα, Germany) and (XRD, Unisantis-XMD-300, CuKα, $\lambda =$ 1.5419 Å, India). Morphologies were observed using transmission electron microscope (TEM, Philips CM 120, Grid 200 Mesh Cu, Voltage 100 kV) by examination of a solution drop, and scanning electron microscopy/energy dispersive X-ray spectroscopy (SEM/EDX) mapping using (SEM: TESCAN, Mira 3, Czech Republic) for the powders. Also, PL of bare, doped and core-shells of ZnSe was recorded using photoluminescence spectrophotometer (PL Avantes spectrometer Ava Spec-2048 TEC). Optical testing of the bare/doped and core–shell ZnSe-containing solutions, glassy blown films and also for GCs were done employing UV-Visible spectrometers of (Agilent Technologies, Cary series, Cary 100 UV-Vis) and double beam spectrophotometer (PerkinElmer, Lambda 25-USA) within the wavelength range of 190–1100 nm. Centrifuging of ZnSe-containing solutions was done by the centrifugal apparatus (Model: PIT320, Iran: 6000 rpm for 6 min) to obtain the needed powders of the samples. Fourier transform Infrared Spectroscopy (FTIR) was done using a Thermo (AVATAR, USA) within the range of 400–4000 cm^{-1}.

2.3 *Antibacterial Bioassays*

In this section, general remarks on the bioassays and two more used assays of disc-diffusion and serial dilution are introduced.

2.3.1 General Remarks of Bioassays

Bioassay stands on a simple, particular, and accurate laboratory technique to evaluate the antimicrobial/antibacterial behavior. Antimicrobial testing may be used for drug discovery, epidemiology, and prediction of therapeutic results [55, 56, 90]. There are numerous laboratory approaches, which can be employed to assess the antimicrobial activity of an agent. As mentioned at the beginning of Sect. 2.3, the most famous, and basic bioassays are the disc-diffusion and broth or agar dilution methods. Some other procedures such as poisoned food method are used particularly for antifungal testing. There are many established and agreed standards issued by the Clinical and Laboratory Standards Institute (CLSI) for bacteria and yeasts testing [91, 92]. So, based on these standards, in spite of the existence of several bioassays, the most commonly used bioassays may be divided into 3 categories [90]. The first one, being diffusion methods include agar disc-diffusion method, antimicrobial gradient method (Etest), agar well diffusion method, agar plug diffusion method, cross streak, and the poisoned food method. The second one, called as thin-layer chromatography (TLC)–bioautography, includes the agar diffusion methods, direct bio-autography, and the agar overlay bioassay. While the third, nominated as dilution methods, include broth dilution method and agar dilution method [90–92].

Disc diffusion bio-assay, as a frequently used method, has been executed, which aided to overcome some limitations and errors [65, 93]. So, serial dilution method can serve as a substituting method which has more rapidity and precision [65, 93]. The serial dilution method has benefits such more precise antibacterial activity reports delivery to find the top samples for biological requests in drug production and delivery, food industry, wound healing tissue engineering and etc. [66].

2.3.2 Disc-Diffusion Bioassay in Agar Media

Disc-diffusion bioassay is considered as one of the simplest and broadly used approaches from 1940s to the present [93]. This method owns some advantages and features such as: repetitive nature, easiness, cheap, the ability to test vast types of microorganisms and antimicrobial materials, and especially the ease to interpret [90]. Different steps of this method are illustrated schematically in Fig. 4.

To perform familiar disc-diffusion assay maintaining the sterile condition of all primary cultures and equipment is mandatory, which in the present work was guaranteed using an autoclave system (Iran Tolid Medical Industries Co: model: 121A, Iran). The first Muller-Hinton-broth (MHB)-agar was prepared and autoclaved solution was poured into sterile petri-dishes under the same conditions and volumes. The content of the petri-dishes turn into a solid gel, due to the existence of agar. At this stage, plates containing MHB-agar are inoculated (seeded) with a standardized inoculum of the test microorganisms of either Gram-negative bacteria (i.e. Escherichia coli, Klebsiella pneumoniae and Pseudomonas aeruginosa) or Gram-positive's (Staphylococcus aureus and Bacillus cereus). To guide for the eyes, inoculation manner of the surface of MHB-agar solid into petri-dishes is shown in Fig. 5.

Then, sterile Whatman filter paper discs of about 6 mm in diameter, containing the test agent such as the solution containing ZnSe NPs at a desired concentration (loading volume) of 70 and 140 μL, were placed on the MHB-agar surface. The petri-dishes (Incubators of Fan Azma Gostar: Model: FG, Iran.) were incubated under

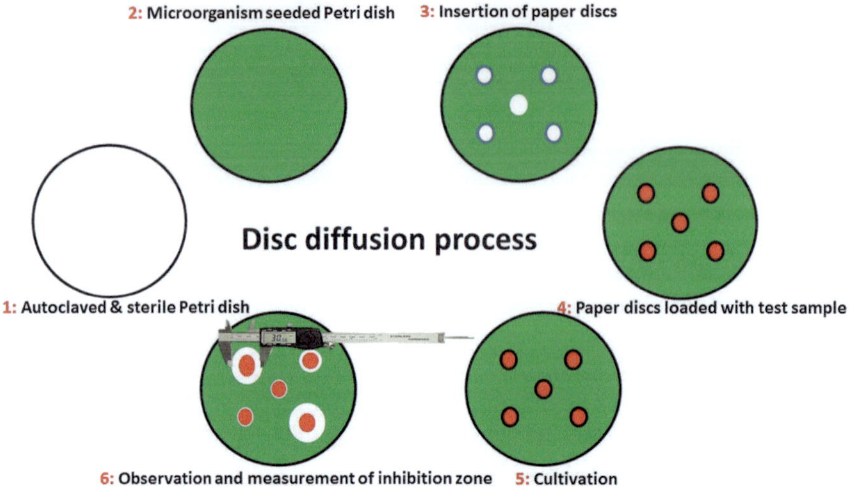

Fig. 4 Schematics of different experimental steps of disc diffusion method

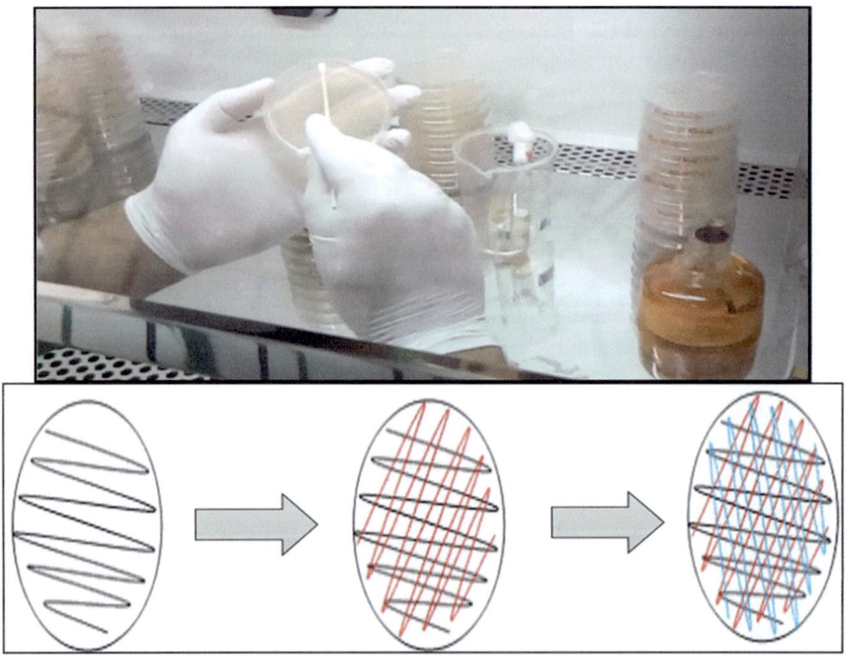

Fig. 5 Swabbing the surface of agar medium (bacteria seeding of the dish): true experimental swabbing image and schematically depiction of correct swabbing, from up to down, respectively

suitable conditions of 37 °C overnight for 18 h. During this time, the antibacterial agent can penetrate/diffuse into the agar and inhibit germination and growth of the test microorganisms and hence the growth inhibited zones (halos) appear. So, after incubation, halos were observed around the paper discs, indicating antibacterial or even probable bactericidal activity. Each experiment was repeated 3 times, and so the average diameter of such halos was measured and reported in millimeter scale. There is a key point in the experiments; negative and positive control tests were performed on the plates containing only inoculated MHB-agar and free ZnSe (or any probable antibacterial agent) respectively. Such control tests were incubated at 37 °C for 18 h, verifying the correct growth performance of bacteria and the sterile nature of probable antibacterial agent accordingly [51, 56]. In short, the appearance of the inhibition zone around the paper disc specifies the antimicrobial activity of the tested agent, the diameter of which is proportional to its antimicrobial potential (i.e. under the same concentration and loading volume of the agent). As will be shown and discussed in Sect. 3, plotting a comparative antibiogram can provide comparative results by categorizing bacteria into susceptible, intermediate, or resistant microorganisms. However, since the bacterial growth inhibition does not mean the bacterial demise, the disc-diffusion assay cannot determine the bactericidal feature [90]. So, sampling of the inhibition and again culturing this media can elucidate the existence of any further bacteria or not.

2.3.3 Serial Dilution Bioassay

Serial dilution assay is a commonly used route [7, 55, 65, 90], which has no specific limitations, is easy and user friendly, inexpensive, reproducible and, has the potential of accurate, rapid and broad-usage for the recognition of antibacterial activity. Furthermore, the main gains of the microdilution method may be stated as its reproducibility and economic features besides the space needed for such miniaturized test method [62, 65, 90]. Serial dilution bioassay is based on assessing the concentration of an unidentified sample (i.e. number of colonies, organisms, bacteria, or viruses). The importance and detailed aspects of serial dilution and colony counting can be followed from the standard operating procedures and regulatory guidelines describing this methodology [65]. Therefore, performance of serial dilution targets to determine the minimum inhibitory concentration (MIC) and minimum bactericidal concentration (MBC) as vital information about the potential of the understudied agents in broth medium [62, 65, 90]. Evaluation of MIC is commonly executed as designated by Silva et al. [94]. Also, MBCs can be acquired as the lowest concentration of agent (glass, GCs or NPs) at which bacterial growth was prevented about 99.9% which mean bacterial death by the way termed by Fernandes et al. [95]. It is worth mentioning that the bacterial growth inhibition (indicated by MIC) does not mean the bacterial death (i.e. MBC). Also, as mentioned in Sect. 2.3.2, comparing diffusion and dilution assays reveals that disc-diffusion approach cannot provide any content about MIC [90].

To explain the steps of serial dilution, an inoculum of 0.5 McFarland (Mc) corresponding to 1.5×10^8 CFU (colony forming unit/mL [96]) of each of G-negative or G-positive bacteria (listed in Sect. 2.3.2) was provided from overnight cultures and inoculated in Luria Bertani (LB) or MHB with initial NPs concentrations of about 12 ppm. McFarland standard means the optical absorbance/optical density (OD) = 0.08–0.13 at radiation wavelength of 630 nm [96]. In the serial dilution method, shaken 96-well non-stick microplates were used, and each microplate was used for one bacteria with three replications of the corresponding agent. In each well, 25 μL of sterilized MHB or LB fluid is poured. For example, in the case of MHB culture, 4.2 g of MHB powder was dissolved in 200 mL of distilled water in an appropriate Erlenmeyer (upon the standard protocol). Also to prepare the bacteria for inoculation and prepare 0.5 Mc, 1.05 g of MHB is dissolved separately into a 100 mL flask. After pouring same volume of MHB or LB liquid in wells, the same values of McFarland solution (microorganism) were added to the wells. The work continued by filling 25 μL of the mother solution of a probable antibacterial agent to the first well of each row of the microplate and its dilution by the factor of 2 from well numbers 1–10. The 11-th and 12-th wells are devoted to negative and positive control tests. Final samples were incubated overnight at 37 °C for 18 h. Figure 6, shows the two-fold dilution array schematically.

After incubation, the resistance of the understudied bacteria to the agents was determined by measuring the OD taken by Elisa reader (ELX800) at a radiation wavelength of 630 nm. Each experiment was taken as triplet on 3 rows of microplate. In addition to what is inferred from the OD data, MIC well number is observable by the unaided eye from the clearness of wells' content. These remarks will be more and more clarified in Sect. 3 by presentation of results on the agents.

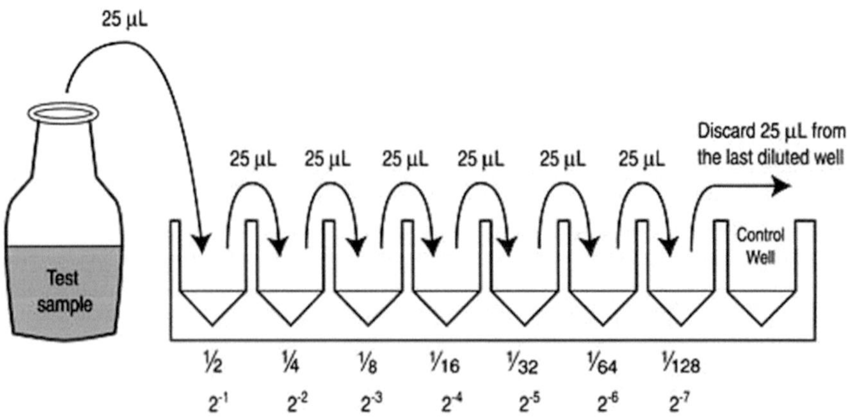

Fig. 6 Two-fold serial dilution schematics [55]

3 Bio-Optical Results and Advantages

During the time, many efforts have been done to expand and improve theories and experimental procedures related to the synthesis and applications of new materials. Any scientific attempt which moves toward removing all or parts of older inadequacies is fundamentally and technologically vital and worthy. Such labors can be surely seen in different scientific fields. Physics and materials science are the fields which play a major role in the past, nowadays, and future developments, and they are able to match with biology, medical sciences, engineering and etc. In other word, in suggesting agents/materials in unique applications, some issues such as environment, ambient reliability, durability, functionality, speed of operation, economical aspects, etc. should be noticed. In other words, there is a competition between materials proposed by researchers. Keeping in mind the mentioned statuses, and what was subjected in Sect. 1, VI-elements based materials specially tellurate glasses and GCs and also ZnSe NPs in bare, shelled and doped states are interesting categories. The mentioned materials are reviewed in terms of their optical and some biological aspects.

3.1 Te-Based Glasses and Glass–Ceramics

Glasses and glass–ceramics are important materials used in different areas such as dental applications [33] and restorative dentistry (due to their biocompatibility, outstanding aesthetic properties, good mechanical strength, and easy using), luminescence [29, 30], optoelectronic applications and optical fibers [32], as well as antibacterial and bactericidal features [7, 24]. Advanced glasses and GCs and glasses may be engineered materials developing for their particular optical, mechanical, electrical, or biological properties.

As mentioned earlier, multicomponent TeO_2-based glasses, known as low melting point, moisture-proof, good optical window, high refractive index, good mechanical strength, and high thermal and ambient stability are valuable materials. Their unique properties make them preferred over other engineering materials [22, 25]. Nowadays, the manufacturing of glassy and GCs structures is further advanced and more understood, hoping to explore new and functional agents. In Sect. 3.1 an attempt is made to manifest some optical and biological features and advances in the field of Te-containing glasses and GCs.

3.1.1 Optical and Structural Aspects

Based on what has been discussed before and keeping in mind the importance of tellurate and tellurate-vanadate Gs, the present subsection aims at the review some of the vital optical features of such multicomponent glasses. Specific applications of a material need to investigate different properties and the correlation between them. Optical applications such as optical fibers, luminescence, solar cells, etc. also need to control the band gap through composition or even engineering by theoretical studies and simulations. So, in experimental works, the search for the optimal optical samples is an advanced area, which also needs a more accurate method to determine the band gap. The major effect of the band gap on the other optical properties and function of the material, reveals its vitality. Therefore, before quantifying the subject, it is helpful to summarize the historical progress in band gap determination approaches. From literatures, three main approaches of Tauc's [97, 98], absorption spectrum fitting (ASF) [99], and derivation of absorption spectrum fitting (DASF) [4] are seen, of which DASF is used by Souri and Tahan in 2015 [4]. Such a revolution in the used methods arose from some inadequacies of the older methods as reported in Refs. [4, 22], The DASF method is considered as an easy-to-use, rapid, reliable and more accurate approach for the evaluation of the optical band gap (E_g), which can be employed for amorphous, crystalline, and nano-scaled semiconductors. To perform this, UV-Visible spectroscopy is needed to study the shape and shift of absorption edge, determination of the optical charge carrier transition types, and the energy band structure beside the band gap [61, 97, 98, 100–102]. Such effort and the necessity of using precise methods are due to the inconsistencies between the reported data for different materials, which can affect the optical device functionality [102]. To the best of our knowledge, a brief review of the mentioned approach comes as follows.

- ***Tauc's approach***

In the Tauc's method [97], absorption coefficient of semiconductors is given by Eq. (4):

$$\alpha\hbar\omega = B\left(\hbar\omega - E_{\text{gap}}\right)^{m} \tag{4}$$

in which, α, $\hbar\omega$, E_{gap}, and B are the absorption coefficient, energy of the photon incident on the sample (where $\hbar$ is Planck's constant divided by 2π, and ω is the frequency of incident radiation), optical band gap, and a constant, respectively [98]. The power m denotes the type of charge carrier transition which takes the values 1/2, 3/2, 2 and 3 for direct allowed, direct not-allowed, indirect allowed and indirect not-allowed transitions, respectively. The absorption coefficient α is given by Eq. (5) [98]:

$$\alpha = \left(\frac{1}{d}\right)\ln\left(\frac{I_o}{I_t}\right) = \frac{2.303A}{d} \tag{5}$$

where d, A, I_o and I_t are thickness of the film sample or dimension of the quartz cell or concentration of the solution, optical absorbance, incident and transmitted beam intensity respectively.

If one performs the Tauc's plot ($(\alpha\hbar\omega)^{1/m}$ vs. $\hbar\omega$) for different m-values and finds its optimal fitting parameters (such as the linear correlation coefficient R^2 for the linear part of graph), E_{gap} can be obtained from the extrapolation of the linear part to get the intercept with $\hbar\omega$ axis at $(\alpha\hbar\omega)^{1/m} = 0$. As is obvious, there are some inadequacies and error sources which limit the accurate assessment of the band gap. Namely, the need to know about d and so the value of α, and error attributed to the correct choice of *m* as well as the correct linear extrapolation. Such difficulties reveal the necessity of method improvement.

- ***Absorption spectrum fitting (ASF) approach***

Based on the aforementioned paragraph about the sources of error in implementing Tauc's approach, Souri and Shomalian [99] proposed a revision of the Tauc formula called as the ASF method, which started from the use of Beer-Lambert Law ($\alpha(\lambda) = (2.303/d)A$) on Tauc's equation Eq. (4); such substitution, with the aim of using the direct experimental values of *A* instead of $\alpha(\omega)$, resulted to Eq. (6) for ASF formalism [99]:

$$A(\lambda) = C\lambda\left(\lambda^{-1} - \lambda_g^{-1}\right)^m \tag{6}$$

in which, $C = [B(hc)^{m-1}\frac{d}{2.303}]$, and λ_g, h and c denote the wavelength corresponding to the band gap $\left(E_{gap}^{ASF} = hc\lambda_g^{-1} = 1239.83 \times \lambda_g^{-1}\right)$, Planck's constant and the speed of light, respectively. Although the least square extrapolation and finding the optimal value (presumption about *m*) of *m* is still needed, one can see the removal of any direct dependence and any need to *d* and $\alpha(\omega)$.

- ***Derivation of absorption spectrum fitting (DASF approach)***

In spite of the widespread use of Tauc's and ASF approaches [28, 97, 98, 103–106], the use of the least squares technique and especially presumption of the power *m* which refers to the type of charge carrier optical transition before determination of the band gap are important inadequacies. DASF approach [4], which avoids any need for sample thickness, and also avoids any need for presumption of transition type, only needs the direct spectral data of sample absorption in the UV-Vis region.

By some simple mathematical manipulations on Eq. (6) as shown in Eqs. (7) and (8), DASF formalism can be written as Eq. (8).

$$\ln\left[A(\lambda)\lambda^{-1}\right] = \ln(C) + m\ln\left(\lambda^{-1} - \lambda_g^{-1}\right) \tag{7}$$

where $C = [B(hc)^{m-1}\frac{d}{2.303}]$, and then:

$$\frac{d\left\{\ln\left[A(\lambda)\lambda^{-1}\right]\right\}}{d\left(\lambda^{-1}\right)} = \frac{m}{\left(\lambda^{-1} - \lambda_g^{-1}\right)} \tag{8}$$

Equation (8) includes a discontinuity that appears in the plot of $\frac{d\{\ln[A(\lambda)\lambda^{-1}]\}}{d(\lambda^{-1})}$ versus (λ^{-1}) at $(\lambda^{-1} = \lambda_g^{-1})$. Therefore, substituting the value of λ_g^{-1} of such plot in the relationship $E_{gap}^{DASF} = 1239.83 \times \lambda_g^{-1}$ (eV), gives the band gap directly. Knowing the exact band gap, one can evaluate the power m from the slope of the linear part of the plot of $\ln\left[A(\lambda)\lambda^{-1}\right]$ versus $\ln\left(\lambda^{-1} - \lambda_g^{-1}\right)$.

In addition, some reviews were performed on the opto-structural features of tellurate glass series of the forms A_yO_z-BaO-TeO_2 (A_yO_z: Eu_2O_3, Dy_2O_3, Er_2O_3 and Nd_2O_3) [30], $40TeO_2$-(60-x)V_2O_5-xA_yO_z (or A) (i.e. the third components as MoO_3 [10, 25, 107–109], NiO [98, 110], Sb_2O_3 [98, 111, 112], Ag_2O [4, 7, 13, 27] and Sb [12, 113]), Bi_2O_3-TeO_2-V_2O_5 [114], TeO_2-Bi_2O_3-ZnO [32], $\{[(TeO_2)_{0.70}(B_2O_3)_{0.30}]_{0.7}(ZnO)_{0.3}\}_{0.95}(Er_2O_3)_{0.05}$ [29]. Tables 3 and 4 represent some of the referred opto-structural properties of the aforementioned compositions.

As can be seen from Tables 3 and 4, some optical and thermo-structural properties of the mentioned systems have been reported, so it seems necessary to present them briefly.

- ***Molar volume***

The molar volume of a multicomponent glassy system is defined by Eq. (9) [13]:

$$V_m = \sum_i M_i / \rho \tag{9}$$

where glass molecular mass is denoted by $M_i = x_i m_i$ (where x_i and m_i denote mole content and the molecular mass of the ith component in the glass formula respectively).

- ***Oxygen packing density (OPD) and oxygen molar volume***

Other informative structural parameters aiding to interpret the optical absorption, are oxygen packing density (OPD) and oxygen molar volume (V_O^*). OPD and V_O^* depict the rigidity (in opposite of fragility) of glass structure, in relation with the content of bridging (BOs) and non-bridging oxygen atoms (NBOs) [13, 115]. These parameters are given by Eqs. (10) and (11).

$$\text{OPD} = \rho N_O / M = N_O / V_m \tag{10}$$

In which ρ, N_O, M, and V_m denote the glass density, the number of oxygen atoms per formula unit, molecular mass, and molar volume respectively.

As well, V_O^* can be defined as below for the ternary oxide glass of the general form of TeO_2-V_2O_5-(A_yO_z or A):

$$V_O^* = \frac{\left\{\left(M_{V_2O_5} - 16C_V\right)x_{V_2O_5} + M_{TeO_2}x_{TeO_2} + M_{A_yO_z\text{ or A}}x_{A_yO_z\text{ or A}}\right\}}{\left\{\rho(5 - C_V)x_{V_2O_5} + 2x_{TeO_2} + zx_{A_yO_z\text{ or A}}\right\}} \tag{11}$$

where M_i is the molar mass, and x_i represents the mole fraction of each of the glass components. Also, $C_V = \left[V^{4+}\right]/V_{tot}$ is the ratio of reduced vanadium ions as $[V^{4+}]/V_{tot}$, which can be obtained by different methods of titration [22], such as cyclic voltammetry, etc.

- ***Refractive index***

There is a specific relationship (see Eq. (12)) between optical band gap (E_g) and refractive index (n) of a semiconductor as suggested by Dimitrov and Sakka [118]:

$$\frac{n^2-1}{n^2+1} = 1 - \left(\frac{E_g}{20}\right)^{1/2} \tag{12}$$

There is an inverse relation between the two mentioned parameters, and so, samples owing lower band gap have higher third order non-linear optical susceptibility $\chi^{(3)}$. $\chi^{(3)}$ plays the important role in the operation and engineering of optical devices, especially in designing the optical fibers [55]. It is worth-mentioning that the high thermal stability and high $\chi^{(3)}$ are strongly accompanied to have successful device operation.

As mentioned before, Tables 3 and 4 represent the optical and thermo-structural data of some tellurate glasses of the forms A_yO_z-BaO-TeO_2 (A_yO_z: Eu_2O_3, Dy_2O_3, Er_2O_3, and Nd_2O_3) [30], 40TeO_2-(60-x)V_2O_5-xA_yO_z (or A) (i.e. the third components as MoO_3 [10, 25, 107–109], NiO [98, 110], Sb_2O_3 [99, 111, 112], Ag_2O [4, 7, 13, 27] and Sb [12, 113]), Bi_2O_3-TeO_2-V_2O_5 [114], TeO_2-Bi_2O_3-ZnO [32], $\{[(TeO_2)_{0.70}(B_2O_3)_{0.30}]_{0.7}(ZnO)_{0.3}\}_{0.95}(Er_2O_3)_{0.05}$ [29]. All the reported data of a sample are strongly correlated with its probable/promising applications. As can be seen from Tables 3 and 4, the glass formation region of each glass series is different from others, and systems such as TVMx and TVAgx have the more diverse glass formation region, which can be related to the higher glass forming ability of silver oxide or molybdenum tri-oxide. Also, there are fluctuations in properties such as optical band gap, OPD and V_O^* and thermal stability in each glass composition series. Such change trends are correlated to each other. Therefore, any possible suggestion for the application should be stated by the simultaneous consideration of all such properties. For example, for TVMx, TVM10-20 samples have the lowest values of band gap, the highest refractive index, and so the highest third order non-linear optical susceptibility $\chi^{(3)}$. Such samples may be good candidates for optical fibers if they can resist crystallization and have good ambient durability. Their ambient reliability is guaranteed as tellurate glasses owing to their high content of TeO_2. As shown in Table 4, their thermal stabilities are the highest among their glass family, and thus they have the excellent resistance to thermal shocks and crystallization. Therefore they can be used for signal transmission with the lowest scattering. Such explanations are true for TVAg20 and 30 in TVAgx system, which means the critical silver oxide content is at $x = 20$–30 mol%. The critical behavior could be observed in all reported data in Tables 3 and 4. Therefore, for TVAg20–30, the lowest band gap and V_O^*, the highest OPD, refractive index, and thermal stability, make it the preferred agent for optical device engineering. The reported TVAg20–30 showed the highest antibacterial/bactericidal activity, as will be discussed in the next sub-section. For comparison of TVAgx glasses with the reported tellurite oxide glasses, the energy band gap of 30SiO_2-50Bi_2O_3-15B_2O_3-5TeO_2 (2.545 eV) [119], 81TeO_2-19P_2O_5-5La_2O_3 (3.03 eV) [120] and 80TeO_2-5TiO_2-10WO_3-5Nb_2O_5 (4.17 eV) [121] can be followed. These mentioned systems have energy band gap much higher than the values devoted to TVAgx glasses, implying a higher third-order non-linear optical susceptibility of TVAgx, indicating their potential in optical applications. Among the reviewed glassy systems, TeO_2-Bi_2O_3-ZnO core-clad system (72.5TeO_2-10Bi_2O_3-17.5ZnO as core, and 70TeO_2-10Bi_2O_3-20ZnO as clad) was reported for its physical and technological benefits [32]. Its core and clad had the optical band gap of $\approx$1.25 and 1.27 eV respectively. The thermal stability of its core was reported to be 178, which excellently passes the criteria of >90 or 100. Therefore, such very low core-optical band gap (accompanied with the sufficiently high refractive index and thermal stability), ensures its high-performance core-clad system as optical fiber (More details can be followed in Ref. [32]). In brief, tellurate glasses have high technological promises. As another application of glasses and glass–ceramics, one can mention their excellent capabilities in photocatalytic agents such as dual-crystalline glass–ceramics containing NASICON-type $KTi_2(PO_4)_3$ and anatase-type TiO_2 [122] and barium-borate glasses including lanthanum oxide (La_2O_3) [123]. It is a worthy note to refer to shielding properties of some materials against the harmful radiations (i.e. gamma rays) [124–130]; lithium telluro-borate [124], Li_2O-K_2O-B_2O_3-TeO_2 [125], Cao-K_2O-Na_2O-P_2O_5 [127], B_2O_3-Bi_2O_3-SrF_2-Na_2F [128] and borotellurate [130] glasses are some materials introduced for γ-ray shielding.

In Tables 3 and 4, the band gap evaluation approach, luminescence region, and wavelength are also reported, respectively. It should be mentioned that in the BaTe system, 1Eu9BaTe and 1Dy9BaTe samples have shown the photoluminescence in the visible region within the wavelength range 571–702 nm [30]. Also, green emission (i.e. 538–552 nm) has been reported for, $\{[(TeO_2)_{0.70}(B_2O_3)_{0.30}]_{0.7}(ZnO)_{0.3}\}_{0.95}(Er_2O_3)_{0.05}$ [29]. Such reported luminescent glasses can be used in optoelectronic and bio-imaging applications.

In the case of GCs, there are few reports on the reviewed Te-based systems in Tables 3 and 4. One of the relatively recent works on GCs is that which deals with the TVAgx system [27]. In the follow-up, such a system has been briefly performed to give more insight into its structural and optical [27] properties, and in the next sub-section for biological [7] properties in comparison with its glass counterpart. Growth of silver NPs in the tellurovanadate glassy background, opto-structural investigation and comparison of the obtained results with the previously reported data for corresponding glasses are followed [27]. Figures 7 and 8 present the UV-Vis absorption spectra of the TVAg0–40 glasses and the corresponding GCs nominated as ATVAg0–40 GCs.

Depending on Eq. (8), DASF plots for ATVAgx GCs are presented in Fig. 9.

The comparative data of the mentioned Gs and GCs are summarized in Table 5; so, it is obvious that the energy gap in GCs show major blue shift which arises from the decrease in NBOs, related to the growth of nano-crystals [4, 5, 27]. The critical behavior of the gap for TVAg20–30 is repeated for their correspond GCs, inferred as the result of their different structures.

The capability of TVAgx glasses and especially for ATVAgx GCs as antibacterial and bactericidal potential, which is the main scope of this chapter are presented in Sect. 3.1.2.

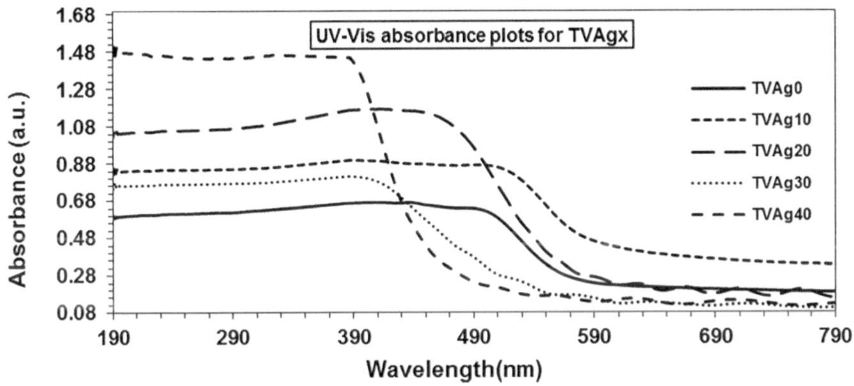

Fig. 7 Plot of the optical absorbance versus wavelength for TVAgx glasses [4]; non-sharp absorption edges refer to the amorphous nature

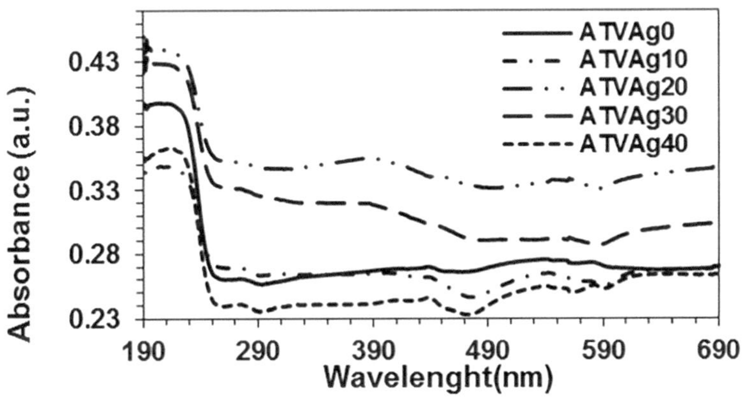

Fig. 8 Optical absorbance of ATVAgx GCs fabricated by heat treatment of their corresponding glasses [27]. The large blue shift of the absorption edge compared to the corresponding glasses in Fig. 7 is apparent

3.1.2 Antibacterial/Bactericidal Activity

Since the main objective of the chapter is antibacterial activity of the selected material, the antimicrobial activity of glasses and GCs is reviewed in this subsection. It would be noted that the nano-scaled agents and NPs can increase their antibacterial or even bactericidal potential. Before any discussion, it should be mentioned that there are several published works on the importance and probable mechanism of antibacterial activity [131–133]. The antibacterial mechanisms of NPs are not well understood and need more and more investigations. However, some of the mechanisms currently thought involve induction of oxidative stress, release of metal ion, and non-oxidizing agents [132]. It is worth-mentioning that most bacteria are categorized into two

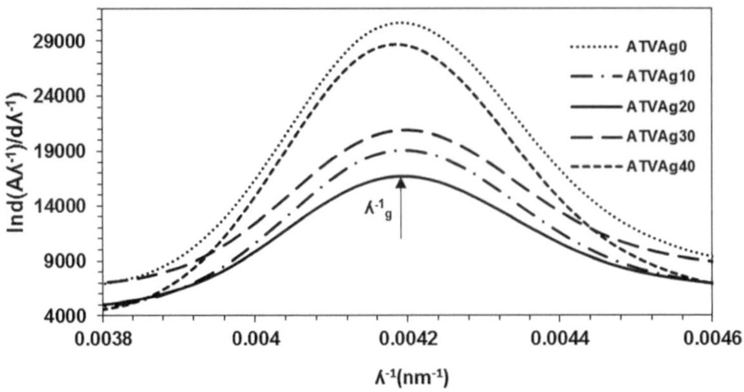

Fig. 9 The DASF plots for ATVAgx GCs, where the discontinuity is indicated by an arrow [27]

categories as Gram (G)-positive and G-negative based on their cell wall structure. The first one contains a thick layer of peptidoglycan in the cell walls, and the second one has a thin peptidoglycan layer with an additional outer membrane consisting of lipopolysaccharide [133–135]. In brief, the multiple and real-time mechanisms of antimicrobial activity is probable and may involve multiple/simultaneous gene mutations in the same bacterial cell [132].

There are few published works on the antibacterial feature of TeO_2-based glasses and GCs. Therefore some works in this field are discussed herein. In the case of investigations related to the materials focused here, one of the published works is that of Elkhoshkhany et al. entitled "Preparation and study of optical, thermal, and antibacterial properties of vanadate-tellurite glass" [24]. In the mentioned work, a part of the work dealt with the antibacterial properties of binary tellurate-vanadate glass using the disc diffusion approach [24]. Only the antibacterial activity of such glass was reported qualitatively and no data on the bactericidal potential was reported. In our recent work (Shahmoradi et al. [7]) on the antibacterial/bactericidal and structural properties of glass–ceramic nanoparticles and glasses in the Ag_2O-TeO_2-V_2O_5 system an extensive work was reported. Figure 10 shows the XRD patterns of TVAgx glasses (a) [4, 6, 13], and their corresponding ATVAgx GCs (b) [7, 27], which confirm the formation of NPs within the glassy matrix.

More extended information about the XRD outputs and their peak analysis by Scherrer's [35, 54] and Williamson–Hall (W–H) [27] methods are tabulated in Tables 6 and 7 for typical ATVAg20 and ATVAg40 GCs, respectively. Tables 6 and 7 involve detected nano-crystalline phases and their crystallographic information (diffraction angles Θ, Bravais lattice type, magnitude of primitive vectors a, b and c and related angles α, β and γ, miller indices (hkl), full width at half maximum (FWHM) of each Bragg's peak, interplanar spacing d_{hkl}, the size of NCs from Scherrer's (D_S) and W–H ($D_{W\text{–}H}$), and related lattice strain ε_S, and $\varepsilon_{W\text{–}H}$).

Also, for better comparison of morphology and structural features of each glass and its corresponding GC, Figs. 11 and 12 represent the microscopic images and

Table 5 Comparative data of TVAgx glasses and their corresponding ATVAgx GCs: absorption edge, band gap, the power (m), and refractive index (n), taken from Refs. [4, 27]

Sample	TVAg0	TVAg10	TVAg20	TVAg30	TVAg40	ATVAg0	ATVAg10	ATVAg20	ATVAg30	ATVAg40
λ_g^{-1} (nm^{-1})	0.00186	0.00177	0.00184	0.00185	0.00236	0.0041838	0.00419	0.00418	0.0041951	0.00417
E_g^{DASF} (eV)	2.35	2.30	2.19	2.30	2.72	5.187	5.204	5.189	5.201	5.179
m	0.623	0.510	0.483	0.869	0.551	0.117	0.083	0.089	0.119	0.126
n	2.598	2.617	2.658	2.617	2.477	1.710	1.708	1.710	1.709	1.711

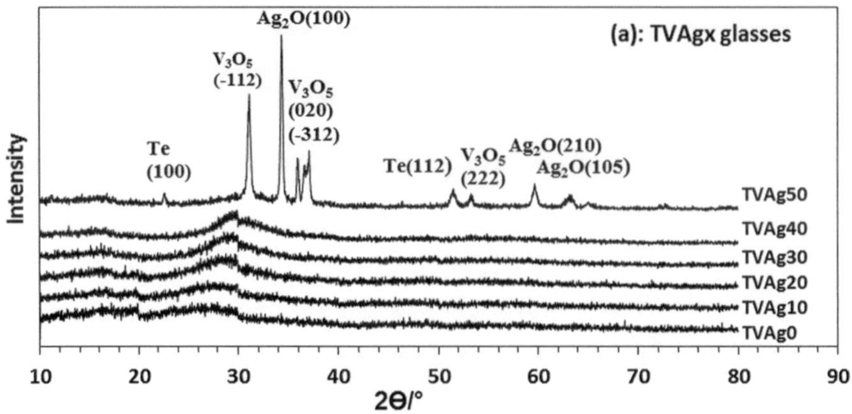

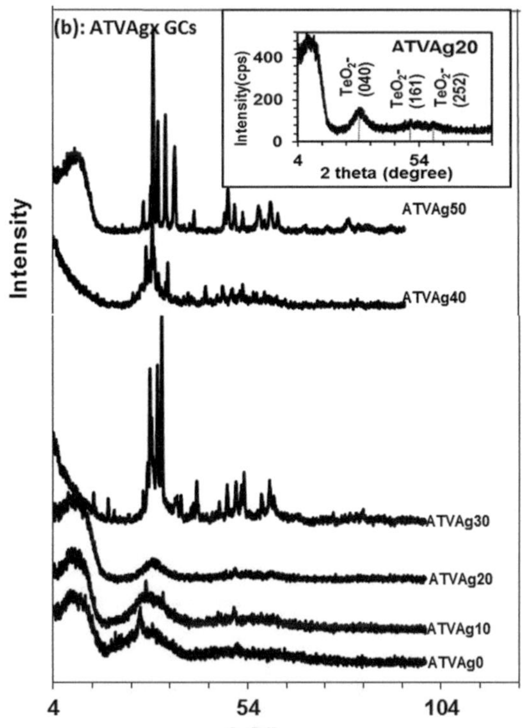

Fig. 10 XRD patterns of **a** TVAgx glasses [4], and **b** ATVAgx GCs [7] (XRD pattern of ATVAg20 is presented as an inset, as guide for the eye)

Table 6 Extended crystallographic information of ATVAg20 GCs

ATVAg20

Related nano crystal phase	Θ (°)	Bravais lattice	a (Å)	b (Å)	c (Å)	α (°)	β (°)	γ (°)	(hkl)	FWHM (°)	d_{hkl} (Å)	NCs D_S (nm) [7, 27]	Average ε_S [7, 27]	D_{W-H} (nm) [27]	ε_{W-H} [27]
TeO_2	14.83	Orthorhombic	5.60	12.04	5.46	90	90	90	(040)	0.31	3.010	26.227	0.013900	47.280	0.00183
	25.56		5.60	12.04	5.46	90	90	90	(161)	0.57	1.785	15.198			
	30.5		5.60	12.04	5.46	90	90	90	(252)	0.49	1.518	18.503			

Table 7 Extended crystallographic information of ATVAg40 GCs

ATVAg40

Related nano crystal phase	Θ (°)	Bravais lattice	a (Å)	b (Å)	c (Å)	α (°)	β (°)	γ (°)	(hkl)
$AgVTeO_5$	14.68	Monoclinic	5.87	11.40	6.81	90	102.00	90	(−131)
	17.62		5.87	11.40	6.81	90	102.00	90	(032)
Ag_3O_4	15.09		3.57	9.20	5.67	90	106.13	90	(−121)
	16.37		3.57	9.20	5.67	90	106.13	90	(121)
	22.77		3.57	9.20	5.67	90	106.13	90	(−213)
	25.10		3.57	9.20	5.67	90	106.13	90	(132)
	27.87		3.57	9.20	5.67	90	106.13	90	(303)
	29.33		3.57	9.20	5.67	90	106.13	90	(−332)
Ag_2TeO_4	15.74	Orthorhombic	10.00	6.94	5.55	90	90	90	(220)
Ag_2O	14.1	Cubic	4.72	4.72	4.72	90	90	90	(110)
	32.36		4.72	4.72	4.72	90	90	90	(311)
AgTe	26.34		6.01	6.01	6.01	90	90	90	(222)

ATVAg40

Related nano crystal phase	FWHM (°)	d_{hkl}—orthorhombic (Å)	d_{hkl}—monoclinic (A°)	d_{hkl}—cubic (Å)	NCs size (nm)—Scherrers [7, 27]	Average ε—Scherrers [7, 27]	$D_{W\text{-}H}$ (nm) [27]	$\varepsilon_{W\text{-}H}$ [27]
$AgVTeO_5$	0.28	–	2.889	–	29.045	0.003913	26.880	0.00083
	0.27	–	2.492	–	30.544			
Ag_3O_4	0.196	–	2.690	–	41.534	0.003141		
	0.23	–	2.249	–	40.831			
	0.31	–	2.492	–	35.617			

(continued)

Table 7 (continued)

ATVAg40								
Related nano crystal phase	FWHM (°)	d_{hkl}—orthorhombic (Å)	d_{hkl}—monoclinic (A°)	d_{hkl}—cubic (Å)	NCs size (nm)—Scherrers [7, 27]	Average ε—Scherrers [7, 27]	D_{W-H} (nm) [27]	ε_{W-H} [27]
	0.23	–	1.447	–	30.544			
	0.31	–	1.603	–	27.148			
	0.50	–	0.865	–	36.779			
Ag_2TeO_4	0.20	2.85084	–	–	40.831	0.003095		
Ag_2O	1.00	–	–	3.337	8.124	0.010987		
	0.70	–	–	1.423	13.294			
AgTe	0.47	–	–	1.734	18.582	0.004157		

elemental analysis of TVAg20 glasses and ATVAg20 GCs. As is evident from Fig. 11, amorphous nature as observed in the FESEM image, stoichiometry, and elemental analysis of TVAg20 glasses are validated. Also, FESEM and TEM images in Fig. 12, confirm the formation of NPs within the glassy matrix of ATVAg20 GCs. Generally, for TVAgx samples, the sizes of NPs were in the range (11–30) nm in agreement with TEM outputs [7, 27].

In the case of biological aspects, Staphylococcus aureus (PTCC 1431) and Bacillus cereus (PTCC 1015) and also two G-negative bacteria of Klebsiella pneumoniae (PTCC 1053) and Escherichia coli (PTCC 1393) were chosen as test strains for the detection of antibacterial and bactericidal activity of TVAgx glasses and their corresponding ATVAgx GCs. To the best understanding, based on the mentioned explanations about the serial dilution method, the results of Elisa red outputs and 96-well plate antibacterial tests of TVAg30 (3 top rows of Table 8) and ATVAg30 (3 bottom rows of Table 8), against G-positive Staphylococcus aureus are presented in Table 8 and Fig. 13, respectively [7]. All the MIC and MBC data of different glasses and GCs are tabulated in Table 9 [7]. The above mentioned figure and table indicate that ATVAg30 sample has the highest (MIC) in well number 5 and TVAg30 has the highest (MIC) in well number 7 against Gram-positive Staphylococcus aureus bacteria.

To investigate the MBC (killing the whole bacteria beside the inhibition of bacterial growth) of each sample, all inhibited wells were tested for MBC. So, in accordance with the complete results of MIC (listed in Table 9), the MBC test was done, in which for ATVAg30 GCs, the results are shown typically in Fig. 14 with Staphylococcus aureus. As can be seen in Fig. 14, no bacteria can grow in the liquid taken from well number 1, which means that the MBC-well number is 1 (liquids taken from MIC well numbers 2–6 show the growth of remained/not-killed bacteria).

As can be understood from Table 9 as the summary of all results of the antibacterial and bactericidal potential, the highest antibacterial potential is related to ATVAg30 (the minimum concentration of the agent). In the cases of Bacillus cereus and Klebsiella pneumonia bacteria, the bacterial inhibition effect is almost the same for glasses and GCs. Furthermore, all agents are effective against G-positive and G-negative microorganisms, indicating that the studied GCs have more antibacterial effect in comparison to the corresponding glasses. Such result can be attributed to the production of nanoparticles having the higher penetration probability to the bacteria cell membrane due to their small particles size. In brief, vanadium-tellurium glasses and GCs containing silver oxide can be used as new and promising agents in bio-optical applications, biosensors, and bio-imaging.

3.2 Bare and Doped Zinc Selenide Nanoparticles

Based on the characteristics mentioned in Sect. 1 about the features and importance of ZnSe NPs in different doped, co-doped and core–shell structures, their opto-structural and antibacterial properties are reviewed in the following sub-sections.

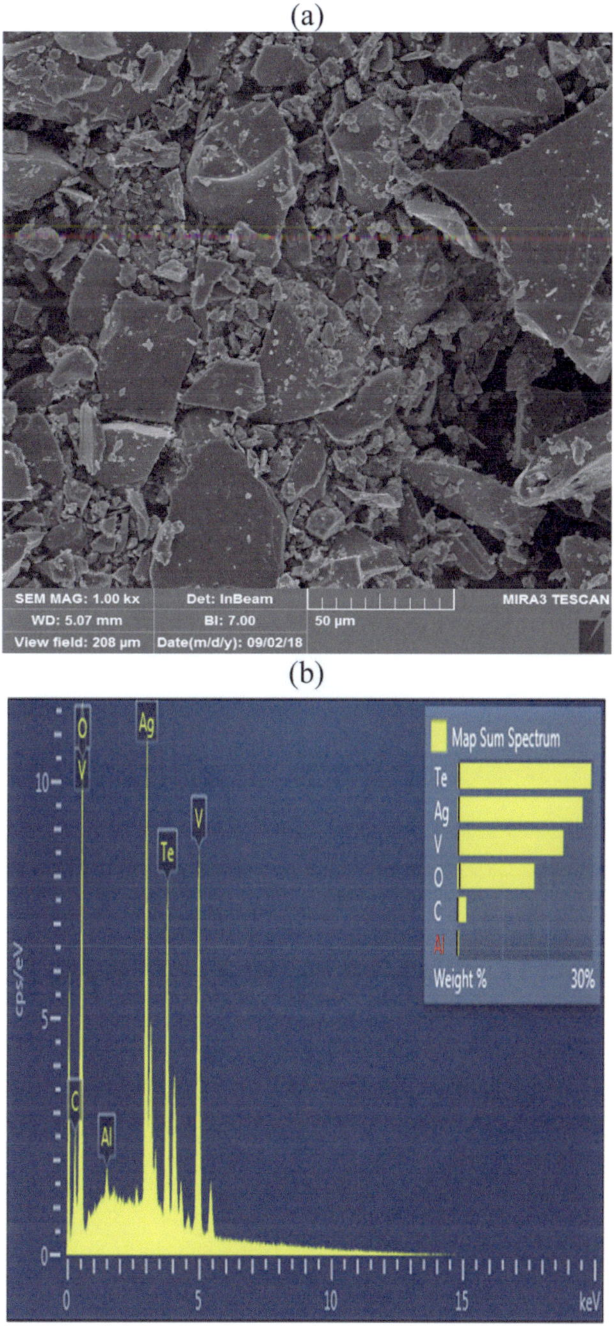

Fig. 11 TVAg20 glasses: **a** FESEM image; **b** its EDX spectrum; **c** elemental map of the detected elements in EDX [27]

(c)

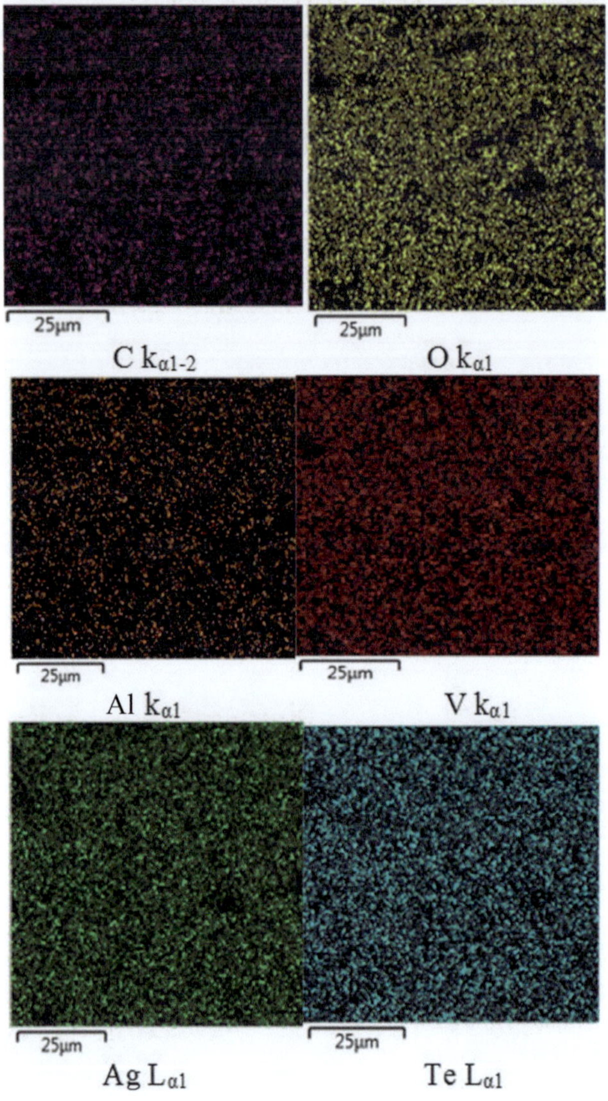

Fig. 11 (continued)

3.2.1 Opto-structural Properties

There are several published works on the different properties of Zn-containing nanoparticles (i.e. ZnO NPs [34, 76]) and especially ZnSe NPs [40, 41, 51–62, 88]. Many have been found in the literatures, such as bare ZnSe [52], Cu-doped ZnSe [53], Sr-doped ZnSe [75], Ni-doped ZnSe [40, 88], Cu-Mn co-doped ZnSe NPs [59,

Fig. 12 ATVAg20 GCs: **a** FESEM image; **b** its EDX spectrum; **c** elemental map of the detected elements in EDX [27]; **d** TEM image

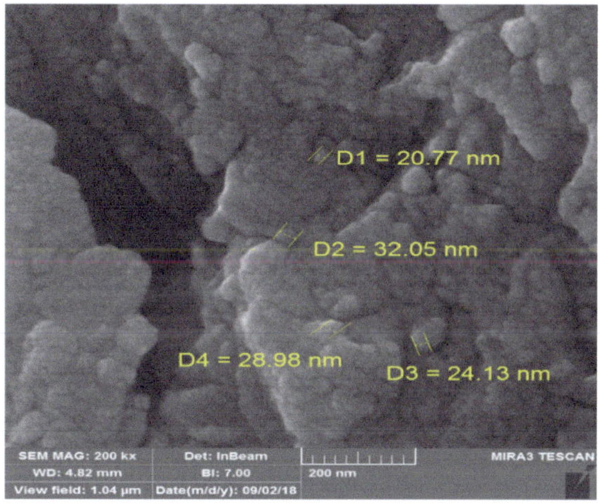

(a)

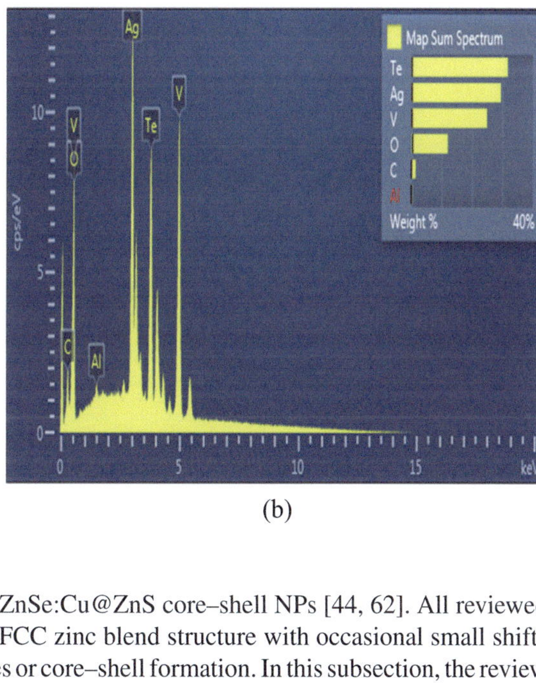

(b)

60] and also ZnSe@ZnS and ZnSe:Cu@ZnS core–shell NPs [44, 62]. All reviewed structures have relatively the FCC zinc blend structure with occasional small shifts related to introduced impurities or core–shell formation. In this subsection, the review is generally focused on structures fabricated by the assistance of MWIR. Figure 15 shows the XRD patterns of ZnSe, ZnSe:Cu 0.1%, ZnSe:Cu 0.1%-Mn 0.1%, ZnSe:Cu 0.1%-Mn 1%, ZnSe:Cu 1.5%-Mn 2% and ZnSe:Cu 0.1%-Mn 4% (synthesized by MWIR for 6 min at fixed pH of 11.2) labeled as Spure, SMn0, SMn1, SMn2, SMN3, SMn4 and SMn5, respectively. The XRD patterns confirm the formation of NCs of size in the range (1.5–3) nm with the FCC zinc blend structure (i.e. main diffraction

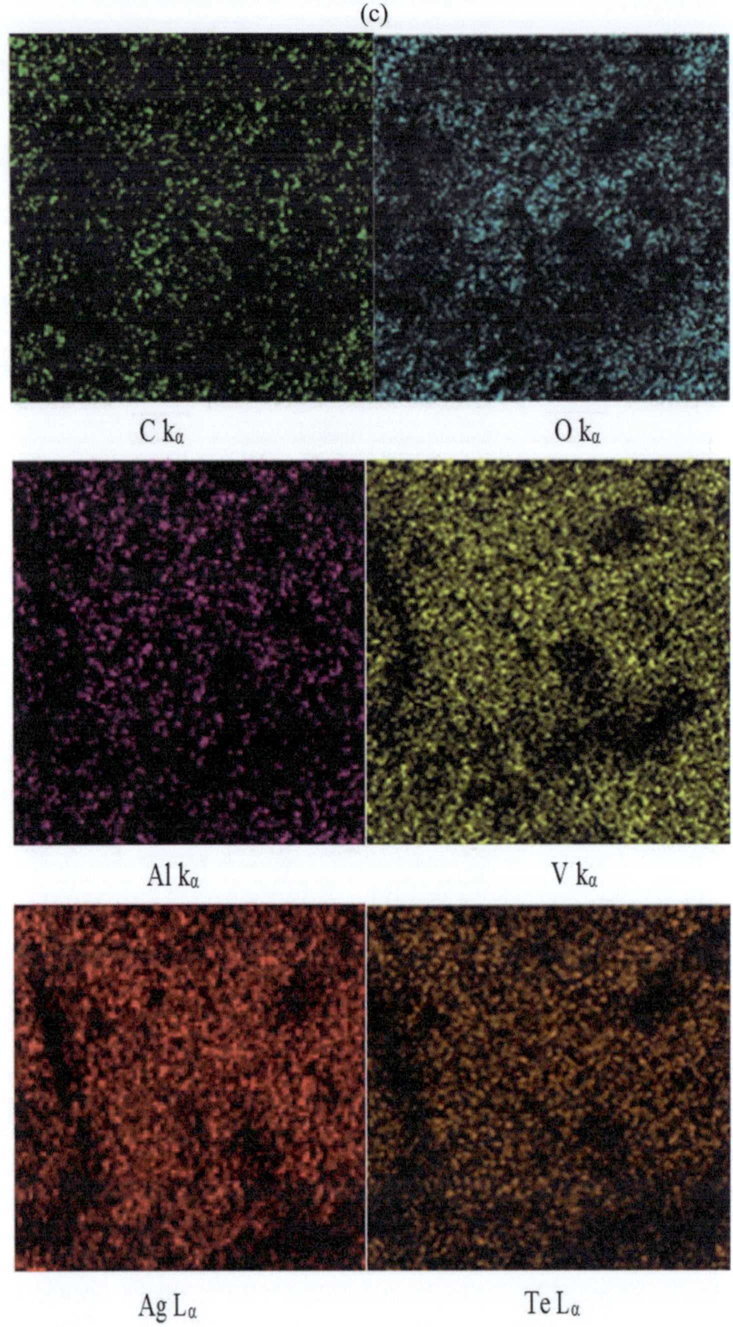

Fig. 12 (continued)

Fig. 12 (continued)

(d)

Table 8 Elisa-read OD-data of TVAg30 glasses (three top rows of the Table: means 3 repetitions) and ATVAg30 GCs (three bottom rows) against Staphylococcus aureus G-positive bacteria (see Fig. 13 corresponding to this table)

Test sample	Well No.											
	1	2	3	4	5	6	7	8	9	10	11	12
A	0.171	0.082	0.065	0.067	0.059	0.07	0.105	**0.405**	**0.376**	**0.388**	**0.297**	**0.702**
B	0.105	0.07	0.066	0.062	0.063	0.158	0.338	**0.365**	**0.324**	**0.356**	**0.245**	**0.428**
C	0.086	0.071	0.067	0.062	0.08	0.239	0.392	**0.347**	**0.376**	**0.378**	**0.251**	**0.441**
D												
E												
F	0.092	0.068	0.069	0.073	0.105	**0.414**	**0.471**	**0.374**	**0.343**	**0.377**	**0.343**	**0.465**
G	0.1	0.067	0.065	0.069	0.082	**0.221**	**0.378**	**0.351**	**0.346**	**0.372**	**0.219**	**0.484**
H	0.104	0.069	0.067	0.094	0.111	**0.395**	**0.464**	**0.348**	**0.386**	**0.388**	**0.271**	**0.664**

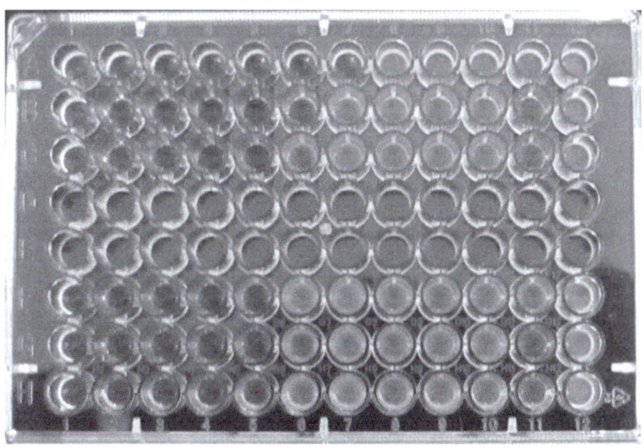

Fig. 13 96-well plate image of serial dilution for TVAg30 (three top rows of the Table: mean 3 repetition) and ATVAg30 (three bottom rows of the Table) against Staphylococcus aureus G-positive bacteria in correspondence with the data of Table 8 [7]

Table 9 Antibacterial/bactericidal activity of TVAgx Gs and ATVAgx GCs in detail: Concentration of tested antibacterial agents, MIC and MBC for different Escherichia coli, Klebsiella pneumonia, Staphylococcus aureus, and Bacillus cereus bacteria [7]

Sample	Ag_2O (mol%)	$D_{S\text{-}H}$ (nm)	Initial concentration (mg/ml)	Escherichia coli				Bacillus cereus			
				MIC		MBC		MIC		MBC	
				Well NO	Concentration (mg/ml)	Well NO	Concentration (mg/ml)	Well NO	Concentration (mg/ml)	Well NO	Concentration (mg/ml)
TVAgx	0	–	40.96	6	1.28	1	40.96	2	20.48	0	0
	10	–		1	40.96	0	0	0	0	0	0
	20	–		3	10.24	1	40.96	2	20.48	0	0
	30	–		3	10.24	1	40.96	2	20.48	0	0
	40	–		4	5.12	1	40.96	3	10.24	2	20.48
	50	–		2	20.48	1	40.96	1	40.96	0	0
ATVAgx	0	11.72	43.92	0	0	0	0	0	0	0	0
	10	24.69		2	21.96	1	43.92	2	21.96	0	0
	20	19.97		2	21.96	0	0	2	21.96	0	0
	30	25.53		4	5.49	1	43.92	2	21.96	0	0
	40	29.4		2	10.98	0	0	2	21.96	0	0
	50	24.74		3	10.98	1	43.92	2	21.96	0	0

Sample	Ag_2O (mol%)	Klebsiella pneumonia				Staphylococcus aureus			
		MIC		MBC		MIC		MBC	
		Well No.	Concentration (mg/ml)	Well No.	Concentration (mg/ml)	Well No.	Concentration (mg/ml)	Well No.	Concentration (mg/ml)
TVAgx	0	4	5.12	1	40.96	3	10.24	2	20.48
	10	1	40.96	0	0	1	40.96	1	40.96

(continued)

Table 9 (continued)

Sample	Ag_2O (mol%)	Klebsiella pneumonia				Staphylococcus aureus			
		MIC		MBC		MIC		MBC	
		Well No.	Concentration (mg/ml)	Well No.	Concentration (mg/ml)	Well No.	Concentration (mg/ml)	Well No.	Concentration (mg/ml)
	20	2	20.48	1	40.96	3	10.24	1	40.96
	30	3	10.24	2	20.48	7	0.64	2	20.48
	40	2	20.48	1	40.96	6	1.28	1	40.96
	50	2	20.48	0	0	5	2.56	1	40.96
ATVAgx	0	6	1.372	4	5.49	0	0	0	0
	10	1	43.92	1	43.92	4	5.49	2	21.96
	20	1	43.92	1	43.92	2	21.96	1	43.92
	30	2	21.96	1	43.92	5	2.745	1	43.92
	40	2	21.96	1	43.92	3	10.98	1	43.92
	50	2	21.96	2	21.96	8	0.686	2	21.96

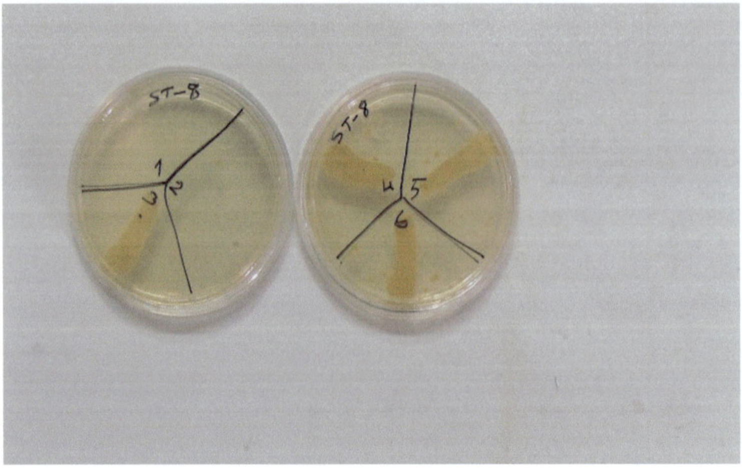

Fig. 14 Representation of MBC test on the solid gel MHB-agar of inhibited wells of Fig. 13 for ATVAg30 GCs. All MIC wells (1–6) have been sampled for MBC test [7]

peaks assigned to Miller indices (111), (220), and (311) at scattering angle $2\,\theta \cong 27°$, $45°$ and $53°$, with lattice constant $a = 5.66882$ Å).

Various works have reported the effect of pH [40, 41, 53], MWIR time [41, 53] and dopant content [41, 52] on the structural, optical, and luminescence properties of ZnSe NPs. In the case demonstrated in Fig. 15, the interplanar spacing (d_{111}) was reported to be about 3.272 Å and 3.267 Å for ZnSe pure and SMn0 samples, respectively. The substitution of Zn^{2+} ions ($r_{Zn^{2+}} = 0.74$ Å) [88] with Cu^{2+} ions with slightly smaller radius of $r_{Cu^{2+}} = 0.73$ Å [136] resulted in a slight decrease (≈0.005 Å) in lattice constant of bare ZnSe. On the other hand, the interplanar distance of SMN1 and SMn2 (where $r_{Mn^{2+}} = 0.66$ Å) was reported as $d_{111} \approx 3.272$ Å similar to that of the bare ZnSe. This result may refer to the incorporation of the dopant ions (Cu and Mn) into the host matrix. The interplanar distance of the co-doped SMn5 sample was reported as $d_{111} = 3.326$ Å showing a clear increase with respect to other samples. This result may be due to the incorporation of the dopant ion (co-doped) into the host matrix and not interstitials, which can be confirmed by PL spectra.

Another certification of the formation of ZnSe structure is Fourier Transform Infrared spectroscopy (FTIR). The FTIR spectra of the samples mentioned in Fig. 15 are presented in Fig. 16. The IR bands are shown in the figure are explained as follows: The bands in the range (3000–3600) cm^{-1}, and the strong band at the 1558 cm^{1} are related to the hydroxyl group (–OH) stretching and bending vibrational modes respectively. The stretching vibrational mode of –NH is seen at 3112 cm^{-1}, and the bands at 2908 cm^{-1} and 895 cm^{-1} are respectively related to the asymmetric stretching vibrations and bending vibrations of CH_2 group. The bands at 1224 and 1159 cm^{-1} are related to C–O stretching vibrational modes, and the band at 930 cm^{-1} corresponds to the stretching vibrations of C=C, while the band at 1697 cm^{-1} is related to C=O stretching modes [137, 138]. The bands at 565 cm^{-1} and (702 and

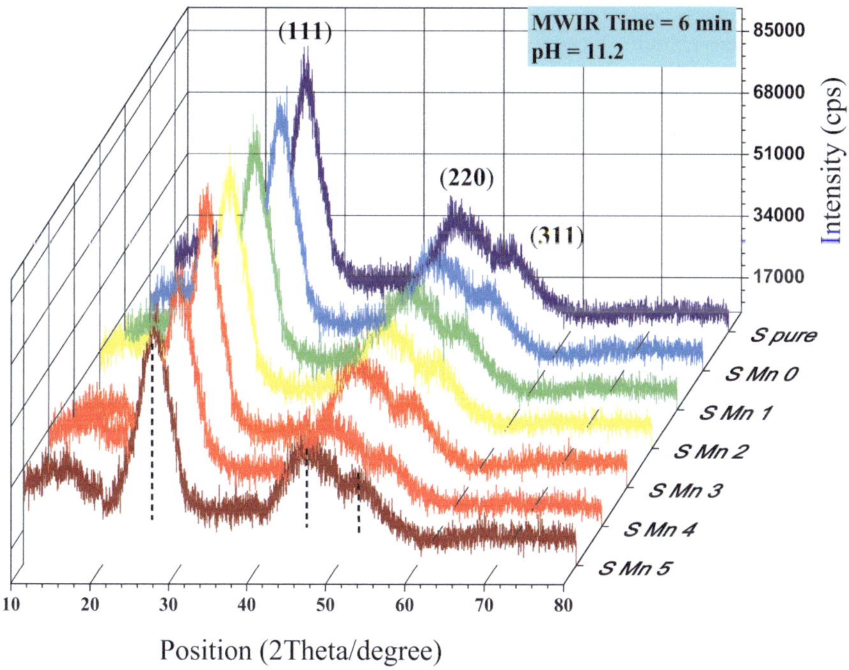

Fig. 15 XRD patterns of ZnSe (Spure) and ZnSe:Cu(0.1)Mn(0, 0.1, 1, 1.5, 2 and 4%) (SMn0–5) NPs prepared by MWIR for 6 min at fixed pH = 11.2 [59]

776 cm^{-1}) are assigned to the stretching mode and bending vibrational modes of Zn-Se bands respectively [139–141], and finally, the bands at 440 and 508 cm^{-1} correspond to ZnSe-Mn and Zn-Se-Cu.

For the different works mentioned at the beginning of this section, it was found that depending on the different synthesis conditions i.e. pH, MWIR time (cause of NPs growth), and dopant content, the optical band gap of the produced NPs ranges from 3.5 to 3.7 eV. The increase in pH up to 11.2 has the role of increasing NCs size, hence decreasing the band gap obtained by DASF. However, for pH = 12.2, trend was opposite [53]. The increase in MWIR time caused a reasonable increase in the size of NCs, and so the band gap obtained was decreased using DASF [39, 53]. The optimal pH (i.e. higher crystalline quality) was found to be 11.2, since this sample had the least lattice strain and dislocation density, and also the least Urbach energy [41, 54, 55].

In the case of Cu-doped ZnSe NCs, based on the trends of optical band gap [52, 59, 60], Urbach energy [52], lattice strain, and dislocation density [41, 44, 54, 55], the optimal sample was found to be ZnSe:Cu 0.1%. Also, for the fabrication of ZnSe:Cu-Mn co-doped NPs [59, 60], and ZnSeCu@ZnS core-shells [44], the optimal Cu content, Mn content and pH were selected as 0.1%, 1 and 11.2, respectively. Such samples have higher optical susceptibilities, which makes them good candidates for

Fig. 16 FTIR spectra of: **a** Spure, **b** SMn0, **c** SMn1, **d** SMn2, **e** SMn3, **f** SMn4, and **g** SMn5 nanoparticles (where MWIR time is fixed at 0 min and pH is fixed at 11.2) [59]

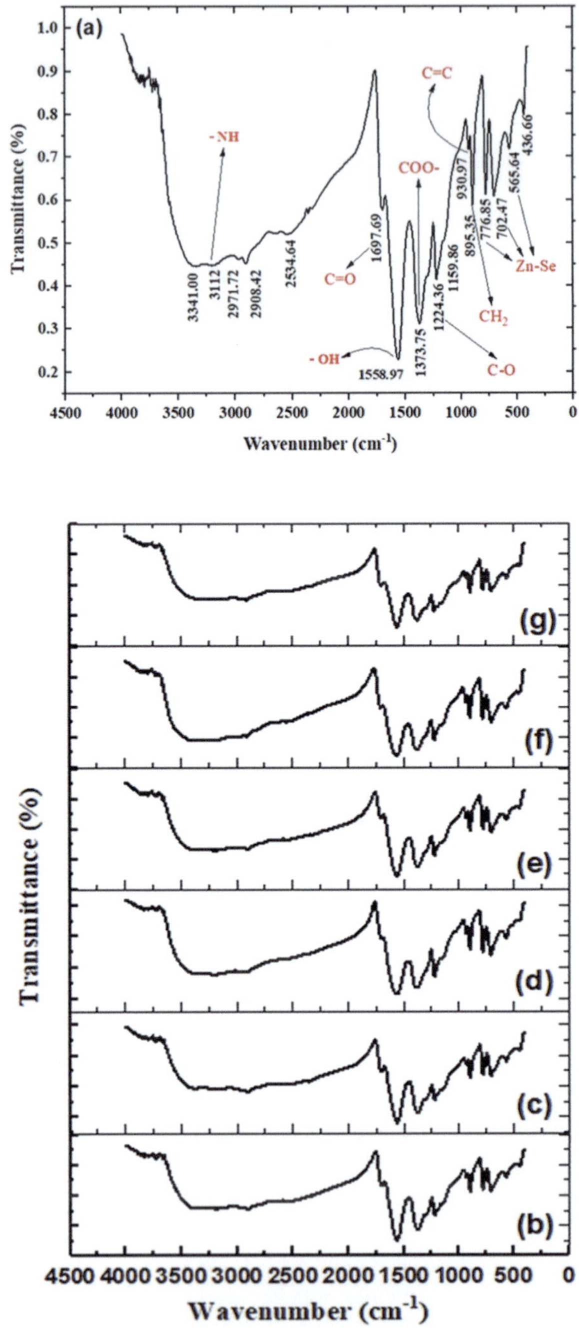

signal transformation. Figure 17, shows the TEM image and EDX spectrum of SMn1 sample (ZnSe:Cu-Mn NPs) confirming the rounded shape [60].

Several works reported the photoluminescence (PL) spectra of nanoparticles of interest in this chapter such as; ZnSe [41, 61], ZnSe:Cu [41, 61] and ZnSe:Cu-Mn NPs [42, 59]. Figure 18, shows the PL spectra of ZnSe:Cu NPs (Fig. 18a) [41] and

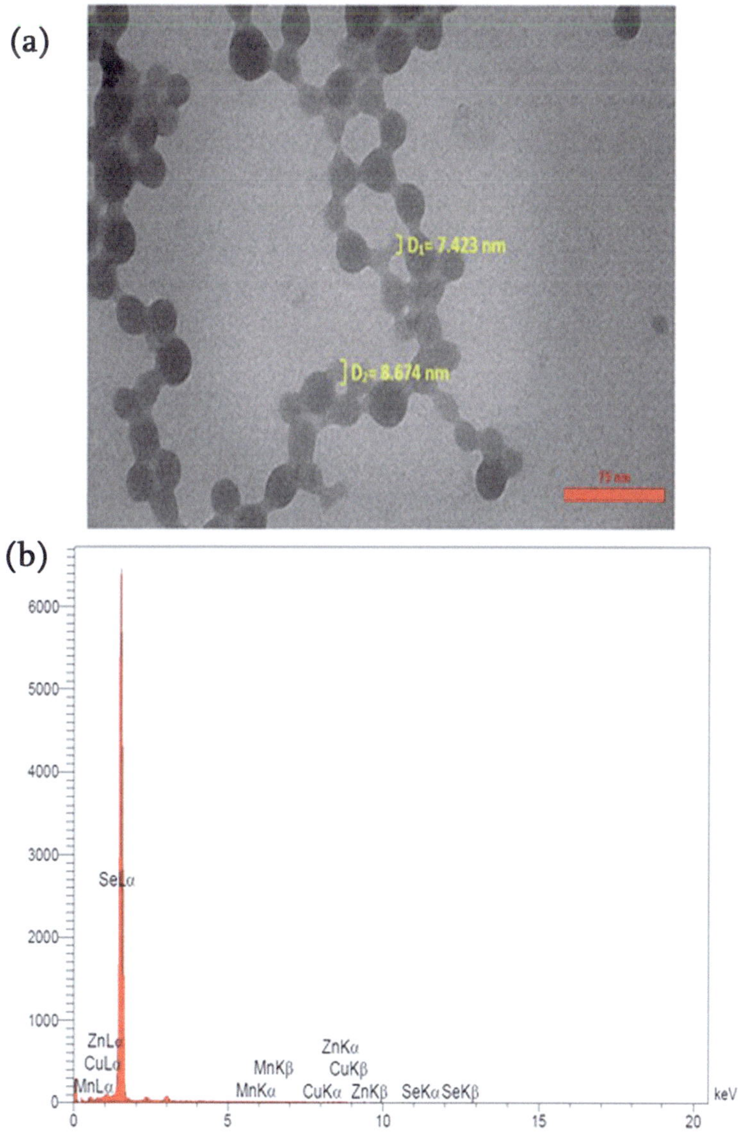

Fig. 17 TEM image (up), and EDX plot (down) of SMn1 NPs (ZnSe:Cu(0.1%)Mn(0.1%)) [60]

ZnSe:Cu-Mn NPs (Fig. 18b) [59] showing their green emission. It may be concluded that these NPs can be used in imaging technology especially in bio-imaging. As depicted in the inset of Fig. 18b, the optimal sample is SMn2 (ZnSe:Cu 0.1%-Mn 1%) [59].

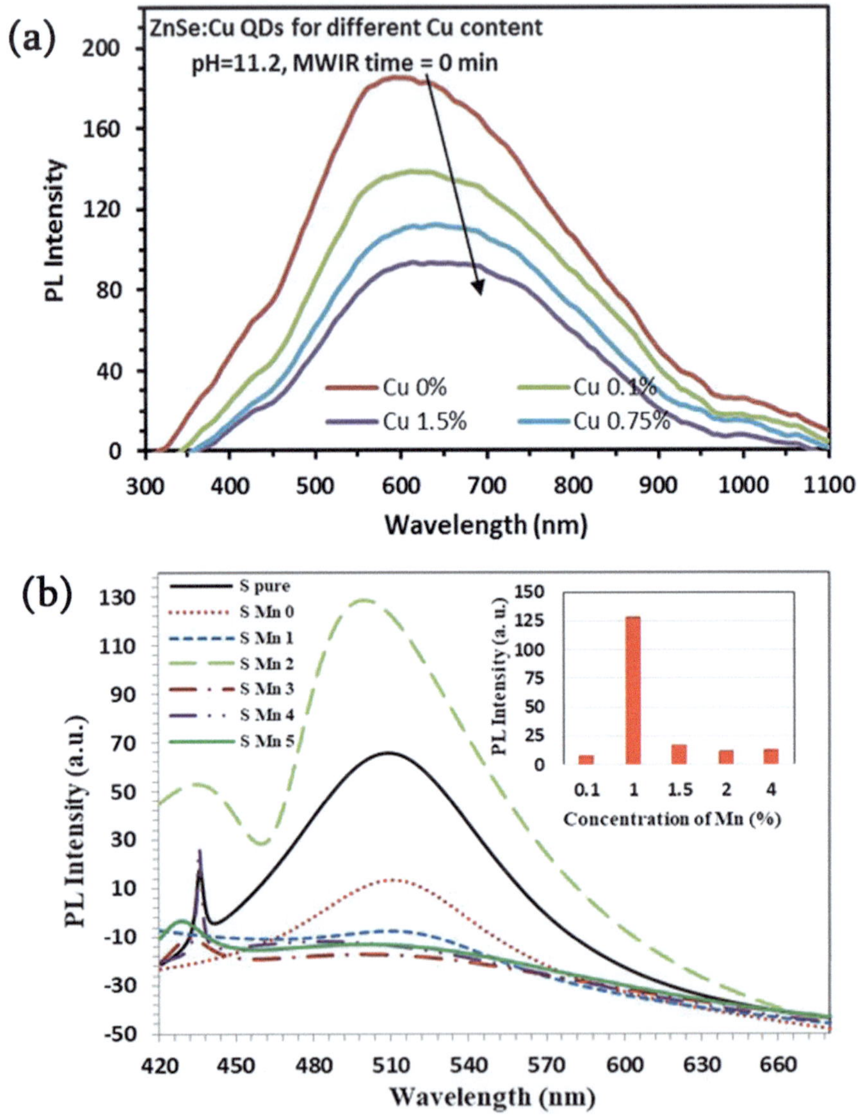

Fig. 18 PL spectra of **a** ZnSe: Cu (= 0, 0.1, 0.75, and 1.5%) NPs [41], and ZnSe:Cu 0.1%-Mn co-doped ZnSe NPs (**b**) [60]

3.2.2 Biological Evaluation: Bacterial Growth Inhibition

In the previous subsections, antibacterial/bactericidal features and their experimental detection approaches were extensively presented. The present subsection aims to review some recent works on the antibacterial or even bactericidal activity of ZnSe NPs. There are several published works on the morphology, biological and antimicrobial potential of Zn-based NPs (e.g. ZnO [34, 76, 142], undoped and doped ZnSe NPs [41, 44, 51–62, 131–133]. ZnSe NPs have good antibacterial/bactericidal ability against the well-known microorganisms of both G-positive's and G-negative's. Doping of ZnSe may affect its biological aspects in terms of modifications in other physical/chemical properties. In one of the works performed by Beena et al. [75], Sr-doped ZnSe NPs, with different molar ratios of Sr^{2+} ions (0.01, 0.05, and 0.1) have been synthesized by co-precipitation method at the presence of $NaBH_4$ and 2-mercaptoethanol as reducing and stabilizing agents respectively. ZnSe:Sr NPs have been examined and evaluated for their successful photocatalytic degradation of methyl orange (MO). Moreover, different Zn- and Ti-based nanostructures have been examined for their photocatalytic activities and applications [143–147]. Also, their antibacterial activity has been investigated against Staphylococcus aureus and Escherichia coli bacteria employing disc diffusion method on the Mueller Hinton Broth (MHB) culture. In another work, Beena et al. [74], produced zinc selenide NPs by co-precipitation approach, and studied their physical, morphological, luminescence and antibacterial features. Disc-diffusion antibacterial activities were investigated using MH-agar (MHA) culture against Bacillus subtilis G-positive and E. coli G-negative strain. The results showed the higher activity against E. coli.

It should be mentioned that the metal NPs such as Ag NPs [133, 148] or even matrices containing Ag ions [7] were reported for their antibacterial and bactericidal features. Also, there are some recent reports on the antibacterial or even bactericidal potential of ZnSe NPs prepared by MW-assisted aqueous method (focused on approach in this chapter). The effect of synthesis conditions such as MWIRT [51, 55, 58], Cu-dopant [55–57], or even pH have been investigated employing different approaches of disc-diffusion [51, 55, 57, 58]. Table 10, tabulates the main information about conditions and some biological aspects of some reviewed works especially on ZnSe NPs.

It is known that under the same conditions, higher antibacterial activity in disc-diffusion assay means a larger inhibition zone diameter. In the case of pure-ZnSe and ZnSe:Cu NPs fabricated by hydrothermal MW-assisted route [51, 56–58] the results showed that the inhibition zone was greater than 6 mm (diameter of paper disc) against E. coli, P. aeruginosa [51, 58] G-negative bacteria and also against S. aureus and Bacillus cereus G-positive bacteria [56, 57]. The results of inhibition zone diameters are presented in Table 10. In the mentioned works, two loading volumes of 140 and 70 μL of solutions containing ZnSe and ZnSe:Cu NPs were selected on agar plates, demonstrating the excellent antibacterial activity of studied NPs. G-negative bacteria in the lower loaded volume of 70 μL appeared to exhibit lower antibacterial activity, which may be due to the potential presence of lower Zn^{2+} ions as a known antimicrobial agent. Another result is the inverse relationship between the

Table 10 Some physical, chemical, and biological information of selected ZnSe and metal NPs. * and ** in columns 7 and 9, represent the concentrations and their corresponding activities respectively

Type of NPs	Synthesis method	Size (nm)	Shape of NPs	Microorganism	Bio-assay	Concentration/ loaded volume	Culture	Activity (MIC, MBC or inhibition diameter)	Refs.
ZnO	Chemical method followed by mechanical and heat treatment	<50 ± 9	Spherical	V. cholerae, E. coli, C. Jejuni, St. aureus (MRSA)	Disc diffusion	0.78; 1.56; 3.125; 6.25; 12.5; 25; 50 mM	LB medium, MHA	N.R.	[149]
ZnSe:Sr	Co-precipitation	21–35	Spherical	E. coli	Disc diffusion	25–100 μL	MHB	0.8–5.1 mm	[75]
				S. aureus				0.5–5 mm	
ZnSe	Co-precipitation	11.8–29.6	Spherical	E. coli	Disc diffusion	25–100 μL	MHA	2–6.5 mm	[74]
				S. subtilis				1–2 mm	
Ag	Coating of Ag NPs	N.R.	Spherical	Bacillus	Serial dilution	N.R.	N.R.	MIC = 600 μg/L	[148]
ZnSe:Ni	Wet chemical precipitation	2–4 nm	Spherical	N.R.	N.R.	N.R.	N.R.	N.R.	[40]
	Solvothermal	≈5 nm							[88]
ZnSe	MW-assisted hydrothermal (MWIRT = 0–6 min)	1.94–2.14	Spherical	E. coli	Disc diffusion	70*, 140 μL**	MHB	*: 6.69–6.80 mm **: 10.43–7.79 mm	[51]
				P. aeruginosa				*: 7.42–10.67 mm **: 10.01–11.58 mm	

(continued)

Table 10 (continued)

Type of NPs	Synthesis method	Size (nm)	Shape of NPs	Microorganism	Bio-assay	Concentration/ loaded volume	Culture	Activity (MIC, MBC or inhibition diameter)	Refs.
ZnSe	MW-assisted hydrothermal (MWIRT = 0–6 min)	1.94–2.14	Spherical	St. aureus	Disc diffusion	140 μL	MHB	9.37–14.34 mm	[58]
				Bacillus cereous				8.96–16.86 mm	
ZnSe:Cu (0–1.5%)	Hydro-microwave (MWIRT = 0–6 min)	2.24–2.27	Spherical	St. aureus	Disc diffusion	70*, 140 μL**	MHB	*: 7.07–11.74 mm **: 9.37–14.34 mm	[56]
				Bacillus cereous				*: 7.10–10.95 mm **: 8.25–18.45 mm	
ZnSe:Cu (0–1.5%)	MW-assisted aqueous (MWIRT = 0, 6 min)	1.59–1.74	Spherical	E. coli	Disc diffusion	140 μL	MHB	10.43–14.78 mm	[57]
				P. aeruginosa				10.01–9.64 mm	
ZnSe:Cu (0–1.5%) at different pH's of 8, 10.2, 11.2, 12.2	MWIR-assisted chemical route (MWIRT = 0–6 min)	1.74–2.18	Spherical	E. coli P. aeroginosa St. aureus Bacillus cereos	Serial dilution	6.172–12.470 ppm	LB	Details are presented in Table 11	[55]

Table 11 ZnSe:Cu NPs produced under synthesis conditions (dopant %, pH, and MWIRT), MIC, and MBC well numbers for different microorganisms (Escherichia coli, Pseudomonas aeruginosa, Staphylococcus aureus and Bacillus cereus) [41, 52, 53, 56]

Type of NPs	Cu-dopant %	Synthesis-pH	MWIRT (min)	Size (nm) obtained by XRD	Concentration of NPs (ppm)	Type of microorganisms							
						Escherichia coli		Pseudomonas aeruginosa		Staphylococcus aureus		Bacillus cereus	
						MIC Well no.	MBC Well no.	MIC Well no.	MBC Well no.	MIC Well no.	MBC Well no.	MIC Well no.	MBC Well no.
Pure ZnSe	0	8.0	0	1.736	12.470	3	3	3	3	3	3	3	3
			6	2.048	10.025	2	2	2	2	2	2	2	2
		10.2	0	1.496	11.206	3	3	3	3	1	1	3	3
			6	1.993	9.072	3	3	2	2	3	3	3	3
		11.2	0	1.942	11.164	3	3	2	2	3	3	3	3
			1	–	10.674	3	3	3	3	3	3	–	–
			2	–	9.973	3	3	3	3	3	3	–	–
			3	–	9.671	3	3	3	3	3	3	–	–
			4	–	9.299	3	3	2	2	3	3	3	3
			6	2.143	8.596	3	3	–	–	3	3	3	3
		12.2	0	1.773	12.387	2	2	2	2	3	3	2	2
			6	2.180	8.404	2	2	–	–	2	2	2	2
ZnSe:Cu	0.1	11.2	0	1.987	8.714	3	3	–	–	3	3	3	3
	0.75			1.878	7.934	3	3	–	–	3	3	–	–
	1.5			1.975	6.172	3	3	–	–	4	4	–	–

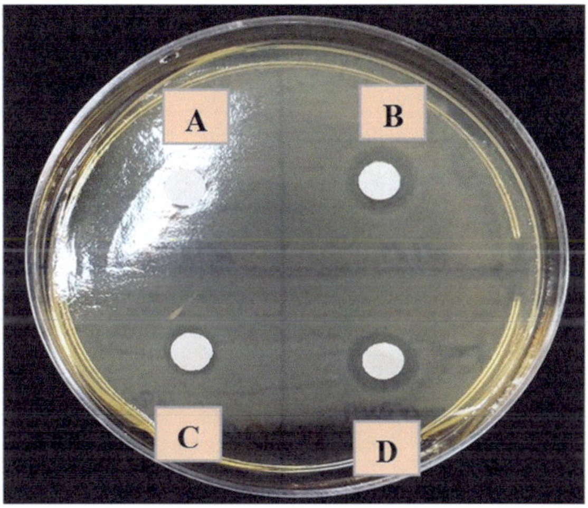

Fig. 19 Depiction of inhibition zone of ZnSe NPs fabricated under different MWIRTs against P. aeruginosa; **A** MWIRT = 1 min; loaded volume = 70 μL; **B** MWIRT = 1 min; Loaded volume = 140 μL; **C** MWIRT = 2 min; loaded volume = 70 μL; **D** MWIRT = 2 min; volume = 140 μL [51]

NPs size and antibacterial activity. The nanoparticles also release ions, which react with the thiol groups of proteins on the bacterial cell membrane. The extension of these proteins across the bacterial cell surface allows the agar to be transported across the cell membrane. As reported by Malarkodi et al. [150], G-negative bacteria have thin and simple multilayered lipid materials in their cell walls. Therefore, suitable antibacterial NPs can somewhat easily penetrate the bacterial cells and then show an inhibition zone. It is worth noting that discrepancies in the sensitivity against different microorganisms may be inferred from their different cell structure, physiology, and coverage of NPs around the bacteria [67, 150]. For further inference about the disc diffusion assay, Fig. 19 shows a typical image of the inhibition zones of ZnSe NCs synthesized under different MWIRTs of 1 and 2 min against P aeruginosa and E. coli.

Figure 20 shows the comparative histogram for antibacterial activity (inhibition zone diameter) of ZnSe NPs against E. coli (each experiment has been repeated three times). This figure obviously shows that the increase in particle size (with increasing MWIRT) results in decreasing the antibacterial sensitivity, which is reasonable. There is a published research [67] on the strong binding of NPs to the outer wall of E. coli, which results in the inhibition of energy transfer and dehydrogenase activity of the enzyme leading to RNA and DNA damages and finally probable bacterial death.

Also, comparing the results mentioned [51] (see Table 10) with the results of G-positive bacteria from S. aureus and Bacillus cereus [58] shows that un-doped ZnSe NPs have higher inhibition against the tested G-negative bacteria. Similar trends for antibacterial activities were reported for Cu-doped ZnSe NPs against the bacteria studied in 4 cases. However, it should be stated that agents containing 0.1 and 0.75% of Cu have higher activity and relatively high $\chi^{(3)}$ values. It has been suggested that the potential electrostatic adsorption of Cu^{2+} impurity ions on the bacterial cell-wall can damage the membrane and change its function. Therefore, such changes can alter the penetrability of NPs to cellular membranes, causing the cell death [151].

Fig. 20 Comparative histogram of the antibacterial potential of ZnSe Nps synthesized under different MWIRTs (times) against E. coli G-negative bacteria; $V_1 = 140$ μL is the loading volume [51]

Considering their optical properties (especially PL) and biological aspects at the same time may indicate that they are agents with vital bio-optical benefits.

The antibacterial/bactericidal potential of ZnSe and ZnSe:Cu NPs prepared under the synthesis conditions of different pH, dopant content, and MWIRTs were reported employing the micro-serial dilution technique, which can provide more comprehensive results besides MIC and MBC [55]. Detailed tabulated results of the mentioned work can be observed in Table 11. In that work (as listed in Table 10), Lysogeny broth (LB) culture was used. Similar to that mentioned in Sect. 3.2.1 for serial dilution testing of TVAgx glasses and ATVAgx GCs [7], 96-well microplates have been used for ZnSe NPs (i.e. Table 11).

As can be seen from Table 11 and Fig. 21, the present NPs exhibited excellent antibacterial/bactericidal activity as promising agents, where the optimal agents were revealed to be those synthesized under the conditions: (pH = 10.2, MWIRT = 0 min), (pH = 12.2, MWIRT = 0 min; and pH = 11.2, Cu dopant percent = 1.5%), and (pH = 11.2, MWIRT = 4 min) [41, 56].

The optical (especially PL) properties and biological aspects of these nanoparticles may indicate that they are agents of bio-optical applications such as use in biosensors. They also have the potential to act as antibacterial/bactericidal materials, photo-catalysts, and especially as agents that can be used in bio-imaging and drug delivery.

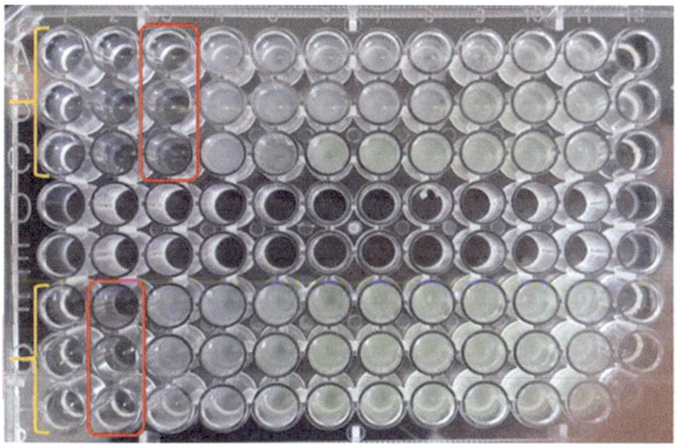

Fig. 21 96-well microplate serial dilution test against Pseudomonas aeruginosa for pure ZnSe NPs synthesized at pH = 8.0, MWITR = 0 min (The 3 top rows of the image mean 3 repetitions) and at pH = 8.0, MWITR = 6 min (The 3 lowest rows of the image mean 3 repetitions); The two mentioned agents show accordingly MIC well numbers 3 and 2 as indicated by red rectangles [55]

4 Conclusions

In this chapter, tellurate transition metal oxide containing glasses (TMOGs) (especially ternary silvered telluro-vanadates), their corresponding GCs, bare, shelled, and doped ZnSe NPs were reviewed due to their special and unique bio-optical properties and benefits. For glasses and GCs, a critical composition was commonly detected, in which for xAg_2O-$(60-x)V_2O_5$-$40TeO_2$ the sample containing 20 and 30 mol% of Ag_2O had the minimum band gaps indicating the highest χ^3. Such result beside their highest thermal and ambient stability makes them as excellent promised agents in optical fibers. The precise and the newest derivative absorption spectrum fitting method (abbreviated as DASF) has been used for precise determination of the band gap. Regarding the recent non (low)-toxic antibacterial/bactericidal materials, glasses and specially nanomaterials (as the bridge between nano-science and biology as nano-biotechnology) are novel agents in this field, and in bio-imaging employing the luminescent agents. Also, their antibacterial/bactericidal potential detected by employing serial dilution and/or agar disc-diffusion bioassays was presented and argued. Results emphasized on the higher antibacterial/bactericidal activity of glasses and glass–ceramics containing 30%-silver than the studied bacteria from both Gram-negative and Gram-positive categories. Also, in the case of zinc selenide nanoparticles synthesized under various conditions, the optical, luminescence and antibacterial potential have been evaluated as dopant content and particle size dependent aspects. Finding 0.1% Cu as the optimum antibacterial-optical dopant content, Mn-Cu co-doped ZnSe NPs were investigated considering a fixed optimum Cu value (i.e. 0.1%). Based on the probable type of impurity substitution in the host ZnSe lattice, the samples owing 1 and 1.5% of Mn were considered as the best optically-promising agents. So, these materials reviewed in this chapter are of great potential for Te-based glasses and GCs. The bared/shelled and doped ZnSe NPs are of potential use in optoelectronic devices, and they are promising agents in optical and biological applications.

References

1. R. El-Mallawany, *Tellurite Glass Smart Materials: Applications in Optics and Beyond*, ed. by R. El-Mallawany (Springer Cham, 2018), pp. 1–16
2. R. El-Mallawany, J. Appl. Phys. **73**(10), 4878 (1993)
3. R. El-Mallawany, A.H. El-Sayed, M.M.H.A. El-Gawad, Mater. Chem. Phys. **41**(2), 87 (1995)
4. D. Souri, Z.E. Tahan, Appl. Phys. B **119**, 273 (2015)
5. D. Souri, Z.E. Tahan, S.A. Salehizadeh, Indian J. Phys. **90**(4), 407 (2016)
6. D. Souri, F. Honarvar, Z.E. Tahan, J. Alloys Compd. **699**, 601 (2017)
7. Y. Shahmoradi, D. Souri, M. Khorshidi, Ceram. Int. **45**, 6459 (2019)
8. R. El-Mallawany, M. Sidkey, A. Khafagy, H. Afifi, Mater. Chem. Phys. **37**(2), 197 (1994)
9. R.K.N. Ningthemcha, D. Biswas, Y.B. Singh, D. Sarkar, R. Mondal, D. Mandal, L.S. Singh, Mater. Chem. Phys. **249**, 123207 (2020)
10. D. Souri, M. Elahi, Phys. Scr. **75**(2), 219 (2007)
11. D. Souri, M. Elahi, M.S. Yazdanpanah, Cent. Eur. J. Phys. **6**(2), 306 (2008)
12. D. Souri, H. Zaliani, E. Mirdawoodi, M. Zendehzaban, Measurement **82**, 19 (2016)
13. D. Souri, J. Non-Cryst, Solids **475**, 136 (2017)
14. G.S. Murugan, Y. Ohishi, J. Non-Cryst, Solids **341**, 86 (2004)
15. I.Z. Hager, R. El-Mallawany, M. Poulain, J. Mater. Sci. **34**(21), 5163 (1999)
16. D. Biswasa, R.K.N. Ningthemcha, A.S. Das, L.S. Singh, J. Non-Cryst, Solids **515**, 21 (2019)
17. D. Biswas, A.S. Das, R. Mondal, A. Banerjee, D. Deb, A. Dutta, S. Bhattacharya, S. Kabi, L.S. Singh, J. Non-Cryst, Solids **542**, 120104 (2020)
18. L.C. Barbosa, C.O. Filho, E.F. Chillcce, *Technological Advances in Tellurite Glasses*, ed. by V. Rivera, D. Manzani, Springer Series in Materials Science, vol. 254 (Springer, Cham, 2017), pp. 131–153
19. S. Jayaseelan, P. Muralidharan, M. Venkateswarlu, N. Satyanarayana, Mater. Sci. Eng. B **119**, 136 (2005)
20. H.M.M. Moawad, H. Jain, R. El-Mallawany, T. Ramadan, M. El-Sharbiny, J. Am. Ceram. Soc. **85**(11), 2655 (2002)
21. R. El-Mallawany, Phys. Stat. Sol. (A) **177**, 439 (2000)
22. D. Souri, *Tellurite glass smart materials*, ed. by R. El-Mallawany (Springer, Cham, 2018). pp. 67–105
23. G. Webb, D. Agata, E.M.C. Magal, P. Ruan, Proc. Natl. Acad. Sci. U. S. A. **102**(37), 13343 (2005)
24. N. Elkhoshkhany, A. Reda, A.M. Embaby, Ceram. Int. **43**, 15635 (2017)
25. S.F. Hosseini, D. Souri, A. Nouri Emamzadeh, J. Inorg. Organomet. Polym. Mater. **31**, 2877 (2021)
26. S.F. Hosseini, D. Souri, E. Mohammadi-manesh, R. El-Mallawany, Ceram. Int. **45**(17(Part B)), 2364 (2019)
27. Y. Shahmoradi, D. Souri, Ceram. Int. **45**, 7857 (2019)
28. A.A. Ali, Y.S. Rammah, R. El-Mallawany, D. Souri, Measurement **105**, 72 (2017)
29. A.M. Noorazlan, H.M. Kamari, S.O. Baki, D.W. Mohamad, J. Nanomater. **16**(1), 407 (2015)
30. A. Kaur, A. Khanna, L.I. Aleksandrov, J. Non-Cryst, Solids **476**, 67 (2017)
31. Y. Hatefi, D. Souri, H. Beigpour, J. Mater. Sci. Mater. Electron. **28**, 939 (2017)
32. J. Massera, A. Haldeman, D. Milanese, H. Gebavi, M. Ferraris, P. Foy, W. Hawkins, J. Ballato, R. Stolen, L. Petit, K. Richardson, Opt. Mater. **32**, 582 (2010)
33. S.J. Saint-Jean, *Advanced Ceramics for Dentistry*, ed. by J.Z. Shen, T. Kosmač (Elsevier, 2014), pp. 255–277
34. V.D. Mote, Y. Purushtham, B.N. Dole, J. Theor. Appl. Phys. **6**, 6 (2012)
35. K. Venkateswarlu, M. Sandhyarani, T.A. Nellaippan, N. Rameshbabu, Procedia. Mater. Sci. **5**, 212 (2014)
36. N.C. Halder, N.C.J. Wagner, Acta Crystallogr. **20**, 312 (1966)
37. V.V. Nikesh, A.D. Lad, S. Kimura, S. Nozaki, S. Mahamuni, J. Appl. Phys. **100**(11), 113520 (2006)

38. N. Oraee, M. Molaei, E.S. Iranizad, Mod. Phys. Lett. B **26**, 1250171 (2012)
39. D. Souri, A.R. Khezripour, M. Molaei, M. Karimipour, Curr. Appl. Phys. **17**, 41 (2017)
40. N. Priyadharsini, M. Elango, S. Vairam, T. Venkatachalam, M. Thamilselvan, Optik **127**, 7543 (2016)
41. A.R. Khezripour, D. Souri, Optik **183**, 294 (2019)
42. C. Rajesh, C. Phadnis, K.G. Sonawane, S. Mahamuni, Generation of white light from co-doped (Cu and Mn) ZnSe QDs. J. Exp. Nanosci. **10**(14), 1082 (2015)
43. I. Ahmad Mir, K. Rawat, H.B. Bohidar, Mater. Res. Express **3**(10), 105014 (2016)
44. S. Ebrahimi, D. Souri, J. Alloys Compd. **840**, 155712 (2020)
45. K.P. Tiwary, S.K. Choubey, K. Sharma, Chalcogenide Lett. **10**, 319 (2013)
46. R. Zhang, Y. Liu, S. Sun, Opt. Mater. **34**, 1788 (2012)
47. H.D. Duong, J. Rhee, Sens. Actuators B: Chem. **134**, 423 (2008)
48. L.S. Devi, K.N. Devi, B.I. Sharma, H.N. Sharma, IOSR-J. App. Phys. **6**, 6 (2014)
49. N. Arshi, F. Ahmed, S. Kumar, M.S. Anwar, J. Lu, B.H. Koo, C.G. Lee, Curr. Appl. Phys. **11**, S360 (2011)
50. B. Kwamboka, W. Omwoyo, N. Oyaro, Ind. J. Nanosci. **4**, 1 (2016)
51. D. Souri, N. Salimi, M. Ghabooli, Inorg. Chem. Commun. **123**, 108345 (2021)
52. D. Souri, M. Sarfehjou, A.R. Khezripour, J. Mater. Sci. Mater. Electron. **29**(4), 3411 (2018)
53. D. Souri, K. Ahmadian, A.R. Khezripour, J. Electron. Mater. **47**(11), 6759 (2018)
54. M. Ghasemi Hajiabadi, M. Zamanian, D. Souri, Ceram. Int. **45**, 14084 (2019)
55. A.R. Khezripour, D. Souri, H. Tavafi, M. Ghabooli, Measurement **148**, 106939 (2019)
56. S. Ebrahimi, D. Souri, M. Ghabooli, J. Clust. Sci. **30**(3), 677 (2019)
57. S. Ebrahimi, D. Souri, M. Ghabooli, Nanoscale **5**(2), 179 (2018)
58. N. Salimi, D. Souri, M. Ghabooli, Iran. J. Mater. Sci. Eng. **7**(3), 1 (2018)
59. M. Sarfehjou, D. Souri, A. Khezripour, Opt. Laser Technol. **145**, 107467 (2022)
60. M. Sarfehjou, D. Souri, Appl. Phys. A **127**, 139 (2021)
61. M. Molaei, A.R. Khezripour, M. Karimipour, Appl. Surf. Sci. **317**, 236 (2014)
62. M. Molaei, A.R. Bahador, M. Karimipour, J. Lumin. **166**, 101 (2015)
63. M. Bagheri, P. Sangpour E. Badiei, M. Pazouki, Int. J. Eng. Trans. C: Aspects **27**, 1803 (2014)
64. R.Y. Pelgrift, A.J. Friedman, Adv. Drug Deliv. Rev. **65**, 1803 (2013)
65. A. Ben-David, C.E. Davidson, J. Microbiol. Methods **107**, 214 (2014)
66. P.W. Su, C.H. Yang, J.F. Yang, P.Y. Su, L.Y. Chuang, Molecules **20**, 11119 (2015)
67. S. Rezaei-Zarchi, A. Aved, M.J. Ghani, S. Soufian, F.B. Firouzabadi, A.B. Moghaddam, S.H. Mirjalili, Iran. J. Pathol. **5**, 83 (2010)
68. C. Chaliha, B.K. Nath, P.K. Verma, E. Kalita, Arab. J. Chem. **12**, 515 (2019)
69. J.P. Ruparelia, A.K. Chatterjee, S.P. Duttagupta, S. Mukherji, Acta Biomater. **4**, 707 (2008)
70. D.M. Ali, N. Thajuddin, K. Jeganathan, M. Gunasekaran, Colloids Surf. B Biointerfaces **85**, 360 (2011)
71. A. Rai, A. Prabhune, C.C. Perry, J. Mater. Chem. **20**, 6789 (2010)
72. B. Bagchi, S. Kar, S.K. Dey, S. Bhandary, D. Roy, T.K. Mukhopadhyay, S. Das, P. Nandy, Colloids Surf. B Biointerfaces **108**, 358 (2013)
73. E.M. Costa, S. Silva, S. Vicente, C. Neto, P.M. Castro, M. Veiga, R. Madureira, F. Tavaria, M.M. Pintado, Mater. Sci. Eng. C **79**, 221 (2017)
74. V. Beena, S. Ajitha, S.L. Rayar, C. Parvathiraja, K. Kannan, G. Palani, J. Inorg. Organomet. Polym. Mater. **31**, 4390 (2021)
75. V. Beena, S.L. Rayar, S. Ajitha, A. Ahmad, M.D. Albaqami, F.A.A. Alsabar, M. Sillanpää, Water **13**, 2189 (2021)
76. S.V. Gudkov, D.E. Burmistrov, D.A. Serov, M.B. Rebezov, A.A. Semenova, A.B. Lisitsyn, Front. Phys. **9**, 64148 (2021)
77. Y. Hatefi, N. Shahtahmasebi, A. Moghimi, E. Attaran, J. Rare Earth **29**, 484 (2011)
78. Y. Hatefi, N. Shahtahmasebi, A. Moghimi, E. Attaran, J. Lumin. **131**, 114 (2011)
79. J. Pisarska, W. Ryba-Romanowski, G. Dominiak-Dzik, T. Goryczka, W.A. Pisarski, J. Alloys Compd. **451**, 223 (2008)
80. D. Deng, S. Xu, S. Zhao, C. Li, H. Wang, H. Ju, J. Lumin. **129**, 1266 (2009)

81. E. Mohaghegh, A. Ne'mati, B. Eftekhari Yekta, S. Bani Jamali, Iran. J. Mater. Sci. Eng. **2**(3), 19 (2013)
82. P. Bindu, S. Thomas, J. Theor. Appl. Phys. **8**, 123 (2014)
83. Z.A. Peng, X. Peng, J. Am. Chem. Soc. **123**, 183 (2001)
84. C.J. Murphy, J. Mater. Chem. **18**, 2173 (2008)
85. Q. Dai, N. Xiao, J. Ning, C. Li, D. Li, B.O. Zou, W.W. Yu, S. Kan, H. Chen, B. Liu, G. Zou, J. Phys. Chem. C **112**, 7567 (2008) 7567
86. R. Xie, Y. Li, X. Zhang, H. Liu, J. Alloys Compd. **621**, 396 (2015)
87. D. Zhu, X. Jiang, C. Zhao, X. Sun, J. Zhang, J.J. Zhu, Chem. Comm. **46**(29), 5226 (2010)
88. K. Yadav, Y. Dwivedi, N. Jaggi, J. Lumin. **158**, 181 (2015)
89. R. Yousefi, F. Jamali-Sheini, A.K. Zak, M.R. Mahmoudian, Ceram. Int. **38**(8), 6295 (2012)
90. M. Balouiri, M. Sadiki, S.K. Ibnsouda, J. Pharm. Anal. **6**, 71 (2016) 71
91. CLSI, *Methods for Dilution Antimicrobial Susceptibility Tests for Bacteria that Grow Aerobically, Approved Standard*, 9th edn., CLSI document M07-A9 (Clinical and Laboratory Standards Institute, Wayne, USA, 2012)
92. CLSI, *Methods for Antimicrobial Dilution and Disk Susceptibility of Infrequently Isolated or Fastidious Bacteria, Approved Guideline*, 2nd edn., CLSI document M45-A2 (Clinical and Laboratory Standards Institute, Wayne, USA, 2010)
93. A.J. Driscoll, N. Bhat, R.A. Karron, K.L. O'Brien, D.R. Murdoch, Clin. Infect. Dis. **54**, S159 (2012)
94. S. Silva, E.M. Costa, M.R. Costa, M.F. Pereira, J.O. Pereira, J.C. Soares, M.M. Pintado, Food Control **51**, 314 (2015)
95. J.C. Fernandes, F.K. Tavaria, J.C. Soares, O.S. Ramos, M.J. Monteiro, M.E. Pintado, F.X. Malcata, Food Microbiol. **25**, 922 (2008)
96. N.R. Bhalodia, V.J. Shukla, J. Adv. Pharm. Technol. Res. **2**, 104 (2011)
97. J. Tauc, A. Menth, J. Non-Cryst, Solids **8**, 569 (1972)
98. D. Souri, S.A. Salehizadeh, J. Mater. Sci. **44**, 5800 (2009)
99. D. Souri, K. Shomalian, J. Non-Cryst, Solids **355**, 1597 (2009)
100. N.F. Mott, E.A. Davis, *Electronic Processes in Non-Crystalline Materials*, 2nd edn. Clarendon (Oxford University Press, 1979), pp.45–58
101. L.E. Alarcon, A. Arrieta, E. Camps, S. Muhl, S. Rudil, E.V. Santiago, Appl. Surf. Sci. **254**, 412 (2007)
102. L. Changshi, L. Feng, Opt. Commun. **285**, 2868 (2012)
103. R. El-Mallawany, Y.S. Rammah, A. El Adawy, Z. Wassel, Adv. Opt. Photonics **5**, 11 (2017)
104. Y. Al-Hadeethi, M.I. Sayyed, Y.S. Rammah, Ceram. Int. **46**, 2055 (2020)
105. Y.S. Rammah, K.A. Mahmoud, M.I. Sayyed, F.I. El-Agawany, R. El-Mallawany, J. Non-Cryst, Solids **534**, 119944 (2020)
106. Y.S. Rammah, G. Kilic, R. El-Mallawany, U.G. Issever, F.I. El-Agawany, J. Non-Cryst, Solids **533**, 119905 (2020)
107. D. Souri, Measurement **44**, 1904 (2011)
108. M. Elahi, D. Souri, Indian J. Pure Appl. Phys. **44**, 468 (2006)
109. D. Souri, Measurement **44**, 717 (2011)
110. D. Souri, S.A. Salehizadeh, J. Therm. Anal. Calorim. **112**(2), 689 (2013)
111. D. Souri, Y. Shahmoradi, J. Therm. Anal. Calorim. **129**, 601 (2017)
112. D. Souri, Eur. Phys. J. B. **84**, 47 (2011)
113. D. Souri, M. Mohammadi, H. Zaliani, Electron. Mater. Lett. **10**(6), 1103 (2014)
114. H.S. Farhan, Aust. J. Basic Appl. Sci. **11**(9), 171 (2017)
115. D. Souri, Phys. B: Condens. Matter **456**, 185 (2015)
116. D. Souri, J. Non-Cryst, Solids **470**, 112 (2017)
117. D. Souri, J. Mater. Sci. **46**, 6998 (2011)
118. V. Dimitrov, S. Sakka, J. Appl. Phys. **79**, 1736 (1996)
119. N. Berwal, S. Dhankhar, P. Sharma, R.S. Kundu, R. Punia, N. Kishore, J. Mol. Struct. **1127**, 636 (2017)
120. A. Abdel Kader, R. El-Mallawany, M. Elkholy, J. Appl. Phys. **73**, 71 (1993)

121. R. El-Mallawany, M.D. Abdalla, I.A. Ahmed, Mater. Chem. Phys. **109**, 291 (2008)
122. X. Liu, Z. Luo, C. Wen, S. Liu, P. Zhang, H. Liang, A. Lu, J. Non-Cryst, Solids **589**, 121661 (2022)
123. H. Eskalan, M. Kavgaci, H. Yaykasli, M. Gok, Mater. Today Commun. **39**, 108737 (2024)
124. S. Karthika, S. Shanmuga Sundari, K. Marimuthu, P. Meena, Chem. Phys. Impact **8**, 100430 (2024)
125. H. Al-Ghamdi, A.H. Almuqrin, M.I. Sayyed, A. Kumar, Results Phys. **29**, 104726 (2021)
126. M.E. Medhat, Ann. Nucl. Energy **36**, 849 (2009)
127. M.I. Sayyed, B. Albarzan, A.H. Almuqrin, A.M. El-Khatib, A. Kumar, D.I. Tishkevich, A.V. Trukhanov, M. Elsafi, Materials **14**, 3772 (2021)
128. A.S. Abouhaswa, M. Almurayshid, F. Almasoud, M.I. Sayyed, K.A. Mahmoud, Sci. Rep. **12**, 3548 (2022)
129. M.S. Al-Buriahi, H. Arsalan, H.O. Tekin, V.P. Singh, B.T. Tonguc, Mater. Res. Express **7**, 025202 (2020)
130. G. Lakshminarayana, M.I. Sayyed, S.O. Baki, A. Lira, M.G. Dong, Kh.A. Bashar, I.V. Kityk, M.A. Mahdi, J. Electron. Mater. **48**, 930 (2019)
131. M.J. Hajipour, K.M. Fromm, A.A. Ashkarra, D.J. de Aberasturi, I.R. de Larramendi, T. Rojo, V. Serpooshan, W.J. Parak, M. Mahmoudi, Trends Biotechnol. **30**, 499 (2012)
132. L. Wang, C. Hu, L. Shao, Int. J. Nanomedicine **12**, 1227 (2017)
133. Y.N. Slavin, J. Asnis, U.O. Häfeli, H. Bach, J. Nanobiotechnol. **15**, 65 (2017)
134. E.D. Cavassin, L.F.P. de Figueiredo, J.P. Otoch, M.M. Seckler, R.A. de Oliveira, F.F. Franco, V.S. Marangoni, V. Zucolotto, A.S.S. Levin, S.F. Costa, J. Nanobiotechnol. **13**, 64 (2015)
135. L. Dorobantu, C. Fallone, A. Noble, J. Veinot, B. Guibin, G. Ma, G.G. Goss, R.E. Burrel, J. Nanopart. Res. **17**, 172 (2015)
136. B.T. Huy, M.H. Seo, P.T. Phong, J.M. Lim, Y.I. Lee, Chem. Eng. Sci. **236**, 75 (2014)
137. K. Nakamoto, *Infrared Raman Spectra of Inorganic and Coordination Compounds*, 5th edn. (John Wiley & Sons, New York, 1997), pp. 97–99
138. S.B. Qadri, E.F. Skelton, D. Hsu, A.D. Dinsmore, J. Yang, H.F. Gray, B.R. Ratna, Phys. Rev. B **60**(13), 9191 (1999)
139. M. Arslan, A. Maqsood, A. Mahmood, A. Iqbal, Mater. Sci. Semicond. Process. **16**(6), 1797 (2013)
140. M.M. Ivashchenko, I.P. Buryk, A.S. Opanasyuk, D. Nam, H. Cheong, J.G. Vaziev, Mater. Sci. Semicond. Process. **36**, 13 (2015)
141. M. Sharma, S.K. Tripathi, J. Phys. Chem. Solids **73**(9), 1075 (2012)
142. K. Swaroop, H.M. Somashekarappa, Res J Recent Sci **4**, 197 (2015)
143. Z. Vaez, V. Javanbakht, J. Photochem. Photobiol. A Chem **388**, 112064 (2020)
144. J. Saien, M. Asgari, A. Soleymani, N. Taghavinia, J. Chem. Eng. **151**(1–3), 295 (2009)
145. A.R. Soleymani, A.M. Tavassoli, H. Rezaei-Vahidian, Chem. Eng. Res. Des. **190**, 759 (2023)
146. A. Haghighizadeh, A. Beni Ali, M. Haghmohammadi, M.S. Samie Adel, S. Farshad, Water Air Soil Pollut. **234**, 366 (2023)
147. H.G. Hameed, N.A. Abdulrahman, Phys. Chem. Res. **11**(4), 771 (2023)
148. A.M. El Badawy, R.G. Silva, B. Morris, K.G. Scheckel, M.T. Suidan, T.M. Tolaymat, Environ. Sci. Technol. **45**, 283 (2011)
149. S. Siddique, Z.H. Shah, S. Shahid, F. Yasmin, Acta Chim. Slov. **60**(3), 660 (2013)
150. C. Malarkodi, S. Rajeshkumar, K. Paulkumar, M. Vanaja, G. Gnanajobitha, G. Annadurai, Biosynthesis and antimicrobial activity of semiconductor nanoparticles against oral pathogens. Bioinorg. Chem. Appl. **2014**, 347167 (2014)
151. M. Raffi, S. Mehrwan, T.M. Bhatti, J.A. Akhter, A. Hameed, W. Yawar, M.M. Hasan, Ann. Microbiol. **60**(1), 75 (2010)

Bioactive Glass and Glass–Ceramics for Bone Healing and Medical Applications

Patrick Munyensanga, Meriame Bricha, and Khalil EL Mabrouk

Abstract This chapter provides an overview of current research on bioactive glasses, glass-ceramics, and their related composites, including their synthesis techniques and processing methods. Bioactive glasses have shown the ability to form strong and stable bonds within the host rapidly. Their components, such as silicon-containing protein serum, enable drug delivery and release. However, unlike load-bearing implants, bioactive glasses are brittle, a challenge that has been addressed through the synthesis and fabrication of nanocomposite matrix forms. One interesting aspect is the easy integration of these glasses with other biomaterials, such as polymers, metals, or ceramic scaffolds with specific topologies. Their scaffolds exhibit excellent mechanical properties, tailored to meet the requirements of different bone tissues. Additionally, biosilicates have been developed focusing on their morphology and combined with glass-ceramic strength. The chapter further discusses the factors involved in the media preparation of bioactive glass and glass-ceramic, highlighting how modifications to their composites contribute to the formation of micro- and macro-sized structures.

Keywords Biomaterials · Porous materials · Scaffolds · Biocompatible · Nanostructures · Bioactive glass · Good health and well-being

1 Introduction

In recent years, biomaterials have gained recognition and attraction in tissue engineering. A reliable example is titanium (Ti) and titanium alloys, which have recently been used for implants as alternatives. Some other biomaterials were selected based on their biofunctional properties, such as zirconia (ZrO_2) for dental implants compared to other ceramics [1, 2]. The demand and selection of biomaterials have resulted in higher risk of insufficient tissues in humans, which has been seen in the elderly population. Therefore, the demands were fast-growing with a significant

P. Munyensanga · M. Bricha · K. EL Mabrouk (✉)
Euromed University of Fes (UEMF), Fes, Morocco
e-mail: k.elmabrouk@ueuromed.org

S. J. Ikhmayies (ed.), *Advances in Biomaterials Research*, Advances in Material Research and Technology, https://doi.org/10.1007/978-3-031-97481-6_8

need for tissue engineering [3]. An example, in the United States alone, nearly 13 million prosthetics and medical devices were implanted each year [4, 5]. This draws attention to the need for biomaterials and helps to understand this rise. Such as those bioinspired devices and tissues/membranes [6] used to replace failing body organs and support physiological activities. Interestingly, these biomaterials were interacted with human blood cells, tissues, or organs, and perform their biological functions [3, 4]. An example of engineered tissues, purposely for restoring failed organ parts or drug delivery with biodegradability or bioresorbable functions [7]. However, despite the diversity of current surgical techniques for repairing bone defects, reconstruction of critical-sized soft and hard tissue defects, remains challenging for surgeons worldwide. Insufficient bone volume, poor anatomical match, bone graft resorption, and rejection are some of the complications [8].

On the other hand, metal implants were typically required to repair extensive damage to the human body [9], and commonly selected for prosthetic operations due to their superiority over traditional ceramics and polymeric biomaterials [8]. Additionally, they were also served as load components when implanted as bone[1] replacements. Several diseases over different complications caused by disordered bone or body tissues contributed to this need, as well as, the accidental fracture that leads to the replacement of the specific part with implants or tissues required to mimic their biological functions [10]. Therefore, it has been recommended that the biomaterials of such prosthetic procedure must have adequate bioactivity regarded as effective in implantation. Thus, the biocompatibility of the material is achieved, with optimum criteria being well thought out before any desired procedures. Several fabrication techniques for biomaterials which improve biocompatibility and mechanical properties have been developed. They involved traditional and novel methods [11] for criterion philosophy in selection of any biomaterial for orthopedic application [8]. The following five characteristics are frequently considered in advanced biomaterials research to meet requirements. Thus, the biomedical researchers tasked the majority of them to optimize the biofunctionalities of any desired implantable material as follows:

(a) The biomaterial must receive a favorable host response such as osseointegration capability and stimulation of biomineralization of the host during service without sacrificing local or systemic responses of the living system or tissue, i.e., bioactivity properties.
(b) Bio-metal implants must meet the approximate distinctive design criteria of metallic implants manufactured for replacement applications, such as load transmission and stress distribution (mechanical properties).
(c) Articulating biomaterials, such as metallic implants or tissue-engineered ones, should allow movement in artificial joint applications. An example of hip and knee joint replacement.
(d) Implantable devices developed based on the first generation of biological performance from the 1940s to the 1980s, must be achievable using chemical reactive

[1] **Bone**: A composite structure material with hierarchical organization, and primarily composed with collagen fibrils and hydroxyapatite (HA).

materials that improve osteointegration and bioactivity, such as bioactive glasses or glass-ceramics [12].

(e) Developing strategies aimed to yield surfaces that promote bio-integration and facilitate osteointegration by providing good layers that colonize bacterial growth [13].

Furthermore, most specifically, the traditional methods used to fabricate biomaterials, ceramics, and ceramic matrix composites were limited to the manufacturability of large parts with complex shapes and component production, in addition to those time-consuming and expensive processes. More specifically, the manufacturing processes of bio-joining or bioprinting of functional hydrogels are novel fabrication processes using additive manufacturing, while the combinations of ceramics and ceramic matrix composites with polymers, metallic, and bioactive glasses for biofunctional hydrogels were not. Since, they were produced to address the drawbacks of traditional methods and not yet regulated [14] or patented in international standards as for other AM techniques found in International Organization for Standardization/American Society for Testing Materials (ISO/ASTM) 52,900 [15]. In contrast, ceramic integration in additive manufacturing has increasingly enabled fabrication of biomedical manufacturing [16, 17] and the utilization of components for high-temperature structural requests, including those in energy, environment, transportation, and aerospace [18, 19].

Joining processes of biomaterials described in the research of Kirihara [20], which explains the composite of two hybrid biomaterials that allows a combination of simpler parts to form large structures or complicated shapes is an example. Thus, the integration processes allow the incorporation of ceramics with glasses and ceramic matrix composite components with metal-based cores act as load-bearing components [20]. Therefore, come apart from the novel advanced fabrication technique, and form a nanohybrid composite material structure with enhanced mechanical and physicochemical properties.

2 Biomaterials in Research

The most recent bioengineered materials research focuses on overcoming the impediments that materials face due to inherent limitations in their integration with the host environment. Such limitations occur due to the host's response to osteointegration and restrict the onsite target. Due to cell toxicity, the physiological function of the immune system and the blood cells transportation are disrupted [21]. Thus, it can be seen that the needs of developing and designing medical devices come from their biocompatibility. This addresses the physiological, mechanical, and physicochemical properties relevant to the site of interest. This host features capture almost all the mandatory considerations of the basic types of biomaterials (Fig. 1). According to Sharma et al. [21], the biomaterials of soft tissues can be synthetic polymers such as polyanhydrides, naturally occurring polymers such as complex sugars (hyaluronan

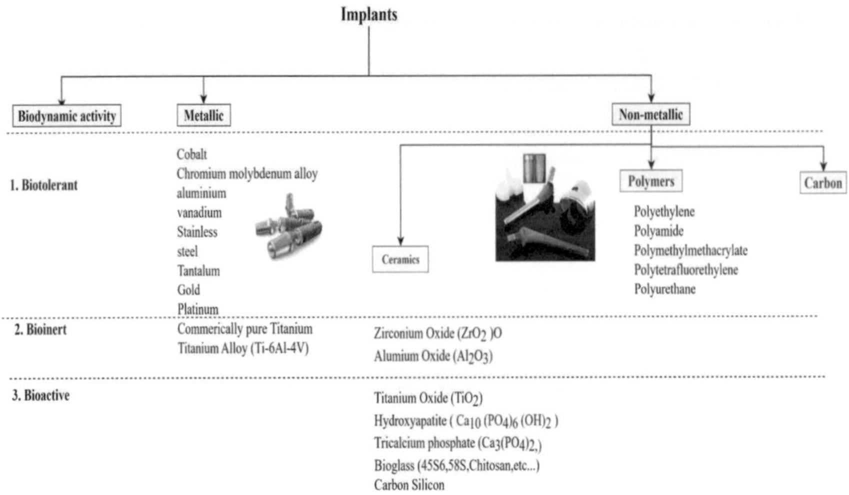

Fig. 1 Biomaterials' categories and classification

and chitosan), and/or inorganics such as hydroxyapatite (HA). Thus, their classification can be based on functions such as hydrogels [22], injectables [23, 24], drug delivery [25, 26], and surface-modifier [27, 28] for healing or replacement, and by other different bio-features [21, 29, 30].

2.1 Bioactive Glasses

Throughout history, mankind has always been interested in glasses, a brittle material that is artistic with fined surface roughness. In terms of engineering gadgets, these are environmental devices with a variety of applications. Thanks to Fernandes et al. [31], who summarized these appliances from compact objects, including bottles, windows, containers, light bulbs, etc., to technical applications, including television tubes, computer screens, spectacles, telescope lenses, spectrometer prisms, laboratory ware, optical fibers, etc., and or artistic purposes. The glasses' appliances were not limited to technical applications only. They were also brought to the biotechnical appliances. At the time of Larry Hench (1938–2015), majority of the researchers focused on the theory of engineering devices that were less reactive to the human body and adhered to the biological response implant integration.

In the early 1970s, researchers were interested in the compatibility of the surface reaction of biomaterials in contact with blood cells. There was no theory about material-tissue bonding that limited the elicitation or inflammatory response. It was beyond the imagination of the theory to consider the possibility of synthetic materials that could have a positive impact on tissue healing when implanted. These biomaterials were being studied and are considered as materials with minimal reactivity in the body. Some of them are ceramics for bone grafts and polymers [32]. The first bioactive glass developed met some skepticism and curiosity from the publication and later studies. After the group of Hench published the verification of direct attachment between bone and the glass implant—except the mechanism incurred and theory phenomenon that was not yet developed—the bioactive glasses gained desirability [32, 33].

Therefore, the term bioactivity[2] was related to other biomaterials used in orthopedic implants. The term bioactive glass was first used in the U.S. army report, appeared as "bio-glass-ceramic". These terms come with the addition of ceramics since the purpose of the bioactive glass innovator was to synthesize a strong enough material that can be load-bearing for bone replacements. This is also peer-reviewed from the following researches, which were constructed on bioactive glass ceramics and the amorphous crystalline glass samples [32]. As written above, the first operational theory mechanism of tissue bonding with the surface of bioactive glass-ceramics implants was described in the research of *Clark, Hench, and Paschall in 1976* (see Fig. 1). They defined different reactivity events that occur after implantation for both in vivo and in vitro. As a result, the bioactive glass developed a chemodynamic surface that depends on glass composition (wt%) in both hard and soft tissues [35]. The confirmation was later explained by researchers worldwide looking at the new bioactive glass composition; new synthesis process, new applications, and more depth for the biological response and reactivity that affects their performance. This theory was gradually developed until the publication of Hench, about the "reaction sequence[3]" occurring after implantation of bioactive glass into bone or soft tissue [34, 36, 37] (Fig. 2).

The sequence is explained by the formation of a new layer of hydroxycarbonate apatite (HCA) on the surface of bioactive glass as a new biologically active layer used as a surrogate to claim any biomaterials' bioactivity. In vitro tests of bioactive glasses based on the kinetics of hydroxycarbonate apatite (HCA) synthesis appeared to demonstrate the bioactivity of bioactive glasses (BGs) and their potential to mimic the formation of the HA layer of the bone [38]. However, the disposition of the HA layers on the surface of BG comes from two components; dissolution of the material surface, and the precipitation of the Ca-P layer [38]. On the other hand, the synthesized bioactive glass with a high crystallization level above 60% lowers the formation of the surface apatite layer. Filho et al. [38] synthesized glass-ceramics to study the formation of the apatite layer after the incubation time, and obtained 1N2C3S ($Na_2Ca_2Si_3O_9$) with a process that includes melting, quenching, and annealing.

During the in vitro testing, the HCA formation increased the crystallization structure as the onset time of immersion increased from 10 to 22 h for 100% crystallized glass-ceramics. However, even when exposed to simulated body fluid (SBF), the existence of less than 5% or less residual glass in glass ceramics composition

[2] *Bioactivity*: "A bioactive material is one that elicits a specific biological response at the inter-face of the material that results in the formation of a bond between tissues and the material" [34].

[3] *Reaction sequence*: "for a bond with tissue to occur a layer of biologically active HCA must form" [32].

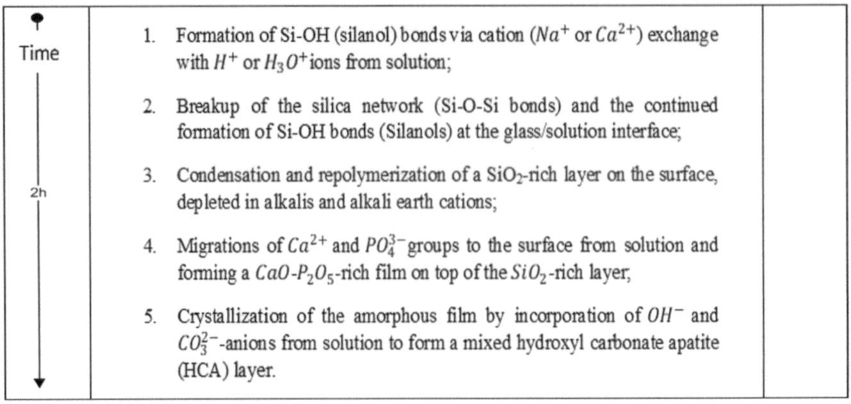

Fig. 2 Reaction sequence formation of hydroxycarbonate apatite (HCA) as stated in the early theory of reaction stages where the bioactive glass in vitro forms a biologically active layer that defines HCA layer as postulated by Professor Hench [31]

shows no evidence of the creation of a surface apatite layer. In these studies, the range of grain size and crystallization degree explain the amorphous structure found at 36% crystallization, through Fourier-transform infrared spectroscopy (FTIR) plot in wavenumber range 700–400 cm^{-1}. Hydroxylation of the glass, the solution's ion concentration, and the connectivity of the residual glassy phase are also incurred in the reaction [38, 39].

As previously stated, ion migration to the surface can sometimes delay dissolution, making deposition difficult and delaying the creation of the surface layer. Furthermore, these two methods for increasing chemical reactivity for the formation of a hydroxyapatite (HA) layer (Fig. 3) on the glass surface are based on the rate of dissolution and the presence of dopants (serum protein) that inhibit the formation of an amorphous Ca-P layer, thereby attaching the cells that interface within this layer disposition [40]. Table 1 and Fig. 3 explain well the different parameters involved in the dissolution-precipitation based on material and solution [41].

Various protocols (as shown in Fig. 3) have been evolved in bioactive glass synthesis from the first discovered in 1969's. Different parameters including pH, temperature, type of precursor and synthesis route, were optimized and adjusted, as they were determined to influence the morphology of synthesized BGNPs [43–45]. In addition, these parameters significantly affected the size, shape, and surface characteristics of the NPs, an example of the case of Stöber process routes [46]. Optimum structure of synthesized BGNPs was obtained by controlling these parameters. During synthesis, the increase in hydrophilicity was accompanied by an increase in hydroxyl groups (OH^+) in the precursor reagents, causing phase transformation and modification of particle aggregation. Also, varying the concentration of reagents, specifically alkoxyl precursor (M-OH) [44], surfactants (e.g., cetyltrimethylammonium bromide CTAB) affect the agglomerations of NPs and determine the mean size of the monodispersed particles [47]. An example was given by Kesse et al. [44],

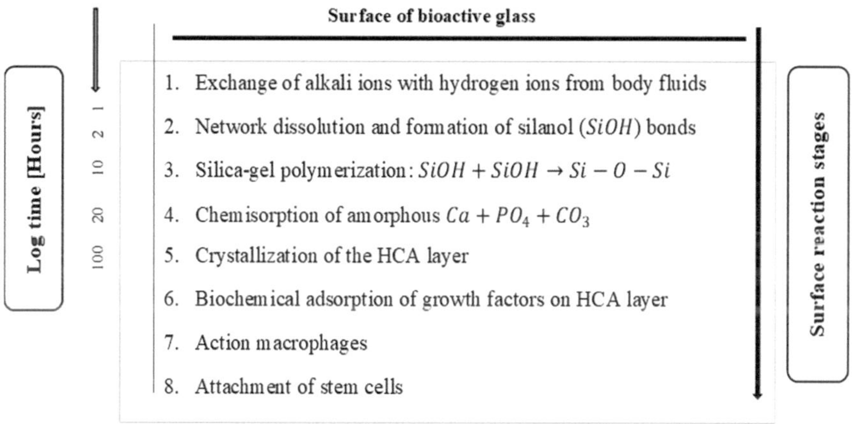

Fig. 3 Sequences of interfacial reactions involved in forming a bond between bone and bioactive glass [31, 42]

Table 1 In vitro parameters involved in dissolution-precipitation of BG [40, 41]

Material related	Solution related
Chemistry composition	Dopants and serum protein content
Crystalline structure	Ionic concentration
Surface charge and area	pH variation
Porosity	Buffer capacity
	Temperature

where the change of mean diameter size of the particles from 71 to 452 nm was illustrated. In contrast, the size distribution remained unchanged, ranging between 6 and 18% due to the variation of ammonium hydroxide concentration. In their study, variation of reagents ratio and dissolution time (when calcium nitrate was added for periods of 0.5–6 h through a different process under stirring or agitation has produced the purely amorphous bioactive glass nanoparticles (BGNPs) as confirmed by X-ray diffraction (XRD) characterization techniques shown in ref. [44] (Fig. 4).

After 30 min of stirring, the precipitate starts to give strong aggregate particles with irregular morphology, contrary with the addition after 1 h; where the transmission electron microscope (TEM) was used to get an image of aggregated particles (Fig. 5) with spherical morphology as well as a homogeneous size of ≈150 nm as shown in Fig. 6 [44]. Therefore, at the end of synthesizing the BGNPs, it was found that temperature parameter, was highly influencing crystallization, and causing other structural and morphological changes. Various studies synthesized bioactive glass under a temperature gradient including the famous one by Hulbert et al. [49], where the temperature of the synthesis protocols was varied through melt quenching process. The study of Zeimaran et al. [50], reported the fabrication of rod-like bioactive glass-ceramics (BGCs) under hydrothermal treatment based on the sol–gel process and confirmed this aforementioned synthesis route hypothesis.

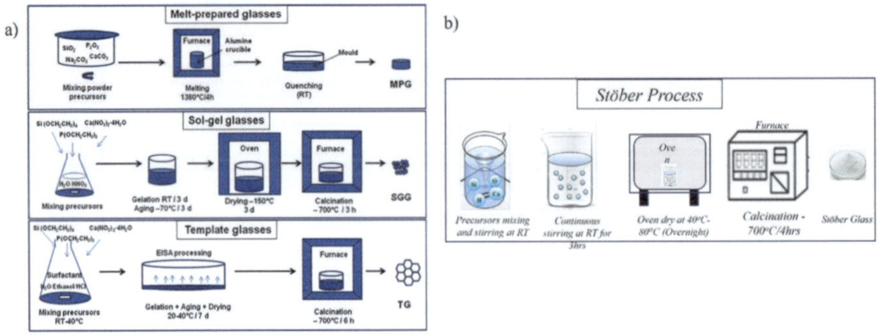

Fig. 4 Schematic diagram of the main stages in the synthesis process of **a** three families of bioactive glasses: melt-prepared (MPG), sol–gel (SGG), and template glasses (TG). Reprinted from Izquierdo-Barba et al. [48], with permission from John Wiley and Sons, Copyright (2013), and **b** the Stöber process found to reduce synthesis time and produce monodispersed bioactive glass nanoparticles with controlled size [46]

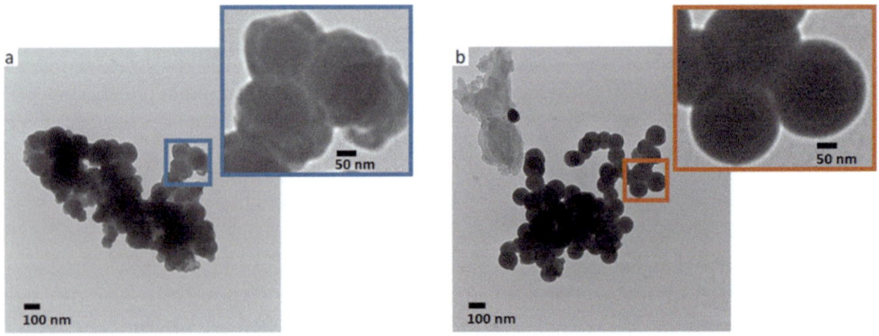

Fig. 5 TEM images of bioactive glass NPs prepared with **a** 30 min, and **b** 1 h calcium nitrate addition time after the beginning of the sol–gel reaction. Reprinted from Kesse et al. [44], with permission from, American Chemistry Society (ACS). Copyright (2019). https://doi.org/10.1021/acsomega.8b03598, Note to readers: Further permissions related to the material excerpted should be directed to the ACS

A bioactive glass of composition SiO, CaO, and P_2O_5 with wt% ratios of 58:33:9 respectively was synthesized under two thermal conditions of 200 and 220 °C to assess the effect of morphology, element composition, and degree of crystallinity, which affect the bioactivity response. The results showed a decrease in rod diameter from micro to nano (6.9 μm to 76–650 nm), decrease of the crystallinity, and an increase in the specific surface area of BGNPs [50]. More explanation will be given in the next section.

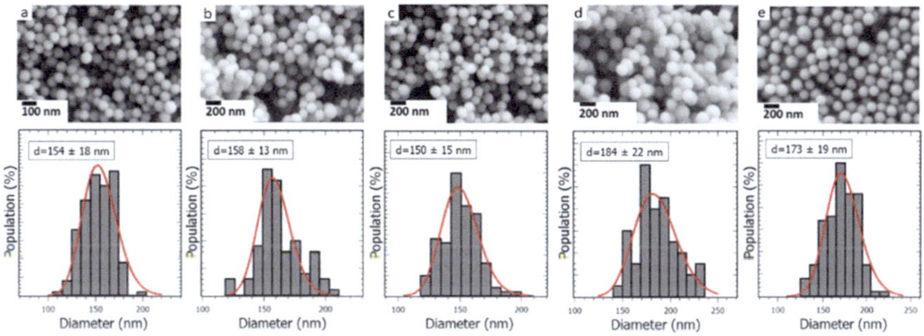

Fig. 6 Scanning electron microscope (SEM) images and particle size distribution reported as histograms of BGNPs prepared with **a** 2 h, **b** 3 h, **c** 4 h, **d** 5 h, **e** and 6 h of calcium nitrate addition time after the beginning of the sol–gel reaction and while other synthesis parameters were kept constant, including NH_4OH concentration which was fixed at 0.21 M and the Ca/Si ratio was fixed at 0.2. Reprinted from Kesse et al. [44], with permission from American Chemistry Society (ACS). Copyright (2019). https://doi.org/10.1021/acsomega.8b03598, Note to readers: Further permissions related to the material excerpted should be directed to the ACS

2.2 *Glass–Ceramics*

The glasses and ceramics have been used for a long time in orthopedic applications, especially for healing soft tissues of body, skeletal repair, and reconstruction. Due to their compatibility within medical care and surgery, they are often referred to as bioactive glasses, bioactive glass-ceramics, and bioactive ceramics, and have a pretty similar synthesis preparation route. The differences come from their indifferences in mechanical properties, as well as their thermal properties such as thermal expansion investigated in refs. [42, 51]. In addition, bioceramics may be almost composed of a small set of whole periodic table elements and compounds such as alumina, zirconia, carbon, silica-contained compounds, and some other elements. Bioceramics are different from bioactive glasses as discussed above, thought to be prepared from silica content minerals. Both may be prepared as powders or granules for porous and dense materials [52]. An example for machinability and workability of bioactive glasses and ceramics joined to form a biosilicate glass-ceramic composite that retains similar properties as bioactive glass itself, such as biocompatibility, antibacterial properties, and facilitates cost-effective production of foam complex shape and scaffold foams. At the same time, bioactive glass that is limited to powder, granule, or small monoliths forms cannot be made by traditional manufacturing methods [53].

In the article of Chevalier and Gremillard, in 2009 [52], the authors reviewed the future trends of bioceramics in medical appliances for the next 20 years. In this review article, the nano-structure of calcium phosphate-made ceramics and porous bioactive glasses were discussed. From what have been projected there, it has been under research for more than ten years ago up to now. Furthermore, the possibility of combining both with organic phases, and the desired properties for bone substitution and engineered tissues for regeneration purposes were presented [54]. The mechanical properties of glass and glass-ceramics structures indicate that their elastic constants, Poisson's ratios, and their atomic packing depend on the Cortical and Cancellous bone (Fig. 7). Consequently, the scaffolds stiffness modulus of glass-ceramics should match that of the replaced tissue; this will reduce bone resorption problems caused by load transfer and stress shielding [42, 55]. Due to a mismatch in elastic moduli between the biomaterial and the nearby bone, the bone experiences "stress shielding," which is undesirable since healthy bones require some level of tensile load stimulation. In the literature, a wide range of values has been reported for the compressive modulus of 0.12–1.1 GPa and 11.5–17 GPa for trabecular bone and cortical bone respectively, depending on the measurement technique, parameters used, the bone source, and the structural variations within the bone from the same source [56, 57].

Therefore, the critical size defect of bone or soft body tissues was frequently used to be treated by auto and allograft transplantation [58]. This was practiced by Hindus in 600 BC when their surgeons repaired organ deformities limited to the body's surface by the aid of anesthesia and asepsis performed to limit excessive damage to living organs [59, 60]. Susrutha, for example, rebuilt a damaged nose with flesh from the cheek area. This type of orthopedic procedure was very popular, especially when

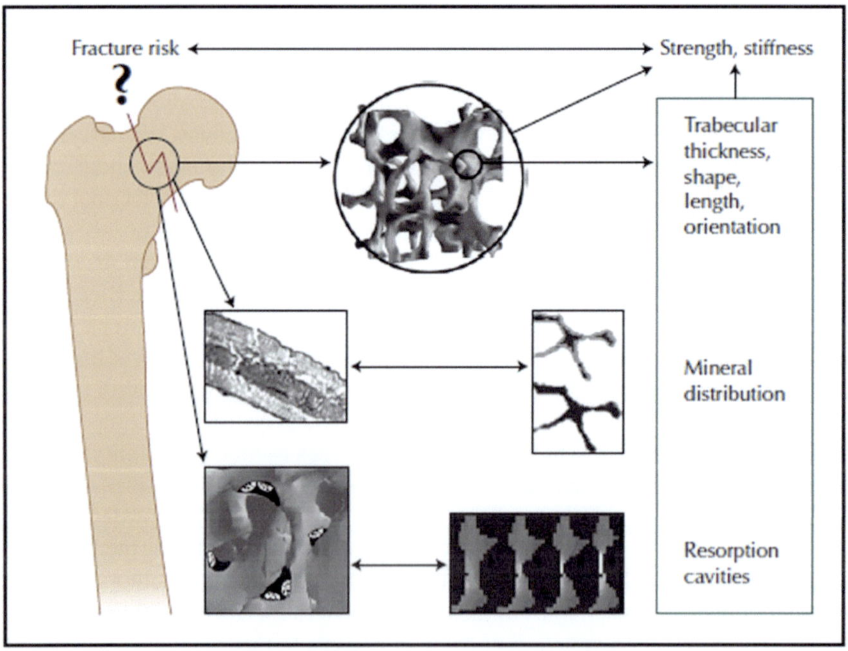

Fig. 7 Bone extraction imaging indicating the resolution of clinical imaging techniques, which gradually become available to enhance the reproducibility of mimicked hard and soft tissues morphology. Reprinted from Van Der Linden and Weinans [57], with permission from Springer Nature. Copyright (2009)

transferring living tissue from one organ to the damaged one [59, 61]. However, such procedures need bone harvesting from patient donor sites, which may be challenging due to tissue morphology scarcity or the danger of surgical complications. Three-dimensionally (3D) printed implants, and biomaterial-based tissue engineering (TE) methods have been promised to be the next generation of bone regeneration therapies because they did not require bone donation [58].

Moreover, in the reviewed article of Fernandes et al. [31], the authors discussed briefly the Zachariasen's ground breaking work [62], which introduced early the concepts in the study of glass structure, and continued to be fundamental in the field of glass and glass ceramic research. According to Zachariasen, both crystals and glasses consist of atoms connected by the same interactions, vibrating around their equilibrium positions. However, the key difference lies in the structural arrangement: where the crystals possess periodicity and symmetry, while glass lacks these properties. Because of this lack of symmetry, the glass unit cell extends infinitely.

To delve deeper into the structure of glass, Zachariasen [62] considered the oxides that form the glass network. In the case of vitreous silica, the structure comprises oxygen tetrahedra surrounding silicon atoms. The corner-sharing linkage of these tetrahedra involves each oxygen atom being connected to two silicon atoms (Fig. 8), which results in avoiding the formation of periodic structures, such as oxygen octahedra or oxygen cubes. Additionally, following their established rules, three components were identified as network formers: SiO_2, B_2O_3, and P_2O_5. These "*Network formers*" have a crucial role in constructing a three-dimensional glass structure, with each component sharing corners through oxygen bridges, commonly referred to as bridging oxygens (BO). Conversely, "*Network modifiers*", known as non-bridging oxygens (NBO), disrupt the glass network by expelling oxygen atoms. Finally, the "*intermediate oxides*" are trace elements or components necessary for acting as either network formers or modifiers. These oxides, such as MgO, Al_2O_3, Ag_2O, ZnO, etc., are incorporated into the glass network as dopants [43, 46, 63, 64].

In traditional glass production, monovalent oxides such as Na_2O are added as fluxing agents to significantly lower the melting temperature and reduce production costs. However, binary compositions of alkali silicate glasses exhibit low chemical durability. For that, to enhance stability and bioactivity, divalent cation oxides such as CaO are introduced into the silicate networks [65]. Consequently, the prepared glass was known to be classified based on the presence of specific network former oxides in the composition, leading to categorizations such as silicate, borosilicate, phosphosilicate, or borophosphate glasses, as documented in the literature [65]. Thus, by controlling the glasses' crystallization process, networks can be controlled to form a glass-ceramic structure through heat treatment, known as sintering. The resulting material consists of residual glass with embedded crystal phases, whose composition depends on the initial glass composition [66]. During crystallization, there are molecular rearrangements within the glass. This process enables the production of crystalline structures, often in the form of metastable polymorphs (Fig. 9). These structures can be transformed into thermodynamically stable crystalline phases through additional and appropriate heat treatment. Glass-ceramic fabrication techniques are cost-effective, offering distinct advantages over other materials. They can

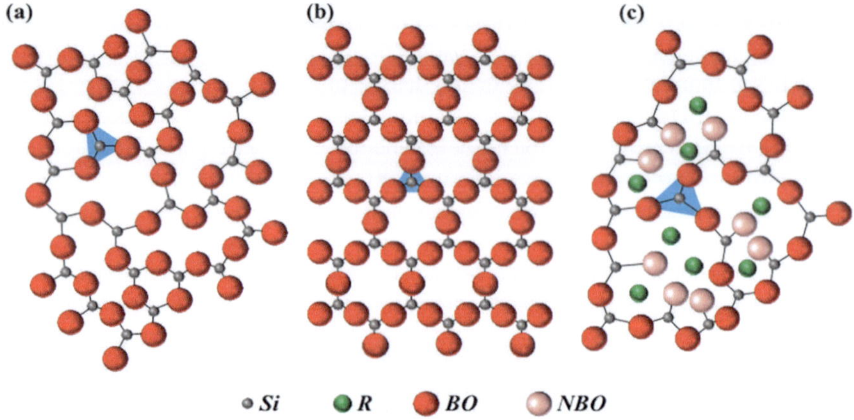

Fig. 8 Schematic representation of 2D network structure for **a** vitreous, **b** crystalline silica, and **c** vitreous silicate, where Si refers to Silicate, R represents a generic network modifier cation, while BO and NBO refer to bridging oxygen and non-bridging oxygen, respectively, Reproduced from Fernandes et al. [31], under the terms of the Creative Commons Attribution (CC BY) license. http://creativecommons.org/licenses/by/4.0/

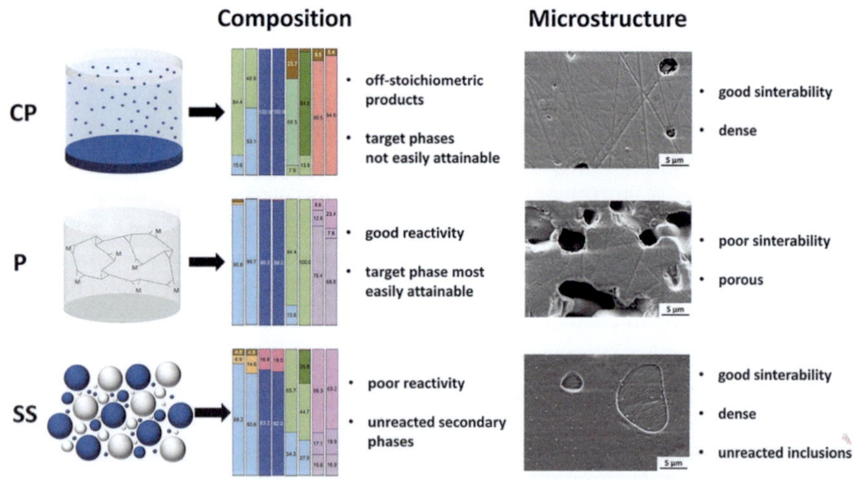

Fig. 9 Quantified crystalline phase wt% composition from XRD patterns: the dashed lines are the targeted phases with SiO_2 phase predominantly quantified as cristobalite, while the microstructure of sintered glass-ceramic pellets is displayed for the selected Zn_2SiO_4. The "CP" stands for Co-precipitation approach, the "P" attributed to sol–gel approach of Maggio Pechini [67], and the SS for solid state approach. Reproduced from Kamutzki et al. [68], under the terms of the Creative Commons license: CC-BY. https://creativecommons.org/licenses/by/4.0/

be shaped into highly complex forms and exhibit superior mechanical properties due to their fine microstructure, which contains minimal or no residual pores.

However, the sintering window, which determines their cellular properties, is crucial for achieving desired characteristics. For instance, 13–93 glass [69] is prone to crystallizing, but remains stable at very high temperatures [30, 70]. On the other hand, the classic Bioglass® (45S5) has a narrow sintering window and tends to crystallize above its glass transition temperature (T_g), resulting in undesirable microstructures [70]. To ensure successful crystallization, various parameters must be carefully controlled during the heat treatment process. These parameters include the heating rate, nucleation temperature, crystallization temperature, and the duration of heating at specific temperatures. Effective crystallization involves initiating the formation of numerous small crystals rather than a small number of large crystals. Following the nucleation stage, crystal growth occurs at elevated temperatures, which is a crucial step in the sintering process [31]. Interestingly, combining glass with ceramics to create a composite material offers initial advantages, especially when utilizing bioceramics. These ceramics can be categorized as relatively bioinert, bioactive (or surface reactive), or bioresorbable, depending on their properties. Moreover, bioceramics can be designed to be porous, facilitating tissue ingrowth [51]. Consequently, they serve as replacement materials for prostheses such as hip or knee joints. This is primarily due to their favorable mechanical properties, which include high strength, toughness, and wear resistance [71].

3 Bioactive Glass and Bioactive Glass–Ceramics Composites for Implants

While working with his associate, the discovery of 45S5 bioactive glass by renowned researcher iconic Larry Hench in the late 1969 was a great start to the exploration of biomaterials for orthopedic appliances. The team developed an implant material that could bond with soft body tissues [32]. During this period, the inert metals and plastic implants were not compatible with the host due to the elicit reaction of the surface morphology of the implants [31]. Due to several attempts from various researchers for the bioactive glass itself, the implants, when implanted, presented loosening and low cytocompatibility, ion release from their constituents, and tend to induce problems to the host. Accordingly, the researchers looked at enhancing bioactive glass' biological and response properties to introduce innovative tissue engineering using a different class of biomaterials ranging from natural to synthetic materials. These are trace elements incorporated in the bioactive glass composition to handle properties overcoming these elicit problems and implant failure [72, 73].

Successful doping of bioactive glass and bioactive glass-ceramics by some specific metal ions is effective in various biofunctionalities such as cancer treatment, drug delivery, and biological properties. It has also affected properties such as angiogenic and osteogenic performance as well as antimicrobial activity. Bioactive glass-ceramics are often used for fabrication of hard tissue recovery and prosthesis coatings due to biocompatibility, biodegradability, and osteoconduction from such composition of calcium phosphate (hydroxyapatites (HA), α- and β-tricalcium phosphate (TCP)), Calcium silicate for wollastonite or hardystonite and most prepared bioactive glass composed of SiO_2, CaO, Na_2O and P_2O_5. Doping of bioactive glass-ceramics by trace elements such as Ag^{2+} and Cu^{2+} helps in the mineralization and metabolic processes to prevent microbial infections [74]. Furthermore, zinc-containing ceramic composites are opening up new possibilities in bone tissue engineering research [75].

Bioactive glass composite is synthesized by incorporating trace elements or combining with ceramics and polymers to make hybrid materials. Therefore, each element in the composite is controlled depending on the biofunction of bioactive glasses and bioactive glass-ceramics. Another route is hybrid materials through joining layer by layer. Electrospun materials are examples of this process. An example is given by Saatchi et al. [76], who synthesized and characterized cerium doped bioactive glass, that is, chitosan (Ce-BG), and used electrospinning to make a bioactive glass polyethylene oxide composite scaffold with the purpose to obtain an engineered tissue. In addition, the theory behind this was the biological lever of these composite tissues with cerium content into BG chitosan, which could significantly enhance the cytocompatibility of bioactive glass scaffolds tissue. The in vitro test proved this, where culturing the tissue with fibroblast cells, showed that the more excellent 8Ce-BG content increases cell attachment and expansion on a fiber mesh. Not only that, but this material has the ability of thorough dispersion of the development of submicrometric clusters in various polymeric matrices (Fig. 10) attached to the composite through electrospinning techniques [76]. Researchers have worked on the synthesis route as a technique that can incorporate different elements and obtain hybrid material with covalent bonds rather than van der Waals forces [58]. Unfortunately, this route is quite challenging for elements with different mechanical properties, such as thermal expansion. The coating is another way for bioactive glasses and bioceramics coated implant cores to opt for this, by changing composite structures [77].

An example is given for the metallic or polymers implants that possess great mechanical properties but low biocompatibility. Coating's process increases their surface bioactivity by joining them with bioactive glasses or bioactive glass-ceramics, as well as their mechanical strength, and toughness are enhanced. Moreover, there have been developments in utilizing polymeric biomaterials to create polymer-glass composites [78, 79]. These composites combine the bioactivity of bioactive glass with flexible polymers. In addition, they provide wear resistance from deposited layers on polymer-bioactive glass. For instance, the application of polyether ether ketone (PEEK)/BG coatings through an electrophoretic deposition process resulted in a tenfold improvement in corrosion resistance for steel specimens [80]. Similar to coating a metallic implant with bioactive glass has advantage of the mechanical strength of the implants bulk material while retaining bioactive glass effects at the surface. In addition to this, numerous techniques are listed in Table 2, indicating the associated procedure used in research to form these composites with bioactive glasses and glass-ceramics. Some of the most noticeable modifications have used various forms of carbon to improve the properties of 45S5 bioactive glass [81].

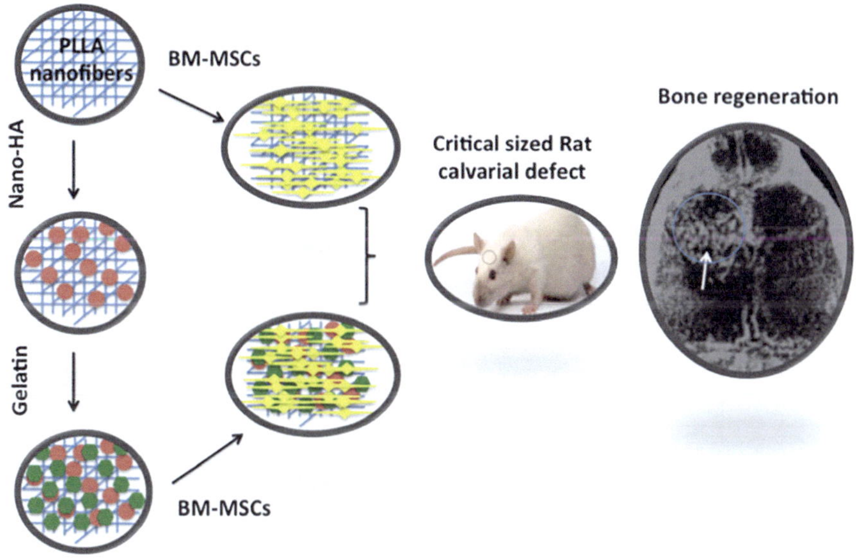

Fig. 10 Polymer fibers electrospun for in vivo application were used in a rat study. Radiographic analysis and macroscopic observation revealed critical-sized rat calvarial defects. Therefore, scaffolds with and without human bone marrow-derived mesenchymal stem cells (BM-MSCs) were implanted at the defect site for eight weeks. The results did not reveal any inflammation, infection, or bleeding. Interestingly, all implanted scaffolds were perfectly integrated into the bone at the defect site without any encapsulation or noticeable undesirable body response. Reprinted from Andalib et al. [6], with permission from Springer Nature. Copyright (2021)

In the study of Touri et al. [82], they developed a method to incorporate carbon nanotubes (CNTs) into the structure without interfering with the materials bioactive properties. The CNTs were chosen based on their large aspect ratio and high strength. This technique incorporates the synthesized bioactive glass 45S5 on a CNT scaffold shape by creating a composite that doubles the compressive strength and the elastic modulus compared to the pure glass [82]. The polymeric coatings technique, a similar process has been explored to improve the mechanical properties of metallic implants. It is a naturally biocompatible process to facilitate the biological fixation of implants with a native extracellular matrix. This coating promotes the osseointegration process and reduces infection while replacing failed tissue or bone [8]. The coating process of polymer on metallic implants improves cell-material interaction and provides structural support to the newly formed tissues. Polymeric coatings can also prevent the body fluid's exposure to the biodegradable metallic implants, such as alloying of magnesium, which presents low corrosion resistance by adding a protective layer of an appropriate polymer coating. In addition, coatings may enhance porous structure as cells thrive through the interconnected pores during regeneration [4, 83].

Even though the metal ions are used with bioactive glass, their ion release and bioactivity degradation rate must be controlled at the synthesis and in vivo applications. An example of this is Mg-based metallic implants that outperform other

Table 2 Strategies for surface and bulk modification

Strategy	Dopants	Composite	Biofunctions	Refs.
Coating's techniques of BG	Mg^{2+}	Calcium orthophosphate magnesium composite	• Bone mineralization • Influencing osteoblast • used as a dopant source of Mg^{2+} to increase mechanical strength and decrease the rate of matrices degradation	[84, 85]
Modified Stöber process	Ag^{+} Sr^{2+}	Ag-Sr MBGNs[a] $70SiO_2$-(30-x)CaO	• Bacteriostatic ions increase and control antimicrobial properties • Enhancing biocompatibility, non-toxicity and exhibiting good thermal and physiological medium stability	[84, 86]
Sol–gel process	Cu^{2+}	CuBG-CS[b] $60SiO_2$, (36-x)CaO, $4P_2O_5$, xCuO	• Stimulating angiogenesis capacity and osteogenesis • Adding antibacterial properties in bioactive glass and bioceramics composite • Reducing bone infection such as osteomyelitis	[87]
Sol–gel derived	Sr^{2+}	Sr MBGNs[c] $60SiO_2$-(36-x)Cao-$4P_2O_5$-xSrO	• Stimulating osteogenesis, angiogenesis • Antibacterial ions • Enhancing osteoporosis treatment	[84, 86, 88]
350 mT magnetic field	$Fe^{2+ \text{ or } 3+}$	Bioactive glass (BG)-based composites reinforced with bismuth ferrite (BF)	• Aiding in the oxygen transport • Drug delivery elements when doped in BG or bioceramics • Enhanced bone tissue formation and accelerated osseointegration under magnetic field exposure	[83]

(continued)

Table 2 (continued)

Strategy	Dopants	Composite	Biofunctions	Refs.
Sol–gel for BG	Zn^{2+}	Glass-ceramic $64SiO_2$-26CaO-$5P_2O_5$-5ZnO	• Antibacterial effect • Aids in bone regeneration and wound healing • Promotes angiogenesis and homeostasis	[84, 89]
*Sol–gel for BG *Electrospinning for nanofibrous scaffolds	Ce^{2+}	8CeBG/chitosan/polymer $60SiO_2$-28CaO-$4P_2O_5$-$8Ce_2O_3$	• Enhance cytocompatibility • Enhance fibroblast cell attachment and cell expansion on a fiber mesh	[76]
Electrospinning Sol–gel	Polycaprolactone	Mesoporous silica layered hybrid nanofibers	• Silica shell provides an active biointerface • The 3D nanoscale fibrous structure provides a cell-stimulating matrix suitable for bone regeneration • Good mechanical properties (tensile strength and elastic modulus) • Promising therapeutic platforms for stimulating stem cells • Wound healing and bone regeneration	[25]
Fused deposition modelling (FDM) for composite printing	Polylactic acid (PLA)	45S5-PLA • d50: 4.0 ± 1.0 μm • d95: ≤20 μm	• BG inclusions lead to increased osteogenic differentiation • PLA-BG composite scaffolds are suitable for bone tissue engineering	[58]
Solvent-based extrusion 3D bioprinting with a modified cartesian 3D printer	Polycaprolactone (PCL), borate	13-93B3 glass-PCL $53B_2O_3$, 20CaO, $12K_2O$, $6Na_2O$, 5MgO, $4P_2O_5$	• Promote angiogenesis • Degradation of composition glass scaffolds possess a controlled release of bioactive glass with ≈23% weight loss in two weeks	[90]

[a] MBGNs: Mesoporous bioactive glass nanoparticles
[b] CuBG-CS: Copper-doped bioactive glass collagen scaffolds
[c] Sr MBGNs: Strontium-doped mesoporous bioactive glass nanoparticles

implanted metals in terms of biofunction. This approach has been used to eliminate the effects of stress shielding and prevent a second surgery from removing the implant. Mg degrades at a quick stage of implantation rate, and it incorporates biodegradable alloys in contact with blood cells in the physiological environment. Such a family of biodegradable metals, also known as bioresorbable or bioabsorbable metals that may degrade relatively safely within the body, has been found to be an attractive biomaterial to use. This is the category of magnesium-based and iron-based alloys. Although zinc and the biodegradable magnesium alloys coated with calcium orthophosphate were investigated [85, 91]. Kraus et al. [92] also investigated Mg alloys for temporary implants intended to be used in osteosynthesis. The in vivo studies have been conducted on a growing rat skeleton for their degradation and interaction with bone remodeling after implant dissolution. The study was performed every 4 weeks over the 24 weeks of the test. The finding was that the earlier degradation of the fast degrading (ZX50) Mg alloy compared to the slowly degrading (WZ21) Mg alloy, that kept their veracity for this period and showed good osteoconductive, enhancing bone accumulation at the surface layer [92].

Furthermore, specific metal biomaterials can spontaneously corrode in the physiological environment, making them excellent candidates for creating temporary implants such as stents [85, 93]. Thus, implants and engineered tissues must be developed with regulated degradation rates to match the kinetics of healing and regeneration. A suitable option appears to be the application of protective elements that are biocompatible and biodegradable, capable of delaying the beginning of biocorrosion.[4] Calcium orthophosphates appear to be ideal candidates for such coatings since they are readily tolerated by living organisms [85].

4 Composites of Bioactive Glass and Glass–Ceramics as Porous Implants

Porous biomaterials are engineered materials that interact with biological systems for medical purposes, such as treating, augmenting, repairing, or replacing a tissue function in the body, or diagnosing a disease. They have also been employed as a matrix for bone regeneration, such as calcium phosphates, calcium sulfates, and bioactive glasses. Metals in various forms for implants have been created to aid human recovery from various bone disorders or fractures [94]. Thanks to manufacturing mechanisms including additive manufacturing methods with their functionality for mass fabrication of porous metallic implants with different geometrical forms and complex objects [95]. This fabrication was prominently adopted for the very promising production of clinical implants. Moreover, enhanced metals with bioactive glasses and ceramics

[4] *Biocorrosion*: A process in which metabolic activities of microorganisms associated with metallic materials supply insoluble products that are able to accept electrons from the base of metal biomaterial. i.e., in the biofilms on the surface of the corroding material.

are still the biomaterial of choice for each machining process to enhance properties based on several clinical applications [11].

Furthermore, the term "porous implant" comes from the biomimicking host organ of the implant. Furthermore, the incorporation of bioactive glasses stands for biocompatibility. One of the advantages of porous structure implant is that it stimulates the in-growth and allows permeability of the blood cells. The distinction between open-cell and closed-cell porous metals, or metal foams, is a key distinction. Individual cells in closed-cell foams are surrounded by a thin metal wall or membrane, but individual cells in open-cell foams are interconnected, allowing the tissue to enter the foam and hold it into place. Partially or porous-coated solid substrates, fully porous materials, and porous metal segments linked to a solid metallic component are all designated as distinct types of porous implants [96]. A unique three-dimensional, multiscale interconnected porous tantalum scaffold implant was investigated to explore a novel trabecular bones biocompatibility and bone ingrowth characteristics. The implant mimicked the bone structure was constructed to hold the potential for bone tissue engineering [97]. A range of tests required to validate the capability of implant need to be specified. The morphology and structure are characterized by scanning electron microscopy (SEM) and some industrial computerized tomography. Bioactivity and biological response are required for the in vivo and in vitro test through culturing and exposing samples to stem cells or using simulated body fluid and buffered saline solution. In that study of Wang et al. [97], hard tissue slice inspection and push-out tests were used to assess the healing effects and bone integration at the given times.

Therefore, the SEM imaging was used to examine cell shape and adherence, and the 3-(4,5-dimethylthiazol-2-yl)-2,5-diphenyl-2*H*-tetrazolium bromide (MTT) test was used to assess cytotoxicity. The bone ingrowth and regeneration start from the third week to six weeks after implantation (Fig. 11). The new porous structures based on the lightweight implant are in the current research as shown in the literature, therefore, interests to the researchers as well as the industrial manufacturer. There is great potential in biomedical applications for three-dimensional (3D) printing [98]. These techniques of forming bioinspired [98, 99] or metamaterials [30] are a response to additive manufacturing of biomaterials through the approaches and philosophy of machinability. Porous structures based on triply periodic minimum surfaces (TPMS) have emerged as potential bone replacement scaffold design options with excellent porosity and strength [100]. An example is given for a modified femoral stem design that substantially improved stress-shielding and the proximal femoral bone. This prospective study assessing total hip arthroplasties, as documented in ref. [101].

Aside from macro implant design and the creation of a porous structure, lowering the overall stiffness of the prosthesis [102], a technique that reduces the effect of stress-shielding by enhancing bone-implant bonding using bioactive coating materials [103, 104]. An example, is bioactive glass coated metal implant. Therefore, the stress shielding effect on the outer layer surface of the implant has been extensively studied and acknowledged in the field of bone research. However, the bonding strength between the coating layer and core in implantology remains controversial [105, 106].

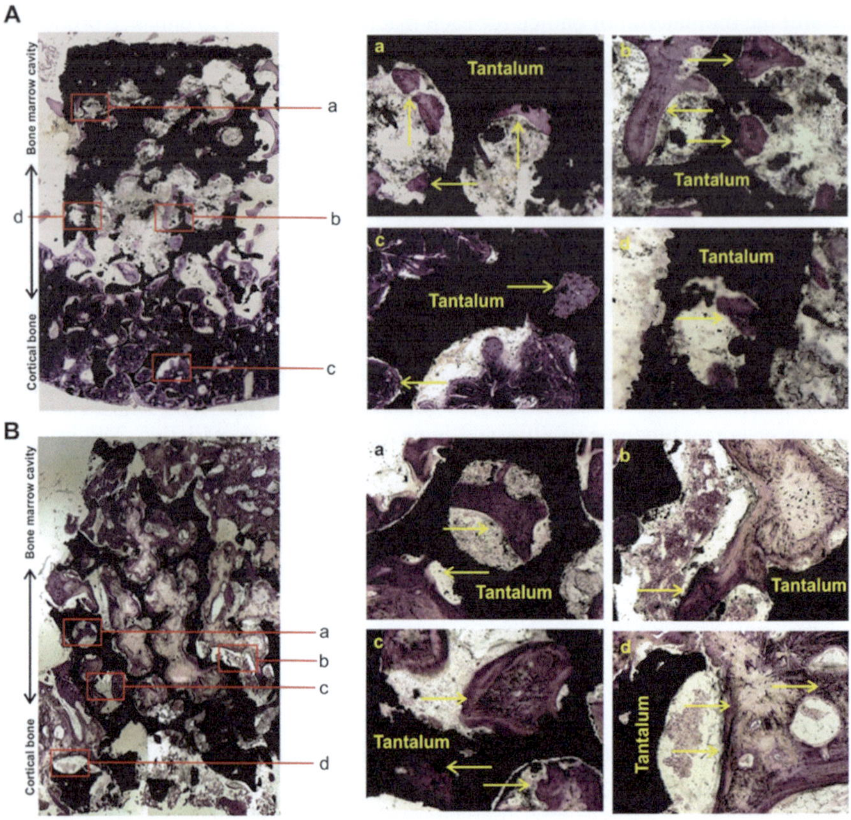

Fig. 11 Methylene blue staining for retrieved tantalum rod-bone samples. **A** Methylene blue staining for retrieved tantalum rod-bone samples at three months after the implantation. A merged panoramic version in the left panel and distinct-separated images (**a**–**d**) are listed in the right panel; arrows indicate the new bone and tantalum rod interfaces. **B** Methylene blue staining for retrieved tantalum rod-bone samples at 6 months after the implantation. A merged panoramic version in the left panel and distinct-separated (**a**–**d**) are listed in the right panel; arrows indicated the new bone and tantalum rod interfaces. Reprinted from Wang et al. [97], with permission from, American Chemistry Society (ACS). Copyright (2020). https://doi.org/10.1021/acsomega.0c03127. Note to readers: Further permissions related to the material excerpted should be directed to the ACS

5 Conclusions

In the current chapter, the synthesis processes, properties, and medical applications of bioactive glasses and glass ceramics, with a particular focus on their relevance in implants and medical settings were discussed. The versatility of bioactive glasses and glass ceramics extends beyond regenerative medicine, finding applications in eyeglasses, diagnostic tools, chemical ware, thermometers, tissue culture flasks, and fiber optics for endoscopy within the healthcare sector. The modification of these biomaterials into micro and macro composites were found utilization in additive manufacturing techniques to create porous implants and scaffold tissues. Significant progress has been observed in the biomedical field, especially ceramic materials tailored for bone repair and reconstruction. This progress reflects a breakthrough in regenerative medicine, with bioactive glasses and composites playing a pivotal role. In the realm of research, the incorporation of trace elements into bioactive glass networks has been explored to enhance biocompatibility, facilitate drug delivery, and exert antibacterial effects in vivo. Achieving meticulous control over the "biodecoration" of implant surfaces at the molecular level, it was identified as crucial for promoting bone tissue regeneration. These advancements have not only improved the physicomechanical properties of bioactive glass but have also influenced the integration ability of implants with the host. In this comprehensive chapter underscored the remarkable potential of bioactive glasses and glass ceramics as versatile biomaterials in regenerative medicine and beyond. The progress made to date serves as a foundation for continued research and innovation, promising further advancements in the treatment and restoration of damaged or diseased tissues and ultimately enhancing patients' healthcare.

References

1. A. Ross, Zirconia Bioceramics in Metal Free Implant Dentistry, https://ceramics.org/event-session/gfmat-2-29. Accessed 17 January 2021
2. J. Han, J. Zhao, Z. Shen, Adv. Appl. Ceram. **116**, 138 (2017)
3. S. José Gobbi, V. João Gobbi, Y. Rocha, Biomed. J. Sci. Tech. Res, **14**, 002554 (2019)
4. S. Bose, D. Ke, H. Sahasrabudhe, A. Bandyopadhyay, Prog. Mater. Sci. **93**, 45 (2018)
5. X. Wang, *Advances in Biomaterials Science and Biomedical Applications*, ed. by R. Pignatello (IntechOpen, 2013). https://doi.org/10.5772/53461
6. N. Andalib, M. Kehtari, E. Seyedjafari, N. Motamed, M.M. Matin, Cell Tissue Bank. **22**, 467 (2021)
7. I.L. Tsiklin, A.V. Shabunin, A.V. Kolsanov, L.T. Volova, Polymers (Basel) **14**, 3222 (2022)
8. J. Adhikari, P. Saha, A. Sinha, *Fundamental Biomaterials: Metals*, ed. by P. Balakrishnan, M.S. Sreekala, S. Thomas (Woodhead Publishing, 2018), pp. 299–321
9. P.N. Kumta, A.R.D.-T. Chou, D.H. Hong, V.S. Gorantla, *Degradable Magensisum-Based Implant Devices for Bone Fixation: WO2017035072A1*, PTC/US2016/048035 (2015)
10. D. Wang, G.Wu, X. Lin, Y. Liu, *Surface Coating and Modification of Metallic Biomaterials*, ed. by C. Wen (Woodhead Publishing, 2015), pp. 345–358
11. M. Saini, World J. Clin. Cases **3**, 52 (2015)
12. D.F. Williams, Biomaterials **29**, 2941 (2008)
13. M. Djošić, A. Janković, V. Mišković-Stanković, Materials **14**, 5391 (2021)
14. Z. Younoszai, M. Dobias, E. McAndrew, L. Richardson, K. Talkington, FDA's Regulatory Framework for 3D Printing of Medical Devices at the Point of Care Needs More Clarity, https://www.pewtrusts.org/en/research-and-analysis/issue-briefs/2022/07/fdas-regulatory-framework-for-3d-printing-of-medical-devices-needs-more-clarity. Accessed 20 September 2022
15. World Trade Organization Technical Barriers to Trade (TBT) Committee, *ISO/ASTM 52900: Standard Terminology for Additive Manufacturing*
16. M.J. Zafar, D. Zhu, Z. Zhang, Materials (Basel) **12**, 3361 (2019)
17. M. Wester, Catal. Undergrad. Sci. Res. **10**, 61 (2017)

18. T.D. Ngo, A. Kashani, G. Imbalzano, K.T.Q. Nguyen, D. Hui, Compos. Part B Eng. **143**, 172 (2018)
19. The American Ceramics Society, Aircraft, Space VehiclesCeramics and Glass in the Aerospace Industry—The American Ceramic Society, https://ceramics.org/about/what-are-engineered-ceramics-and-glass/ceramics-and-glass-in-the-aerospace-industry. Accessed 24 June 2023
20. S. Kirihara, Ceramic Additive Manufacturing and Integration Technologies, https://ceramics.org/event-session/gfmat-2-13. Accessed 18 January 2021
21. K. Sharma, M.A. Mujawar, A. Kaushik, Front. Mater. **6**, 172 (2019)
22. W. Rosamond, K. Flegal, G. Friday, K. Furie, A. Go, K. Greenlund, N. Haase, M. Ho, V. Howard, B. Kissela, S. Kittner, D. Lloyd-Jones, M. McDermott, J. Meigs, C. Moy, G. Nichol, C.G. O'Donnell, V. Roger, J. Rumsfeld, P. Sorlie, J. Steinberger, T. Thom, S. Wasserthiel-Smoller, Y. Hong (American Heart Association Statistics Committee and Stroke Statistics Subcommittee, Heart disease and stroke statistics—2007 update: a report from the American Heart Association Statistics Committee and Stroke Statistics Subcommittee. Circulation **115**(5), e69–171 (2007). https://doi.org/10.1161/CIRCULATIONAHA.106.179918. Epub 2006 Dec 28. Erratum in: Circulation **115**(5), e172 (2007). Erratum in: Circulation **122**(1), e9 (2010). Kissela, Bret [corrected to Kissela, Brett]. PMID: 17194875)
23. B. Baroli, J. Pharm. Sci. **96**, 2197 (2007)
24. J.R. Venugopal, M.P. Prabhakaran, S. Mukherjee, R. Ravichandran, K. Dan, S. Ramakrishna, J.R. Soc, Interface **9**, 1 (2012)
25. R.K. Singh, G.-Z. Jin, C. Mahapatra, K.D. Patel, W. Chrzanowski, H.-W. Kim, A.C.S. Appl, Mater. Interfaces **7**, 8088 (2015)
26. J.D. Caplin, A.J. García, Acta Biomater. **93**, 2 (2019)
27. Z. Ma, Z. Mao, C. Gao, Colloids Surf. B Biointerfaces **60**, 137 (2007)
28. H.J. Chung, T.G. Park, Adv. Drug Deliv. Rev. **59**, 249 (2007)
29. L. Yuan, S. Ding, C. Wen, Bioact. Mater. **4**, 56 (2019)
30. A. Zadpoor, Int. J. Mol. Sci. **18**, 1607 (2017)
31. H.R. Fernandes, A. Gaddam, A. Rebelo, D. Brazete, G.E. Stan, J.M.F. Ferreira, Materials (Basel) **11**, 2530 (2018)
32. D. Greenspan, Biomed. Glas. **5**, 178 (2019)
33. C.R. Arciola, D. Campoccia, G.D. Ehrlich, L. Montanaro, Adv. Exp. Med. Biol. **830**, 29 (2015)
34. L.L. Hench, Biomed. Glas. **1**, 1 (2015)
35. A.-M. Pobloth, M.J. Mersiowsky, L. Kliemt, H. Schell, A. Dienelt, B.M. Pfitzner, R. Burgkart, R. Detsch, D. Wulsten, A.R. Boccaccini, G.N. Duda, Sci. Rep. **9**, 16692 (2019)
36. L.L. Hench, J.R. Jones, Front. Bioeng Biotechnol. **3**, 194 (2015)
37. L.L. Hench, *An Introduction to Bioceramics*, 2nd edn., ed. by L.L. Hench (Imperial College Press, 2013), pp. 71–85
38. O.P. Filho, G.P. La Torre, L.L. Hench, J. Biomed. Mater. Res. **30**, 509 (1996)
39. M.R. Filgueiras, G. La Torre, L.L. Hench, J. Biomed. Mater. Res. **27**, 445 (1993)
40. A. El-Ghannam, P. Ducheyne, *Comprehensive Biomaterials*, ed. by P. Ducheyne (Elsevier Ltd., 2011), pp. 157–179
41. A.R. El-Ghannam, J. Biomed. Mater. Res. **69**, 490 (2004)
42. L.-C. Gerhardt, A.R. Boccaccini, Materials (Basel) **3**, 3867 (2010)
43. Q. Ju, T. Zenji, A.L.B. Maçon, E. Norris, G. Poologasundarampillai, A. Obata, J.R. Jones, T. Kasuga, Biomater. Adv. **134**, 112561 (2022)
44. X. Kesse, C. Vichery, J.M. Nedelec, ACS Omega **4**, 5768 (2019)
45. I.E. Pinto e Souza, S.M. de Carvalho, T. Martins, M.d.M. Pereira, Mater. Res. **23**, 20200070 (2020)
46. P. Munyensanga, M. Bricha, K. El Mabrouk, Mater. Chem. Phys. **305**, 127969 (2023)
47. K. Zheng, A. Solodovnyk, W. Li, O.M. Goudouri, C. Stähli, S.N. Nazhat, A.R. Boccaccini, J. Am. Ceram. Soc. **98**, 30 (2015)
48. I. Izquierdo-Barba, A.J. Salinas, M. Vallet-Regí, Int. J. Appl. Glas. Sci. **4**, 149 (2013)

49. S.F. Hulbert, J.C. Bockros, L.L. Hench, J. Wilson, G. Heimke, *High Tech Ceramics* (Elsevier, Amsterdam, 1987), p.189
50. E. Zeimaran, S. Pourshahrestani, S.F.S. Shirazi, B. Pingguan-Murphy, N.A. Kadri, M.R. Towler, J. Non Cryst. Solids **443**, 118 (2016)
51. S.V. Dorozhkin, Biomaterials **31**, 1465 (2010)
52. J. Chevalier, L. Gremillard, J. Eur. Ceram. Soc. **29**, 1245 (2009)
53. J. Foreman, Additive Manufacturing Offers Fast and Simple Way to Fabricate Bioactive Glass-Ceramics, https://ceramics.org/ceramic-tech-today/biomaterials/additive-manufacturing-offers-fast-and-simple-way-to-fabricate-bioactive-glass-ceramics. Accessed 20 January 2021
54. S. Pina, R. Rebelo, V.M. Correlo, J.M. Oliveira, R.L. Reis, *Osteochondral Tissue Engineering*, ed. by J. Oliveira, S. Pina, R. Reis, J. San Roman, Part of the Advances in Experimental Medicine and Biology, vol. 1058 (Springer, Cham, 2018), pp 53–75
55. H.M. Frost, Bone **23**, 395 (1998)
56. S.C. Francisco, *Mechanical Properties of Glass-Ceramics: Public Lecture*, Center for Research, Technology and Education in Vitreous Materials (Federal University of São Carlos, Brazil, 2015), https://drive.google.com/file/d/0Bwbe0R6uLmNOZHkxcDZSTTgxc2M/view?pli=1&resourcekey=0-MAXa7jPhj-xOXKuvbe_naQ. Accessed 25 June 2024
57. J.C. Van Der Linden, H. Weinans, Curr. Osteoporos. Rep. **5**, 56 (2007)
58. T. Distler, N. Fournier, A. Grünewald, C. Polley, H. Seitz, R. Detsch, A.R. Boccaccini, Front. Bioeng Biotechnol. **8**, 552 (2020)
59. A. Balamurugan, S. Rajeswari, G. Balossier, A.H.S. Rebelo, J.M.F. Ferreira, Mater. Corros. **59**, 855 (2008)
60. S. El Baakili, P. Munyensanga, M. Bricha, K. El Mabrouk, Porous metallic implants from additive manufacturing to biocorrosion: a review. Johnson Matthey Technol. Rev. **68**, 71 (2024)
61. Sushruta: Sushruta Samhita, *Bhishagratna: An English Translation of the Sushruta Samhita, Based on Original Sanskrit Text* (Calcutta, Bose, 1907), pp. 156–158
62. W.H. Zachariasen, J. Am. Chem. Soc. **54**, 3841 (1932)
63. S. Kapoor, A. Goel, A.F. Correia, M.J. Pascual, H.Y. Lee, H.W. Kim, J.M.F. Ferreira, Mater. Sci. Eng. C Mater. Biol. Appl. **53**, 252 (2015)
64. M. Khoeini, S. Hesaraki, A. Kolahi, Ceram. Int. **47**, 31666 (2021)
65. J.E. Shelby, *Introduction to Glass Science and Technology*, 2nd edn. (The Royal Society of Chemistry, New York, NY, 2005), pp.7–25
66. R.F. Pettifer, *Glasses and Glass-Ceramics*, ed. by M.H. Lewis (Springer Netherlands, Dordrecht, 1989), pp. 41–58
67. R. Mohassel, A. Sobhani, M. Salavati-Niasari, M. Goudarzi, Spectrochim Acta Part A Mol. Biomol. Spectrosc. **204**, 232 (2018)
68. F. Kamutzki, M.F. Bekheet, S. Schneider, A. Gurlo, D.A.H. Hanaor, Open Ceram. **9**, 100241 (2022)
69. A.A. Mehatlaf, S.B.H. Farid, A.A. Atiyah, I.O.P. Conf, Ser. Mater. Sci. Eng. **1067**, 012136 (2021)
70. E.L. Bourhis, *Glass: Mechanics and Technology*, 2nd edn. (Wiley, 2014), pp. 55–84
71. A.N. Cormack, A. Tilocca, Phil. Trans. R. Soc. A **370**, 1271 (2012)
72. M.N. Rahaman, D.E. Day, B. Sonny Bal, Q. Fu, S.B. Jung, L.F. Bonewald, A.P. Tomsia, Acta Biomater. **7**, 2355 (2011)
73. Q. Chen, G.A. Thouas, Mater. Sci. Eng. R Rep. **87**, 1 (2015)
74. V. Modelski Schatkoski, T. Larissa do Amaral Montanheiro, B. Rossi Canuto de Menezes, R. Monteiro Pereira, K. Faquine Rodrigues, R. Guimarães Ribas, D. Morais da Silva, G. Patrocínio Thim, Ceram. Int. **47**, 2999 (2021)
75. I. Bagherpour, S.M. Naghib, A.H. Yaghtin, IET Nanobiotechnol. **12**, 895 (2018)
76. A. Saatchi, A.R. Arani, A. Moghanian, M. Mozafari, Ceram. Int. **47**, 260 (2021)
77. B.D. de Greñu, R. de los Reyes, A.M. Costero, P. Amorós, J.V. Ros-Lis, Nanomaterials **10**, 1092 (2020)
78. A. Nawaz, M.A. Ur Rehman, J. Appl. Polym. Sci. **138**, 50220 (2021)

79. X. Song, D. Shi, P. Song, X. Han, Q. Wei, C. Huang, High Temp. Mater. Process. **40**, 1 (2021)
80. M. Atiq Ur Rehman, F.E. Bastan, B. Haider, A.R. Boccaccini, Mater. Des. **130**, 223 (2017)
81. Z. Li, N.W. Khun, X. Tang, E. Liu, K.A. Khor, J. Mech. Behav. Biomed. Mater. **65**, 77 (2017)
82. R. Touri, F. Moztarzadeh, Z. Sadeghian, D. Bizari, M. Tahriri, M. Mozafari, BioMed Res. Int. 465086 (2013)
83. C. Khatua, S. Bodhak, B. Kundu, V.K. Balla, Materialia **4**, 361 (2018)
84. S.M. Rabiee, N. Nazparvar, M. Azizian, D. Vashaee, L. Tayebi, Ceram. Int. **41**, 7241 (2015)
85. S.V. Dorozhkin, Acta Biomater. **10**, 2919 (2014)
86. S. Bano, M. Akhtar, M. Yasir, M. Salman Maqbool, A. Niaz, A. Wadood, M.A. Ur Rehman, Gels **7**, 34 (2021)
87. E.J. Ryan, A.J. Ryan, A. González-Vázquez, A. Philippart, F.E. Ciraldo, C. Hobbs, V. Nicolosi, A.R. Boccaccini, C.J. Kearney, F.J. O'Brien, Biomaterials **197**, 405 (2019)
88. Z. Goudarzi, A. Ijadi, A. Bakhriari, S. Eskandarinezhad, N. Azizabadi, M. Asgari Jazi, J. Compos. Compd. **2**, 105 (2020)
89. A. Balamurugan, G. Balossier, S. Kannan, J. Michel, A.H.S. Rebelo, J.M.F. Ferreira, Acta Biomater. **3**, 255 (2007)
90. C. Murphy, K. Kolan, W. Li, J. Semon, D. Day, M. Leu, Int. J. Bioprinting **3**, 53 (2017)
91. Wikipedia, Bioresorbable Metal—Wikipedia, https://en.wikipedia.org/wiki/Bioresorbable_metal. Accessed 18 February 2021
92. T. Kraus, S.F. Fischerauer, A.C. Hänzi, P.J. Uggowitzer, J.F. Löffler, A.M. Weinberg, Acta Biomater. **8**, 1230 (2012)
93. M.P. Staiger, A.M. Pietak, J. Huadmai, G. Dias, Magnesium and its alloys as orthopedic biomaterials: a review. Biomaterials **27**, 1728 (2006)
94. S. Arabnejad, R. Burnett Johnston, J.A. Pura, B. Singh, M. Tanzer, D. Pasini, Acta Biomater. **30**, 345 (2016)
95. J. Ni, H. Ling, S. Zhang, Z. Wang, Z. Peng, C. Benyshek, R. Zan, A.K. Miri, Z. Li, X. Zhang, J. Lee, K.-J. Lee, H.-J. Kim, P. Tebon, T. Hoffman, M.R. Dokmeci, N. Ashammakhi, X. Li, A. Khademhosseini, Mater. Today Bio **3**, 100024 (2019)
96. F. Matassi, A. Botti, L. Sirleo, C. Carulli, M. Innocenti, Clin. Cases Miner. Bone Metab. **10**, 111 (2013)
97. X. Wang, Z. Zhu, H. Xiao, C. Luo, X. Luo, F. Lv, J. Liao, W. Huang, ACS Omega **5**, 22520 (2020)
98. J. Liu, N.B. Mohd Rafiq, L.M. Wong, S. Wang, Front. Chem. **9**, 768007 (2021)
99. M.A. Meyers, P.-Y. Chen, A.Y.-M. Lin, Y. Seki, Prog. Mater. Sci. **53**, 1 (2008)
100. V. Rati, N. Singh, S. Rai, S. Kumta, *Advances in Computational Methods in Manufacturing*, ed. by R. Narayanan, S. Joshi, U. Dixit, Part of the Lecture Notes on Multidisciplinary Industrial Engineering book series (LNMUINEN) (Springer, Singapore, 2019), pp. 955–966
101. A. Herrera, J. Mateo, A. Lobo-Escolar, J.J. Panisello, E. Ibarz, L. Gracia, J. Arthroplasty **28**, 1160 (2013)
102. M. Dallago, V. Fontanari, E. Torresani, M. Leoni, C. Pederzolli, C. Potrich, M. Benedetti, J. Mech. Behav. Biomed. Mater. **78**, 381 (2018)
103. H.S. Hedia, N. Fouda, Comput. Mater. Sci. **87**, 83 (2014)
104. A. Vladescu, C.M. Cotrut, F. Ak Azem, M. Bramowicz, I. Pana, V. Braic, I. Birlik, A. Kiss, M. Braic, R. Abdulgader, R. Booysen, S. Kulesza, T.K. Monsees, Biomed. Mater. **13**, 025011 (2018)
105. G. Brunello, H. Elsayed, L. Biasetto, Materials **12**, 2929 (2019)
106. R. Korabi, K. Shemtov-Yona, D. Rittel, Clin. Implant Dent. Relat. Res. **19**, 935 (2017)

Advances in Microbial-Based Electrochemical Biosensors for Heavy Metal Detection

Ab Rahman Marlinda, Norazriena Yusoff, Abu Hashem, and Suresh Sagadevan

Abstract A more current development among environmental management is to use cutting-edge novel technologies based on synthetic biology and genetic engineering that are programmed to offer precise signal outputs, sensitivity, and selectivity. Microbial biosensors are analytical devices capable of sensing specific substances present in the environment due to the microorganisms' particular biological reaction, which are later converted into electrical signals by an electrochemical transducer. Microbial biosensors have a wide range of uses, including environmental monitoring and energy production. As a result, unique developments made by combining microbial biosensors with electrochemical detection techniques have the potential to improve the selectivity and sensitivity of the created sensor for monitoring heavy metal toxicity in the environment. In the future, these microbial biosensors will provide broad applications in monitoring environmental metal pollution and sustainability for energy production.

Keywords Biomaterials · Electrochemical · Heavy metal · Bacteria · Yeast · Microalgae

A. R. Marlinda (✉) · A. Hashem · S. Sagadevan
Nanotechnology and Catalysis Research Centre (NANOCAT), Institute for Advanced Studies, Universiti Malaya, Kuala Lumpur, Malaysia
e-mail: marlinda@um.edu.my

A. Hashem
e-mail: hashemnibbd@gmail.com

S. Sagadevan
e-mail: sureshsagadevan@gmail.com

N. Yusoff
Physics Department, Faculty of Science, Universiti Malaya, Kuala Lumpur, Malaysia
e-mail: norazrienayusoff@um.edu.my

A. Hashem
Microbial Biotechnology Division, National Institute of Biotechnology, Ashulia, Savar, Dhaka, Bangladesh

S. J. Ikhmayies (ed.), *Advances in Biomaterials Research*, Advances in Material Research and Technology, https://doi.org/10.1007/978-3-031-97481-6_9

1 Introduction

A microbial biosensor is an analytical device consisting of a transducer and immobilized viable or non-viable microorganisms [1]. The combination of these two important components formed a device that detects, transmits, and records information regarding a physiological or biochemical change rapidly with high accuracy and sensitivity. Technically, the transducer is an electronic probe that integrates a biochemical signal from biological components, thereby converting the signal into quantifiable electrical response such as current, voltage potential, or intensity of light through electrochemical or optical point of view [2–4]. Later, these responses can be further amplified, processed, and stored for later analysis [5–7]. The biochemical signal can be obtained from a change in protons of analytes concentration, release or uptake of gases, light emission, absorption [8–12], and others, brought about by the metabolism of the target compound by the biological recognition element. For instance, glucose is an 'analyte' in a biosensor designed to detect glucose [13]. Meanwhile enzyme glucose oxidase (GOD) is bioreceptor. When analyte reacts with GOD, it gives gluconic acid by producing two electrons and two protons. During the reduction of GOD, particles such as electrons and protons, as well as oxygen, will interact to produce hydrogen peroxide and oxidised GOD, allowing additional GOD for glucose to react with. As the glucose content is increasing, a higher amount of oxygen is required. Meanwhile, as glucose content is decreasing, the more hydrogen peroxide is generated. This scenario implies that a rise in hydrogen peroxide or a decrease in oxygen can be monitored to determine the concentration of the analyte, glucose. The detection of target analytes is variously applied in many diverse fields such as medicine [14, 15], environmental monitoring [16–19], defense system [20, 21], food processing [22], fermentation [23–25], and food safety [26, 27]. Figure 1 presents components and the important aspects for developing an effective biosensor.

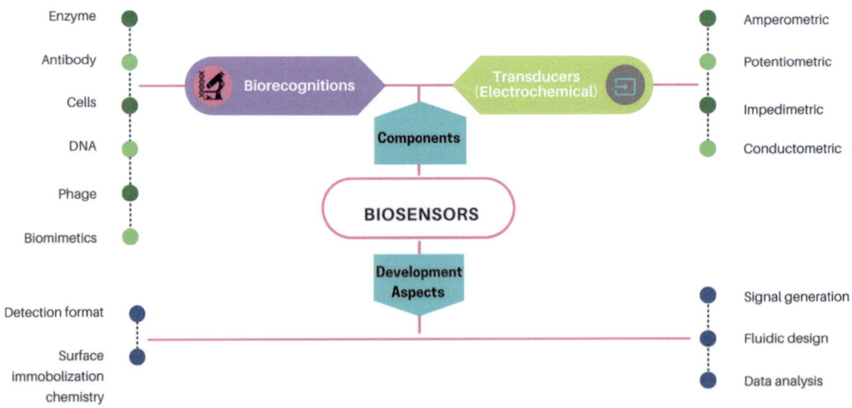

Fig. 1 The components and the important aspects for developing an effective biosensor

Metals are essential elements in living organisms as they act as co-factors for enzyme activities such as growth and cellular metabolism [28]. However, certain metals are toxic and persistently inorganic pollutants to the environment since they cannot be degraded such as silicon, vanadium, nickel, copper, zinc, arsenic, selenium, molybdenum, chromium, manganese, iron, cobalt, and iodine [29]. This metal group is known as heavy metals, which have a density larger than 4 ± 1 g cm^{-3} or has density five times higher than that of water [19, 30]. Heavy metals occur naturally in the outermost layer of the earth and can be released into the environment as a result of manmade activity such as harmful compounds disseminated by mining and processing companies around the world.

Nowadays, due to various human activities such as industry, mining, and agriculture, large amounts of heavy metals pollution are accumulated. Heavy metals, such as zinc, nickel, cadmium, copper, lead, arsenic, chromium, and mercury, are crucial environmental pollutants that easily enter the food chains but are challenging to discharge [31–33]. The excessive concentration of different types of heavy metals may contaminate the soil and disturb the ecosystem. Heavy metals pollution is also detrimental to human health that can cause short and long-term adverse health effects after months or years of exposure. This chapter focuses on selecting appropriate heavy metals detoxification mechanisms of microbial cells and the further uses of microbial cells such as bacteria, yeast, and microalgae in biosensors to detect heavy metals. This chapter also discusses the advantages, limitations, and challenges of biological sensing components in biosensors technology which offer an invaluable and essential overview for individuals interested in microbial sensing technologies.

2 Electrochemical Microbial Biosensor

Many conventional mediation methods aim to investigate and eliminate heavy metal elements that contaminated the soil and water supply. The conventional mediation methods either by mechanical or physio-chemical approach such as landfill thermal treatment, excavation, acid leaching, chemical precipitation, ion exchange, filtration, etc. are the most common approaches to mediate heavy metals from the environment [34, 35]. However, these conventional methods have drawbacks such as generating large amounts of toxic by-products, high cost, and time-consuming as the biological reaction requires more reaction time. It also requires more complicated techniques such as high-temperature treatment and condensation to identify the heavy metal elements. During heat treatment process, a temperature of 300–400 °C is required to completely degraded metals such as cadmium and copper [36]. A very high temperature of 900–1000 °C is also required to recover metals generated by thermochemical methods treated with KCl and $MgCl_2$ [36]. These methods require very high temperature and tedious operating procedures and produce large quantities of sludge and other toxic products which demand careful disposal. As a result, in order to prevent heavy metal pollution from endangering ecosystems and human health, more efficient, cost-effective, and environmentally friendly procedures must be implemented. In this regard, the detection using microbes or other biological components to measure the state or levels below concentration limits constituted by the regulatory regime is compulsory.

Many types of biological molecules are commonly used as receptors of biosensors, such as animal and plant cells, tissues, enzymes, nucleic acids, receptors, organelles, and microorganisms to recognize the imbalance target of heavy metal ions. These bio-recognition components will be combined with transduction elements to construct the biosensor devices. Meanwhile, among these bio-recognition components, the microorganisms are remarkable components that provide many advantages of detecting a various range of chemical substances, operating pH values, and operating temperatures [2–5, 37]. With a low cost of purification of many enzymes and cofactors in microorganisms, the cells can ingest and detect a vast amount of chemicals [1, 38, 39]. Additionally, in parallel with Deoxyribonucleic acid (DNA) technologies, the possibility of genetic modification of microorganisms to upgrade the existing enzyme in host cells has increased [40, 41]. The interaction of pollutant molecules or heavy metal ions with microorganisms will produce biological responses that can be detected and converted into a physicochemical signal by a transducer. The transducer element can be categorized depending on their transducers such as electrochemical, optical, and piezoelectric, or thermometric [28]. Later, the physicochemical detection signal can be amplified and converted into numeric records.

The electrochemical microbial biosensors can be categorized into three transducers: amperometric, potentiometric, and conductometric [2–6]. Electrochemical microbial biosensors may provide particular quantitative or semi-quantitative analytical information when a biological recognition element is in direct spatial contact with an electrochemical converting element [42]. These techniques were also offering high sensitivity and rapid detection of targets. In 1962, the first biosensor was introduced by Leland C. Clark for glucose detection has opened a new vision to the electrochemical biosensor technology [43]. Figure 2 displays various electrochemical techniques that are used for microbial sensors in diagnostic applications, such as chronoamperometry (CA), differential pulse voltammetry (DPV), cyclic voltammetry (CV), square wave voltammetry (SWV), electrochemical Impedance Spectroscopy (EIS), and its advantages. Few main advantages of electrochemical techniques, including reliable technique, perform an excellent sensitivity, highly selective, simple fabrication, requires inexpensive instrumentations, fast response, environmentally friendly, and portable compared to other analytical methods [44]. Furthermore, the selection of bio-recognition components is heavily influenced by the analyte targets' redox potential, which has characterised the rivalry for the fabrication of better sensors and biosensors.

2.1 Amperometric Microbial Biosensors

The amperometric microbial biosensors works at a set voltage concerning a reference electrode [12]. The biochemical output is a current signal that corresponds

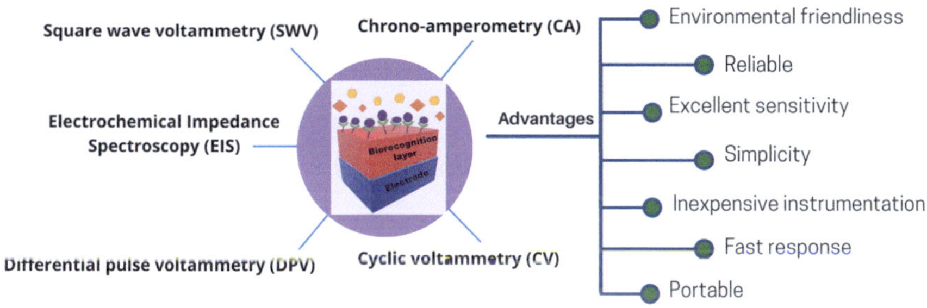

Fig. 2 Various electrochemical techniques that are used for microbial sensors in diagnostic applications and their advantages

to the oxidation or reduction of electroactive species that interact with the modified electrodes surface. The most extensively utilised amperometric microbial biosensors were employed to evaluate the biochemical oxygen demand (BOD) for biodegradable organic contaminants in wastewater monitoring [45, 46]. In general, BOD microbial sensors are created in the forms of biofilm-whole cell-based coupled to a suitable transducer to detect the signal on the microbial metabolic rate in near environmental circumstances [40].

2.2 *Potentiometric Microbial Biosensors*

In General, the potentiometric microbial biosensors can be categorized based on ion-selective electrode such as pH, ammonium, chloride, and others that coated with an immobilized microbe layer. The consumption of analyte from ion accumulation or depletion by the microbe layer will result in the change in potential values [12]. Potentiometric techniques evaluate the potential difference between a reference electrode (or grounded) and a working electrode [47, 48]. Potentiometric electrochemical microbial sensors have the benefit of being both selective and very sensitive to any changes of the target chemicals. However, the requirement toward the reference electrodes stable potential has limited their sensing application as a transducer [49].

2.3 *Impedimetric Microbial Biosensors*

In impedimetric microbial biosensor the system's potential always stays close to the equilibrium state of the potential redox reaction. Thus, this mechanism gives few advantages including:

1. Stable time response and accurate measurement over the average time range.
2. Linear *I*–*V* relationship (current versus voltage) around the equilibrium potential [50].

3. Wide detection range over broad frequency range such as to 10^{-4}–10^{6} Hz [51].
4. Ability to miniaturize into a chip format [52].

2.4 Conductometric Microbial Biosensors

The Conductometric biosensors are classified based on variations in electric conductivity in the environment resulting from target analytes, such as the consumption or creation of ions by transducers [12, 53]. For example, Bhatia et al. [54] used a one-time-use conductivity and microbial sensor to evaluate the influence of species including chloride, nitrate, and sulphate, as well as ion concentration/osmolarity on *E. coli* metabolism. It was found that conductometric biosensor is able to detect target chemicals with high sensitivity in a short time.

3 Microorganism Detoxification Mechanisms for Heavy Metals Removal

Microorganisms own remarkable metabolic pathways to detoxify heavy metal compounds via respiration, fermentation, and co-metabolism to maintain their development and growth [55]. The adaptation of the microorganisms toward the heavy metals occurs when they create various mechanisms to maintain their homeostasis and resistance in the ecosystem [56, 57]. The production of degradative enzymes enables it to resist contaminated compounds. There are few mechanisms that the microorganisms have developed to detoxify the heavy metals in the environment, including bioaccumulation [58], biomineralization [59–61], biosorption [58, 62–64], and biotransformation [65, 66]. Bioaccumulation and biosorption show effective mechanisms to decay heavy metals [58, 67]. There are two approaches by utilizing living and non-living biomass [62], where the heavy metals can be transported into the bacterial cell. The heavy metals are accumulated intracellularly by the bioaccumulation mechanism [58, 68, 69].

Meanwhile, non-living cells will adsorb heavy metals via cell walls known as biosorption mechanism [70]. Thus, both living and non-living microbial adsorbents can be utilized to remove heavy metal ions from wastewater [71]. However, the living biomass is identified as effective biosorbents due to the capability of their cellular surfaces to adapt to various environments for heavy metal detoxifying and recovery [72–74]. Many types of heavy metal compounds may destroy microbial cell membranes; however, microorganisms resist the heavy metals since they can evolve a defense mechanism. Thus, the response of microorganisms to heavy metal toxicity is essential for re-establishing polluted sites and removing them from their ecosystem. Several microorganisms can detoxify heavy metal compounds such as bacteria, algae, yeast, fungi, weeds, and plants [75–78]. The economic aspect and availability of biological sources of microorganisms have resulted in the production of biosorption technologies. These microbial biosorbents will be cultured or developed from various food processing or waste products and identified and categorized according to their capability for heavy metals removal properties. Previous research proved that biosorbents such as algae and bacteria performed very high adsorption capacity for heavy metal removal [79–87]. The selection of bacteria for detection of the bio-available concentration of heavy metals is dependent on the gene coding of the bacteria where the ability of its chemical sites to bind with heavy metals. Then, the detoxifying process is followed by complex mechanisms such as ion exchange, chelation, and adsorption [47, 88, 89].

4 Electrochemical Microbial-Based Biosensors for Heavy Metal Ions Detections

In microbial electrochemical sensors, electrode-biofilm has been utilized as the sensing element, and it has been widely used to trace heavy metal ions by tracking the electron transfer between electroactive microorganisms and electrodes in an appropriate environment. Bacteria, yeast, and microalgae are some of the examples of microorganisms that can be employed as the sensing element to sense specific heavy metal ions. In this section, we will review some of the previous work reported on the development of microbial-based electrochemical sensors using microorganisms for heavy metal ion detection. Table 1 shows the types of microorganisms that have been utilized for detection of heavy metal ions from several studies.

4.1 Using Bacteria to Sense Specific Heavy Metal Ions

Bacteria possess a capability to degrade organic substrates and provide power for fermentation. Essentially, the technique employs a bio-electrochemical system that regulates the strength of microbial respiration in order to transform organic material instantly into electrical energy. Based on the published manuscript by Abu-Ali et al. [90], they have successfully developed the bacteria-based inhibition sensor array for the detection of heavy metal ions mercury (Hg^{2+}). Two types of bacteria namely *Escherichia coli* and *Shewanella oneidensis* were used to immobilize them on the surface of gold electrode via poly L-lysine, thus allowing for the electrochemical reactions with heavy metal ions to occur. Figure 3 presents fluorescence microscopy images of *Shewanella oneidensis* in Luria–Bertani (LB) broth (1:1 ratio) before and after treatment with 1 M $HgCl_2$, along with the corresponding cyclic voltammograms of the sensing probe exposed to varying $HgCl_2$ concentrations. It can be seen that there is increase in the recorded current due to the increase in the concentration of $HgCl_2$ salt, thus revealing the ability of the proposed electrochemical bacteria-based sensor to detect Hg^{2+} ions within the linear concentration range of 1 mM to 1 M. The

Table 1 Types of microorganisms that have been utilized for detection of heavy metal ions

Type of microorganism	Sensor element	Detection elements	Detection limit	Linear range	Refs.
Bacteria	*Escherichia coli*	Hg^{2+}	0.1 mM	0.1 mM–1 M	[90]
	Shewanella oneidensis		0.1 mM	0.1 mM–1 M	
	Rhodotorula-ALT72 *Rhizobium-MAP7*	Cr(VI) Cd(II)	0.1 mg/L 1 mg/L	–	[92]
	Escherichia coli	Cd^{2+}	25 nM	25 nM–1000 nM	[91]
		Cu^{2+} $Cr_2O_7{}^{2-}$	0.14 mg/L 0.025 mg/L	–	[96]
		Cd^{2+} Pb^{2+}	95.5 μM 207.0 μM	–	[97]
Microalgae	*Phormidium* sp.	Pb^{2+}	2.5×10^{-8} M	5.0×10^{-8} to 2.0×10^{-5} M	[93]
	Chlorella vulgaris	Cd^{2+}	1 ppb	–	[98]
	Chlorella vulgaris *Anabaena* sp.	Pb^{2+} Cu^{2+} Cd^{2+}	–	0–5 ppm	[99]
	Chlorella sp.	Cd^{2+} Fe^{3+} Cr^{3+} Mn^{2+}	50 nM	55 nM–5.45 mM 62 nM–6.2 mM 38 nM–3.75 mM 79 nM–7.95 mM	[100]
Yeast	*Saccharomyces cerevisiae*	Cu^{2+}	0.5 mM	0.5 mM–2 mM	[94]
		Cu^{2+} Cd^{2+} Ni^{2+} Pb^{2+}	10.12 mg/L 13.88 mg/L 17.06 mg/L 34.56 mg/L	–	[95]

ppb = parts per billion; ppm = parts per million

Escherichia coli was also reported by Biran et al. [91] to investigate the accumulation of Cd^{2+} via online and in-situ monitoring method. They employed the *Escherichia coli* as cadmium-responsive promoter to promoter less *lacZ* gene. The monitoring was applied using an electrochemical assay of β-galactosidase activity. The whole-cell biosensor was able to detect concentration of Cd^{2+} in nanomolar ranges within periods of minutes.

Furthermore, Alfadaly et al. discovered that *Rhodotorula-ALT72* and *Rhizobium-MAP7* can support the detection and removal of heavy metals Cr(VI) and Cd(II) with a higher percentage at concentrations of 0.1 and 1 mg/L [92]. In their study, the tested Rhizobium strain shown a greater capacity to remove heavy metals and was more resistant to metal ions, as its cell viability was not significantly impacted by hazardous metal ions at varied doses. Thus, this study can promise a huge potential technique for a cost-effective, simple, and reproducible assay, which can allow quick screening

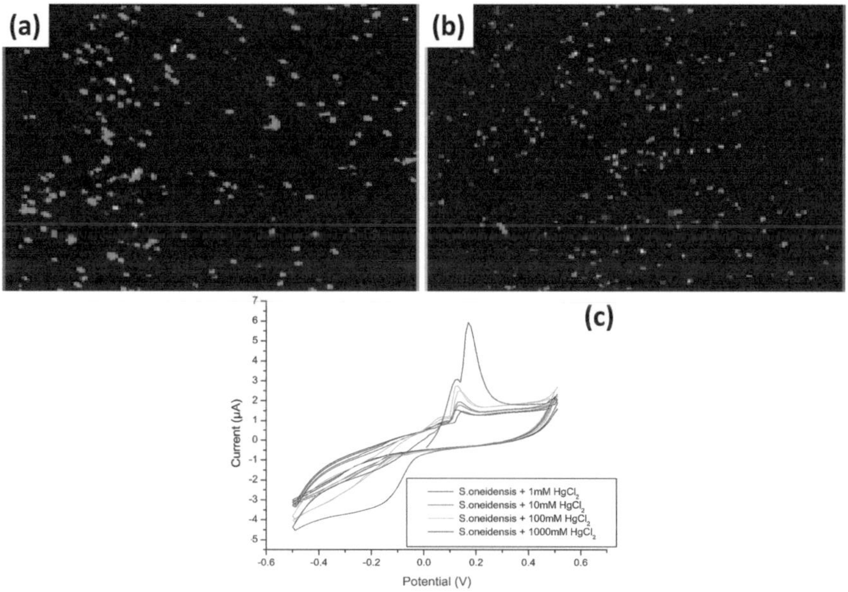

Fig. 3 Fluorescence microscopy images of *Shewanella oneidensis* before (**a**), and after treatment with $HgCl_2$ salt (1 M) (**b**), and cyclic voltammograms of *Shewanella oneidensis* bacteria solutions in LB broth (1:1 ratio) which was treated with different concentration of $HgCl_2$ (**c**). Reprinted from Abu-Ali et al. [90], with permission from Elsevier. Copyright (2017)

of heavy metals and decrease pollution concentration, especially in water-based environments.

4.2 *Using Microalgae to Sense Specific Heavy Metals*

Apart from using bacteria as one of the sensing elements, the potential of microalgae in determining the concentration of heavy metal ions has also been proven. A study carried out by Yüce and co-workers [93] has revealed the ability of using a strain of filamentous Cyanobacterium known as *Phormidium* sp. to trace lead (Pb^{2+}) ions in aqueous solution. This result was attributed to the presence of carboxyl and sulphoxide groups in *Phormidium* sp. which became the predominant contributors in Pb^{2+} ions uptake and thus enhanced the electrochemical response. Figure 4 demonstrates the calibration graph obtained under the optimum experimental conditions where a linear equation given by $y = 1.1878x$ (where y corresponds to current and x corresponds to Pb(II) concentration) with $R^2 = 0.9901$ is obtained in the Pb(II) concentration range 5.0×10^{-8} to 2.0×10^{-5} M (Pb(II) corresponding to 0.01–4.0 mg/L. Based on this graph, detection limit was calculated using the formula $3S_b/$

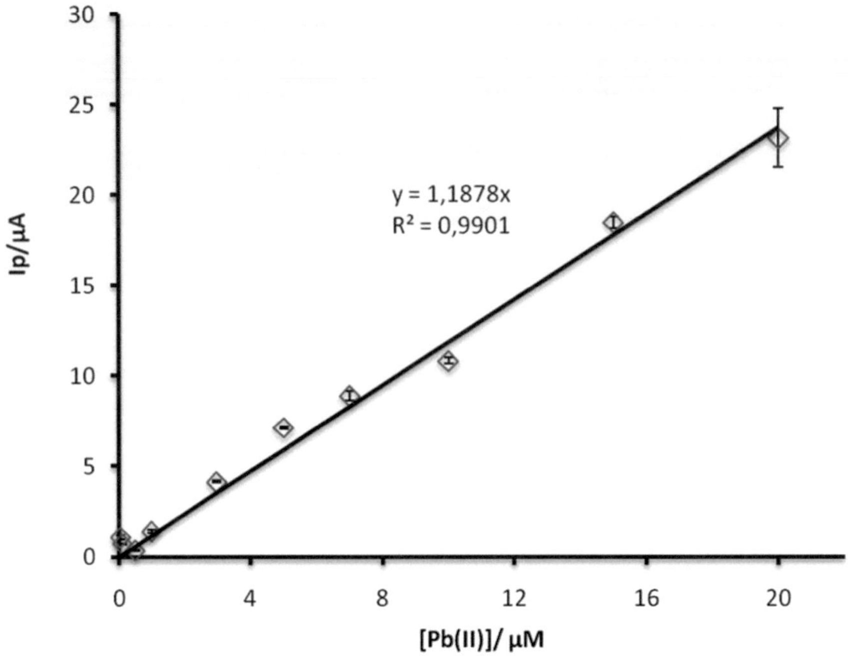

Fig. 4 The calibration curve of the developed microalgae sensor for the detection of Pb^{2+} ions. Reprinted from Yüce et al. [93], with permission from Elsevier. Copyright (2010)

m (where S_b is the standard deviation of the responses and *m* is the slope of the calibration curve) and found to be about 2.5×10^{-8} M.

4.3 *Using Yeast to Sense Specific Heavy Metals*

An increase in the development of microbial based electrochemical sensors has motivated several researchers such Lehmann et al. [94] who produced a microbial biosensor to trace copper (Cu^{2+}) using an amperometric technique. In this work, *Saccharomyces cerevisiae* strains which is a species of yeast has been recombined with the promoterless *lacZ*-gene from *E. coli* to acts as the sensing element in this microbial sensor. It is suggested that the fusion of the corresponding metal-inducible gene elements with a suitable bacterial reporter gene luciferase (lux-gene) will increase the performance of the microbial based sensor to detect not only heavy metal ions, but also other chemical compounds. Figure 5a shows the scanning electron microscope (SEM) image of *Saccharomyces cerevisiae* 19.3C/YEp352-FUSI cells immobilized with 2.5% polyvinyl alcohol (PVA) on a capillary membrane 'RoTrac' and Fig. 5b presents the graph for indirect amperometric measurements for the detection of Cu^{2+}. It is evident that the proposed sensor could measure Cu^{2+} in a concentration range between 0.5 and 2 mM $CuSO_4$. It is found that 2.6-fold and sixfold increases in the indirect amperometric signals were achieved with different Cu^{2+} concentrations of 1 mM and 1.5 mM, respectively.

Meanwhile, according to Gao et al., the feasibility of an electrochemical biosensor integrated with the same yeast, *Saccharomyces cerevisiae*, can be used to assess heavy metal ions in real wastewater samples was verified by using four types of acute biotoxicity analytes (Cu^{2+}, Cd^{2+}, Ni^{2+}, and Pb^{2+}) [95]. The toxicity detection of acute biotoxicity ions was monitored via chronoamperometry. The results shown that heavy metal ions Cu^{2+}, Cd^{2+}, Ni^{2+}, and Pb^{2+} had toxicity values of 10.12 mg/L, 13.88 mg/L, 17.06 mg/L, and 34.56 mg/L, respectively.

5 Advantages, Limitations, and Challenges of Biosensor Technology

The utilization of microorganisms such as bacteria, algae, and yeast for developing microbial biosensors technology offers various advantages from different aspects compared to conventional sensing methods. The microbial biosensor-based offer a high sensitivity measurement of the bio-available concentration of heavy metals compared to enzymes or antibody biosensors that are unstable and sensitive to pH and temperature change [101–103]. The microbial biosensors can be continuously performed in a real-time in-situ monitoring environment with highly selective to specific compounds [102]. The excellent characteristics of microbial biosensors are also performed by detecting a rapid and accurate detection of multiple compounds simultaneously [102, 104]. The ability to grow large numbers of microorganisms is also crucial to consider, since it may be accomplished through a simple culture procedure employing inexpensive liquid nutrition medium. As a result, analysis costs can be significantly lowered as compared to traditional sensing methods that need more expensive and sophisticated instruments such as gas chromatography, liquid chromatography, mass spectrophotometry, and other approaches [105].

Although microbial biosensors technology has numerous benefits over traditional sensing technologies, the utilization of microbes remains constrained by various limitations, such as intrinsic cellular heterogeneity both genotypically and phenotypically [106]. The real-world limitation on the legislation of genetic alteration of organisms also serves as a barrier to the creation of microbial biosensors. Furthermore, the unpredictable nature of environmental factors, such as nutrient availability, pH, temperature, and water availability, may disturb the output signal and viability of the microbial biosensors [103, 107–109]. Several bacteria also converted toxic heavy metals into their non-toxic forms due to the slow absorbance of substrates and products across the cell membrane into cells, contributing to the limitation of microbial biosensors technology [110]. Although bacteria can deal with heavy metals by evolving many internal automated systems, large amounts of heavy metals present in nature could harm or kill the biosensor structure completely [111].

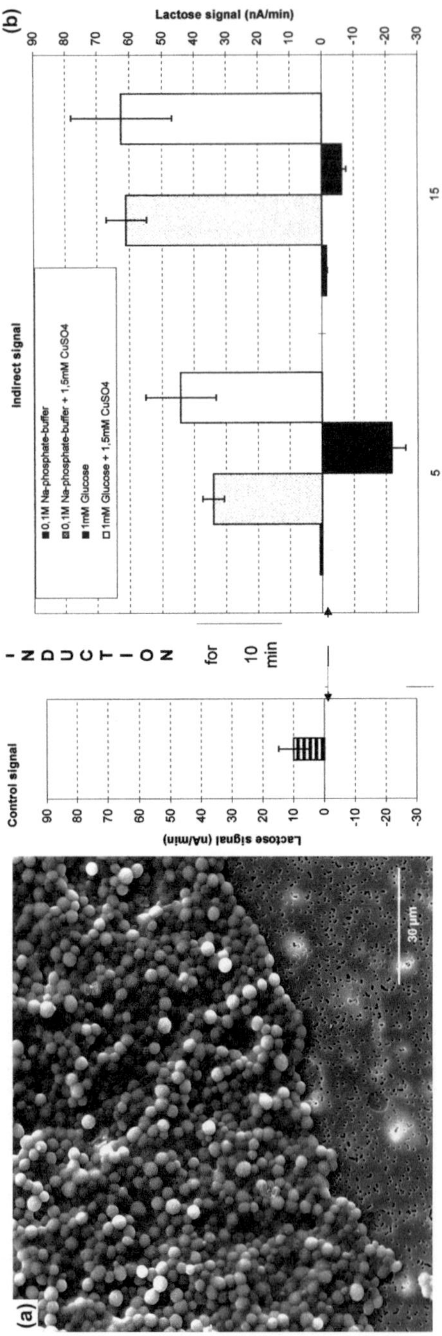

Fig. 5 a SEM image of *Saccharomyces cerevisiae* 19.3C/YEp352-FUSI cells immobilized with 2.5% PVA on a capillary membrane 'RoTrac'. **b** Detection of Cu^{2+} with the indirect amperometric measurement. Reprinted from Lehmann et al. [94], with permission from Elsevier. Copyright (2000)

Furthermore, the risk of dangerous mutations also needs to be considered while designing and modifying the microorganisms to suit the detection of heavy metal components. The potential mutations could occur accidentally and spread to the environment, posing a risk to the safety systems [79, 109, 112]. Besides the accidental risk of potential mutation spread to the natural environment, the transportation process for transferring the microbial biosensor technology from the lab to the field may contribute to another challenge for this biosensor technology. Several investigations have been carried out to exhibit the effect of transportation conditions on the efficiency of the detectable signal for the long-term period [79, 110]. Specific microbial biosensors require activation at the test site along with the transportation factor as they are constructed via air-drying and freeze-drying processes [110, 111]. However, this chapter is intended to highlight the relevant issues so that microbial biosensors can be utilized practically in real-world applications.

6 Conclusion

Despite all limitations and challenges for constructing the microbial sensors, the current progress in developing the microbial biosensors for heavy metals sensing is well planned. Many research groups progressively start to study alternative pathways for constructing microbial-based sensing to increase the robustness of biosensors technology. However, a slight movement towards developing a standard procedure to determine optimal microbial biosensing should be constructed. Few standard criteria for proposing a novel microbial biosensor construction must be provided to ensure the standard framework shortly. The standard criteria including specificity, detection amounts as set by the world health organization (WHO) or equivalent, reproducibility, stability of signal responses, facile detection methods, longevity over time, quick responses and recovery times, biosafety switches, risk-free application, detection, and ease of assembly at the contaminated sites.

Furthermore, the advancement of portable detectors and related technologies has made previously impossible applications feasible. Accurate miniaturization and integration of microbial biosensors with different micro or nanomaterials have the ability to minimize resource consumption, boost sensitivity and selectivity, and shorten response time while detecting metals with high resolution and precision. This development of the series of standards will ensure steady progress for heavy metals and standardize the framework for the microbial biosensors designs and construction.

Acknowledgements The authors would like to thank the Ministry of Higher Education (MoHE) Malaysia for funding this work via research grant numbers (JPT(BKPI)1000/016/018/28 Jld.3(2) & NANOCAT-2024C) and (FRGS/1/2024/STG05/UM/02/5) and the Universiti Malaya International Collaboration Grants for the funding of this research via grant numbers IMG006-2023 and IMG010-2023.

References

1. S.F. D'Souza, Biosens. Bioelectron. **16**, 337 (2001)
2. G. Bardeletti, F. Sechaud, P. R. Coulet, in *Biosensor Principles and Applications*. ed. by P.R. Coulet, L.J. Blum (CRC Press, Boca Raton, 2019), pp.7–26
3. W. R. Seitz, *Biosensors: Fundamentals and Applications*, 1st edn., ed. by A. Turner, I. Karube, G.S. Wilson (Oxford university Press, 1987), pp. 599–617
4. D. Ivnitski, I. Abdel-Hamid, P. Atanasov, E. Wilkins, Biosens. Bioelectron. **14**, 599 (1999)
5. B.D. Malhotra, *Biosensors: Fundamentals and Applications* (Smithers Rapra, 2017), pp. 3–21
6. A. Mulchandani, *Enzyme and Microbial Biosensors: Techniques and Protocols*, ed. by A. Mulchandani, K.R. Rogers (Humana Press, Totowa, NJ, 1998), pp. 3–14
7. D.G. Buerk, *Biosensors: Theory and Applications* (CRC Press, Boca Raton, 2014), pp.21–39
8. S. Mikkelsen, E. Cortón, *Centrifugation Methods, Bioanalytical Chemistry* (John Wiley & Sons, New Jersey, 2004), pp.247–267
9. I. Karube, J. Biotechnol. **15**, 255 (1990)
10. P.A. Siskos, E.B. Bakeas, M.P. Baya, *Biosensors for Direct Monitoring of Environmental Pollutants in Field*, ed. by D. Nikolelis, U.J. Krull, J. Wang, M. Mascini (Springer Science & Business Media, 1997), pp. 18–20
11. H. Nakamura, I. Karube, Anal. Bioanal. Chem. **377**, 446 (2003)
12. Y. Lei, W. Chen, A. Mulchandani, Anal. Chim. Acta **568**, 200 (2006)
13. W.R. Heineman, W.B. Jensen, Biosens. Bioelectron. **8**, 21 (2006)
14. A. Bhan, N. Sarkar, Rev. Environ. Health **20**, 39 (2005)
15. D.W. Boening, Chemosphere **40**, 1335 (2000)
16. Z. Cui, X. Zhang, H. Yang, L. Sun, J. Environ. Chem. Eng. **5**, 3616 (2017)
17. Z. Chen, X. Pan, H. Chen, X. Guan, Z. Lin, J. Hazard. Mater. **301**, 531 (2016)
18. B. Liu, Y. Lei, B. Li, Biosens. Bioelectron. **62**, 308 (2014)
19. A.S. Mohammed, A. Kapri, R. Goel, in *Biomanagement of Metal-Contaminated Soils*. ed. by M. Khan, A. Zaidi, R. Goel, J. Musarrat (Springer, Dordrecht, 2011), pp.1–28
20. R.L. Gallo, T. Nakatsuji, J. Invest. Dermatol. **131**, 1974 (2011)
21. R. Nishad, T. Ahmed, V.J. Rahman, A. Kareem, Front. Microbiol. **11**, 1298 (2020)
22. S. Amaya-González, N. De-los-Santos-Álvarez, A.J. Miranda-Ordieres, M.J. Lobo-Castañón, Sensors **13**, 16292 (2013)
23. I. Karube, E. Tamiya, K. Sode, K. Yokoyama, Y. Kitagawa, H. Suzuki, Y. Asano, Anal. Chim. Acta **213**, 69 (1988)
24. J. Tkac, I. Vostiar, P. Gemeiner, Bioelectrochemistry **56**, 127 (2002)
25. J. Tkac, I. Vostiar, L. Gorton, P. Gemeiner, E. Sturdik, Biosens. Bioelectron. **18**, 1125 (2003)
26. F. Rahmati, S.S. Hosseini, S.M. Safai, B.A. Lajayer, M. Hatami, 3 Biotech **10**, 1 (2020)
27. C. Ranadheera, P. Prasanna, J. Vidanarachchi, R. Mcconchie, N. Naumovski, D. Mellor, *Nanotechnology Applications in Food*, ed. by A.E. Oprea, A.M. Grumezescu (Elsevier, 2017), pp. 245–265
28. J.C. Gutiérrez, F. Amaro, A. Martín-González, Front. Microbiol. **6**, 48 (2015)
29. M.K. Hill, *Understanding Environmental Pollution*, 3rd edn. (Cambridge University Press, Cambridge, 2010), pp.327–344
30. M. Hookoom, D. Puchooa, Curr. Res. Microbiol. Biotechnol. **1**, 119 (2013)
31. M. Diop, R. Amara, Environ. Sci. Pollut. Res. **23**, 11975 (2016)
32. M.S. Islam, M.K. Ahmed, M. Raknuzzaman, M. Habibullah-Al-Mamun, M.K. Islam, Ecol. Indic. **48**, 282 (2015)
33. Y. Zhang, C. Zhang, R. Ma, X. Du, W. Dong, Y. Chen, Q. Chen, Mater. Sci. Eng. C **75**, 175 (2017)
34. J. Singh, A.S. Kalamdhad, Environ. Sci. Pollut. Res. **20**, 8974 (2013)
35. H. Zhang, D. Ma, R. Qiu, Y. Tang, C. Du, Chem. Eng. J. **313**, 157 (2017)
36. W. Shi, C. Liu, D. Ding, Z. Lei, Y. Yang, C. Feng, Z. Zhang, Bioresour. Technol. **137** (2013)

37. Y. Arikawa, K. Ikebukuro, I. Karube, in *Enzyme and Microbial Biosensors: Techniques and Protocols*. ed. by A. Mulchandani, K.R. Rogers (Humana Press, Totowa, NJ, 1998), pp.225–235
38. K.T. Leung, O. Tresse, D. Errampalli, H. Lee, J.T. Trevors, FEMS Microbiol. Lett. **155**, 107 (1997)
39. X. Xu, Y. Ying, Food Rev. Int. **27**, 300 (2011)
40. S. Belkin, Curr. Opin. Microbiol. **6**, 206 (2003)
41. C. Rensing, R.M. Maier, Ecotoxicol. Environ. Saf. **56**, 140 (2003)
42. D.R. Thévenot, K. Toth, R.A. Durst, G.S. Wilson, Biosens. Bioelectron. **16**, 121 (2001)
43. L.C. Clark Jr., C. Lyons, Ann. N. Y. Acad. Sci. **102**, 29 (1962)
44. M.A. Brooks, E.W. Tsai, *Laboratory Techniques in Electroanalytical Chemistry, Revised and Expanded*, ed. by P. Kissinger, W.R. Heineman (CRC Press, Pub. Boca Raton, 2018), pp. 769–806
45. J. Liu, B. Mattiasson, Water Res. **36**, 3786 (2002)
46. P. Melidis, E. Vaiopoulou, A. Aivasidis, Bioprocess Biosyst. Eng. **31**, 277 (2008)
47. J.W. Lim, D. Ha, J. Lee, S.K. Lee, T. Kim, Front. Bioeng. Biotechnol. **3**, 61 (2015)
48. A.L. Simonian, E.I. Rainina, J.R. Wild, *Enzyme and Microbial Biosensors: Techniques and Protocols*, ed. by A. Mulchandani, K.R. Rogers (Humana Press, Totowa, NJ, 1998), pp. 237–248
49. L. Su, W. Jia, C. Hou, Y. Lei, Biosens. Bioelectron. **26**, 1788 (2011)
50. X. Yang, J. Kirsch, A. Simonian, J. Microbiol. Methods **95**, 48 (2013)
51. A.J. Bard, L.R. Faulkner, *Fundamentals and applications: Electrochemical Methods*, 2nd edn. (John Wiley & Sons, 2000), pp. 580–632
52. M. Pedrero, S. Campuzano, J.M. Pingarrón, Sensors **9**, 5503 (2009)
53. S.R. Mikkelsen, G.A. Rechnitz, Anal. Chem. **61**, 1737 (1989)
54. R. Bhatia, J. Dilleen, A. Atkinson, D. Rawson, Biosens. Bioelectron. **18**, 667 (2003)
55. A.S. Ayangbenro, O.O. Babalola, Int. J. Environ. Res. Public Health **14**, 94 (2017)
56. S.K. Brar, M. Verma, R. Surampalli, K. Misra, R. Tyagi, N. Meunier, J. Blais, J. Hazard Toxic Radioact. Waste **10**, 59 (2006)
57. W. Wei, X. Liu, P. Sun, X. Wang, H. Zhu, M. Hong, Z.-W. Mao, J. Zhao, Environ. Sci. Tech. **48**, 3363 (2014)
58. A. Malik, Environ. Int. **30**, 216 (2004)
59. F. Kang, X. Qu, P.J. Alvarez, D. Zhu, Environ. Sci. Technol. **51**, 2776 (2017)
60. M. Li, X. Cheng, H. Guo, Int. Biodeterior. Biodegrad. **76**, 81 (2013)
61. J.P. Maity, G.-S. Chen, Y.-H. Huang, A.-C. Sun, C.-Y. Chen, Geomicrobiology J. **36**, 612 (2019)
62. R. Gabr, S. Hassan, A. Shoreit, Int. Biodeterior. Biodegrad. **62**, 195 (2008)
63. H.K. Alluri, S.R. Ronda, V.S. Settalluri, J.S. Bondili, V. Suryanarayana, P. Venkateshwar, Afr. J. Biotechnol. **6**, 2924 (2007)
64. T.A. Davis, B. Volesky, A. Mucci, Water Res. **37**, 4311 (2003)
65. G. Girma, Egypt. Acad. J. Biol. Sci., G. Microbiol. **7**, 29 (2015)
66. H. Xu, C. Hong, Y. Yao, L. Liu, W. Wang, W. Zhu, L. Hong, J. Weng, Y. Zhou, F. Zhu, Sci. Total. Environ. **776**, 145864 (2021)
67. A.A. Juwarkar, S.K. Yadav, *Bioremediation Technology*, ed. by M.H. Fulekar (Springer, Dordrecht, 2010), pp. 266–284
68. K. Arunakumara, X. Zhang, J. Ocean Univ. China **7**, 60 (2008)
69. F. Huang, Z.-H. Wang, Y.-X. Cai, S.-H. Chen, J.-H. Tian, K.-Z. Cai, Ecotoxicol. Environ. Saf. **157**, 216 (2018)
70. M. Fomina, G.M. Gadd, Bioresour. Technol. **160**, 3 (2014)
71. W. Zhu, X. Xu, L. Xia, Q. Huang, W. Chen, Geomicrobiology J. **33**, 274 (2016)
72. W.S. Tan, A.S.Y. Ting, Bioresour. Technol. **123**, 290 (2012)
73. L. Velásquez, J. Dussan, J. Hazard. Mater. **167**, 713 (2009)
74. B. Volesky, Hydrometallurgy **59**, 203 (2001)

75. T. Nevita, P. Pandey, D.K. Maheshwari, *Bacteria in Agrobiology: Crop Productivity*, ed. by D. Maheshwari, M. Saraf, A. Aeron (Springer, Berlin, Heidelberg, 2013), pp. 439–461
76. R. Dhankhar, A. Hooda, Environ. Technol. **32**, 467 (2011)
77. J.M. Jacob, C. Karthik, R.G. Saratale, S.S. Kumar, D. Prabakar, K. Kadirvelu, A. Pugazhendhi, J. Environ. Manag. **217**, 56 (2018)
78. O.B. Ojuederie, O.O. Babalola, Int. J. Environ. Res. Public Health **14**, 1504 (2017)
79. L.T. Bereza-Malcolm, G.L. Mann, A.E. Franks, ACS Synth. Biol. **4**, 535 (2015)
80. V.K. Gupta, A. Nayak, S. Agarwal, Environ. Eng. Res. **20**, 1 (2015)
81. Y. Jiang, P. Liang, P. Liu, Y. Bian, B. Miao, X. Sun, H. Zhang, X. Huang, Int. J. Mol. Sci. **17**, 1392 (2016)
82. H. Kinoshita, Y. Sohma, F. Ohtake, M. Ishida, Y. Kawai, H. Kitazawa, T. Saito, K. Kimura, Res. Microbiol. **164**, 701 (2013)
83. M.U. Mustapha, N. Halimoon, J. Microb. Biochem. Technol. **7**, 253 (2015)
84. C. Tibazarwa, P. Corbisier, M. Mench, A. Bossus, P. Solda, M. Mergeay, L. Wyns, D. Van der Lelie, Environ. Pollut. **113**, 19 (2001)
85. S.Y. Cheng, P.-L. Show, B.F. Lau, J.-S. Chang, T.C. Ling, Trends Biotechnol. **37**, 1255 (2019)
86. R. Muñoz, M.T. Alvarez, A. Muñoz, E. Terrazas, B. Guieysse, B. Mattiasson, Chemosphere **63**, 903 (2006)
87. S. Xin, Z. Zeng, X. Zhou, W. Luo, X. Shi, Q. Wang, H. Deng, Y. Du, J. Hazard. Mater. **324**, 365 (2017)
88. Y. Jiang, X. Yang, P. Liang, P. Liu, X. Huang, Renew. Sustain. Energy Rev. **81**, 292 (2018)
89. J. Igwe, E. Nwokennaya, A. Abia, Afr. J. Biotechnol. **4**, 1109 (2005)
90. H. Abu-Ali, A. Nabok, T. Smith, M. Al-Shanawa, Sens. Biosensing Res. **13**, 109 (2017)
91. I. Biran, R. Babai, K. Levcov, J. Rishpon, E.Z. Ron, Environ. Microbiol. **2**, 285 (2000)
92. R.A. Alfadaly, A. Elsayed, R.Y. Hassan, A. Noureldeen, H. Darwish, A.S. Gebreil, Molecules **26**, 9 (2021)
93. M. Yüce, H. Nazır, G. Dönmez, Biosens. Bioelectron. **26**, 321 (2010)
94. M. Lehmann, K. Riedel, K. Adler, G. Kunze, Biosens. Bioelectron. **15**, 211 (2000)
95. G. Gao, D. Fang, Y. Yu, L. Wu, Y. Wang, J. Zhi, Talanta **167**, 208 (2017)
96. Y. Yang, Y. Liu, Y. Chen, Y. Wang, P. Shao, R. Liu, G. Gao, J. Zhi, Chemosphere **255**, 126964 (2020)
97. V. Adam, D. Chudobova, K. Tmejova, K. Cihalova, S. Krizkova, R. Guran, M. Kominkova, M. Zurek, M. Kremplova, A.M. Jimenez, M. Konecna, D. Hynek, V. Pekarik, R. Kizek, Electrochim. Acta **140**, 11 (2014)
98. H. Guedri, C. Durrieu, Microchim. Acta **163**, 179 (2008)
99. M. Darder, P. Aranda, L. Burgos-Asperilla, A. Llobera, V.J. Cadarso, C. Fernández-Sánchez, E. Ruiz-Hitzky, J. Mater. Chem. **20**, 42 (2010)
100. D.N. Roxby, H. Rivy, C. Gong, X. Gong, Z. Yuan, G.-E. Chang, Y.-C. Chen, Biosens. Bioelectron. **165**, 112420 (2020)
101. K.M. Aref'ev, M. Guseva, B. Khomchenkov, High Temp. **25**, 174 (1987)
102. B.F. Brehm-Stecher, E.A. Johnson, Microbiol. Mol. Biol. Rev. **68**, 538 (2004)
103. L. Peixoto, B. Min, G. Martins, A.G. Brito, P. Kroff, P. Parpot, I. Angelidaki, R. Nogueira, Bioelectrochemistry **81**, 99 (2011)
104. Y. Zhang, I. Angelidaki, Biotechnol. Bioeng. **108**, 2339 (2011)
105. Y. Jiang, P. Liang, C. Zhang, Y. Bian, X. Sun, H. Zhang, X. Yang, F. Zhao, X. Huang, Appl. Energy **165**, 670 (2016)
106. S. Verma, A. Kuila, Environ. Technol. Innov. **14**, 100369 (2019)
107. Q. Hu, L. Li, Y. Wang, W. Zhao, H. Qi, G. Zhuang, J. Environ. Sci. **22**, 1469 (2010)
108. K.M. Nielsen, P.J. Johnsen, D. Bensasson, D. Daffonchio, Environ. Biosafety Res. **6**, 37 (2007)
109. M. Schmidt, BioEssays **32**, 322 (2010)
110. H. Harms, J. Rime, O. Leupin, S.J. Hug, J.R. van der Meer, Microchim. Acta **151**, 217 (2005)
111. H.-C. Tao, Z.-W. Peng, P.-S. Li, T.-A. Yu, J. Su, Biotechnol. Lett. **35**, 1253 (2013)
112. M. la Farré, M.a.-J. García, L. Tirapu, A. Ginebreda, D. Barceló, Anal. Chim. Acta **427**, 181 (2001)

Therapeutic Effects of Metallic Ions in the Physiological Environment: A Review on Recent Developments

Shaher Bano Zaidi, Syeda Ammara Batool, and Muhammad Atiq Ur Rehman

Abstract Recent studies encouraged the regeneration of living tissue, which enhances the quality of life for those who require surgery because of traumatic accidents, bone disorders, bone deformities, or reconstructions. Bio ceramics (BCs) including bioactive glasses (BGs), hydroxyapatite (HA), bone cement etc. can aid bone regeneration/reconstruction. Bone cement, scaffolds, and nanoparticles can be tailored to mimic natural bone in the physiological environment. They have good biocompatibility and biodegradability that can promote osteoconduction on the bone tissue in the immediate vicinity. BCs can be doped with various ions which are important for bone metabolism and are expected to have a multifunctional role after being implanted, directing the process of tissue regeneration, and giving the body the ability to self-heal. The entry of ions into the network of BGs can increase osteointegration, osteoinduction, antibacterial effect, and vascularization. The incorporation of metallic ions in BCs led to a new family of bone regeneration materials, which can overcome the drawbacks of primitive materials and favor their usage in various bone tissue engineering (BTE) applications. This chapter summarizes recent studies on ionic substitutions in BCs commonly found in human bodies. Furthermore, co-substituted BCs being synthesized from various precursors and their effect on bone regeneration and remodeling are being discussed. The challenges associated with the toxicity of inorganic ions and their therapeutic effects are discussed in detail. The recent advancements of BCs doped with inorganic ions in bone tissue regeneration and their associated therapeutic effects are also highlighted. Finally, conclusions are made with perspectives and future research directions based on identified challenges.

Keywords Bioactive glasses · Bone regeneration · Bone tissue engineering · Antibacterial · Bioactive ceramics

---

S. B. Zaidi · S. A. Batool · M. A. U. Rehman (✉)
Department of Materials Science and Engineering, Institute of Space Technology Islamabad, Islamabad, Pakistan
e-mail: atique1.1@hotmail.com

S. J. Ikhmayies (ed.), *Advances in Biomaterials Research*, Advances in Material Research and Technology, https://doi.org/10.1007/978-3-031-97481-6_10

1 Introduction

Bone is a specialized connective tissue with the ability to regenerate after injuries and during normal skeletal development. Bone homeostasis is maintained through the processes of bone production and resorption, which are mediated by osteoblasts and osteoclasts, respectively Fig. 1A, B. Bone production and resorption are currently in a state of balance. Bone anomalies can be caused by injury, pathological disease (bone loss due to breast cancer or prostate cancer), or surgical operations.

The traditional methods of treatment of bone defects, namely autografts [1], allografts [2], and xenografts. All of them have limitations such as graft rejection, pathogen transmission, and other constraints such as a scarcity of donors, and an immunogenic response by the host of the event [3]. Tissue engineering (TE) has emerged as a potential field of study. It is a multidisciplinary approach to the repair, restoration, and regeneration of injured tissues beyond their natural healing capabilities emerged as a viable alternative to grafting procedures [4]. Bone tissue engineering (BTE) is a new field of biomedical science that provides a local environment for bone regeneration by employing cells, 3-D biomimetic matrices, and growth factors alone or in combination. BCs used for BTE should be biocompatible, biodegradable, mechanically strong, chemically inert, osteoconductive, and osteoinductive with the ability to osseointegrate at the site of implantation [5–9]. BCs are the most commonly employed materials among the many synthetic biomaterials studied for hard tissue recovery and prosthesis-covering calcium phosphate (CP) [e.g., hydroxyapatite (HAp),—and α and β-tricalcium phosphate (α + β TCP),

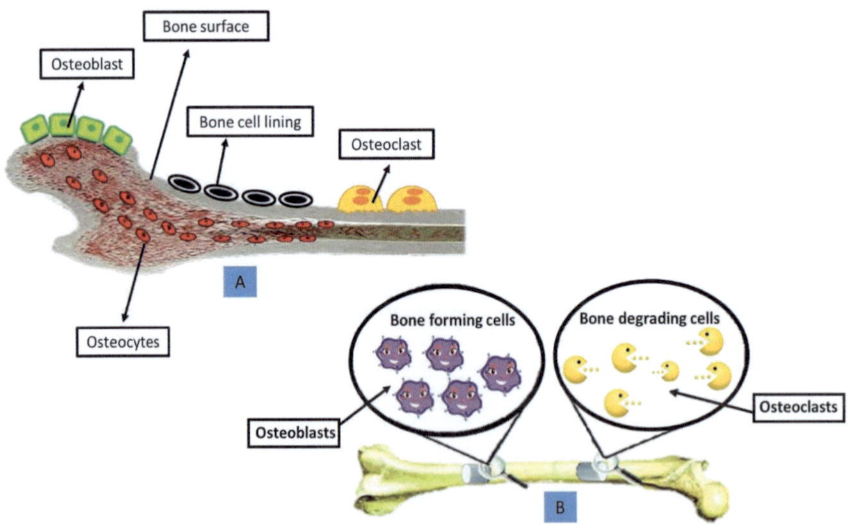

Fig. 1 **A** Surface of the bone containing osteoblasts, osteoclasts, and bone lining cells. Osteoclasts pervade the mineralized matrix interior. **B** Bone-forming osteoblasts and bone-degrading osteoclasts that make up bone. Bone homeostasis is maintained by a delicate balance of these two processes

calcium silicates (CS) [e.g., wollastonite ($CaSiO_3$) and larnite (–Ca_2SiO_4)], BGs with a composition based mostly on the SiO_2–Na_2O–CaO–P_2O_5 system [e.g., 45S5, 58S, 60S3, mesoporous bioactive glass nanoparticles (MBGNs)], and hybrid scaffolds of these materials are some examples of biomaterials (BMs) widely investigated for replacement and repair of hard tissue [10–13].

In this chapter, we have focused on the substitution of inorganic ions in the network of BGs for advanced BTE applications. An overview of various metallic ions doping in BGs is provided. The discussion calls attention to the effect of individual ions present in the BG matrix. The ionic doping improves the surface properties of the biomaterial, making them more reliable and safer for clinical applications. Inorganic metal ions, in tandem with the bone substitutes or on their own, are thus of particular interest for bone tissue regeneration. Moreover, the chapter highlights what is known about the role of inorganic ions in cellular processes and bone regeneration, with an emphasis on research that has already been done for the regenerative potential of bone substitutes. However, there is a lot to be done to completely understand the action mechanism of these metallic ions-doped BGs inside living bodies. Predicting the true outcome of designed BGs for a specific clinical trial is still impossible based on in vitro and in vivo studies only. Providing an idea of how these materials will interact with the body when implanted in animal models. Investigations conducted in vitro and in vivo only offer limited insights; therefore, investigations using animal models are necessary to fully comprehend material interactions in living systems. Beyond these basic tests, clinical studies are necessary to obtain more knowledge. To improve readers' comprehension, the chapter contains instances of clinical experiments on BGs.

2 Ionic Doping of Bioceramics

BCs have the intrinsic ability to produce a bone-like apatite layer on the surface of the implant which can create a chemically stable contact with surrounding tissues, promoting bioactivity. The development of apatite in vivo can be predicted by placing the material in an aqueous media (enriched with minerals) with a chemical composition that mimics physiological liquids [10, 11].

Specific metal ions with therapeutic effects (e.g., calcium (Ca^{2+}) [14], magnesium (Mg^{2+}) [15], copper (Cu^{2+}) [16, 17], zinc (Zn^{2+}) [18, 19], strontium (Sr^{2+}) [20, 21], iron (Fe^{3+}/Fe^{2+}) [22], cobalt (Co^{2+}/Co^{3+}) [23], and manganese (Mn^{2+}) [24]) are incorporated into BCs chemical structure to overcome the disadvantages of pristine material and favor their use in different biomedical applications. Many metal ions are required for regular cell metabolism, but in high enough concentrations, they can be harmful. Inorganic ions and organic molecules/ions are essential ingredients of life and are active in cell-matrix functions [25]. These essential trace elements are involved in multiple metabolic functions and are required for the growth, development, and maintenance of healthy bones. Inorganic ions are commonly spread in nature, so they can be categorized as toxicological or biochemically functional

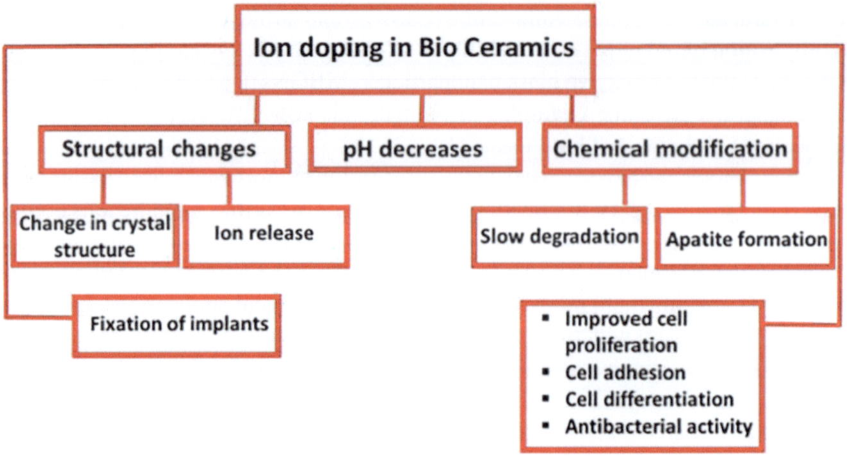

Fig. 2 Doping of different metallic ions in BCs and their functions

in biological processes [26]. The fact that a significant fraction of enzymes needs metals for catalysis is a major determinant of their functional significance in biological systems [27]. The electrochemical law governs the action of metal ion release into biofluid. Inorganic ions are used to treat various ailments such as Ca^{2+} and Mg^{2+} that take part in critical regulatory functions of the cells [28]. Angiogenesis and osteogenesis can be stimulated by their integration into BMs. As a result, it is vital to understand how ions interacted with cells in vitro and how certainly ions could be included in biomaterials in the right doses to benefit both in vitro as well as in vivo [29].

Metallothionins are metal-ion-rich proteins present in living organisms. Zn^{2+}, Ca^{2+}, and Mg^{2+} are divalent cations that inhibit cytotoxicity and antagonism [30]. The inorganic ions are being incorporated in various classes of BMs. For example, Fig. 2 depicts the functioning of different ions substituted in BMs.

Divalent ions such as Zn^{2+} and Mg^{2+} have been demonstrated to play a key role in the activation of signaling cascades linked to bone regeneration in osteoprogenitor cells as well as osteoblasts [30]. Osteoporosis is caused by a lack of Mg^{2+}, which results in a decrease in bone development, an acceleration in bone resorption, as well as the loss of trabecular bone. BGs made of borosilicate have the potential for bone tissue engineering [31]. Dwarfism, osteoporosis, as well as delayed bone development are all symptoms of Zn deficiency in humans. Zn enhances stem cell osteogenesis and mineral deposition when it is incorporated into synthetic polymers [32]. Mg is required for the normal functioning of our cells. Mg is, in particular, the most plentiful mineral within each cell, second only to potassium (K) [33]. In numerous in vitro experiments, the Mg^{2+} created via Mg metal breakdown can promote osteogenic stem cell differentiation and improve endothelial cell migration, ultimately promoting osteogenesis as well as angiogenesis. Because of their ability to enhance osteoblast bone production, Mg^{2+} boosts alkaline phosphatase activity and inhibits osteoclast differentiation. Zn and its alloys have significant benefits in encouraging bone regeneration. The underlying mechanism of Zn in these activities, however, has yet to be discovered [34]. Ca is well-known for its role in the development of healthy bones and teeth [35]. Ca, on the other hand, tends to relieve muscle cramps and triggers a variety of responses in the human body [36]. Along with its osteoconductive as well as osteoinductive characteristics, CP is more commonly used for bone regeneration purposes. Cu^{2+} can promote cell proliferation, increase osteoblast activity as well as proliferation, and augment BG osteostimulation and corresponding antibacterial properties. As a result, Cu has a lot of potential for boosting bone repair [37]. In a rat model, Mn^{2+} can speed up fracture repair by enhancing early angiogenesis [38]. Rare-earth metal ions (such as cerium (Ce^{3+}), terbium (Tb^{3+}), erbium (Er^{3+}), europium (Eu^{3+}), lanthanum (La^{3+}), and dysprosium (Dy^{3+})) have been studied as dopant ions in BCs, resulting in new promising category of luminescent optical nanomaterials that could serve as an alternative to organic fluorophores. An overview of the potential applications of rare-earth-doped BCs was published by the researchers of references [39–41].

The most stable CP phase, hydroxyapatite (HA) with chemical formula $Ca_{10}(PO_4)_6(OH)_2$, offers numerous physiological benefits [42], including direct bone bonding and enhanced new bone production [43]. HA has the potential to be used as a reinforcing filler in composites, insulating agents, and chromatomedium for simple and efficient protein and nucleic acid fractionation [44]. HA is found in a variety of compositions that can be classified based on the Ca/P ratios. The Ca/P ratio of stoichiometric HA is 1.67, although stable compositions can have Ca/P ratios up to 1.5, assuming no ionic substitution in the apatite structure [45]. HA is chemically comparable to hard tissue and has great biocompatibility and bioactivity [44, 46–49]. As a result, it has found a lot of applications in medication delivery systems and tissue engineering. On orthopedic devices, HA has been identified as a bioactive ceramic coating that improves the integrity and durability of the implanted prosthesis [50]. In the realm of bone tissue engineering, nanostructured biomaterials may be able to substitute hard and soft skeletal tissue as well as biocompatible materials for tissue genesis [51]. A lot of effort has been made to improve the mechanical characteristics of synthetic CP. Doping BCs with metal ions such as Mg, Mn, Zn, and Sr [52] is one of the ways used. The lattice parameters, crystallinity, dissolution kinetics, and other physical properties of apatite can all be affected by these residues of ions. These metal enforcers are also thought to play a role in boosting cell-material interactions and strengthening the mechanical properties of CP [10, 53]. The interaction of cells with materials is extremely dynamic, resulting in an outstanding tissue response at the biomaterials interface. Minerals and traces of metal elements may also help grow and accelerate bone formation and resorption in bone cells and bone minerals in vivo and in vitro. Certain metal ions [Ag^{1+}, Zn^{2+}, Cu^{2+}, Fe^{3+}, gallium (Ga^{3+})] [16, 28, 54–57] have also been observed to penetrate bacteria, inactivate their enzymes, and produce hydrogen peroxide: all these processes can kill bacteria and fungi [25].

Calcium silicate (CS) is an appealing BC because it degrades to release Ca^{2+} and silicon (Si^{4+}) ions which are inductive for osteogenesis and angiogenesis, and improves their physical and biological qualities [14]. Incorporating a variety of metal ions into CS ceramics always results in complex structures and varying ionic bond strengths, affecting ion release. Such complex structures may affect the stability and pace of dissolution. A high pH value and dissolution of Ca and Si ions impede cell differentiation and growth; however, osteoblast proliferation and differentiation are aided by these factors. Gene expressions are also stimulated by low Ca and Si ion concentration and a proper pH value [58, 59]. Ion-doped CS ceramics give rise to novel materials with a different crystal structure. The features of pseudowollastonite (α-$CaSiO_3$) ($CaSiO_3$), also known as calcium silicate (α-CS), were first investigated in this field, and it was discovered to be bioactive due to its capacity to produce a surface apatite layer when immersed in a simulated bodily fluid (SBF) [60, 61]. However, CS's progress as a clinically useful bone substitute has been hampered by two key drawbacks. For starters, it has a high dissolving rate, which results in an alkaline environment, causing alkaline-induced toxicity and excessive release of Ca and Si ions to levels that limit cell development. Second, when compared to the ranges of values required for the regeneration of cortical and cancellous bone, the mechanical strength of α-CS as dense monoliths and porous scaffolds [62] is extremely low, limiting the materials ability to maintain mechanical stability [63] under physiological loads [64–66]. Researchers [67–69], have been looking into ways to improve the mechanical properties of Ca silicate ceramics by reducing the rate of degradation. To counter this challenge, various metal ions and/or metal oxides are doped into the CS crystal structure to generate doped-Ca silicate ceramics (DCSCs). The DCSCs do not control the degradation rate, instead, they promote bone regeneration rate due to the presence of osteoinductive metallic ions. The tissues are reformed before the CSCs are completely degraded. BCs contain a dopant that is either an element incorporated for ionic Ca substitution a metal oxide integrated into the $xCaO$–$ySiO_2$ structure [70, 71], or a mixture of both techniques that has been assessed for biocompatibility/bioactivity in at least one in vitro or in vivo investigation. These DCSCs were created in one of three ways: (1) ionic substitution of Ca with divalent ions such as Sr and Cu, (2) integration of metal oxides into the $xCaO$–$ySiO_2$ structure, or (3) a mixture of both procedures [65, 68, 72]. Except for aluminum oxide which produces gehlenite after doping, the oxides used to dope Ca silicate normally contain a divalent (Sr, Zn, Mg, Cu, Co) or quadrivalent [titanium (Ti), zirconium (Zr)] metal ion [73].

BGs have controlled particle size and porous structure that are important for biological use. Ionic substitutions in the bioactive glasses often perform key functions in the biological systems [74] that can be prepared either by the sol–gel method or the melt-quench method. Bioactive glasses synthesized by melt quench or sol–gel method generally exhibit irregularly shaped morphology, non-uniform particle size, and amorphous structure [75, 76]. However, the right choice of precursor and appropriate pH value during processing can result in a controlled morphology [77]. Sol–gel-derived glasses have benefits over melt-derived glasses, including a large surface area as well as nanoporosity inside a lamellar material that may aid dissolution and bioactivity. Among sol–gel derived glasses MBGNs are considered as a preferred option due to their ability to carry the drugs at the desired site in addition to rapid

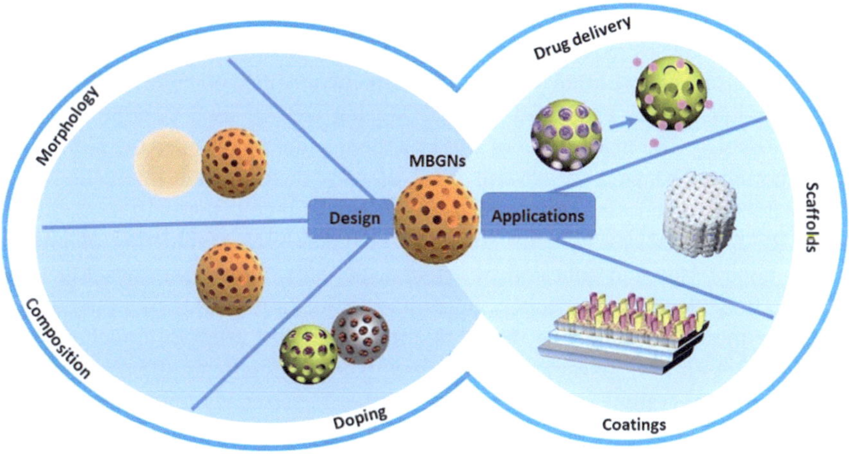

Fig. 3 Design and applications of metal ion-doped mesoporous bioactive glass nanoparticles (MBGNs)

ionic dissolution. MBGNs can thus promote bone healing due to small pores (6–7 nm) inside the particles [77, 78]. MBGNs generally have SiO_2, CaO, and P_2O_5, in a mol% composition of around 60%, 36%, and 4% values, respectively [79]. CaO can be substituted with inorganic ions thus leading to specific functionality and application [80, 81]. Though it may take the therapeutic ions longer to be released, the constant interchange of ions, as well as the liquid milieu, may cause corrosion difficulties that must be managed [82]. One of the advancements in tissue engineering is the synthesis of porous scaffolds obtained from BCs and their applications in bone implantation [81], drug delivery [78], dentistry [83–85], etc. Figure 3 explains the design and applications of metal ion-doped MBGNs. However, significant obstacles remain such as developing a biocompatible and biodegradable nanoparticle imaging probe suitable for clinical usage.

3 Role of Individual Ions in BGs

3.1 Magnesium (Mg)

Magnesium cation is the fourth most abundant cation in the body and the second most abundant intracellular cation. Mg is a trace element found in the body, including bone and extracellular matrix, and is linked to cell differentiation, calcined tissue mineralization, and mineral metabolism in an indirect way [86, 87]. It is known to be essential for several enzymatic activities in many biological functions of the human body [88]. Mg is involved in over 300 chemical reactions in the human body, as well as playing a key part in cellular functions [89]. It can increase osteoblast proliferation, an important mechanism in bone growth, maintenance, and repair [78]. Furthermore, Mg deficiency causes reduced bone formation, increased trabecular bone loss, and bone resorption, underscoring Mg's important involvement in bone metabolism [33]. It has previously been examined whether Mg ion may be used to dope bioactive glasses [90–92]. The importance of Mg as a therapeutic ion has been explored and it is reported to be vital for several types of cells owing to the interaction with phosphate ions (adenosine triphosphate (ATP) exists in cells normally as a chelate with Mg^{2+}), it acts as a cofactor for many enzymes [93], stimulation of growth of new bone tissue and adhesion of osteoblastic cells. Other than this, Mg is an excellent coagulant used especially in cardiac applications [38]. Due to the ability to change their resorption rate, Mg^{2+} is one of the most effective dopants for β-TCP. Gallo et al. [94] investigated the effect of Mg^{2+} doping on the resorption of β-TCP. Mg-doped samples (1 and 6 mol%) were synthesized via solid-state reaction, while non-doped samples were generated by precipitation in an aqueous medium. The doping of β-TCP with Mg^{2+} was validated by a considerable drop in lattice characteristics with an increase in Mg^{2+} content. The grain size of the β-TCP samples, on the other hand, didn't differ significantly. Osteoclasts break down the inorganic components of bones by lowering the pH locally, causing CP crystals to dissolve. To recreate the environment at the surface of an implanted material during cellular resorption, each sample was immersed in acidic solutions ($pH = 4.4$) and compared to osteoclastic-mediated resorption. Mg doping reduced the solubility of the β-TCP phase and delayed acid-etching as well as cell-mediated resorption. Grain orientation (crystal direction [0 0 0 1] for pure β-TCP) led to an early-stage breakdown, which resulted in a spike-like sub-structure. These findings pave the way for new ways to create materials with tailored resorbability [95].

Kanasan et al. [87] investigated an Mg-doped hydroxyapatite/sodium alginate biocomposite (Mg-HAp/SA) intending to improve the physical, mechanical, and bioactivity aspects of bone implants. The precipitation method was used to dope HAp at different Mg^{2+} concentrations ranging from 0.5 to 1.5 wt.% Mg-HAp and sodium alginate (SA) powder were combined (99.5:0.5), uniaxially crushed, and sintered at 1300 °C. The density and porosity of the sample were dramatically affected by the presence of Mg^{2+}. When Mg^{2+} concentration was increased to 1.0 wt.%, the relative density increased by roughly 5%. As a result, the hardness of the sample increased. In comparison to pure HA/SA, increasing Mg^{2+} concentration improved sample hardness (4 GPa for the sample doped with 1.0 wt.% Mg^{2+}) compared with pure HA/SA (3 GPa). Because of the grain expansion and pore reduction during the sintering process, both density and hardness were lowered at 1.5 wt.% Mg. Mg-HA/SA has a higher potential to generate apatite than pure HAp, making it a better candidate for use as a bone substitute during the implantation process [10].

Due to its many biological benefits, Mg ion has been used as a dopant and in the form of complexes with many different matrices, among which glass-ceramic (49.13 wt.% SiO_2–7.68 wt.% CaO–43.19 wt.% MgO) is the most common [80]. The preferred synthesis technique is the sol–gel method. Other than glass ceramics, some quaternary glass systems (64% SiO_2, 26% CaO, 5% MgO, and 5% P_2O_5 in mol.%) are also used. The synergistic effect of bioactive glasses with Mg ions promotes osteoblast cell proliferation and differentiation [96]. Dietrich et al. [97], looked at the effect of magnesium doping on the in vitro bioactivity of melt-derived glasses 46S6. They discovered that magnesium influences the development and evolution of newly created layers and that the degree of these changes is highly dependent on the amount of magnesium added to the glass matrix. According to Dietrich et al. [97], MgO promotes the bioactivity of alkali-mediated glasses while inhibiting them in acid-mediated glasses. Mg has two main mechanisms of interaction; it can either bind to the enzyme-substrate, forming a complex with which the enzyme interacts, or it binds directly to the enzyme altering its structure. All in all, its function is related to ATP utilization, as Mg is found in almost all biological cells as Mg-ATP [30]. Therefore, the inability of Mg to perform its function either due to deficiency or uncontrolled release from the scaffolds can result in the hindrance of Mg-activated functions. A possible mechanism for Mg-induced bone-loss is presented in Fig. 4. It shows that Mg plays a crucial role in cellular function. In the absence of proper Mg release and activation, cell proliferation can be hindered due to reduced deoxyribonucleic acid (DNA), ribonucleic acid (RNA), and protein synthesis [33]. The addition of Mg ions in scaffolds increases the bioactivity and compatibility of the system by promoting bone cell activity. Mg is mitogenic for osteoblasts in cell culture and its depletion causes cell growth inhibition [98, 99].

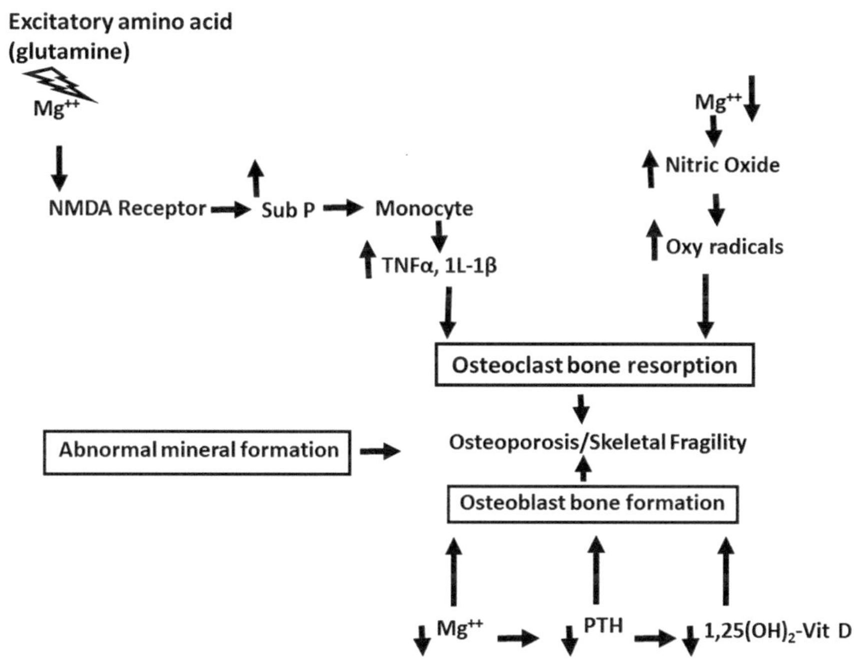

Fig. 4 Mechanism of Mg-induced bone loss. Reprinted from Rude and Gruber [33], with permission from Elsevier. Copyright (2004)

3.2 Strontium (Sr)

Strontium belongs to group II-A of the periodic table same as that of Ca and thus has comparable chemical properties. Due to these comparable properties, Sr can displace Ca in the metabolic process of hard tissues. 0.335% of the human skeleton consists of Sr, as well as its biological importance. Sr is already known to be utilized in pharmaceutical medications to treat osteoporosis, and it stimulates cell proliferation and differentiation [100]. The incorporation of Sr^{2+} into bones improves osteoporosis therapy outcomes. Injecting Sr^{2+} into BGs may preserve the acellular bioactivity of the glasses while promoting osteoblast activity and inhibiting osteoclast differentiation. Due to its chemical and physical similarity to Ca, Sr is frequently used in bone substitute matrices including HAp and BGs [15]. So, Sr can be incorporated as a therapeutic ion in MBGNs for bone formation and tissue regeneration applications [40, 41]. During Sr incorporation in BGs, ions may be present as inclusions on the surface of BGs that slow down the release of ions at the site of defect, which is therapeutically beneficial. Sr-doped BG also has high-lightened a-cellular bioactivity [101, 102].

A highly porous structure with interconnected pores is the most promising material for the healing of bone defects. In this connection, Yin et al. [103] developed borate glass scaffolds (13-93B2) containing Si, B, Ca, Na, K, Mg, and P ($18SiO_2$–$36B_2O_3$–$22CaO$–$6Na_2O$–$8K_2O$–$8MgO$–$2P_2O_5$, in mol%) for BTE doped with various amounts of SrO (0–9 mol%). The polymer foam replication technique was used to construct scaffolds with pore diameters ranging from 200 to 500 microns and a porosity of about 80%. The melting procedure was utilised to incorporate Sr into the glass composition. Fourier-transform infrared spectroscopy (FTIR) was used to confirm that Sr was present in the glass. After soaking in SBF for a week, the scaffold doped with 6 mol% SrO showed good bioactivity since nanometer-sized HAp crystals were present. As the SrO content rose, the scaffolds compressive strength somewhat improved. While the release of boron (B^{3+}) from the BG structure into the SBF solution steadily decreased, the addition of Sr boosted the release of Si^{4+} and Ca^{2+} ions from the BG structure into the SBF solution. Although levels above 0.65 mM have been demonstrated to inhibit cell growth and proliferation, B is essential for bone health. Because Sr ions positively affect bone formation and limit bone resorption through direct or indirect effects on bone cells or bone minerals, they can enhance the biocompatibility and physicochemical qualities of HA for use in implants. As a result, Sr has drawn notice. Alkaline earth elements Sr and Ca are utilised as oxides in BCs because of their high spatial resolution [103, 104]. It's a terrific idea to utilise nanoindentation to examine the mechanical behavior at the implant/bone interface. The nanoindentation method can be used to measure the mechanical properties of individual bone structure units. This novel method was used to investigate the nanomechanics of the interface between cancellous and cortical bones with strontium-containing hydroxyapatite (Sr-HA) bone cement under weight-bearing conditions. After six months, Young's modulus and hardness at the interface between Sr-HA cement and cancellous bone were significantly higher than those at the interface between Sr-HA cement and cancellous bone, while Young's modulus and hardness at the interface between Sr-HA cement and cortical bone were significantly different. These results were supported by histological analysis and chemical composition. Osteointegration of Sr-HA cement with cancellous bone. The interface between cortical bone and Sr-HA cement did not have an apatite layer with a higher Ca and P content than cortical bone, but there was an apatite layer with a higher Ca and P content between cancellous bone and Sr-HA cement. These two interfaces mechanical characteristics indicated that they were diffused interfaces, and six months after implantation, cancellous or cortical bone developed into Sr-HA cement [20].

3.2.1 Mechanism of Action of Sr Incorporation in Bone

Commonly known effects of Sr on bioglass bioactivity are (1) formation of CP structure on BG surface, and (2) chemical stability, strength, and bioactivity of BG. The size of Sr ion (118 pm) is marginally larger than Ca ion (100 pm) which causes the increment in d-spacing in HA crystals [48, 49]. As previously said Ca and Sr can be replaced due to their comparable properties, and this incorporation leads to the expansion of HA crystal. The impact of Sr on rat calvariae organ culture, a model that controls the incorporation and activity of osteoblastic cells, was investigated by Canalis et al. [105]. Histo-morphometric analysis, which was used to analyse this model, revealed that Sr mainly boosted pre-osteoblastic cell multiplication and increased bone matrix development. They also studied the Sr effect on DNA synthesis and bone collagen synthesis. This effect was specific to Sr salts, not Na salts or Ca salts. Response of bone cells to Sr and Ca showed that these cells can discriminate between these two ions. The mechanism of pre-osteoblastic replication and synthesis of collagen by Sr for enhanced bone formation is shown in Fig. 5 [100].

3.3 Silver (Ag)

Silver—Group 1B includes the transition metal element silver. The silver monovalent ion (Ag^+) is typically utilised in metallo-pharmaceuticals and is well known for its potent inhibitory and antibacterial qualities against a range of pathogens [106]. Additionally, this metallic ion performs better than organic antimicrobial agents in terms of heat resistance, antibacterial activity endurance, and safety. Due to its high reactivity, Ag^+ reduces infection during implantation [25, 26]. The rate of infections during knee replacements is around 1%, while after implantation it is between 0.3 and 0.6% for hip replacements [107]. The patients suffering from an infection must get revisionary surgery [108]. A biofilm that forms on the implants surface is the cause of the infection [107, 109]. Several strategies, including medication delivery and surface topology modification, can be used to prevent the formation of biofilms [77]. Ag^+ particles are significant because of their beneficial antibacterial properties

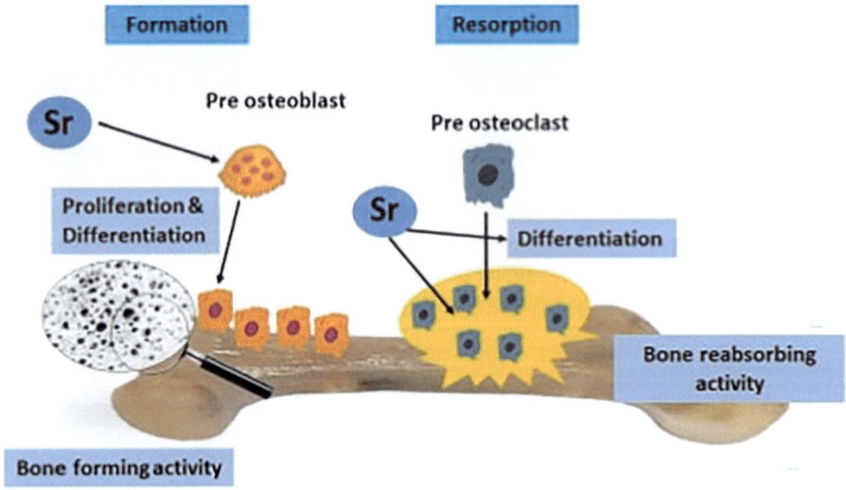

Fig. 5 Mechanisms of action of Sr in bone cells. Strontium stimulates preosteoblast replication, leading to an increase of matrix synthesis

in tissue culture processes [110]. Additionally, studies have demonstrated that the growth of Ag tissue societies in bone does not interfere with ontogenesis indicators, such as the formation of HAp [110]. Therefore, adding Ag to bone repair filler tissues may prove beneficial by helping to prevent diseases with negligible negative effects.

However, Ag has several drawbacks, including an unstable ionic state that can be hazardous if directly consumed. Several studies were conducted to overcome deficiencies such as binding with substrates (HA [111], BG [112], CS [113]) to increase ionic state stability and release of metal ions over an extended time period [95]. Ag's addition to HAp has become more and more common in recent years, indicating the materials' enormous potential. The literature on the topic does not go into detail regarding the synthesised materials or the methods of Ag inclusion, despite the fact that the results are consistent in terms of the good biological benefits. An Ag^{+} doped HAp compound or an Ag/HAp nanocomposite can be formed by a variety of synthesis techniques, including co-precipitation possibly followed by a thermal or reducing treatment, electrochemical, hydrothermal [114, 115], chemical precipitation [116, 117], sol–gel [118, 119], and impregnation methods. While both options are acknowledged and discussed in the literature, little is known about the mechanisms that result in one of them. The first scenario involves the replacement of Ca^{2+} in the HAp crystal structure by Ag^{+} cations, while the second scenario involves the intimate intermingling of metallic Ag particles in a multi-phase composite. Paterson et al. [120] and Pajares et al. [121] studied Ag doped HA and Ag doped BGs, respectively. Electrospun scaffolds containing Ag-HA proved antibacterial properties of Ag in orthopedic and dental bone surgeries. Moreover, Ag-BG microparticles exhibited MRSA-specific antibacterial action. Furthermore, Ag-BG microparticles bioactive and regenerative qualities were shown in a calvarial defect model (in vivo) and co-culture with hBMSCs (in vitro). Ag-BG increased the rate of cell proliferation and osteoblastic differentiation in hBMSCs, which in turn increased bone formation. These characteristics along with their special ability to combat bacteria even in growth-arrested environments show that Ag-BG microparticles are a highly effective regenerative material for treating bone deformities and reducing the likelihood of recurrent *Staphylococcus aureus* infections.

It has been discovered that adding Ag to a bioactive glass framework increases the crystallisation of quartz and metallic Ag in the bioactive glass. Large quantities of Ag ions have been found to be cytotoxic, although their release from bioactive glass implants can help treat bone infections. It was demonstrated that glass holding 2 wt.% Ag was harmful to cells, however, glass having 0.75 and 1 wt.% Ag did not harm the cells.

According to Shirazi et al. [113], CS doped with Ag and tantalum (Ta) has superior mechanical and biological qualities. The addition of 5 wt.% Ta to the CS lattice produced hardness and fracture toughness values that were noticeably higher than those of 5 wt.% Ag. Increasing the amount of Ag did not affect the mechanical properties. Both Ag-doped and Ta-doped CS were capable of producing surface apatite upon immersion in simulated physiological fluid. Ta incorporation into the CS enhanced human fetal osteoblast (hFOB) cell growth, with greater levels of Ta resulting in increased proliferation. However, compared to smaller quantities of Ag and doping silica, raising the concentration of Ag to 15 wt.% inhibited cell growth.

3.3.1 Mechanism of Action and Therapeutic Effect of Ag

In the 1890s, the term "oligodynamic action" was introduced which explains the toxic effect of metal ions on pathogens. It also includes the antibacterial action of Ag. Ag combines with many cellular components. Attachment of Ag with bacterial cells or membranes is more important than nucleic acid. Ag-ions (Ag^{2+}) enter through the proteinous cell wall, cause precipitation of deoxyribonucleic acid (DNA), and destroy bacteria [122]. Incorporation of Ag in BGs enables controlled, sustainable release of Ag^{2+}, so the antibacterial activity of Ag ions is preserved in the body fluids that contain a different type of proteins and chloride ions that can form complex compounds with Ag, which inhibit the antibacterial action of Ag [36, 37]. It was found that Ag ions interact with bacterial cells and with enzymes containing thiol groups. The interaction of Ag^{+} with bacterial cell walls causes inhibition of various functions of cells, which causes bacterial cell death. Free radicals are created as a result of the production of reactive oxygen species (ROS). Strong bactericidal action is exhibited by these radicals. The proposed mechanism is inhibition of respiratory enzymes by Ag ions and when these ions enter the bacterial cell wall ribosomal denaturation causes inhibition of protein synthesis. DNA transcription and translation are also inhibited due to the binding of genetic material with bacterial cells [123–125]. However, a high concentration of Ag in BGs can be toxic. It was found that 2 wt.% Ag shows toxicity for osteoblast cells and 0.75–1 wt.% Ag did not influence osteoblast cell activity [126]. Ag has been shown to act through a variety of methods, including

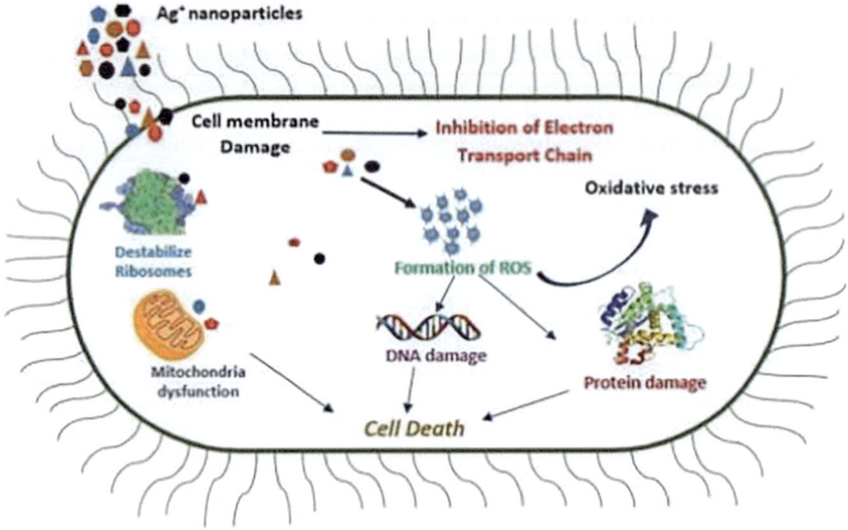

Fig. 6 Proposed mode of action of Ag nanoparticles against bacterial growth. Reprinted from Rahman et al. [126], with permission from Springer Nature. Copyright (2019)

(1) interfering with electron transport, (2) attaching itself to DNA, and (3) interacting with cell components. Among pathogenic microbes, silver is extremely poisonous. Previous research has examined its biocidal activity on 12 distinct bacterial species [122]. Ag nanoparticles mode of action against bacterial growth is described in Fig. 6 [126].

3.4 *Zinc (Zn)*

Zinc is another transition metal element and a vital micronutrient for the body. It is a cofactor of numerous enzymes that, via promoting protein synthesis, are crucial to DNA replication. Since Zn is necessary for the production of bones, a lack of it is linked to a delay in the growth of the skeleton [127–129]. Zn ion (Zn^{2+}) has been used as a doping agent (thin films or composites) in matrices including BGs [130] and HAp [131]. In addition to promoting angiogenesis and homeostasis, Zn^{2+} insertion in these matrices possesses antimicrobial qualities. Furthermore, because Zn^{2+} can increase the specific activity of alkaline phosphate (ALP), it is crucial for the growth, mineralisation, and production of bones [129].

Balamurugan et al. [131] and Hall et al. [132] separately added Zn into BGs that does not affect the bioactivity of bioglass. Controlled Zn additions can either increase or preserve bioglasss bioactivity rather than decrease it. The concentration of Zn in the bioglass and its interaction with the overall composition and structure of the glass determine its bioactivity and effectiveness. Zn can have beneficial effects on bone growth and regeneration, but its bioactivity must not be adversely affected by improper formulation. Zn is larger and thus has a higher surface area than common ions present in bioglass or hydroxyapatite like Ca^{2+} and Mg^{2+}. In some circumstances, Zn's ionic radius (74 pm) is greater than Ca's (100 pm). Because Zn is larger than these other ions, when it is introduced into a material, its larger size might affect the materials structure and properties, increasing the number of sites accessible for nucleation sites and new phase formation. In vitro studies of Zn-doped BGs show enhanced formation of the CP phase. Increasing the concentration of Zn in BGs leads to an increase in the concentration of ZnO, It's concentration dictates whether it will function in a glass structure as an intermediate oxide or as a network modifier. ZnO reduces the glass bioactivity due to a decrease in non-bridging oxygen atoms. As the doping of therapeutic ions getting innovative and is remarkably used for bone regeneration and improvement of biological properties. Among other therapeutically impacting ions, Zn serves as an imperative mediator in bone matrix mineralization [133]. Along with the bone formation properties and upgraded a-cellular formation of CaP layer, Zn also possesses antibacterial effects [134]. It is found in the literature that bioactivity [135, 136]. Zn-doped BGs became promising materials for bone regeneration as well as antibacterial effects at the nanoscale. The release of metal ions at higher concentrations is cytotoxic for human osteoblast cells. It has been found that more than 0.4 wt.% Zn is cytotoxic and 0.4 wt.% Zn oxide in 58S bioglass improved bone proliferation [53, 54]. However, Aina et al. [18] pointed out that the cellular bioactivity results of Zn-BG are not persistent and showed that hydroxycarbonate apatite (HCA) formation on BG glass is prohibited due to the presence of Zn content which lessens the overall release of BG ion. Similarly, O'Connor et al. [137] reported identical results of delayed development of HCA on the BG surface. On the other hand, the release of Zn ions (ZnO-BGNs) is dependent upon pH (pH-sensitive) [130]. Synthesis of Zn-MBGNs are an upcoming building block in the field of nanocomposites and biomaterials applications. Given that Zn is essential for bone production and growth in the physiological environment, a great deal of research has been done on how it affects osteoblast activity, as seen in Fig. 7 [137, 138].

Researchers need to develop materials that mimic the structure and composition of bone and have antibacterial properties to prevent post-operative infection in implantable devices. Zn-doped HAp (Zn-HAp) thin films on silicon discs (SiD) were produced by Predoi et al. [138] using a sol–gel spin coating technique. The material was annealed at two different temperatures—500 and 700 °C—after spin coating. The surfaces of both materials are smooth and undamaged. Proliferating cell lines (HeLa cells) were used for the cytotoxicity testing, and the outcomes demonstrated that the materials were safe. However, there was a slight impact of heat treatment temperature in antibacterial tests against *Candida albicans* and *S. aureus*. The inhibitory effect on microbial cells was greater when SiD-coated Zn-HAp layers treated at 700 °C were compared to SiD-coated Zn-HAp layers treated at 500 °C. Bandyopadhyay and colleagues [139] investigated the effects of ZnO in CPs. They discovered that adding ZnO to CPs with customised sintering improves their densification and microhardness (the maximum microhardness was attained at 2.5% ZnO sintered at 1250 °C).

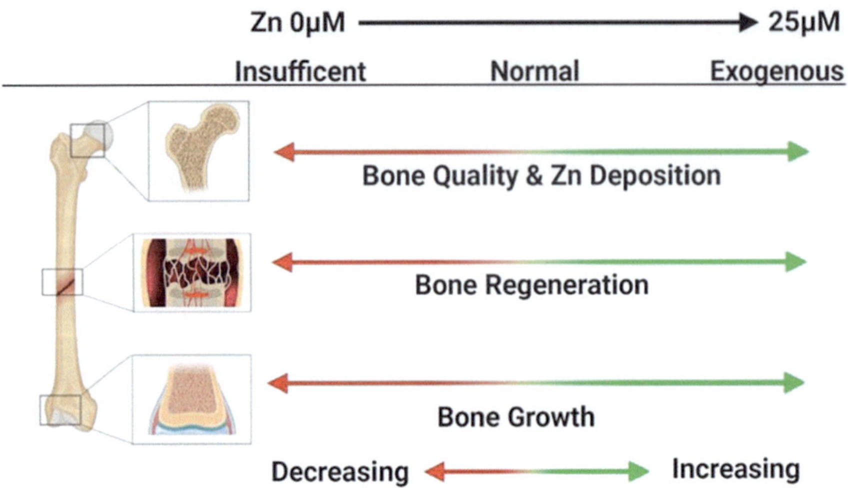

Fig. 7 Effect of Zn on skeletal cells. Reproduced from O'Connor et al. [137], under the terms and conditions of the Creative Commons Attribution (CC BY) license. http://creativecommons.org/licenses/by/4.0/

Furthermore, investigations on cells employing the osteoprecursor cell line OPCI demonstrated that ZnO in substrates is non-toxic.

Zn-releasing substances, such as Zn-TCP, have shown that Zn effectively increases bone density or stops bone loss in vivo and promotes greater bone growth in vitro. A biphasic layer of fluoride hydroxyapatite (FHA) and Zn-containing tricalcium phosphate (ZnTCP) was created using the sol–gel method [131]. The ZnTCP/FHA layer increased cell viability statistically significantly and maintained cell growth as compared to another biphasic layer (TCP/FHA) deficient in Zn. ZnO-doped TCP and HA were synthesised in a ball mill in 5 h and 1.5 min. After five days, a cell material interaction study using an osteoprecursor cell line showed that ZnO-doped HA and TCP were safe. However, it was shown that the increasing ZnO content inhibited cell interactions and spreading. High crystalline nanoparticles of CP containing Zn (ZnP) were also created by wet chemical processes, using precursor particles that ranged in size from 300 to 800 nm. ZnP nanopowder and porogen were mixed to create the porous scaffolds, which were then compressed into a thin disc and sintered for an hour at 1100 °C. Following in vitro analyses of adhesion and spreading onto these matrices for 24 and 5 days, an examination of the cells on these surfaces showed that osteoblast cells were, albeit in small amounts, attached to the ZnP surface.

3.5 Copper (Cu)

Copper is a transition metal with great conductivity and corrosion resistance. It is a key micronutrient in the body and is required in small amounts by most living species for metabolic functions [140]. It has an important catalytic role in the initial stage of collagen maturation, which results in the formation of stable fibrils [141, 142]. Cu has been shown to improve endothelial cell proliferation and angiogenesis in vivo, both of which are critical in the repair of massive bone lesions [141]. Cu co-works with Fe for the formation of red blood cells and helps in iron absorption. Its deficiency weakens muscles which leads to the problems of fragile bones di (osteoporosis) [143, 144]. The importance of Cu in the body originates from the synthesis of Cu-doped biomaterials. Mg–Cu and Zn–Cu biomaterials are studied [65, 66] and Cu-doped nanoparticles [145] are topics of interest in this era. It has been reported that Cu-BGs play an essential role in cardio vascularization, bone transformation, and in the curing of wounds [146–148]. It has been proven that the replacement of Cu by Ca promotes HAp formation upon immersion in SBF. Moreover, it doesn't show the cytotoxic effect on human mesenchymal stem cells (hMSCs) and fibroblast cells in the concentrations of 10 and 1 μg/mL. When compared to traditional melt-derived bioglass, which has a BET (Braunauer–Emmett–Teller model) surface area of 0.15–2.7 m^2/g, 5Cu-BGN's surface area (16.3 m^2/g) is noticeably larger. Significant zeta potential value of 5Cu-BGN, which is −27.5 mV represents the negatively charged nature of particles that indicate their stability, and it is acceptable for the fabrication of nano-composites [149]. Direct ingestion of Cu ions is considered dangerous due to their unstable ionic states. After deactivating proteins by binding, Cu ions interact with microbial membranes to alter their structure and permeability, and finally with microbial nucleic acids to restrict microbial reproduction [150]. Using Cu-doped HA nanopowders, Stanić et al. [149] observed a 95% decrease in microorganisms for all three strains of *Escherichia coli*, *S. aureus*, and *Candida albicans* in 2010.

3.6 Other Miscellaneous Ions

A method to enhance some of these biomaterials qualities without compromising advantageous ones is to add dopant ions to the BC matrix. In addition to the ions, other ions serve the same function. For example, Shi et al. [151] studied Fe doped octa-calcium phosphate and poly (lactic-co-glycolic acid) scaffolds. They have studied the Fe^{3+} release and its angiogenic effect on bone grafts. Experimental studies (in vitro and in vivo) revealed that these scaffolds facilitated the adhesion and migration of human endothelial cells (HUVECs) and enhanced the development of new blood vessels. Increased levels of vital markers critical for blood vessel formation, such as HIF-1, CD31, and VEGF-A, were associated with this improvement. It was discovered that the release of Fe from the scaffold increased VEGF synthesis and HIF-1 levels, both of which contribute to the development of new blood vessels. Furthermore, the release of Fe^{3+} boosted the generation of eNOS (endothelial nitric oxide synthase) and nitric oxide, two substances that are essential for controlling blood vessel function. Boron (B) doped BCs also were studied by Uysal et al. [31], and they found that a higher concentration of B fastens the HA formation but cell-reproduction is prohibited. *Borate-based bioactive glass scaffolds were also evaluated* in vitro [152]. His work revealed 0.65 mol concentration of B accelerates cell proliferation. After five days of immersion, the stimulation simulated body fluid (SBF) test showed that many bone-like HA appeared or precipitated on HA particles, and the permanent magnet (neodymium (Nd)-Fe-B based magnet) also responded favorably to Fe-HA, moving easily when a magnetic field was applied [22, 152]. For tissue engineering applications, co-doped polyvinyl alcohol polymer/bioactive glass hybrid scaffolds (PVA/Co-BG) were created. For the best angiogenesis promotion and to prevent any negative effects of Co^{2+} within the body, 5 wt% substitution is usually advised [153]. Furthermore, bone regeneration therapy has made use of lithium (Li^{+}). Lithium is frequently used in the treatment of mental illnesses like mania and depression. As a side effect of lithium maintenance medication, Zamani et al. [154] found an increase in bone mass of patients suffering from these illnesses.

Fluorides are recognized to trigger the proliferation of osteoblast cells, but concentration should be less than 500 ng mol^{-1}, otherwise it will be cytotoxic for osteoblast cell activity [79]. Cytotoxicity facilitates oxidative cell damage as determined by cytotoxicity indicator lactate dehydrogenase (LDH) and accrues extracellular malonyl-di-aldehyde (lipoperoxidation index) when studied in osteoblast cells (MG 63) [79].

In the literature, TiO_2 has been introduced (P_2O_5–CaO–Na_2O–TiO_2) to slow down the solubility of BG to enhance cellular activity [155]. The reported concentration of TiO_2 is 5 mol.% which is responsible for relatively lower dissolution in SBF and water [110]. In BC matrices, Mn^{2+} is commonly employed as a trace metal transition element. The osteoinductivity of biomaterials is linked to the release of active ions during breakdown. Therefore, by enhancing osteoblast responsiveness, cell adhesion, and the production of osteogenic chemicals, the addition of Mn^{2+} ions to BGs, HAp, and β-TCP [153, 156, 157] matrices is meant to promote bone regeneration.

The literature also suggests that early vascular regeneration at the implantation site is essential for the success of BTE-based therapies. New bone development and vascularization are frequently limited to the area around the implant. The development of new BCs with other ions as dopants e.g., nickel (Ni), Ga, lead (Pb), chromium (Cr), bismuth (Bi), yttrium (Yt), Ce, Tb, Er, Eu, La, Dy with higher angiogenic and osteogenic potential is gaining a lot of attention as a way to improve the in vivo performance of synthetic biomaterials [25, 39, 40, 158, 159].

Therefore, the use of metal ions to dope biomaterials as potential therapeutic agents for enhanced bone regeneration has been reported in this chapter. Ion-doped BCs have been studied for a range of therapeutic uses, including medication administration, bone regeneration, and hyperthermia, because of the combinatory properties of these materials. The impact of metal ions on the biological properties of the aforementioned matrices is compiled in Table 1.

Table 1 Different dopants of metal ions in BC matrices and their biological effects

Ions	Type of matrix	Biological effects	Refs.
Mg^{2+}	Biocomposite	• Controlled drug release • Promoted angiogenesis and cell proliferation	[15, 160]
Ag^{+}	HA, CS, BGs	• Improved antibacterial activity • Absorption of blood plasma	[124–126]
Cu^{2+}	BGs, β-TCP, HA	• Enhanced antibacterial properties • Angiogenesis • Enhanced cell differentiation	[36, 127]
Sr^{2+}	HA, BGs	• Adhesion of protein • Improved ALP activity • Enhanced cell viability, angiogenesis, osteogenesis, cell proliferation • Ability to improve formation of new bone	[20, 161]
Li^{+}	CP, HA, BGs	• Enhanced cell proliferation • Helps to improve differentiation of osteoblast cells • Triggers osteoinduction	[154, 162, 163]
Fe^{2+}/Fe^{3+}	CP, HA, BGs	• Cell proliferation • Protein adsorption • Angiogenesis • Antibacterial in nature • Superparamagnetic in nature • Anti-cancerous	[22, 25, 156, 164]
Zn^{2+}	BGs, β-TCP, HA	• Improved cell viability • Antibacterial activity • Enhanced adsorption of protein	[114, 139, 143, 147]
Mn^{2+}	HA, BGs, β-TCP	• Cell proliferation • Protein adsorption • Antibacterial in nature • Enhanced gene expression • Improved osteogenic properties • Bioactivity of substance increases	[170–172]
Co^{2+}	HA, BGs	• Improved osteogenesis • Antibacterial activity	[153, 165, 166]
Ga^{3+}	HA, BGs	• Osteoporosis and cancer treatments • Regarding their toxicity and suitability for the patient's physiological state, a number of issues are being considered	[25, 158, 167]

(continued)

Table 1 (continued)

Ions	Type of matrix	Biological effects	Refs.
K^{1+}	HA, BGs, β-TCP	• Plays key role in bio-mineralization • Control of biochemical process • Contributes to the nucleation of the apatite mineral	[168, 169]
Na^{1+}	HA, BGs, β-TCP	• Improves cell adhesion • Enhanced bone metabolism and resorption process	[170, 171]
Rare earth metals Bi, In, Tb, Er, Eu, La, Dy	BGs, HA	• Great bioactivity • Luminescence properties • photoluminescence properties	[39–41, 158]

4 In Vivo Testing of Ion-Doped BGs

To validate the clinical administration of ion-doped BCs in human body, in vivo study is a pre-requisite to carry out. A lot of studies were conducted on various animal models administered with ion-doped BGs [172–175]. The readers are referred to the cited studies for detailed information on animal studies conducted for ion-doped BGs. One example of such investigation is provided below.

Baheiraei et al. [176] studied the osteogenic potential of Sr-ion doped BG incorporated in gelatin (Gel) matrix in a rabbit calvarial bone defect over a period of 12 weeks. The Gel-BG/Sr scaffold was investigated as compared to the Gel-BG scaffold. Unfilled bone defects served as the control. At the end of the study, it was found that both types of scaffolds promoted bone formation at the bone defect site. However, the new bone area was higher for Gel-BG/Sr scaffold. This sample showed early bone formation in histological results as shown in Fig. 8. Hence, Sr ion is approved to be a promoter of bone formation in vivo.

5 Future Perspectives and Challenges

The scientists have just started to understand the mechanism behind tissues and BCs interaction. Few BG ceramics are already approved and commercialized to treat bone-related diseases. Examples of such products include Bioglass® 45S5, Bioverit®, Biosilicate®, Ceravital®, and A/W Cerabone® [177]. Metallic ion doped BG-ceramics have huge potential to offer in regenerative medicine and targeted drug delivery approaches. However, metal ion doped BG-ceramics have a long way to go until commercialization. One of the main concerns is the small number of in vivo and countable research that has been done for clinical trials. Considering the benefits of ion-doped BG-ceramics, researchers need to speed up their efforts to work towards the availability standardized products in the market.

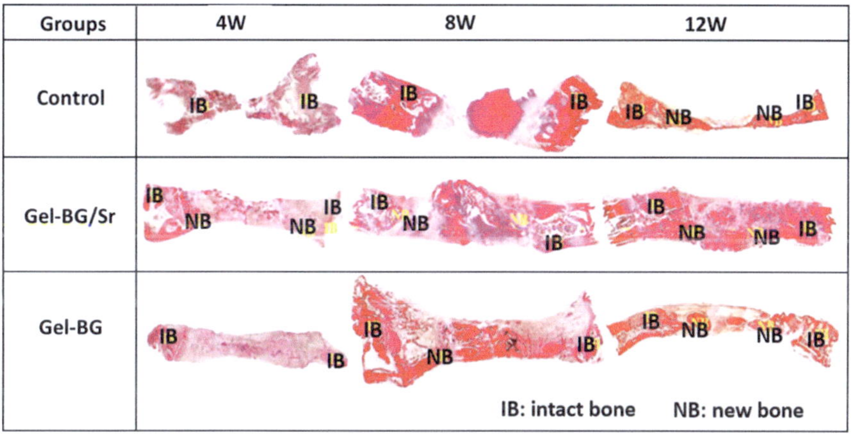

Fig. 8 Histological examination of rabbit bone defects at 4, 8 and 12 weeks. Reproduced from [176], under the terms and conditions of the Creative Commons Attribution 4.0 International License. http://creativecommons.org/licenses/by/4.0/. *Note* The color of the text inside the figure was changed from yellow to black

Fabrication of BG-ceramics into scaffolds or nano-fibers embedded in the polymer matrix is a success, still, there are challenges to be faced while performing surface coating of ion-doped BGs. The difference between the thermal expansion coefficient (TEC) of BGs and the substrates does not allow a well-adherent coating, hence there is a need to find ways to overcome the mismatch of materials for a specific application. Furthermore, there are concerns about handling the degradation of such materials. The type of dopant, glass composition, and quantity that needs a constant pH environment all have a significant impact on BG's biodegradability. While product research and development is one problem, there are still few trustworthy test methods available to forecast long-term mechanical and biological durability. We lack standardized computational models and tools to accelerate the research on multi-component BG-ceramics systems.

6 Conclusions

One crucial aspect of bone tissue engineering (BTE) is the development and fabrication of intelligent scaffolds that can initiate cellular interactions and osteoprogenitor cell differentiation. This can be achieved by injecting therapeutic chemicals or ions into the scaffold, which will enable it to function as a carrier for controlled in-situ drug delivery as well as a bed for the creation of new tissue. Metallic ions as therapeutic agents are one of those additives that have received much attention recently. Some metallic ions, as explained in this chapter, are capable of guiding stem cell development along the desired path, hence there is a lot of potential for using them in regenerative medicine. As cofactors for several enzymes, metal ions play crucial roles in the body's cellular metabolism, signaling pathways, ionic channels, and other biological processes. Metallic ions regulate osteoclast-mediated bone resorption and aid in the differentiation of mesenchymal stem cells into osteoblasts. The hybrid partner of the metal ion is very important. The ion can interact with any molecule and become active, which readily reacts with a molecule of water, or an anion instantly based on the amount and mass of the molecule. Inactive ions, for example, Ni and Cu ions, on the other hand, do not interact with molecules of water or metal ions quickly and can remain in an ionic environment for a longer period. As a result, such ions have a higher likelihood to interact with biomolecules, thus, revealing toxicity. The following factors should be taken into account when evaluating the toxicity of metallic biomaterials: (1) the resistance of the material to corrosion; (2) ions released as a result of corrosion and wear; (3) the behavior of the released metallic ions; and (4) the toxicity of the ions.

Metals and their ions were not integrated with biomolecules throughout the evolution of life. To put it another way, such metals might not exist as ions in a primordial ocean. This demonstrates how difficult it is to combine these ions with biomolecules. Cu, for example, is a detectable component that is lacking in a variety of human diseases. In addition, it participates in several redox enzymatic reactions that are a part of numerous metabolic processes as a catalytic cofactor. When Cu consumption reaches a tolerable level, toxic effects occur, resulting in cell death.

Chemical species in the body normally accelerate metallic ions emitted from metallic materials implanted into the living system. The surface oxide acts as a passive film that prevents metal ion release from metallic implants in aqueous solutions by preventing partial dissolution and re-precipitation. Metallic ions are released when the equilibrium of the film is disrupted. Metal ions that are released do not always affect the human body. To address the protection of metallic biomaterials, partners for combining with released ions, along with their volume as well as toxicity, must be considered.

References

1. V.S. Yadav, S.C. Narula, R.K. Sharma, S. Tewari, R. Yadav, J. Oral Sci. **53**(4), 481 (2011)
2. H.J. Mankin, F.J. Hornicek, K.A. Raskin, Clin. Orthop. Relat. Res. **432**, 210 (2005)
3. V.T. Athanasiou, D.J. Papachristou, A. Panagopoulos, A. Saridis, C.D. Scopa, P. Megas, Med. Sci. Monit. Int. Med. J. Exp. Clin. Res. **16**(1), BR24 (2010)
4. Q.Z. Chen, I.D. Thompson, A.R. Boccaccini, Biomaterials **27**(11), 2414 (2006)
5. C. Wu, J. Chang, Interface Focus. **2**(3), 292 (2012)
6. V. Mouriño, A.R. Boccaccini, J.R. Soc, Interface **7**(43), 209 (2010)
7. M. Karadjian C. Essers, S. Tsitlakidis, B. Reible, A. Moghaddam, A.R. Boccaccini, F. Westhauser, Int. J. Mol. Sci. **20**(2), 305 (2019)
8. E. Fiume, J. Barberi, E. Verné, F. Baino, J. Funct. Biomater. **9**(1), 24 (2018)
9. M.N. Rahaman, D.E. Day, B.S. Bal, Q. Fu, S.B. Jung, L.F. Bonewald, A.P. Tomsia, Acta Biomater. **7**(6), 2355 (2011)
10. V.M. Schatkoski, T.L.d.A. Montanheiro, B.R.C. de Menezes, R.M. Pereira, K.F. Rodrigues, R.G. Ribas, D.M. da Silva, G.P. Thim, Ceram. Int. **47**(3), 2999 (2021)

11. T. Kokubo, H. Takadama, Biomaterials **27**(15), 2907 (2006)
12. S. Gomes, C. Vichery, S. Descamps, H. Martinez, A. Kaur, A. Jacobs, J.-M. Nedelec, G. Renaudin, Acta Biomater. **65**, 462 (2018)
13. A. Tilocca, A.N. Cormack, Proc. R. Soc. A Math. Phys. Eng. Sci. **467**(2131), 2102 (2011)
14. H. Mohammadi, M. Hafezi, N. Nezafati, S. Heasarki, A. Nadernezhad, S.M.H. Ghazanfari, M. Sepantafar, J. Ceram. Sci. Technol. **5**(1), 1 (2014)
15. D. Bellucci, A. Sola, I. Cacciotti, C. Bartoli, M. Gazzarri, A. Bianco, F. Chiellini, V. Cannillo, Mater. Sci. Eng. C. Mater. Biol. Appl. **42**, 312 (2014)
16. A. Moghanian, A. Ghorbanoghli, M. Kazem-Rostami, A. Pazhouheshgar, E. Salari, M.S. Yazdi, T. Alimardani, H. Jahani, F.S. Jazi, M. Tahriri, Int. J. Appl. Glas. Sci. **11**(4), 685 (2020)
17. S. Kargozar, M. Mozafari, S. Ghodrat, E. Fiume, F. Baino, Mater. Sci. Eng. C **121**, 111741 (2021)
18. V. Aina, G. Malavasi, A. Fiorio Pla, L. Munaron, C. Morterra, Acta Biomater. **5**(4), 1211 (2009)
19. M. Furko, Y. Jiang, T. Wilkins, C. Balázsi, Ceram. Int. **42**(4), 4924 (2016)
20. V. Aina, L. Bergandi, G. Lusvardi, G. Malavasi, F.E. Imrie, I.R. Gibson, G. Cerrato, D. Ghigo, Mater. Sci. Eng. C. Mater. Biol. Appl. **33**(3), 1132 (2013)
21. N.D. Ravi, R. Balu, T.S. Sampath Kumar, J. Am. Ceram. Soc. **95**(9), 2700 (2012)
22. M.A. Morsi, A.E.M. Hezma, J. Mater. Res. Technol. **8**(2), 2098 (2019)
23. A. Hoppe, V. Mouriño, A.R. Boccaccini, Biomater. Sci. **1**(3), 254 (2013)
24. Q. Nawaz, M. Atiq Ur Rehman, A. Burkovski, J. Schmidt, A.M. Beltrán, A. Shahid, N.K. Alber, W. Peukert, A.R. Boccaccini, J. Mater. Sci. Mater. Med. **29**(5), 64 (2018)
25. C.R. Chitambar, Pharmacol. Res. **115**, 56 (2017)
26. E. Baye, B. Ukropcova, J. Ukropec, A. Hipkiss, G. Aldini, B. de Courten, Amino Acids **48**(5), 1131 (2016)
27. R. Giampietro, F. Spinelli, M. Contino, N.A. Colabufo, Mol. Pharm. **15**(3), 808 (2018)
28. Y. Ando, H. Miyamoto, I. Noda, N. Sakurai, T. Akiyama, Y. Yonekura, T. Shimazaki, M. Miyazaki, M. Mawatari, T. Hotokebuchi, Mater. Sci. Eng. C **30**(1), 175 (2010)
29. H.R. Bakhsheshi-Rad, E. Hamzah, A.F. Ismail, M. Aziz, M. Daroonparvar, E. Saebnoori, A. Chami, Surf. Coatings Technol. **334**, 450 (2018)
30. I. Cacciotti, J. Mater. Sci. **52**(15), 8812 (2017)
31. T. Uysal, A. Ustdal, M.F. Sonmez, F. Ozturk, Angle Orthod. **79**(5), 984 (2009)
32. S. Wang, R. Li, D. Xia, X. Zhao, Y. Zhu, R. Gu, J. Yoon, Y. Liu, Stem Cell Res. Ther. **12**(1), 123 (2021)
33. R.K. Rude, H.E. Gruber, J. Nutr. Biochem. **15**(12), 710 (2004)
34. S.A.E.-M. Bashandy, E.A.A. Omara, H. Ebaid, M.M. Amin, M.S. Soliman, Asian Pac. J Trop. Biomed. **6**(12), 1056 (2016)
35. H. Zhou, J. Wei, X. Wu, J. Shi, C. Liu, J. Jia, C. Dai, Q. Gan, J. Mater. Sci. Mater. Med. **21**(7), 2175 (2010)
36. E. Tokuda, Y. Furukawa, Int. J. Mol. Sci. **17**(5), 636 (2016)
37. G. Cárdenas-Triviño, C. Elgueta, L. Vergara, J. Ojeda, A. Valenzuela, C. Cruzat, Int. J. Biol. Macromol. **104**(Part A), 498 (2017)
38. R.K. Rude, H.E. Gruber, H.J. Norton, L.Y. Wei, A. Frausto, B.G. Mills, J. Nutr. **134**(1), 79 (2004)
39. T.S.H. Perera, Y. Han, X. Lu, X. Wang, H. Dai, S. Li, J. Nanomater. **2015**, 705390 (2015)
40. I.A. Neacsu, A.E. Stoica, B.S. Vasile, E. Andronescu, Nanomaterials **9**(2), 239 (2019)
41. Q. Fan, X. Cui, H. Guo, Y. Xu, G. Zhang, B. Peng, J. Biomater. Appl. **35**(2), 237 (2020)
42. L.C. Chow, Dent. Mater. J. **28**(1), 1 (2009)
43. K. Ishikawa, P. Ducheyne, S. Radin, J. Mater. Sci. Mater. Med. **4**(2), 165 (1993)
44. K. Prabakaran, S. Rajeswari, Trends Biomater. Artif. Organs **20**(1), 20 (2006)
45. S. Ramesh, C.Y. Tan, M. Hamdi, I. Sopyan, W.D. Teng, *Proceedings of SPIE 6423, International Conference on Smart Materials and Nanotechnology in Engineering*, vol. 64233A (2007, November 01)

46. T. Leventouri, Biomaterials **27**(18), 3339 (2006)
47. L. Cristea, R. Piticescu, B. Popescu, M. Teleanu, Metalurgija **42**(2), 99 (2003)
48. U. Vijayalakshmi, A. Balamurugan, S. Rajeswari, Trends Biomater. Artif. Organs. **18**(2), 101 (2005)
49. M.P. Mahabole, R.C. Aiyer, C.V. Ramakrishna, B. Sreedhar, R.S. Khairnar, Bull. Mater. Sci. **28**(6), 535 (2005)
50. Y. Ahmed, M. Yasir, M. Atiq, U. Rehman, Surfaces **3**, 237 (2020)
51. C. Gao, S. Peng, P. Feng, C. Shuai, Bone Res. **5**, 17059 (2017)
52. A. Hoppe, B. Sarker, R. Detsch, N. Hild, D. Mohn, W.J. Stark, A.R. Boccaccini, J. Non Cryst. Solids **387**, 41 (2014)
53. S. Adzila, M. Murad, I. Sopyan, Recent Patents Mater. Sci. **5**, 18 (2012)
54. X. Qin, R. Cao, J. Zheng, G. Shi, L. Ji, A. Zhu, H. Yao, RSC Adv. **10**(73), 44835 (2020)
55. V. Uskoković, M.A. Iyer, V.M. Wu, J. Mater. Chem. B **5**(7), 1430 (2017)
56. A. Jacobs, G. Renaudin, N. Charbonnel, J.-M. Nedelec, C. Forestier, S. Descamps, Materials **14**(9), 2393 (2021)
57. Y. Li, L. Liu, P. Wan, Z. Zhai, Z. Mao, Z. Ouyang, D. Yu, Q. Sun, L. Tan, L. Ren, Z. Zhu, Y. Hao, X. Qu, K. Yang, K. Dai, Biomaterials **106**, 250 (2016)
58. T. Subramaniam, M.B. Fauzi, Y. Lokanathan, J.X. Law, Int. J. Mol. Sci. **22**(12), 6486 (2021)
59. M.S.N. Shahrbabak, F. Sharifianjazi, D. Rahban, A. Salimi, Silicon **11**(6), 2741 (2019)
60. P. De Aza, F. Guitián, S. Aza, Scr. Metall. Mater. **31**, 1001 (1994)
61. S. Ni, J. Chang, L. Chou, J. Biomed. Mater. Res. A **76**(1), 196 (2006)
62. G. Wang, Z. Lu, D. Dwarte, H. Zreiqat, Mater. Sci. Eng. C **32**(7), 1818 (2012)
63. K. Lin, W. Zhai, S. Ni, J. Chang, Y. Zeng, W. Qian, Ceram. Int. - CERAM INT. **31**, 323 (2005)
64. Y. Iimori, Y. Kameshima, K. Okada, S. Hayashi, J. Mater. Sci. Mater. Med. **16**(1), 73 (2005)
65. H. Zreiqat, Y. Ramaswamy, C. Wu, A. Paschalidis, Z. Lu, B. James, O. Birke, M. McDonald, D. Little, C.R. Dunstan, Biomaterials **31**(12), 3175 (2010)
66. S. Ni, J. Chang, J. Biomater. Appl. **24**(2), 139 (2009)
67. K. Lin, L. Xia, H. Li, X. Jiang, H. Pan, Y. Xu, W.W. Lu, Z. Zhang, J. Chang, Biomaterials **34**(38), 10028 (2013)
68. T.C. Schumacher, A. Aminian, E. Volkmann, H. Lührs, D. Zimnik, D. Pede, W. Wosniok, L. Treccani, K. Rezwan, Biomed. Mater. **10**(5), 55013 (2015)
69. N. Kong, K. Lin, H. Li, J. Chang, J. Mater. Chem. B **2**(8), 1100 (2014)
70. Y. Ramaswamy, C. Wu, A. Van Hummel, V. Combes, G. Grau, H. Zreiqat, Biomaterials **29**(33), 4392 (2008)
71. S.-I. Roohani-Esfahani, Y.J. No, Z. Lu, P.Y. Ng, Y. Chen, J. Shi, N.J. Pavlos, H. Zreiqat, Biomed. Mater. **11**(3), 35018 (2016)
72. T. Tian, Y. Han, B. Ma, C. Wu, J. Chang, J. Mater. Chem. B **3**(33), 6773 (2015)
73. Y.J. No, J.J. Li, H. Zreiqat, Materials **10**(2), 153 (2017)
74. E. Boros, P.J. Dyson, G. Gasser, Chem. **6**(1), 41 (2020)
75. R. Li, A.E. Clark, L.L. Hench, J. Appl. Biomater. **2**(4), 231 (1991)
76. M. Lombardi, L. Gremillard, J. Chevalier, L. Lefebvre, I. Cacciotti, A. Bianco, L. Montanaro, Key Eng. Mater. **541**, 15 (2013)
77. K. Zheng, B. Sui, K. Ilyas, A.R. Boccaccini, Mater. Horizons **8**(2), 300 (2021)
78. Z. Tabia, K. El Mabrouk, M. Bricha, K. Nouneh, RSC Adv. **9**(22), 12232 (2019)
79. L. Bergandi, V. Aina, S. Garetto, G. Malavasi, E. Aldieri, E. Laurenti, L. Matera, C. Morterra, D. Ghigo, Chem. Biol. Interact. **183**(3), 405 (2010)
80. X. Chen, X. Liao, Z. Huang, P. You, C. Chen, Y. Kang, G.-F. Yin, J. Biomed. Mater. Res. Part B Appl. Biomater. Part B **93**(1), 194 (2010)
81. C.-H. Fang, Y.-W. Lin, J.-S. Sun, F.-H. Lin, J. Orthop. Surg. Res. **14**(1), 162 (2019)
82. M. Atiq, U. Rehman, Prosthesis **2**(3), 211 (2020)
83. N. Zhu, X. Chatzistavrou, L. Ge, M. Qin, P. Papagerakis, Y. Wang, J. Dent. **83**, 18 (2019)
84. I. Izquierdo-Barba, A.J. Salinas, M. Vallet-RegíInt, J. Appl. Glas. Sci. **4**(2), 149 (2013)
85. M. Erol-Taygun, K. Zheng, A.R. Boccaccini, Int. J. Appl. Glas. Sci. **4**(2), 136 (2013)
86. J.H. Beattie, A. Avenell, Nutr. Res. Rev. **5**(1), 167 (1992)

87. N. Kanasan, S. Adzila, C.T. Koh, H.A. Rahman, G. Panerselvan, Adv. Appl. Ceram. **118**(7), 381 (2019)
88. G.S. Theodorou, E. Kontonasaki, A. Theocharidou, A. Bakopoulou, M. Bousnaki, C. Hadjichristou, E. Papachristou, L. Papadopoulou, N.A. Kantiranis, K. Chrissafis, K.M. Paraskevopoulos, P.T. Koidis, Int. J. Biomater. **2016**, 3858301 (2016)
89. W. Jahnen-Dechent, M. Ketteler, Clin. Kidney J. (Suppl. 1), i3 (2012)
90. M. Shoaib, A. Bahadur, S. Iqbal, M.M. AL-Anazy, A. Laref, M.A. Tahir, P.A. Channar, S. Noreen, M. Yasir, A. Iqbal, K.W. Ali, J. Alloys Compd. **866**, 159013 (2021)
91. S.K. Nandi, A. Mahato, B. Kundu, P. Mukherjee, *Advanced Techniques in Bone Regeneration*, ed. by A.R. Zorzi, J.B. de Miranda (IntechOpen, 2016). https://doi.org/10.5772/63266
92. M.T. Souza, M.C. Crovace, C. Schröder, H. Eckert, O. Peitl, E.D. Zanotto, J. Non Cryst. Solids **382**, 57 (2013)
93. S. Dhivya, J. Ajita, N. Selvamurugan, J. Biomed. Nanotechnol. **11**(10), 1675 (2015)
94. M. Gallo, B.L.G. Santoni, T. Douillard, F. Zhang, L. Gremillard, S. Dolder, W. Hofstetter, S. Meille, M. Bohner, J. Chevalier, S. Tadier, Acta Biomater. **89**, 391 (2019)
95. R.G. Ribas, V.M. Schatkoski, T.L.d.A. Montanheiro, B.R.C. de Menezes, C. Stegemann, D.M. Leite, G.P. Thim, Ceram. Int. **45**(17(Part A)), 21051 (2019)
96. A. Saboori, M. Rabiee, F. Moztarzadeh, M. Sheikhi, M. Tahriri, M. Karimi, Mater. Sci. Eng. C **29**(1), 335 (2009)
97. E. Dietrich, H. Oudadesse, A. Lucas-Girot, M. Mami, J. Biomed. Mater. Res. A **88**(4), 1087 (2009)
98. A. Sgambato, F.I. Wolf, B. Faraglia, A. Cittadini, J. Cell. Physiol. **180**(2), 245 (1999)
99. F. Sekiya, M. Yoshida, T. Yamashita, T. Morita, J. Biol. Chem. **271**(15), 8541 (1996)
100. P.J. Marie, Bone **38**(2(Suppl. 1)), S10 (2006)
101. V. Mouriño, J.P. Cattalini, A.R. Boccaccini, J.R. Soc, Interface **9**(68), 401 (2012)
102. J. Lao, E. Jallot, J.-M. Nedelec, Chem. Mater. **20**(15), 4969 (2008)
103. H. Yin, C. Yang, Y. Gao, C. Wang, M. Li, H. Guo, Q. Tong, J. Alloys Compd. **743**, 564 (2018)
104. H. Lan, M. Zhang, X. Chen, Z. Huang, G. Yin, Smart Mater. Med. **3**, 94 (2022)
105. E. Canalis, M. Hott, P. Deloffre, Y. Tsouderos, P.J. Marie, Bone **18**(6), 517 (1996)
106. A. Balamurugan, G. Balossier, D. Laurent-Maquin, S. Pina, A.H.S. Rebelo, J. Faure, J.M.F. Ferreira, Dent. Mater. **24**(10), 1343 (2008)
107. L. Montanaro, P. Speziale, D. Campoccia, S. Ravaioli, I. Cangini, G. Pietrocola, S. Giannini, C.R. Arciola, Future Microbiol. **6**(11), 1329 (2011)
108. D. Pavithra, M. Doble, Biomed. Mater. **3**(3), 34003 (2008)
109. C.R. Arciola, D. Campoccia, P. Speziale, L. Montanaro, J.W. Costerton, Biomaterials **33**(26), 5967 (2012)
110. A. Hoppe, N.S. Güldal, A.R. Boccaccini, Biomaterials **32**(11), 2757 (2011)
111. A. Jacobs, M. Gaulier, A. Duval, G. Renaudin, Crystals **9**(7), 326 (2019)
112. A.C. Vale, P.R. Pereira, A.M. Barbosa, E. Torrado, N.M. Alves, Mater. Sci. Eng. C **94**, 161 (2019)
113. S.F.S. Shirazi, S. Gharehkhani, H.S.C. Metselaar, B. Nasiri-Tabrizi, H. Yarmand, M. Ahmadi, N.A. Abu Osman, RSC Adv. **6**(1), 190 (2016)
114. M. Li, X. Xiao, R. Liu, C. Chen, L. Huang, J. Mater. Sci. Mater. Med. **19**(2), 797 (2008)
115. T.-G. Kim, B. Park, Inorg. Chem. **44**(26), 9895 (2005)
116. E. Landi, S. Martorana, A. Tampieri, S. Guicciardi, C. Melandri, Key Eng. Mater. **361–363**, 547 (2008)
117. H. Yun, S. Kim, S. Lee, I. Song, Mater. Lett. **64**(16), 1850 (2010)
118. A. Grandjean-Laquerriere, P. Laquerriere, E. Jallot, J.-M. Nedelec, M. Guenounou, D. Laurent-Maquin, T.M. Phillips, Biomaterials **27**(17), 3195 (2006)
119. G. Bezzi, G. Celotti, E. Landi, T.M.G. La Torretta, I. Sopyan, A. Tampieri, Mater. Chem. Phys. **78**(3), 816 (2003)
120. T.E. Paterson, R. Shi, J. Tian, C.J. Harrison, M. De Sousa Mendes, P.V. Hatton, Z. Li, I. Ortega, J. Funct. Biomater. **11**(3), 58 (2020)

121. N. Pajares-Chamorro, Y. Wagley, C.V. Maduka, D.W. Youngstrom, A. Yeger, S.F. Badylak, N.D. Hammer, K. Hankenson, X. Chatzistavrou, Mater. Sci. Eng. C. Mater. Biol. Appl. **120**, 111693 (2021)
122. A.D. Politano, K.T. Campbell, L.H. Rosenberger, R.G. Sawyer, Surg. Infect. (Larchmt.) **14**(1), 8 (2013)
123. F. Martinez-Gutierrez, P.L. Olive, A. Banuelos, E. Orrantia, N. Nino, E.M. Sanchez, F. Ruiz, H. Bach, Y. Av-Gay, Nanomedicine Nanotechnol. Biol. Med. **6**(5), 681 (2010)
124. J. Singh, G. Kaur, P. Kaur, R. Bajaj, M. Rawat, World J. Pharm. Pharm. Sci. **6**, 730 (2016)
125. C. Greulich, D. Braun, A. Peetsch, J. Diendorf, B. Siebers, M. Epple, M. Köller, RSC Adv. **2**(17), 6981 (2012)
126. S. Rahman, L. Rahman, A.T. Khalil, N. Ali, D. Zia, M. Ali, Z.K. Shinwari, Appl. Microbiol. Biotechnol. **103**(6), 2551 (2019)
127. F. Westhauser, S. Decker, Q. Nawaz, F. Rehder, S. Wilkesmann, A. Moghaddam, E. Kunisch, A.R. Boccaccini, Mater. **14**(8), 1864 (2021)
128. J.O. Eniwumide, H. Yuan, S.H. Cartmell, G.J. Meijer, J.D. de Bruijn, Eur. Cell. Mater. **14**, 30 (2007)
129. K. Zheng, M. Lu, B. Rutkowski, X. Dai, Y. Yang, N. Taccardi, U. Stachewicz, A. Czyrska-Filemonowicz, N. Hüserb, A.R. Boccaccini, J. Mater. Chem. B **4**(48), 7936 (2016)
130. V. Usković, G. Abuna, P. Ferreira, V.M. Wu, L. Gower, F.C.P. Pires-de-Souza, R.M. Murata, M.A.C. Sinhoreti, S. Geraldeli, J. Sol-Gel Sci. Technol. **97**(2), 245 (2021)
131. A. Balamurugan, G. Balossier, S. Kannan, J. Michel, A.H.S. Rebelo, J.M.F. Ferreira, Acta Biomater. **3**(2), 255 (2007)
132. S.L. Hall, H.P. Dimai, J.R. Farley, Calcif. Tissue Int. **64**(2), 163 (1999)
133. O.-M. Goudouri, E. Kontonasaki, U. Lohbauer, A.R. Boccaccini, Acta Biomater. **10**(8), 3795 (2014)
134. S. Sánchez-Salcedo, S. Shruti, A.J. Salinas, G. Malavasi, L. Menabue, M. Vallet-Regí, J. Mater. Chem. B **2**(30), 4836 (2014)
135. E. Saino, S. Grandi, E. Quartarone, V. Maliardi, D. Galli, N. Bloise, L. Fassina, M.G.C.D. Angelis, P. Mustarelli, M. Imbriani, L. Visai, Eur. Cell. Mater. **21**, 59 (2011)
136. S. Haimi, G. Gorianc, L. Moimas, B. Lindroos, H. Huhtala, S. Räty, H. Kuokkanen, G.K. Sándor, C. Schmid, S. Miettinen, R. Suuronen, Acta Biomater. **5**(8), 3122 (2009)
137. J.P. O'Connor, D. Kanjilal, M. Teitelbaum, S.S. Lin, J.A. Cottrell, Materials **13**(10), 2211 (2020)
138. D. Predoi, S.L. Iconaru, M.V. Predoi, N. Buton, M. Motelica-Heino, Zinc doped hydroxyapatite thin films prepared by sol-gel spin coating procedure. Coatings **9**(3), 156 (2019)
139. A. Bandyopadhyay, E.A. Withey, J. Moore, S. Bose, Mater. Sci. Eng. C **27**(1), 14 (2007)
140. X. Qu, Z. He, H. Qiao, Z. Zhai, Z. Mao, Z. Yu, K. Dai, J. Orthop. Transl. **14**, 34 (2018)
141. W. Opsahl, H. Zeronian, M. Ellison, D. Lewis, R.B. Rucker, R.S. Riggins, J. Nutr. **112**(4), 708 (1982)
142. A. Bari, N. Bloise, S. Fiorilli, G. Novajra, M. Vallet-Regí, G. Bruni, A. Torres-Pardo, J.M. González-Calbet, L. Visai, C. Vitale-Brovarone, Acta Biomater. **55**, 493 (2017)
143. Z. Tang, J. Niu, H. Huang, H. Zhang, J. Pei, J. Ou, G. Yuan, J. Mech. Behav. Biomed. Mater. **72**, 182 (2017)
144. W. Gao, Y. Sun, M. Cai, Y. Zhao, W. Cao, Z. Liu, G. Cui, B. Tang, Nat. Commun. **9**(1), 231 (2018)
145. C. Stähli, M. James-Bhasin, A. Hoppe, A.R. Boccaccini, S.N. Nazhat, Acta Biomater. **19**, 15 (2015)
146. S. Zhao, L. Li, H. Wang, Y. Zhang, X. Cheng, N. Zhou, M.N. Rahaman, Z. Liu, W. Huang, C. Zhang, Biomaterials **53**, 379 (2015)
147. X. Wu, G. Meng, S. Wang, F. Wu, W. Huang, Z. Gu, Mater. Sci. Eng. C Mater. Biol. Appl. **52**, 242 (2015)
148. K. Zheng, X. Dai, M. Lu, N. Hüser, N. Taccardi, A.R. Boccaccini, Colloids Surf. B Biointerfaces **150**, 159 (2017)

149. V. Stanić, S. Dimitrijević, J. Antić-Stanković, M. Mitrić, B. Jokić, I.B. Plećaš, S. Raičević, Appl. Surf. Sci. **256**(20), 6083 (2010)
150. I.-S. Lee, C.-N. Whang, K.-S. Oh, J.-C. Park, K.-Y. Lee, G.-H. Lee, S.-M. Chung, X.-D. Sun, Nucl. Instrum. Methods Phys. Res. Sect. B, Beam Interact. Mater. Atoms **242**(1–2), 45 (2006)
151. H. Shi, S. Yang, S. Zeng, X. Liu, J. Zhang, J. Zhang, T. Wu, X. Ye, T. Yu, C. Zhou, J. Ye, Appl. Mater. Today **15**, 100 (2019)
152. H. Fu, Q. Fu, N. Zhou, W. Huang, M.N. Rahaman, D. Wang, X. Liu, Mater. Sci. Eng. C **29**(7), 2275 (2009)
153. P. Dorkov, I.N. Pantcheva, W.S. Sheldrick, H. Mayer-Figge, R. Petrova, M. Mitewa, J. Inorg. Biochem. **102**(1), 26 (2008)
154. A. Zamani, G.R. Omrani, M.M. Nasab, Bone **44**(2), 331 (2009)
155. A. Baranowska, M. Kochanowicz, J. Żmojda, P. Miluski, A. Wajda, M. Leśniak, D. Dorosz, *Proceedings Volume 11456, Optical Fibers and Their Applications 2020*, ed. by R.S. Romaniuk, J. Dorosz (Bialowieza, Poland, 2020, January 27–31)
156. A. Nawaz, S. Bano, M. Yasir, A. Wadood, M. Atiq Ur Rehman, Mater. Adv. **1**(5), 1273 (2020)
157. S. Cañaveral, D. Morales, A.F. Vargas, Mater. Lett. **255**, 126575 (2019)
158. P. Melnikov, A.R. Teixeira, A. Malzac, M.d.B. Coelho, Mater. Chem. Phys. **117**(1), 86 (2009)
159. Z. Du, H. Leng, L. Guo, Y. Huang, T. Zheng, Z. Zhao, X. Liu, X. Zhang, Q. Cai, X. Yang, Compos. Part B Eng. **190**, 107937 (2020)
160. K. Kaur, K.J. Singh, V. Anand, G. Bhatia, S. Singh, H. Kaur, D.S. Arora, Ceram. Int. **42**(11), 12651 (2016)
161. F. Erdem Bastan, M.A. Ur Rehman, F. Ustel, Mater. Chem. Phys. **258**, 123910 (2021)
162. L. Li, R. Wang, B. Li, W. Liang, H. Pan, X. Cui, J. Tang, B. Li, J. Biomed. Mater. Res. Part B Appl. Biomater. **105**(5), 944 (2017)
163. Y. Luo, D. Li, J. Zhao, Z. Yang, P. Kang, Biomed. Mater. Eng. **29**, 699 (2018)
164. J.L. Heath, J.M. Weiss, C.P. Lavau, D.S. Wechsler, Nutrients **5**(8), 2836 (2013)
165. Z. Deng, B. Lin, Z. Jiang, W. Huang, J. Li, X. Zeng, H. Wang, D. Wang, Y. Zhang, Int. J. Biol. Sci. **15**(6), 1113 (2019)
166. K. Czarnek, S. Terpiłowska, A.K. Siwicki, Cent. J. Immunol. **40**(2), 236 (2015)
167. F. Kurtuldu, N. Mutlu, M. Michálek, K. Zheng, M. Masar, L. Liverani, S. Chen, D. Galusek, A.R. Boccaccini, Mater. Sci. Eng. C **124**, 112050 (2021)
168. S. Kannan, J.M.G. Ventura, J.M.F. Ferreira, Ceram. Int. **33**(8), 1489 (2007)
169. S.M. Rabiee, N. Nazparvar, M. Azizian, D. Vashaee, L. Tayebi, Ceram. Int. **41**(6), 7241 (2015)
170. R.-Y. Wang, S.-H. Yang, W.-H. Xu, Chin. Med. J. (Engl.) **129**(5), 594 (2016)
171. H. El Feki, J.M. Savariault, A. Ben Salah, J. Alloys Compd. **287**(1), 114 (1999)
172. E.J. Ryan, A.J. Ryan, A. González-Vázquez, A. Philippart, F.E. Ciraldo, C. Hobbs, V. Nicolosi, A.R. Boccaccini, C.J. Kearney, F.J. O'Brien, Biomaterials **197**, 405 (2019)
173. L.P.L. de Souza, J.H. Lopes, F.V. Ferreira, R.A. Martin, C.A. Bertran, J.A. Camilli, J. Biomed. Mater. Res. Part A **108**(3), 446 (2020)
174. A.M. Deliormanlı, H. Seda Vatansever, H. Yesil, F. Özdal-Kurt, Ceram. Int. **42**(10), 11574 (2016)
175. Y. Lin, W. Xiao, B.S. Bal, M.N. Rahaman, Mater. Sci. Eng. C **67**, 440 (2016)
176. N. Baheiraei, H. Eyni, B. Bakhshi, R. Najafloo, N. Rabiee, Sci. Rep. **11**(1), 8745 (2021)
177. M.C. Crovace, M.T. Souza, C.R. Chinaglia, O. Peitl, E.D. Zanotto, J. Non Cryst. Solids **432**, 90 (2016)

Advancements and Applications of Biomaterials in Medicine, Artificial Organs, and Biology

Wardha Zafar, Sajjad Hussain Sumrra, Muhammad Zubair, Muhammad Faizan Nazar, and Zahid Hussain Chohan

Abstract The human body experiences many wears and tears with aging, and sometimes human organs and tissues fail to function properly because of certain genetic defects, diseases, trauma, injuries, disorders, and eventually with age. The day-to-day medication by drugs can handle some of these conditions, but sometimes the administration of medicines is not sufficient for complete repair/recovery of these disorders. In that case, biomaterials that have the features of facile fabrication, mechanical stability, biocompatibility, and inertness are used to maintain the regular functioning of the body. Biomaterials are chemical entities, which are designed and developed to interact effectively with biological systems, thus providing a platform for the improvement of bio-artificial implantations. These materials are either synthesized using the nanostructured technology, chemical and physical approaches or engineered with the cells and tissues for their biocompatible applications for various diagnostic and therapeutic purposes. The advancement in biomaterials is a continually rising research field signifying the introduction of definite materials for supplementing or substituting human tissues. Biomaterials are being used throughout the body and have a great impact on the treatment of different types of diseases and wounds. Owing to their eco-friendly nature and biodegradability, they are widely used for biological applications. The chapter surveys and discusses some of the biomaterials that have been used or are under consideration for use in biomedical applications.

Keywords Biomaterials · Biomedical applications · Bioartificial implantations · Cardiovascular devices · Orthopedic implants

W. Zafar · S. H. Sumrra (✉) · M. Zubair
Department of Chemistry, University of Gujrat, Gujrat 50700, Pakistan
e-mail: sajjadchemist@uog.edu.pk

M. F. Nazar
Department of Chemistry, Division of Science and Technology, University of Education, Lahore (Multan Campus) 60700, Pakistan

Z. H. Chohan
Department of Chemistry, University of Southern Punjab, Multan 60000, Pakistan

S. J. Ikhmayies (ed.), *Advances in Biomaterials Research*, Advances in Material Research and Technology, https://doi.org/10.1007/978-3-031-97481-6_11

1 Introduction

With globalization and continuous rise in the world population, a tremendous increase in people's living standards is being noted, which highlights the possibility of lifestyle-related physical disorders such as orthopedic and cardiovascular diseases. Day after day, many human tissue or bone fracture cases, probably caused by car accidents, falls, and sports injuries, are handled by medical professionals around the world [1]. However, not all types of wounds or injured tissues may require surgical treatments due to the regenerative and self-renewing capabilities of the human body to carry out the healing process. For example, minor injury to bones or epithelial skin can be easily healed without any surgery, unlike kidney glomeruli, lung, cardiac muscle, neural-retina, articular-cartilage, and spinal cord which requires reconstructive/reconstructive surgery or transplantation for complete recovery [2].

Moreover, the human body's self-healing ability is lost or reduced over time due to aging, and many other reasons such as inflammation due to injured tissues, weakened tissue repair capacity, and decreasing progenitor population within the body. Consequently, in this regard, medical science in collaboration with engineering and materials science is continuously introducing new and smart materials also known as biomaterials that are biologically compatible with the human body to overcome such challenges, anatomical defects or fracture fixation/arthrodesis [3]. Biomaterials also can interact with the body in different ways such as direct application onto the skin, injection into blood vessels, implantation within the body, or extracting components of the body for further analysis [4].

The following is a summary of the development and evolution of biomaterials developed and used since the 1940s; the first generation of biomaterials focused primarily on bio-inertness for near-field applications. Subsequently, the second generation was aimed specifically at achieving bioactivity leading to further advances in renewable biomaterials, controlled reactions with nearby application areas, along with bonding and sustained drug delivery. Then the third generation of biomaterials mainly developed in 1980 where the primary goal of these biomaterials was functional tissue regeneration including proliferation, integration, functionality, healing process, or growth differentiation [5].

With the rapid development of biomedical research and practice, medical devices and biomaterials are widely investigated and utilized, which significantly improved the quality of human life. With fast-tracked global aging, the need for medical devices and biomaterials has increased dramatically. The biomaterial market is growing rapidly worldwide with the introduction of novel biomaterials including implanted detection sensors, artificial limbs, joint replacement materials, drug-carrying stents for regenerative medicine, and is predicted to expand at a compound annual growth rate of 13.69% per year, reaching a net worth of $540.5 billion by the end of 2032 [6].

There has been a range of biomaterials that are being utilized for the applications of imaging/diagnostics, tissue engineering, drug delivery, and device coatings. Over the past few years, the highest growth area has been for polymeric materials although the major fraction of commercialized biomaterials has been for metal-based materials followed by ceramic materials [7]. Conventionally, metals have long been used as biomaterials owing to their relative inertness and excellent mechanical strength [8]. Ceramic biomaterials are interesting because dissolution products can stimulate regeneration and bone growth and are also widely used in the dental field. Polymeric biomaterials are promising soft materials, which can effectively mimic natural tissue and have a resemblance with typical extracellular matrix components such as polysaccharides [9].

Over the past decades, many artificial implants have been designed and developed as substitutes for injured tissues and various implantable biosensors have been practiced for monitoring the human body including but not limited to tissue cultures, dental applications, skull reconstruction, total artificial joint implants, pins, wires, plates and screws for bone treatments, cardiac pacemakers, artificial hearts, synthetic blood vessels, hybrid organs, drug delivery systems, and synthetic skin [10]. Malfunctioned and damaged tissues can be substituted by artificial implants for maintaining basic physiology such as vitreous joints and bodies, incessant blood glucose detectors for monitoring blood glucose level of diabetic patients in real time [11], nerve probes for monitoring the electroencephalogram and treatment of brain-related diseases [12], stents to address arterial stenosis [13], pacemakers to regulate abnormal heart rhythms [14], and orthopedic implants for promoting osseointegration and fracture healing [15]. Furthermore, implanted cell encapsulations to regulate physiological functions are said to deliver novel therapeutic strategies shortly, and relevant research developments have also been continuously reported [16].

Overall, biomaterial-based medical implants play an important role in contemporary biomedical practice and human health. Their distinctive features and greater application than other medical instruments, artificial organs, and implants have increased their demands. The quality of life and prolonged existence in its nature are the two aspects that have increased their use to save a human being in terms of curing the fractures, re-settlement of the bones, skin repairing, and wound healing. These diagnostic methods are helpful in temporarily maintaining physical conditions, mimicking the mechanism of normal tissue regeneration, and repairing various parts of the human body. In this manner, we are being introduced to the considerations that set biomaterials apart from most materials explored in material science. In this chapter, we provided an overview of the field of biomaterials and described many of the biomaterials that interface with biological systems.

This chapter highlights different types of biomaterials that have been utilized for several medicinal applications including surgical devices in numerous therapeutic areas, development of scaffolds for tissue engineering, controlled drug delivery systems, replacement and augmentation of body tissues, and providing key strategies for better interventions in a wide-ranging medical condition. Apart from this, advances in these biomaterials are also described so that the readers can gain background and get encouraged to discover their more applicable uses. This chapter also includes the host interactions following the implantation of biomaterials. As shown in Fig. 1, many different kinds of biomaterials are widely used for the reconstruction or repair of various types of organs in the human body. And because

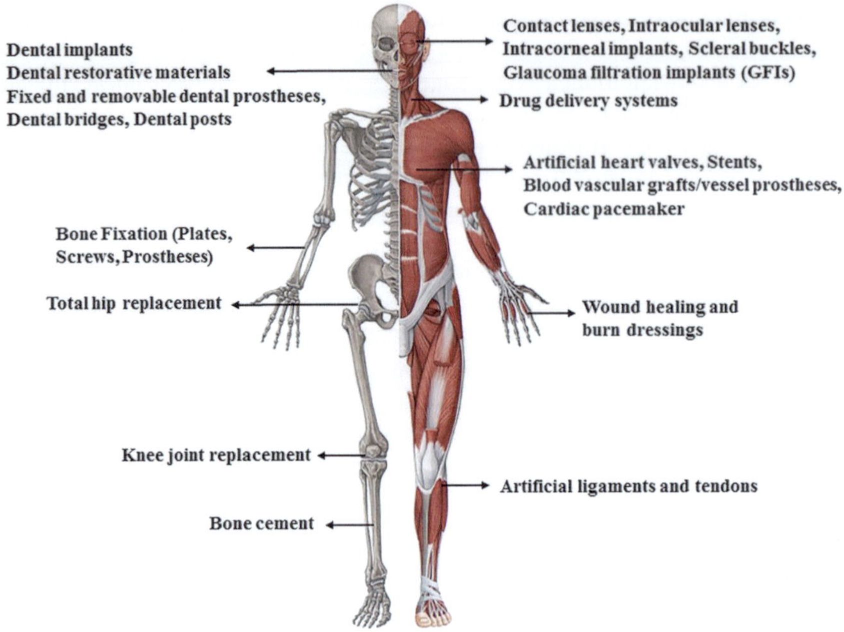

Fig. 1 Significance of biomaterial-based implants for biomedical applications. Reproduced from Tokpah et al. [17], under the terms and conditions of the Creative Commons Attribution (CC BY) license. http://creativecommons.org/licenses/by/4.0/

different human organs require different materials, biomaterials with specific performance characteristics have been designed; each of which has its disadvantages and advantages. We specifically focus on biomaterials used for implantation because they represent several successful innovations in the field of medicine over the past century. This chapter is devoted to a detailed discussion of biomaterials for orthopedic implants, ophthalmologic implants, cardiovascular devices, dental fixation, drug delivery systems, wound healing, and burn dressings. The specific categories discussed were preferred because they are widely used in humans, with particularly good success. In the final part of this chapter, potential future developments for biomaterials are presented.

2 Chemistry of Biomaterials

The development of desirable biomaterials with the capability to regenerate, repair, and improve human tissue is ongoing, and the chemistry of biomaterials significantly contributes to their effective implementation in the clinical setting. Chemistry plays a pivotal role for the design and development of final products of biocomposite-, polymer-, bioceramic- and metal-based structures, which are recently used in clinical practice [18]. Microemulsions are considered definite systems that provide an interesting approach in biophysical research to mimic bio-membranes at the molecular level [19]. Furthermore, there are various commercially available metal-based implants specifically designed and manufactured to replace damaged hard tissues, particularly for load-bearing applications. Their durability, fracture toughness, elongation, and high strength are some of the foremost reasons for employing them in orthopedics [20].

In recent times, polymeric chemical scaffolds are also used to replace many injured soft and hard tissues. But for hard tissue engineering, the use of polymeric substitutes is limited due to poor or insufficient mechanical properties [21]. The chemical composition of bioceramics correlates their performance with the in vivo environment (biological tissues, cells, and fluids) and the consequent outcomes. Coatings, porous scaffolds, bone fillers, and joint prostheses bearings are a few examples of biomedical applications of bioceramics [22]. Titania (TiO_2), zirconia (ZrO_2), and alumina (Al_2O_3) are the most commonly used bioinert ceramics in biomedicine [23].

Composites play a central role in wide-ranging biomedical applications because of their host tissue compatible properties including graded physical properties, strength, and compliance [24]. They are usually prepared by mixing polymeric matrices with ceramics or metals. As with other materials, such as polymers, ceramics, and metals, there are several chemical methods for producing composites, including green synthesis and wet chemical synthesis [25]. Since most human tissues are real composites, synthetic composites are widely used for reconstructive methods such as silk fibroin/hydroxyapatite (Hap) composites being used for bone tissue engineering. Composites made of polymers and metals, known as metallopolymers or organometallic polymers, have been applied for several biomedical applications in interaction with biologically relevant substrates. There are multiple advantages associated with metallopolymer composites, which can be defined as combining the processing and mechanical properties of polymers simultaneously with the radioactive, magnetic, catalytic, and electronic properties of metal building blocks [26].

In reality, the natural bone itself is a composite consisting of water, an organic component (noncollagenous proteins, lipids, type I collagen), and the main inorganic component (hydroxyapatite) [27]. Scaffolds for bone regeneration are being synthesized by adopting a variety of approaches, using a wider range of biomaterials, both in injectable forms and on solid (porous) mediums. Because these are not permanent implants, they play a key role in facilitating the formation of an extracellular matrix that will replace the scaffold over time. A composite material consisting of dispersed hydroxyapatite powder in natural polymers increases the biomimetic actions of the system regarding structural similarity and composition to the extracellular matrix of bone [28].

It was indicated from the reported studies that when bovine bone ash is incorporated into hydroxyapatite containing composite scaffold, an even higher degree of biocompatibility is expected to occur owing to the presence of tricalcium phosphate (TCP) [29]. Unlike artificially synthesized hydroxyapatite, one developed from natural sources is non-stoichiometric, as it may contain other elements such as sodium, zinc, magnesium, potassium, silicon, barium, and fluorine in addition to only calcium and phosphorus. The presence of such elements in the composition of hydroxyapatite makes it very similar to the one which is found in natural human bone [30]. Starch-mediated hydrogels from natural sources are also considered proficient smart materials for pharmaceutical and biomedical applications, providing stimulating approaches in biophysical research at the molecular level [31].

3 Material–Host Interactions

The development of physicochemical interactions between biomaterials and living tissues after implantation is insightful. The timing of specific interactions and the species length scales are of critical importance. The biological nature of a biomaterial defines its positive interaction or effect on the living tissue [32]. The biocompatibility or clinical accomplishment of a biomaterial depends directly on the disturbance induced by the foreign material of the host tissue [33]. Biocompatibility is highly dependent on the size and function of the implant, the duration and site of implantation, with a critical issue being the time scale required to establish material–host tissue interactions [34].

The primary response of any introduced biomaterial surface in a biological environment is the influx of water molecules to the surface and the formation of a water shell around the material on a time scale in the order of nanoseconds. This step can be omitted if the material has been cultured for use in a suitable medium before implantation. The surface properties of the material decide the specific pattern and extent of interaction of water molecules with the surface. This property also regulates which molecules and proteins will act upon formation of the hydration shell. In the second phase, from seconds to a few hours after implantation, the material is mainly covered by an adsorbed layer of proteins found in the extracellular matrix.

In the third step, cells finally get to the “surface” by interacting through the protein coat, consequently cell–surface interactions can be defined as interactions between proteins bound to the surface and cells. This step takes minutes to days after the implant is placed. As the material implantation time increases from a few hours to a few days, cell adhesion, migration, and differentiation occur. This third phase is influenced by its micro-structure (porosity), produced soluble products from the material, the developing physicochemical characteristics of the materials at the surface (chemistry, micro- and nano-topographies), biophysical environment, biological molecules (proteins of cytoskeleton, cell membrane, and extracellular matrix) [35].

The lifespan of implants can range from a few days (short-term biodegradable materials such as sutures) to decades (long-term durable implants such as total hip replacements). The initial implant steps are responsible for setting the phase for their continuous development. Processes that are induced by mineralization or material degradation may occur such as device failure and adverse reactions (formation of fibrous capsules and clots). Future developments of new and improved biomaterials include controlled implant integration healing, thus promoting rapid healing and reducing such effects. The persistent success of any implanted biomaterial in the body is based on the nano- and micron-scale physicochemical properties of the material and controlled macro-functional properties such as mechanical as well as tissue match at the implantation site. The interaction of implanted biomaterials with cellular functions and its influence on living tissue is usually investigated by in vitro experiments designed to mimic in vivo experiments but in more easily controllable simplified style [36].

4 Orthopedic Implants

Orthopedics is the subdivision of clinical medicine that is specified for the diagnosis as well as treatment of musculoskeletal diseases and trauma in the spine and extremities. The development of biomaterial science is attributed to its current state of modern care beyond any other medical specialty. Biomaterials can be used in orthopedics to restore or enhance the physiological function of damaged or diseased tissues through tissue regeneration or transplantation [37]. Orthopedic implants are the medical devices that are used to fix the broken bones. In this technique, implants are used to replace the surfaces of joints in the human body. When other treatments fail to cure the patient or to reduce joint pain and to make the joints move easily, the surgeons choose the procedure of orthopedic implant replacement to strengthen, replace, and fix the fractured joints.

Orthopedic injuries and associated pathologies are recognized worldwide as a serious public health problem that causes the greatest disability burden across the globe. One study estimates that around 1.71 billion people worldwide are affected by muscle conditions [38]. In 2019, the Global Burden of Disease (GBD) study reported a 30.7% rise in Disability-Adjusted Life Years (DALY) rates associated with musculoskeletal disorders as among the ten major causes of increasing burden from teenaged individuals to grownup individuals for the period 1990 to 2019 [39]. These disorders or orthopedic trauma often require surgery and the use of biodegradable/temporary and permanent medical devices containing numerous synthetic as well as natural biomaterials for repairing as well as substituting different tissues (including bone, cartilage, tendons, and ligaments).

4.1 Types of Orthopedic Implants

Orthopedic implants can be classified into two main categories:

(1) Temporary replacement of joints.
(2) Permanent replacement of joints.

Temporary implants are wires, pins, screws, and plates. Whereas permanent orthopedic implants are the implants for wrists, hips, fingers, shoulders, knees, and ankles. Various orthopedic instruments such as crania, cannulated screws, plates, external fixators, wires, and pins are also used to place and insert implants to their accurate points inside the body.

The three main types of orthopedic implants that can be either non-removable or removable are plates, screws, and prostheses.

4.1.1 Plates

There are many ways to get the plates perfect and effective. The width size and thickness of the plate should be appropriate and it must be able to hold both sides of the broken bone. The plate must neutralize and control all forces including compression and bending. These plates are successfully practiced in different surgeries such as pectus bar and excavatum correction. These can also be utilized in femur and tibia.

There are five most important types of plates:

(1) **Compression Plates**: These are metal plates that are practiced for repairing a bone, when forceful pressure is applied between broken parts of bone and to support its healing.
(2) **Buttress Plates**: These are utilized to hold the broken parts together at the end of long bones, especially at fracture sites that experience a variety of deforming forces with high stress such as ankle and knee. These plates are usually designed to move freely with the body, while some might be T- or L-shaped.
(3) **Bridging Plates**: These plates provide stability along with alignment and length to the fractured area. Additionally, they do not interfere with the injured area and at the same time preserve the supply of blood to the crack fragments thus promoting faster and safer bone healing.
(4) **Neutralization Plates**: These plates fall under the category of plates, which function to span the broken area. They provide counterweight so that screws or other devices can stabilize and secure the injured area.
(5) **Tension Plates**: These are generally wires which are utilized to protect an area while it heals.

4.1.2 Screws

A screw is a type of orthopedic implant that is very similar to a common screw available at any hardware store. But an orthopedic screw may have either a crosshead or a flat head. It can be implanted permanently. The most important role of screws is to create compression, which facilitates bone healing at the damaged site. Screws are used to tighten fracture sites such as the torn labrum or the rotator cuff. In addition, they can also be used to restore stability in a weakened area or to repair a broken bone. Because they are used for a variety of purposes, they are available in a variety of sizes and forms, such as the thickness of a reduction screw ranges from 4.5 to 8.5 mm. The screw length is measured from the top to the bottom, whereas the head is normally utilized to insert the screw in the bones.

4.1.3 Prostheses

A prosthesis is an artificial part of the body that is used to replace the lost joints or bones. Prosthetic implants can also be used to support the broken bones. This implant is mostly practiced for hips and knees. It enables the patients to perform their normal activities and regain their physical strength in a relatively short time frame. Sometimes, the doctors may combine healthy bones with the prosthetic materials to either completely replace certain parts of bones or substitute diseased or broken bones [40].

4.2 History of Biomaterials in Orthopedics

Orthopedics originated in the eighteenth century as a set of techniques that work by employing non-surgical approaches to correct the abnormalities in growing children. Surgical treatment of joint and bone diseases was not possible at that time owing to the lack of anesthesia and antiseptic methods which would make the operation manageable and safe. In the nineteenth century, the advancement in antiseptic methods and inhalation anesthesia made general surgery accessible for all the infected patients. But until the early twentieth century, joint and bone surgery was prevented due to the lack of adequate biologically compatible material, possibly the best way to correct a defect or fracture. In the last century, the development of metallurgical engineering has resulted in the establishment of various biologically compatible materials, including alloys such as stainless steel. Nails, screws, and plates have modernized fracture care because of the reason that they can be effectively used to fix the bones. Injuries that used to be treated by wrapping the injury in a heavy cast or by hanging the limb in traction for a month are now treated by just internal fixation. More advanced, biologically compatible metal alloys, such as cobalt-chrome alloys, are now used as the primary material for artificial joints that require permanent implantation [41].

4.3 Characteristics of Biomaterials Used in Orthopedics

Some basic factors define the medical success of the orthopedic intervention such as bioresponse, functional properties of biomaterials, and the design of the medical device. In addition to their chemical, microstructural, and mechanical characteristic features, biomaterials are also tested for their efficiency, safety, biocompatibility, and innocuity for the clinical efficacy [42]. The key properties of orthopedic biomaterials are significant mechanical abilities such as chemical stability, biocompatibility, appropriate microstructural features, and high resistance to wear and corrosion (Fig. 2) [22]. Table 1 depicts characteristic features, advantages and disadvantages of some important metallic, ceramic, and polymeric biomaterials used as orthopedics implants.

The structures of freely moving (synovial) joints including the elbow, ankle, shoulder, knee, and hip are affected by both rheumatoid arthritis and osteoarthritis. Pain in these joints, especially the weight-bearing joints, for example, the knee and hip joints, can be very severe, and can have major effects on the ambulatory functions which can be devastating. With the advent of antibiotics, antiseptics, anesthesia, and biomaterials, it is now satisfactory to substitute damaged or broken joints with prostheses. The restoration of mobility and relief of pain is familiar to hundreds of thousands of patients.

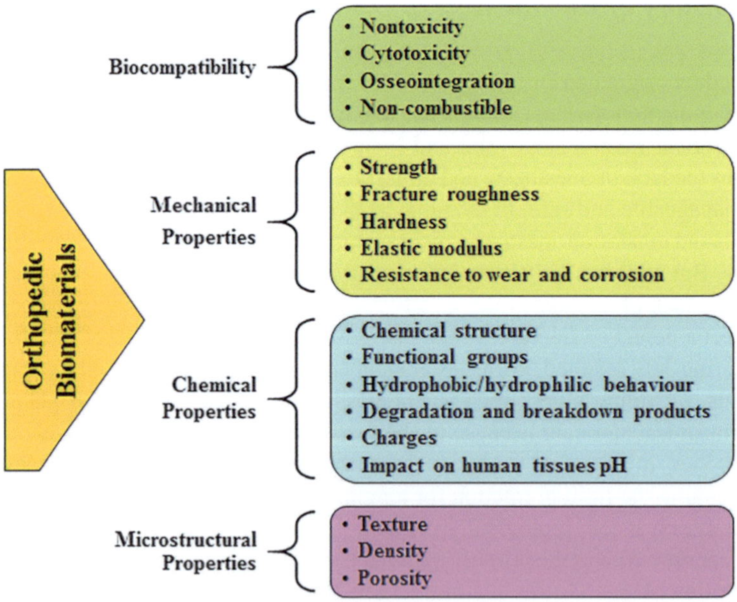

Fig. 2 Essential characteristics of biomaterials in orthopedics

Table 1 Summary of some important metallic, ceramic, and polymeric biomaterials used for orthopedic applications

Biomaterials	Characteristic features	Applications	Disadvantages	References
Zirconia	High biocompatibility, greater strength, and significant fracture toughness	Endosseous implants, bone reconstruction, and hip replacement	–	[43]
Stainless steel	Easy processing, outstanding corrosion resistance, cost-effective and excellent biocompatibility	Fracture plates, cerclage cables, and screws	Allergic reactions (nickel alloys), rigidity, and high modulus of elasticity	[44]
Co–CrMo alloys	High surface smoothness, low ductility, higher fatigue strength, excellent wear, and corrosion resistance	Load bearing components in total joint replacement, prostheses stems	Limited intra-operative manipulation, expensive to manufacture, immunogenic, and reduced biocompatibility	[45]
Titanium and its alloys	Excellent biocompatibility, good corrosion and immunological resistance, osseointegration, excellent mechanochemical properties, moderate elastic modulus	Prostheses stems	Low wear resistance, high hardness, and bacterial adherence	[46]
Alumina	Surface smoothness, low friction/wear coefficients, good corrosion resistance	Femoral head implants	–	[47]
Tricalcium phosphate (ß-TCP)	Significant osteoconductive properties, composition resemblance with bone, biocompatible, bioresorbable	Bone plates and screws, tibial components, hip prostheses and femoral knee	Brittleness, poor mechanical properties	[48]
Hydroxyapatite (HA)	Osseointegration, osteoconduction, high immunotolerance, excellent biocompatibility, and compressive strength	Bone grafts and coating of orthopedic implants	Lesser fatigue and tensile strength	[49]
Polylactic acid (polylactide)	High thermoplastic properties, long-lasting biocompatibility, and biodegradability	Porous scaffolds for growth of neo-tissue, fixation devices and bioabsorbable	Poor ductility, toughness, and thermal stability	[50]

(continued)

Table 1 (continued)

Biomaterials	Characteristic features	Applications	Disadvantages	References
Polyethylene (PE), ultrahigh molecular weight polyethylene (UHMWPE)	Greater availability, less density, and high toughness	Bone joint replacement, liner of acetabular cups in hip arthroplasties	Release of particles, occurrence of osteolysis/inflammatory reactions, and poor wear resistance	[51]
Polymethyl methacrylate (PMMA)	Economical, easy to handle, significant optical and mechanical properties, non-porous structure	Space filler for the proper anchorage of the implant to bone surface, vertebroplasty, and orthopedic surgeries	Low osseointegration, occurrence of systemic reactions (embolism, inflammatory), thermal degradation, and poor tensile resistance	[52]

4.4 Artificial Ligaments and Tendons

Ligaments and tendons are the thin panels of connective tissue which are responsible for providing stability and support to the musculoskeletal system. Injuries to these ligaments and tendons because of any accident in physical activity or sports require urgent implantation with an artificial ligament or tendon. The biomaterial to be used in the musculoskeletal system must have similar functional, biochemical as well as structural properties to native ligament or tendon. The desired characteristic features of any biomaterial to be used as ligament or tendon include sufficient mechanical strength to withstand bodyweight immediately after implantation as well as a degradation rate equivalent to the rate of developing tissue and cellular ingrowth.

The utilization of polymers for ligament or tendon tissue engineering has been highlighted by their degradation properties and their replacement by regenerated ligaments or tendon matrix. Polyhydroxyesters have received greater attention for use in ligament or tendon implants due to their property of hydrolytic degradation. To yield the optimum properties, their degradation rates have been modified [53].

4.5 Bone Cement

Bone cement is a type of biomaterial, which is used to develop an elastic region by filling the vacant space between fractured bone and prosthesis. It comprised of two constituents, a liquid of methyl methacrylate (MMA) monomer, acting as an inhibitor and a amorphous powder/copolymer bead either pre-polymerized polymethyl methacrylate (PMMA) or polymethyl methacrylate (PMMA), acting as initiator. Free radical polymerization begins when the initiator is mixed with the accelerator. Afterward, an evident transformation occurs in the thickness of the cement because it becomes a rigid material. It is important that the gaps must be filled with some suitable bioresorbable material to allow for effective bone growth in the defects. When adequate bioresorbable material is not available, the fibrous tissue ingrowth can obstruct the bone formation surrounding the deformities.

Bioceramics containing calcium phosphate (CaP) have been widely utilized for restoration of the bone because of their remarkable osseointegration, osteoconduction as well as biocompatibility. CaPs are mostly used for hip implants. By coating the surface of stem with CaP, the bonding between osteoconduction of the implant and bones is increased. This practice is called interface bioactive bone cement. Likewise, CaP promotes growth of bones in the osteonecrotic positions in the body. CaP has also been employed to fill the bone defects in reconstruction surgery. Hyaluronic acid is an effective biomaterial, which is recognized for its high mechanical strength and its aptitude to physicochemically bind with the bones [54].

4.6 Joint Replacements

Arthroplasty is a surgery for replacement of joints, where an orthopedic prosthesis replaces a dysfunctional joint. It can be performed for surgery of fingers, ankles, knees, hips, and shoulders. It is used in some cases, such as rheumatoid arthritis and osteoarthritis, when other treatments fail to work. Total Joint Replacement (TJR) is intended to cure arthritis, generally when it interferes with regular actions because of severe pain. Biocompatibility and robustness are the required properties for biomaterials that are going to be used for joint replacement. Implants are well tolerated and biologically compatible if they are not affected by microorganisms and have mechanical stability inside the bone. Composites, ceramics (zirconia and alumina), polymers, metals, and their alloys (titanium alloys, cobalt-chromium alloys, and stainless steel) are the frequently used examples of biomaterials for joint replacement.

4.6.1 Knee Joint Replacement

In replacement surgery of knee joints, also called total knee arthroplasty (TKA), prosthesis made of polymeric materials and metal alloys is used to replace the injured cartilage surfaces of the patella (kneecap), the tibia (shinbone), as well as lower femur (thighbone). While the connecting ligaments and other structures of the knee remain intact. The knee is renowned as a simple hinge joint on account of its aptitude for bending as well as staying straighten similar to a hinged door. But the knee is actually a very complicated joint for the reason that the knee bends while the surfaces glide and roll. During TKA, replacement can be done for up to three levels of bone surfaces: back surface of the kneecap, topmost surface of the shinbone, and the bottom ends (condyles) of thighbone [55].

4.6.2 Total Hip Replacement

Total hip replacement is considered to be one of the major breakthroughs of the twentieth century in orthopedic surgery. Hip arthroplasty usually involves the replacement of the thigh bone (upper femur) along with the replacement or resurfacing of the mating hip (pelvis) bone area. Figure 3 shows a symbolic hip prosthesis consisting of the femoral ball, femoral stem, and a polymeric socket (cup) which is an ultra-high molecular weight polyethylene (UHMWPE) with or without metallic backing. Table 2 lists some combinations for femoral head to socket which are successfully utilized in total hip replacement arthroplasty. For recent metal-on-polymer implants, alloys of cobalt metal are identified as the most commonly used materials [56].

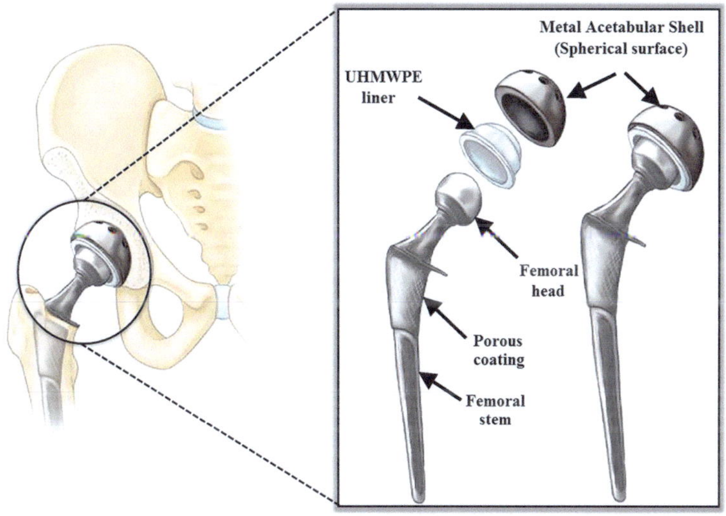

Fig. 3 Typical constituents of implant used in total hip replacement (THR). Reproduced from Devi et al. [57], under the terms and conditions of the Creative Commons Attribution (CC BY) license. http://creativecommons.org/licenses/by/4.0/

Table 2 Materials used as total hip replacement (THR) prostheses

Femoral component	Socket component	Results
Co–Cr–Mo	UHMWPE	Minute wear, commonly used
Co–Cr–Mo	Co–Cr–Mo	Initial limited use and high loosening rate; reduced wear rate of new developments
Alumina	Alumina	Least wear rate, no pain in medical use
Zirconia/alumina	UHMWPE	Greater impact resistance to zirconia, lowest wear rate
Titanium–6-aluminum–4 vanadium (Ti–6Al–4V)	UHMWPE	Reports of greater UHMWPE wear due to failure of titanium surface
Surface-coated Ti–6Al–4V	UHMWPE	Higher wear resistance to abrasion, simply thin treated layer attained

Reprinted from Long and Rack [56], with permission from Elsevier. Copyright (1998)

5 Ophthalmologic Applications

The eye is considered to be the most sensitive and important of all the sensory organs. It is not only a sensitive, but also a complex system whose primary role is in vision and its main responsible parts are retina, lens, and cornea. Any kind of refractory defect can result in blurred vision or even blindness. The transplant of the part is reported, but the supply is less than the requirement. As a result, artificial support for systems is required. Various biomaterials have been used for fabricating the ocular devices to resolve the functional deficits created by ocular trauma, age, and diseases, and many artificial transplants have been employed effectively.

5.1 Considerations for Ophthalmic Materials

The properties that must be considered for the ophthalmic materials are similar to those materials that are used in other parts of the body such as manufacturability, sterilizability, degradation, stability, and biocompatibility. However, the optical properties of the materials to be used in ophthalmic applications should be taken into consideration for effective results. In particular, the optical properties are more critical for the material of intraocular lenses, corneal onlays or inlays, and contact lenses. The ophthalmic materials need to be transparent having a refractive index that is either equal to or greater than that of the tissue being replaced. Sometimes to shield the delicate retina of eye from the exposure to ultraviolet (UV) light, these ophthalmic materials are combined with special UV-blocking agents. Furthermore, any material to be used in the cornea including a corneal inlay or a contact lens must have adequate nutrient and oxygen permeability to maintain the health of cornea owing to the deficiency of vascularization of this tissue. Table 3 lists some biomaterials commonly used in ophthalmic applications.

5.2 Contact Lenses

To correct mild ametropia and obtain satisfactory vision, majority of the people choose contact lenses over spectacles. There are various factors, circumstances, and reasons why people prefer contact lenses instead of using glasses such as:

(1) Irregular astigmatism, in which certain abnormalities in the surface of the cornea damage the vision of the patient, which may be due to corneal dystrophies such as keratoconus and corneal scarring.
(2) Systematic post-operative astigmatism which develops after transplanting of cornea.
(3) Acute anisometropia, when the individual has various refraction powers in each eye, as is the case of patients with one-sided aphakia.
(4) Extreme ametropia, where the use of eyeglasses to compensate a patient's very short or long sight is optically and/or mechanically unbearable.

Contact lenses can be categorized as either soft or hard contact lenses depending on their elastic modulus. For covering the unsightly or damaged eyes, colored contact lenses are often used, while medical contact lenses can serve as a drug delivery vehicle to treat eye syndromes.

Table 3 Types of biomaterials commonly used in ophthalmic implants

Ophthalmic implants	Commonly used biomaterials
Contact lenses	Polymethyl methacrylate (PMMA), polyhydroxyethyl methacrylate (PHEMA), copolymers, polydimethylsiloxane, silicone, trimethylsiloxysilane, hydrogels, *N*-vinylpyrolidone
Inlays or onlays	Collagen, hydrogels, permeable membranes
Intraocular lenses (IOL)	Haptic: Polyvinylidene fluoride (PVDF), polypropylene, PMMA, polyimide Optic: Hydrophobic acrylic, PMMA, PHEMA, silicone, hydrophilic acrylic
Glaucoma shunts	Tubing: Silicone Plates: Polypropylene, silicone (impregnated with barium)
Scleral buckle	Methyl acrylate, hydroxy ethyl acrylate, silicone sponge, soft silicones, poly hydroxyl ethyl acrylate, porcine skin, polyglyceryl methacrylate
Ophthalmic viscosurgical device (OVD)	Collagen, polyacrylamide, hydroxypropyl methylcellulose (HPMS), hyaluronic acid, sodium hyaluronate, chondroitin sulfate
Corneal implants	Polysulfones, PMMA, PHEMA, polyvinyl alcohol, hydroxyapatite, graphite, bacterial cellulose
Vitreous replacements	Gases, silicone oil

Reprinted from Sheardown et al. [58], with permission from Elsevier. Copyright (2020)

In 1508, Leonardo de Vinci first presented the perception of contact lens. But its medical practice was initially detailed in the 1880s [59]. These types of devices were initially effective glass shells made of cadaver or rabbit eye molds that covered the entire front of the eye. Even though these "corneoscleral" lenses succeeded in providing some optical improvement, their drawback was that they were difficult to make and, above all, difficult to wear.

In the 1940s, the discovery of materials containing polymers, mainly polymethylmethacrylate (PMMA), led to the idea of using some other materials to fabricate the ocular devices and give rise to the formation of existing contact lenses. PMMA-based contact lenses were light in weight having exceptional durability, adequate surface wettability, and good optical properties, but their low oxygen permeability (Dk) restricted their long-term use as they were very difficult to wear due to the hard and stiff nature of PMMA [60]. Thus, another alternative material was needed at that time. To solve this issue, in the 1970s rigid gas permeable (RGP) contact lenses were made by copolymerizing methacryloxypropyltris(trimethyl siloxy silane) (TRIS) with methyl methacrylate (MMA), which were licensed to be used for 7 days. The wettability, hardness, elastic modulus, and oxygen permeability of these sorts of materials are controlled by the cross-linking ratio of MMA/TRIS [61].

Then in 1961, the soft contact lenses were made following the synthesis of polyhydroxyl ethyl methacrylate (PHEMA). It has more water content (>38% of water), higher elastic modulus, and provides excellent comfort. While various companies have successfully developed super-thin PHEMA-based contact lenses, people find difficulty in their handling. Some more lenses from hydrogel were also prepared from polymers such as *N*-vinylpyrrolidinone and glyceryl methacrylate (GM). These hydrogel-based lenses were easy to use, but their biggest disadvantage is that these lenses should be permeable to oxygen and very thin, but the reduction in thickness results in a decrease in their strength. Soft contact lenses with high oxygen permeability were also made using silicone elastomer, polydimethylsiloxane (PDMS) [62].

Though this material has high oxygen permeability, excellent tear resistance, and optical properties, the drawback of PDMS is that their reduced surface energy causes poor tear wetting, a tendency of binding with tear lipids and eventually adhesion of contact lens with the cornea. Initially, efforts were made to increase their surface wetting capacity through plasma treatment. However unfortunately, these types of improved surfaces are subjected to instability and do not provide any lasting benefit. The discussion continued over many years whether the manufacturers have to focus on developing soft or rigid stretched wear contact lenses. Research is being done to improve the wettability of contact lenses by embedding them with hydrophilic polymers. To date, PHEMA is the most widely used material for making contact lenses with water content greater than 90% for long-term usage. The structures of some monomer units used for making contact lenses are given in Fig. 4.

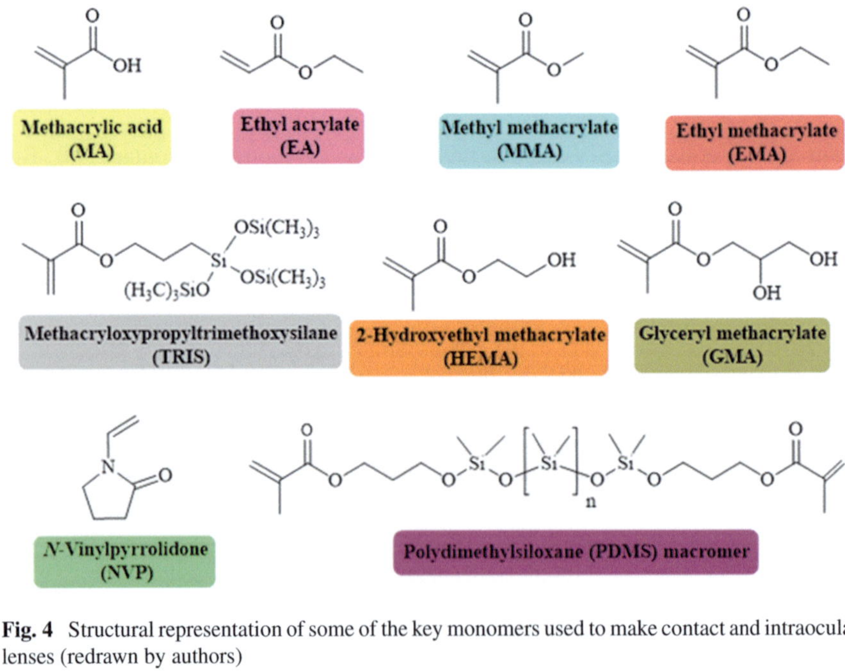

Fig. 4 Structural representation of some of the key monomers used to make contact and intraocular lenses (redrawn by authors)

5.3 Intraocular Lenses

A cataract is a terrible disease that causes blindness by clouding the lens, but, at the same time, it is also categorized as one of the regular treatable types of impaired vision worldwide. This cloudiness can be eliminated by removing the natural lens. Afterward, an intraocular lens (IOL) is placed to re-establish the vision. Cataract extraction and intraocular lens implantation, both are frequently used for ophthalmic surgery.

A typical IOL consists of two parts: haptic and optic. The central optic part is held by haptics and projections evolving out of the core body of the lens, thereby supporting the eye. These haptics can be fabricated using either different or same material, but PMMA or the base material of the optic most frequently used material for the development of IOL. There are many reasons for choosing PMMA as it is light in weight and has high biological compatibility [63].

Although PMMA has been in use for over 40 years, it has some drawbacks, such as it does damage to the endothelium of the cornea due to its low surface energy. Polymeric coatings of HEMA and N-vinylpyrrolidone (NVP) are used to increase the surface energy and bio-compatibility. Furthermore, acrylic polymers, collagen fibers, and PHEMA hydrogels are also used to make foldable IOL. In recent times, poloxamer hydrogel and thermo-sensitive polymers have been tested for IOL and the experimental results were satisfactory [64]. The structures of some monomeric units used for making intraocular lenses are shown in Fig. 4.

5.4 Intracorneal Implants

Occasionally the direct mediation surgery is required for adjusting the refractive shape of the eye. These types of surgeries may include permanent implants such as intracorneal implants or keratoprosthesis. The former is implanted inside the cornea and as a result simply works to enhance the functions of the natural cornea, whereas latter devices involve penetration of the full width of the cornea and thus causing complete replacement of the function of the cornea. The intracorneal lenses are embedded in the main stroma of the cornea to correct conditions such as hyperopia and myopia.

In earlier times, implants completely changed the overall thickness of the cornea, but with the latest developments, only carefully selected layers can be replaced. Early intracorneal lenses were made of flint glass, however today implants are manufactured from polymethyl methacrylate (PMMA) that combines several important mechanical and optical properties with considerable biological compatibility. Although PMMA is a good polymer, the inherent issue with rigid PMMA lenses includes the weakening of epithelial layer over the cornea and the obstruction of the transport of nutrients all over the cornea [65].

Subsequently, polysulfones were used to create a high refractive index and thin film to replace the natural cornea, but polysulfones cause poor nutrient transport because they are resistant to aqueous solutions. Therefore, their surfaces were made porous by increasing the pore size to 10 μm. Hydrogel-based corneal implants were also employed to advance their design, but they did not match the natural corneal refractive index [66]. Their design was further enhanced by the core-skirt approach to improve the bio-adhesion of the artificial cornea and mechanical features. Graphite, PHEMA and polyvinyl alcohols, and some other hydrogel systems such as bacterial cellulose were the materials that were tested during their design improvements, but clinical trials are needed to prove their effectiveness. Myopia could also be cured by modifying the corneal surface area. In 2004, Food and Drug Administration (FDA) approved PMMA rings for the treatment of keratoconus [67].

5.5 Scleral Buckles

The detachment of retina is a critical situation when retina is flaked off from its basement membrane. The sclera is pushed through the scleral buckle and the choroid is brought back into connection with the retina. Some common problems that arise during the use of scleral buckles are infection, reduced extraocular motility, changes in the intraocular pressure of the buckled globe, chronic irritation, and implant extrusion [68]. A wide range of materials is used to formulate the scleral buckles. Initially, the tendons of the same individual were utilized; although they eradicate the reactions by the foreign body, they need two surgeries. The elastic strength is the main disadvantage of implants as the exact desired sizes of the buckle cannot be obtained. Of all the biodegradable materials used, gelatin derived from porcine skin is the most accepted and commonly used substance [69]. The inefficient buckling effect is the main problem of the absorbable buckles. In recent years, extensive design work has been carried out for non-absorbent devices and nowadays these non-absorbent materials provide prominent advantages over absorbent materials. A wide series of polymers were used to fabricate non-absorbable implants, but especially the non-absorbable buckles consisting of soft silicone sponges and silicones are mostly used [70]. The buckles remain soft even after implantation, thus avoiding the destruction of underlying sclera. Other than these soft silicones, several other materials such as methyl acrylate hydroxy ethyl acrylate (MAI), polyhydroxyl ethyl acrylate (PHEA), and polyglyceryl methacrylate (PGMA) are also being used as base materials for scleral buckles. From these three materials, MAI holds superior bulk properties, forms a rigid adjoining capsule, and because of these characteristic features, it is the best widely used hydrogel. Whereas PGMA has weak tensile strength, and PHEA breaks on swelling.

5.6 Glaucoma Filtration Implants (GFIs)

Glaucoma is considered the most typical reason for acute visual handicap, in which elevated intraocular fluid pressure (IOP) results in loss of visual field and injury to optic nerve. Glaucoma filtration surgery (GFS) is considered the principal surgical treatment, involving the establishment of a drainage pathway through the cornea/sclera joining the subconjunctival space to the anterior cavity of the eye. Although most often when a fabricated drainage implant is needed, GFS usually serves as a highly successful option. Initial methods involved the use of various forms of seton, which are made from a wide range of materials such as polymers (nylon), metals (magnesium, platinum, and gold), silk, and horse hair.

Filtration failure rates were higher in these primary devices, mainly as a result of the inflammatory response to chronic foreign bodies initiated by the devices, leading to the closure of fistulas and subconjunctival fibrosis. Some of the other methods which were used to get over this problematic situation involve several ligation processes and short-term blockage of the tube. Several alternate systems have also been developed such as the use of various sorts of valves and micro-tubules, to evade post-operative hypotony and to control the flow of fluid. Recently, in vitro exploration of these devices from the same manufacturer has revealed significant deviations in their resistance values, signifying potential discrepancies between implanted devices and the lack of quality control of manufacturers [64].

6 Cardiovascular Devices

Cardiovascular disease is the leading cause of death worldwide, causing billions of deaths annually. The disease causes irreversible damage to the heart, which necessitates new methods to repair the organ. Treatments for heart disease range from medication to surgery. Most surgical procedures work either by replacing the damaged tissues, such as in heart transplant or circumventing them, as in bypass graft. But there is always a chronic shortage of sources for human donor tissues. With the development of alternative therapies, patient care options will be significantly expanded. The use of effective means for regenerating, reconstructing, and repairing the damaged tissues would reduce the need for limited donor organs. Biomaterials have shown great potential to be used as a tool for such procedures [71]. In heart disease, these biomaterials work effectively through implanting and constructing vascular grafts and heart valves. Cardiovascular biomaterials (CB) are biomaterials that are used as implants in these types of cardiac syndromes and successfully interact and work together with biotic components.

6.1 Biocompatibility of Cardiovascular Biomaterials

There are two main ways to use cardiovascular biomaterials (CB): temporary and permanent. Cardiovascular devices are characterized as permanent internal, temporary internal, and temporary external devices based on how they are used [72]. Table 4 lists the most commonly used synthetic biomaterials for cardiovascular applications. The process of testing the level of compatibility for cardiovascular devices depends primarily on how they are being used. All CBs come in direct interaction with blood and their extent of interaction further explains the test parameters. The extent of interaction may be permanent (>30 days), extended (>24 h to 30 days), and limited (<24 h). Blood compatibility should be tested for all the blood interconnecting devices, regardless of the extent of their contact. According to International Organization for Standardization ISO guidelines (ISO-10993 standard), all these devices should also be evaluated for immunology (complement activation), hemolysis, and thrombogenicity testing. Hemolysis may affect red blood cells (RBC) and cause a decrease in their oxygen-carrying capacity. Hemolysis happens when the RBC directly interacts with the material or its degraded products, which are formed as a result of generated shear stress caused by the relative movement between the surface of the material and blood [73].

Table 4 Applications and advantages of cardiovascular biomaterials

Biomaterial composition	Applications	Advantages
Expanded polytetrafluoroethylene (e-PTFE)	Valve repair, atrial septal defects, extracardiac Fontan procedure, vascular graft, reconstruction, and shunts	Correct or palliate interrupted aortic arch, improved tissue approximation for use in neonates, better handling, kink resistance, prevention of excessive shunt flow, good patency, resterilizable, conformable, nonfraying, high strength, low thrombogenicity, nonbiologic
Polyethylene terephthalate (PET)	Vascular graft	Biocompatibility, hemostasis, high patency, tissue ingrowth
Polyurethane	Valve repair, pacing leads	Better quality handling, resists infection, transparency, elasticity, strength
Platinum–chromium alloy, cobalt–chromium alloy, nitinol, stainless steel, titanium	Stents, mechanical heart valve	Excellent hemodynamics, weak thrombogenicity, durability, strength

6.2 *Artificial Heart Valves*

An artificial heart valve is a machine that is inserted into the heart of a patient with valvular heart disorder when one from these four valves of the heart fails to function properly. Depending on the nature of material from which they are synthesized, cardiac valve prostheses can be classified into two types: biological and mechanical heart valves. Their main differences are illustrated in Fig. 5.

6.2.1 Biological Heart Valves

Based on the construction of a biological valve, it is either with or without stent. Stent-based valves are also known as valvular bioprothesis for the reason that they merge the properties of biological material of different origins with fabricated prosthesis (occurrence of a frame or stent). The first biological valve—homograft—was presented to humans in 1962, even though restricted accessibility constrained its use.

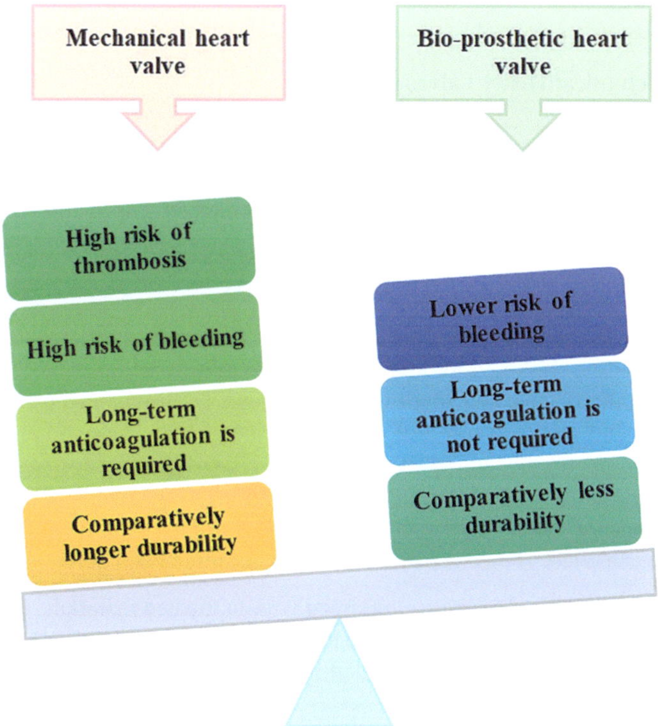

Fig. 5 Different aspects of mechanical (left side) and bio-prosthetic (right side) heart valves. Reproduced from Kumar et al. [75], under the terms and conditions of the Creative Commons Attribution (CC BY) license. http://creativecommons.org/licenses/by/4.0/

In 1965, another material zoonotic-heterograft was also explored, but because of its limited durability, it was also not as effective. The ultimate revolution in the accessibility and design of biological valves took place in 1968 when an advanced technique for the implementation of biological material was explored using glutaraldehyde. This effort resulted in a significant reduction in the antigenic properties, mechanized strengthening by attachment of collagen existing in the tissue, and loss of sensitivity to proteolytic enzymes. Those valves were called bioprothesis, as they combined the concept of artificial and biological valves. The major drawback of all biological valves is their restricted durable nature, which causes them to lose their functionality. It is determined by various features, mainly where the valve is being implanted, and the age of the patient. Recently implanted biological valves display a 15-year shortage of structural imperfections, representing 66% in the mitral point and 88% in the aortic point [74]. Due to the advanced expertise and ongoing research for the preservation of biological material along with the characteristics, operation and construction of blood flow across the aortic valve of the heart succeeded to attain biological valves having remarkably sophisticated hemodynamic characteristics and increased durability, which allows a constant rise in effectiveness in contrast to mechanical valves.

6.2.2 Mechanical Heart Valves

Mechanical heart valve prostheses are composed of fabricated materials such as synthetic polymers, metals, and their alloys. The mechanism of closing and opening is considered as one of the important parameters of their structural difference. All the designs of mechanical heart valves are dependent on the concept of varied pressures throughout the heart cycle.

These valves have the ability that they can differentiate two kinds of blood flow: transverse functioning in ball valves and central functioning in disk valves. Mechanical valves can be classified into the following groups based on their construction: double petal, disk shape, ball shape, and cage shape. Unfortunately, patients having mechanical heart valves require anticoagulation treatment as these valves do not have enough biocompatibility, so they result in hemolysis of blood, causing its coagulation. Regardless of this fact, the mechanical valves are frequently used in cardiac surgeries because their stability decreases the risk of re-operation as a result of valvular dysfunction. A mechanical heart valve has three main components:

(1) Base, normally in the ring style framed with fabricated material.
(2) Cover—the cage style, in which the locking element moves, metal- or graphite-based ring for which it is connected by wires (flakes or disks) of the locking element.
(3) Locking element, generally one or more moving fragments in the shape of semi-circular or circular set of petals, single disk, or a ball [76].

Mechanical heart valves are made up of various materials including ceramics, polymers, and metals, for example, silicone, titanium, pyrolytic carbon, and stainless steel. Stainless steel is the most commonly used material for fabricating heart valves, particularly in constructing struts for assisting leaflets, thus providing mechanized support to the valves and preventing corrosion. A new alloy named L-605, synthesized from cobalt-chromium-tungsten-nickel, is used to make heart valves. Out of different polymers, segmented polyurethanes, polytetrafluoroethylene (PTFE), and polypropylene (PP) are utilized for fabricating the structures of heart valves. Specifically, PTFE is mostly used in making installable rings of prosthetic heart valve.

6.3 Stents

Cardiovascular stents are used to open/bypass the obstructed blood vessels which transport the oxygenated blood to the heart and to improve blood flow in the case of narrowing of blood vessels. These stents are implanted into the lumen of the blood vessels and they may be made up of plastic or metal. During angioplasty process, these coils are implanted to the narrowed artery to improve the light of the blood vessels [74]. Types of vascular stents are shown in Fig. 6.

The most important characteristic of all stents is that they do not form a clot when they come into contact with the surface of the material, as a clot formation at the place of implantation may cause obstruction of the artery. If the stent material does

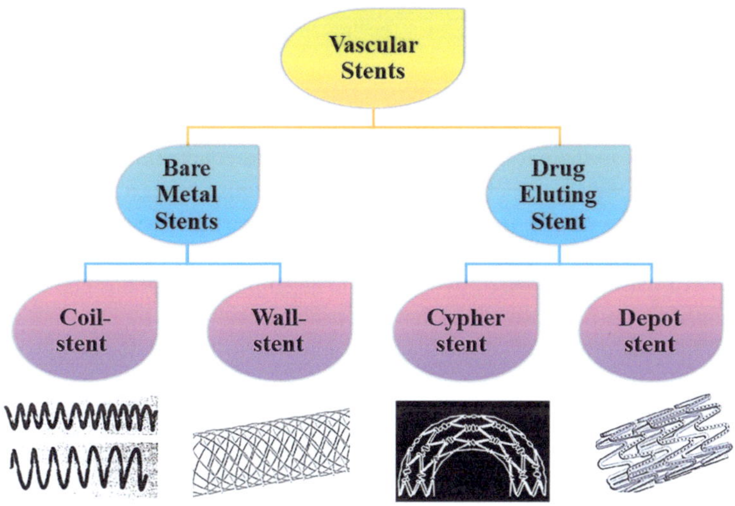

Fig. 6 Types of vascular stents. Reproduced from Pan et al. [77], under the terms and conditions of the Creative Commons Attribution (CC BY) license. http://creativecommons.org/licenses/by/4.0/

not have this property, then special coating of carbon or calcium phosphate is applied on the surface of stent to prevent blood clotting. Some other requirements are ease of implantation, availability, resistance to infection, biocompatibility, sustaining a patent, high fatigue strength, the lack of cross-sectional changes in tension, flexibility, the small diameter of the stent, and resistance to changes in blood pressure [78].

6.4 Cardiac Pacemakers

Cardiac pacemakers are mostly utilized to regulate irregular or slow heartbeats. A pacemaker consists of two pacing leads that connect the pacemaker to the heart, at least one electrode and one pulse generator. The covering of the pulse generator works like a jacket for the circuits and battery that supplies power. It is generally inserted between the pectoral muscle and skin. The pacemaker system specifically applies electrical signals over time to induce contraction of the heart muscle and to encourage the heart to beat naturally as a regular heart rhythm. The pacemaker gets energy from sealed lithium iodine battery. The circuit involves the conversion of electrical energy to minor electrical signals.

The timing of the electrical signals supplied to the heart was also controlled by these circuits. A polyurethane-based connector block, located at the topmost part of the pacemaker, is used to connect the pacemaker with pacemaker lead. Glass-based connector blocks were used in earlier times. The pulse generator is enclosed in titanium. Titanium has substituted ceramics and epoxy resin, which were previously used to encapsulate the pacemakers, with silicone rubber. This advancement to titanium has endorsed the patients to securely use electric devices, such as microwave ovens, for the reason that titanium decreases the interference of external electromagnetic radiations and shields the internal components.

The pacing lead acts as a key element of the pacemaker system, as it efficiently carries electrical signals from the pacemaker to the heart and transfers heart activity information back to the pacemaker. The lead is a non-conducting metal cable with a lead body, at least one electrode, a fixation mechanism, and a connector pin. It transmits electrical energy from pacemaker to the heart. The connector pin is the part of the lead through which it is introduced into the connector block. On the basis of the type of pacemaker, one or two leads can be employed. The one end of the lead is positioned in the right atrium or right ventricle of the heart, by inserting through a vein. While the other end of the lead is connected to the connector block of the pacemaker.

The lead should be stable enough to tolerate the flexing caused by the heart contractions in the corrosive and warm surroundings of the body. The lead body must be good conductor for electric signals, durable, noncorrosive, and flexible. Therefore, those materials should be selected for lead body which are durable, nontoxic, and inert. Nowadays, more resilient materials such as polyurethane and silicone rubber tubes have been used to insulate the lead body, replacing the previously used polyethylene. Polyurethane has more strength than silicone rubber, which enables the use of thinner lead in the pacemaker having superior lead flexibility. Polyurethanes have very low coefficient of friction when wet. On the other hand, oxidation by metal ions can damage polyurethanes, although this method of degradation did not affect the silicone.

The mechanism of fixing holds the head of the lead in the heart. Nowadays, either an electrically active platinum-iridium helix or a nickel-cobalt alloy having a helix of silver core is mostly used to attach the lead electrodes to the heart surface. The electrode is positioned at the topmost of the lead. Electrodes are composed of stainless steel, titanium, platinum, cobalt, and nickel alloys. Titanium also shows excellent durability, high resistance to corrosion, and greater modulus of elasticity. Iridium oxide is used to coat the surface of the electrodes to prevent the formation of insulating layers. The coated electrodes may also offer chronic and less acute thresholds because of their reduced local inflammation [79].

6.5 *Blood Vascular Grafts/Vessel Prostheses*

The diseased or injured blood vessels can be successfully repaired by blood vessel prosthesis, while the vascular bypass, also called vascular graft, is a medical surgery that is used for reconnecting blood vessels by readdressing the blood flow from one part to another. It is mostly maintaining the normal flow rate of blood between two healthy parts by bypassing diseased or injured blood vessels. It can be used effectively during hemodialysis, for providing assistance in organ transplantation and in ischemic conditions of atherosclerosis. Even though, the best and utmost chosen material for vascular graft is either donor vein (allograft) or the patient's own vein (autograft), but some other materials such as polytetrafluoroethylene (Teflon) and polyethylene (PE) terephthalate (Dacron) are also used.

Protein polymers are a new category of polymers that have the capability of embracing the features of the arterial walls and can simulate native proteins. Examples of protein polymers include gels, fibrin fibers, and collagen. These biopolymers provide a new supporting system for fabricating vascular grafts. These polymers also remarkably contribute to the development of artificial blood vessels by attaching to proteins which are involved in directing the cell fate. In the human vasculature, the role of elastin is stimulated by elastic, while the cross-linked collagen fibers provide integrity and mechanical strength to the vascular grafts. Synthetic biomaterials are considered as the primary option for designing the vascular graft because of their various advantages such as comfort and flexible nature of adjusting their mechanical features.

Blockage of the blood vessels and thrombosis rates should be checked to consider material during bypass. Polytetrafluoroethylene (Teflon), polyurethane, and polyethylene terephthalate (Dacron) are the most commonly used non-biodegradable biomaterials for vascular grafts. Patency rate is characterized by the ability of blood vessel to stay open. Polyurethane is more similar to natural vasculature, but its vascular grafts have relatively weak patency rates. Polytetrafluoroethylene (PTFE) is effectively practiced as a vascular graft, when the device is inserted in large diameter arteries having high-flow rate such as the aorta. One more form of PTFE, called elongated-PTFE (e-PTFE), has also been examined. Due to its biochemically inert properties, weak calcification, low rates of restenosis and hemostasis, and low thrombogenicity, it is successfully used for small diameter arteries having low flow rates [80].

7 Dental Implants

Dental treatment is considered to be one of the most common medical treatments for humans. It can range from replacing decayed or broken teeth to filling cavities (also called "dental caries").

7.1 *Dental Restorative Materials*

A huge range of biomaterials is being used in the dental treatments including periodontal, orthodontic, preventive, prosthetic, crown and bridge, fluting, cavity filling, cavity lining, and endodontic treatment of teeth. The biomaterials to be used in the oral cavity must have definite capabilities. It should be non-toxic having better aesthetic appeal, high resistance to tarnish, wear and corrosion, good reproducible quality, excellent osseointegration ability, high mechanical strength, superior biocompatibility, high purity, sterilizability, and hydrolytic stability. The choice of the material is based on its aptitude to mimic the esthetic, mechanical, and physical properties of original tooth structure.

Composite materials have substituted the polymethylmethacrylate (PMMA) fillings, and used in medical practice to reinstate posterior and anterior teeth. In variety of applications, dental composites have a polymeric methacrylic or acrylic matrix reinforced with ceramic particles. The marketable constructions of matrices are mostly constructed using bisphenol-A-glycidyldimethacrylate (Bis-GMA), whereas triethylenglycoldimetachrylate (TEGDMA) has been introduced to decrease the viscous nature. The utmost serious features for long-lasting functioning of dental composites include their mechanical and wear properties, wear resistance, polymerization shrinkage, and dimensional stability [81].

7.2 *Fixed and Removable Dental Prostheses*

Fixed bridges and removable dental prostheses have been practiced for replacement of one or more teeth to restore the aesthetic, phonetic, and masticatory functioning. Both have some drawbacks such as removable dental prostheses do not transmit the occlusal stresses to the bone and at the same time being uncomfortable, whereas the fixed bridges obstruct with the nearby teeth through their attachment points.

Some composite materials such as PMMA reinforced with glass fibers (GFs), ultrahigh molecular weight polyethylene (UHMWPE), Kevlar, and carbon are used as bridges, while PMMA, polyamides, polypropylene (PP), and polycarbonate (PC) reinforced with glass fibers (GFs) are used as removable dental prostheses [82]. Combining the two different types of composite materials, methacrylic matrix strengthened with glass fibers (GFs) and inorganic particles, results in enhanced aesthetic appearance and mechanical properties when they were used as bridges.

7.3 Dental Implants

A dental implant (also known as endosseous implant/fixture) acts as an interface between the skull and jawbone and functions as an orthodontic anchor by supporting facial prosthesis or dental prosthesis including a denture, crown, and bridge. It is considered a widespread tooth replacement option. An artificial permanent implant (tooth root) replaces the lost or damaged tooth. It was designed to combat the defects of removable dental prostheses and fixed bridges. It is surgically implanted into the jawbone, similar to the shape of a screw, and bonds with the natural bone. It forms a base in order to support artificial teeth, called dental crowns. Dental implants are made of various types of materials such as composites (CF/carbon and SiC/carbon), polymers (PET, Polystyrene (PS), PTFE, PMMA, Ultra-High Molecular Weight Polyethylene (UHMPE)), ceramics (carbon, glass, alumina, zirconia), metals (silver, platinum), and their alloys (stainless steel, Ti alloys, Co–Cr–Mo alloys) [83].

The body of a dental implant contains three parts that are capable of performing a variety of functions (Fig. 7).

(1) The implant (screw) functions like the artificial root of teeth.
(2) The abutment connects the crown with the implant screw. Abutment is also used to connect crowns to partial dentures, removable dentures, and dental bridges.
(3) The crown (artificial tooth) is attached at top of the abutment. The crowns are composed of porcelain, which is a tooth-colored material, so it looks like a natural tooth and resembles its shape.

Dental implants exist in two different forms:

(1) **Subperiosteal Implant**: Subperiosteal implant is not surgically implanted into the jawbone, instead it is placed under the gums in such a way that it is either above or on the jawbone. It is not very common. But it can be a good choice for patients who do not have sufficient natural jawbone to support an endosteal implant.
(2) **Endosteal Implant**: An endosteal implant is considered as the leading type of implant used today. It is surgically introduced into the jawbone and with the

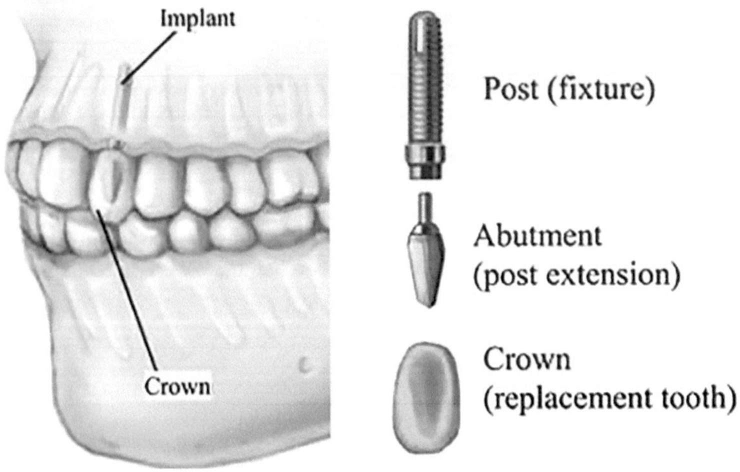

Fig. 7 Components of the dental implants. Reprinted from Praveen et al. [84], with permission from Elsevier. Copyright (2022)

passage of time, it forms a bond with the natural jawbone. It can be composed of alloplastic material, small screws, and titanium, referring to a fake tissue graft.

7.4 Dental Bridges

A dental bridge is a fabricated tooth (partial denture) that is used for filling the empty space between the teeth. In the beginning, the gold bridges were used because of their esthetical characteristic feature. But their time-consuming preparation and high cost has led the researchers to develop easy-to-use and cost-effective materials. And the use of polymer-based biomaterials has fulfilled these conditions and the composites have replaced the gold bridges. Commercially available dental bridges were prepared by using polymer composites such as GF/PMMA, UHMWPE/PMMA, Kevlar fiber KF/PMMA, and carbon fiber CF/PMMA [85].

7.5 Arch Wires and Brackets

Orthodontic arch wires are practiced for correcting the dental arrangement. This is assisted by attaching orthodontic brackets onto the teeth. Brackets should have the accurate strength and hardness so as to provide the particular force from the wire to the teeth. They should have a smooth surface to reduce the accumulation and deposition of plaque, with the smooth opening for the arch wire to reduce the resistance due to friction. They should have a good biocompatibility and high resistance to corrosion.

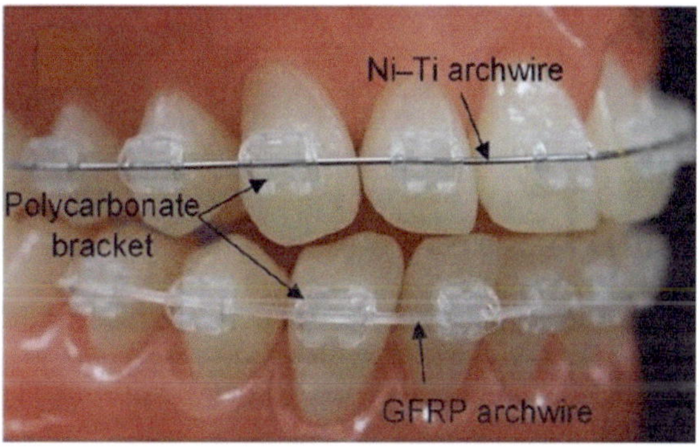

Fig. 8 Comparison of glass fiber reinforced plastic (GFRP) arch wire and nickel-titanium arch wire. Reprinted from Krishnakumar and Senthilvelan [88], with permission from Elsevier. Copyright (2021)

As most orthodontic brackets are made with three-dimensional prescriptions for each tooth, they must be prepared to reflect it. The brackets and arch wire were made from titanium alloys, chromium-cobalt alloy, and stainless steel [86]. Polymeric biomaterials have replaced metallic biomaterials because of their aesthetic problems. Bracket implants and arch wire contain mostly glass fiber (GF) reinforcement material for their dental polymer matrix of polymethylmethacrylate (PMMA), nylon, polypropylene (PP), and polycarbonate (PC) [82]. The polycarbonate-GF provides continual force to teeth and also it is transparent, which gives it an esthetical appearance as compared to Ni–Ti alloys (Fig. 8) [87].

7.6 Dental Posts

The structural integrity of totally damaged teeth can never be attained either by restoration or by the cavity filling. Then in this situation, it is necessary to support the teeth with the help of dental post. The selection of the biomaterial for the dental post is dependent on the fracture strength property. Metal posts and the resin fiber posts provide the needed fracture strength, despite the fact that metal posts have been outpaced by resin fiber posts with reference to fracture resistance in comparison to zirconia or other metal posts. Posts containing titanium alloys have outstanding resistance to corrosion as compared to other metal posts. Carbon fiber and glass fiber are the leading reinforcing materials, while polyamide, epoxy, and Bis-GMA are frequently used resin materials for dental posts. Epoxy-carbon fiber and Bis-GMA-glass fiber composites have been identified as broadly used post materials. Epoxy resin reinforced with carbon fibers exhibits superior features as compared to Bis-GMA glass fiber composites [89]. But various health risks such as asthma, allergies, and skin irritation lead to the discovery of new efficient materials for dental posts.

8 Drug Delivery Systems

In order for any drug formulation to be effective, its active ingredient must be bioavailable in optimum quantities at the targeted position of the human body. In the pharmacological industry, the drug delivery systems are based on the drug dosage form, a device used for effectual delivery, an active ingredient of the drug and its release at the targeted point of action. The use of traditional prescribed amounts of drugs comprising capsules and tablets is restricted due to the requirement for high doses and their related high toxic profiles. Novel drug delivery systems have been developed to counter these problems. These advanced drug delivery systems include excipients and active components having the characteristic features that control the release of active components, thus providing the best benefit at the desired location with reduced drawbacks.

Some of the widely used advanced systems are nanoformulations, microspheres, liposomes, and osmotic drug delivery systems (as listed in Table 5). To achieve effective drug delivery to the targeted site without disturbing healthy cells, the prepared biomaterials must be enriched with directing ligands. The presence of directing ligands confers cytotoxic properties and prevents adverse effects on healthy cells. Moreover, the biomaterials should be designed with the purpose of reducing the drug release into the blood, thereby preventing drug resistance and increasing blood circulation [90].

They have the ability to be used for parenteral use or oral administration. Mucoadhesive drug delivery system is a perfect example of a system specific to transdermal applications. In addition to the development of these novel drug delivery systems, improvements have been made to the traditional dosage forms to maximize the release of active components and eliminate their side effects. These techniques contain compression coatings, sugar coatings, and film coatings (sustained release coatings or enteric coatings) [91]. Table 5 lists some modified and advanced drug delivery systems and the used biomaterials for their manufacture.

Polymeric delivery systems are considered one of the most valuable approaches in which the drug is encapsulated in a polymer matrix or otherwise contained within a polymer membrane. These systems are generally used to accomplish either spatial or temporal control of drug delivery. After implantation, the drug diffuses out of the membrane into the tissues. In some cases, dissolution or erosion of the polymer plays a contributing role in the drug release mechanism. Mostly degradable polymers including poly orthoesters and poly lactic acid are employed in the drug delivery systems. Basically, the polymeric vehicles support the drugs to be effectively delivered at the targeted point of action over a prolonged time period. They are designed to increase patient compliance and improve drug efficacy and safety. Polymers are used in their designs to enable the drug delivery with short in vivo half-lives, reduce the dosage frequency and the quantity of drug molecule, decrease side effects, and sustain therapeutic levels of the drug [92].

Chitosan has been efficiently utilized in drug delivery as a biodegradable release system, drug conjugate, hydrogel system, and polyelectrolyte complex (PEC)

Table 5 Advanced drug delivery systems

Drug delivery systems	Characteristic features	Biomaterials
Tablet/capsule coatings	Compression coating: coating surrounds the intact core surface to ensure precise release of the core materials	Enteric coatings: sodium alginate, zein, shellac, methyl methacrylate-methacrylic acid (MAA) copolymers, hydroxypropyl methylcellulose acetate succinate (HPMCAS), polyvinyl acetate phthalate (PVAP), hydroxypropyl methylcellulose phthalate (HPMCP), cellulose acetate phthalate (CAP), cellulose acetate succinate (CAS)
	Sustained release coatings: drug release is achieved on a fixed frequency, thus maintaining the drug concentration for a fixed period of time	
	Enteric coating: Prevents excretion of active ingredients in the intestine produced as a results of drug degradation in gastric pH of the stomach	
	Film coating: involved in taste masking and aesthetic appeal	Film coating: polymer solution
	Sugar coating: to mask the taste	Sugar coating: sugar syrup
Mucoadhesive systems	Dosage forms: solutions, gels, patches, and tablets	Penetration enhancers: polysorbate 80, sodium salt of ethylenediaminetetraacetic acid (EDTA), menthol, dextran sulfate, benzalkonium chloride, polyethylene glycol (PEG)
	Offer mucoadhesion and greater residence time of controlled release systems in the mucosa of buccal, nasal, and ocular paths	Mucoadhesive agents: hydrogen bonding groups containing polymers: Chitosan, gelatin, agarose, hydroxypropyl cellulose, hydroxypropyl methylcellulose, hydroxyethyl cellulose, sodium alginate, sodium carboxymethyl cellulose, polyvinyl alcohol, polyacrylic acid
Osmotically controlled drug delivery systems	Formulation (tablet) having one small laser-drilled hole and an outer semi-permeable membrane. Osmosis results in absorption of water across the semi-permeable membrane; osmotic pressure forces the active agent from the aperture in the formulation so that the drug can be released effectively	Wicking agents: sodium lauryl sulfate, polyvinylpyrrolidone, titanium dioxide, kaolin, colloidal silicon dioxide
		Semipermeable membrane: Ethyl cellulose (EC), cellulose acetate butyrate (CAB), cellulose propionate (CP), cellulose acetates
	Osmotic pressure acts as the driving energy for the controlled release of drug	Osmotic agents: dextrose, citric acid, fructose, xylitol, potassium chloride, sodium chloride, sorbitol

(continued)

Table 5 (continued)

Drug delivery systems	Characteristic features	Biomaterials
Microspheres	Free-flowing powders (artificial polymers/natural proteins); particle size less than 200 μm; biodegradable	Natural polymers: starch, carrageenan, agarose, gelatin, chitosan, collagen, albumin
		Synthetic polymers: poly anhydrides, poly alkyl cyanoacrylates, lactides and copolymers, glycolides, epoxy polymers, glyceryl methacrylate, polymethyl methacrylate (PMMA)
Nanoformulations	Nano-sized formulations; Solid lipid nanoparticles, nanoparticles	Nanoparticles: iron oxide, polystyrene, magnetite, carbon, zinc oxide, titanium oxide
	Quantum dots	Quantum dots: cadmium telluride, cadmium selenide, cadmium sulfide
	Nanowires, nanotubes	Nanotubes and nanowires: boron nitride, silicon, titanium dioxide, carbon black
Liposomes	To improve the bioavailability of the drug, it is encapsulated in a phospholipid bilayer	Steroids: cholesterol
		Phospholipids: phosphatidylethanolamine, phosphatidylcholine

for various components. Chitosan-based delivery systems are commonly used in bioimaging applications; gene therapy; and to deliver antibiotics, growth factors, peptides/proteins, and anti-inflammatory drugs. Chitosan polyelectrolyte complexes or particles have been tested for nasal delivery of therapeutic proteins. The results revealed that insulin-containing chitosan nanoparticles significantly improved nasal absorption of proteins in comparison to chitosan solutions [93].

9 Wound Healing and Burn Dressings

The skin is the largest organ of the human body which prevents the loss of electrolytes and water from the body. But while protecting the organs and tissues from external environment, it can feel pressure, heat, and cold. The skin comprises subcutaneous, dermis, and epidermis tissues. It also contains nerves, lymph vessels, blood vessels, and cutaneous appendages including sebaceous glands, hair follicles, and sweat glands. Wounds due to diabetes, cuts, burns, or other factors interrupt the physiological functions as well as appearance of the skin. A large portion of the injury causes a massive loss of protein and water from the body with an increase in metabolism, and in more severe cases can be fatal. In case of any trauma or damage to the skin, initially the stem cells of the epidermis provide healing and self-renewal ability. However, this method of self-regeneration is not sufficient in the case of a serious injury, and, in this case, additional therapy is required for treatment. If full-thickness skin of more than 4 cm diameter is lost, then grafting is required for its cure.

On the basis of the biomaterial to be used, the replacements of skin can be categorized as synthetic (non-biodegradable and biodegradable) or biological (xenogeneic, autologous, and allogeneic). Fabricated replacements contain acellular origins and their main functioning is to prevent microbial contamination and to check for lost fluid. Whereas biological replacements (also known as tissue-engineered skin) consist of cultured cell sheets/suspensions, which can be utilized individually or in combination with a dermal matrix [94].

9.1 Skin Grafting

Skin grafting is the conventional method used to treat wounds. Each type of skin has its advantages as well as its own disadvantages. Xenografts and allografts can be obtained from various types of sources, even though they can trigger the immune system and cause immunoreaction and transmit diseases. Although the autografts do not cause immunoreactions, they have the potential to increase the possibility of infection and can cause additional damage to the donor sites.

9.2 Tissue Engineering

With advances in biomedicine and materials science, tissue engineering has been established as an effective way to treat skin defects. Tissue-engineered skin is the equivalent of biological skin, which is made by combining the seed cells with suitable scaffolds to restore the morphology and functions of injured skin. It has significant biological activity and abundant sources, and does not cause damage to the donor sites. Up till now, different cell-free skin substitutes have also been constructed for the reason that cell-based substitutes are costly, need a longer cell culture duration to manufacture, and are more likely to be rejected [95]. Skin substitutes are categorized as acellular and skin substitutes. Tissue-engineered skin substitutes can resemble with either one layer (dermis or epidermis) or both layers of the skin. For that reason, they may also be classified as dermo-epidermal, dermal, and epidermal skin substitutes (Fig. 9) [96].

In addition to wound protection, in tissue engineering, scaffolds also provide a suitable environment for seed cells to grow and prevent cell death initiated by direct cell injection. The optimal scaffold should be advantageous to cell proliferation and adhesion along with having significant biocompatibility. In addition to the elimination of secretions, it should maintain a moist environment favorable for healing, and prevent the loss of body fluids to avoid bacterial infection. Furthermore, scaffolds must have

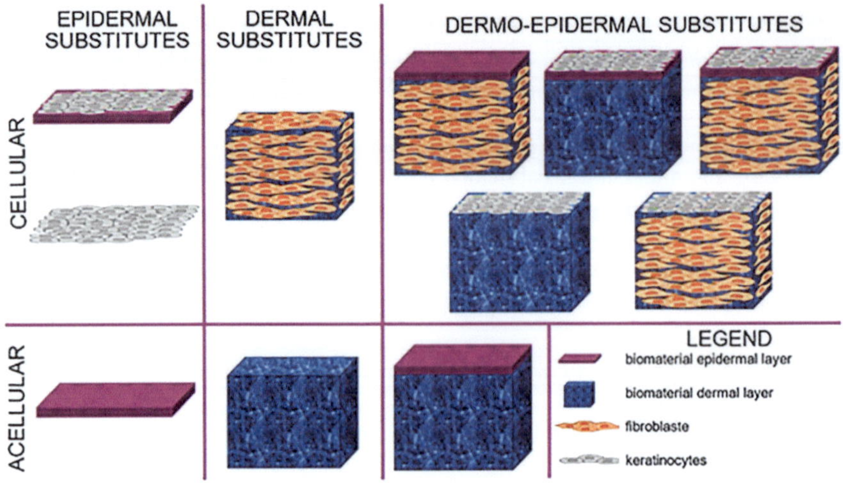

Fig. 9 Different variations of tissue-engineered skin substitutes. Reproduced from Przekora [96], under the terms and conditions of the Creative Commons Attribution (CC BY) license. http://creativecommons.org/licenses/by/4.0/

appropriate pore sizes and porosities to enable the cell migration and nutrient transportation. The most common forms of scaffolds include hydrogels, acellular matrix, sponges, membranes, and nanofibers. Biomaterials affect the characteristic features of scaffolds, such as mechanical properties and biocompatibility. Notably, scaffold materials guide cell differentiation, proliferation, and growth. Scaffold materials can be divided as composite, synthetic, and natural materials (Table 6).

9.3 Biomaterial-Based Wound Dressing

Wound dressing is usually an external healing aid provided to the wound for preventing the flow of proteins, exudates, and blood. Wound dressing materials are the scaffolds that give relief at the wound site, for example, films, hydrogels, or cellulosic materials are constructed either from synthetic or natural or materials [97]. In addition to the dressing of wounds, the antiseptic wound dressings also provide hygienic environment around the wounds. Furthermore, therapies based on various types of biomaterials have been developed for wound management. Wound dressings containing biomaterials are best for loading antibiotics or other drugs because of their release kinetics and tunable properties. The main objective of active wound dressings is to increase the natural healing process. These dressings work to control many aspects that affect poorly healing wounds such as wound infection, scarring, ischemia, and excessive inflammation [98]. Figure 10 shows some of the biomaterials currently in medicinal use.

Table 6 Various synthetic and natural biomaterials used as tissue-engineered skin substitutes

Biomaterial composition	Structural type of substitute	Cellular content
Fibrin-coated polylactic acid (PLA), collagen	Dermo-epidermal	Cellular (keratinocytes, fibroblasts)
Bovine tendon collagen, silk fibroin	Dermal	Cellular bone marrow stromal cells (BMSCs)
Chitosan, collagen	Dermal	Cellular (fibroblasts)
Fish collagen, chitosan, basic fibroblast growth factor(bFGF)-loaded polylactic-co-glycolic acid (PLGA) microspheres, chondroitin sulfate	Dermal	Acellular
Human like collagen, carboxymethylated chitosan	Dermal	Acellular
Polyvinylpyrrolidone, polyglycerol sebacate (PGS)	Dermal	Acellular
Collagen, polycaprolactone (PCL), elastin	Dermal	Acellular
Gelatin, poly(L-lactic acid)-co-poly(ecaprolactone)	Dermal	Acellular
Cellulose, poly(globalide)	Epidermal	Acellular

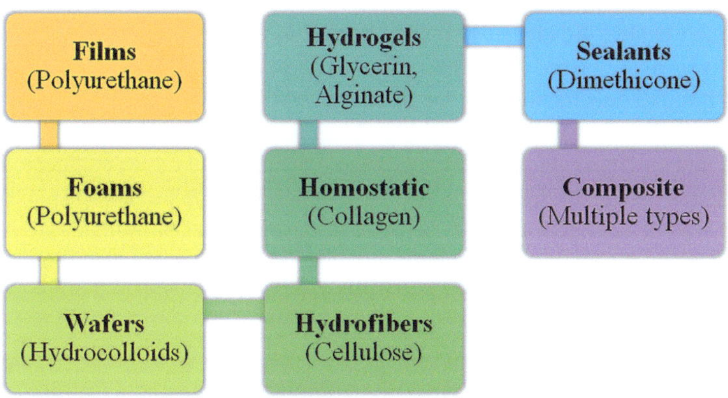

Fig. 10 Some common forms of biomaterial-based wound dressings

10 Future Trends for Biomaterials

Biomaterials are acknowledged as the backbone of the medicinal device industry, an important part of healthcare. These materials have a significant effect in the field of medicine. Nowadays, there are thousands of disposable materials, biomedical devices, soft and hard tissue products, and diagnostic and pharmaceutical products, which are commercially available in the pharmaceutical market for human benefit.

However, there are still many challenges that need to be addressed. In recent times, advanced medical strategies, such as prevention and treatment of leading genetic diseases, have become important, and the need for new treatments has led to the modification of biomaterials. Therefore, the future trend has to link the remarkable biofunctionality and biocompatibility of polymers and ceramics with the mechanically superior metals to achieve most desired medical functioning of the implants. For this, the biomaterial engineers and scientists need to expand the domain of promising applications by combining the biomaterials with physics, genetics, biochemistry, molecular biology, and other fields of science. It is to be expected that in the future there will be a combination of devices containing a bioengineered substrate along with a biological component to break through the intrinsic limitations of entirely artificial prosthetics. This connection assists the tissue engineers, material scientists, and researchers to propose the products from molecular level to cells and tissues [99].

11 Conclusion

Biomaterial science is a multidisciplinary success story, having promoted from developments in physical sciences, as well as biomedical knowledge and engineering. Biomaterials are acting as an indispensable factor for improving the quality of life and human health. Advanced biomaterials can allow people to live healthier and longer lives, as described in the technologies and detailed in this chapter. Applications of biomaterials include therapeutic treatments (medical devices and implants), medical supplies (surgical tools and blood bags), diagnostics (biosensors and gene arrays), and emerging regenerative medicine (cartilage and tissue-engineered skin). The detailed study on the biomedical application of biomaterials displayed here provides the basics needed for the design and development of biomaterial implants and devices by the emerging material scientists, tissue engineers, and researchers.

The development and application of biomaterials is predicted to increase dramatically in the coming decades due to the aging population worldwide. The medical device industry depends on the design and development of advanced composites, polymers, metals, and ceramics to create effective replacements for diseased and damaged body tissue. For instance, the development of resorbable polymers capable of releasing pharmacological agents with the intention of preventing artery occlusion has markedly expanded the clinical use of drug-eluting cardiovascular stents; flexible polymers that dilate the artery and flexible wire meshes that maintain arterial patency. The past few decades have seen rapid development in biomaterial research, influencing almost all areas of dentistry and medicine.

Many ethical concerns associated with the use of synthetically fabricated biomaterials, including ceramics, polymers, and metals, have been raised in the past. Majority of these include potential harmful effects and safety for the living system. The development of non-biological and more traditional material cells such as biologically integrated biomaterials will mean that new ethical questions will arise. With significant advances in nanotechnology, molecular and cell biology, the need for effective and safe treatments will also create unique ethical situations in the near future. The practice of animals in biological research has been hampered by animal rights groups, creating new challenges for researchers and scientists to allow further action. Accountability of biomaterial researchers in the future will require the integration of many new regulations and rules into existing codes of conduct for research. These will be necessary if advanced biomaterials from emerging fields of technology and science are to be ethically and morally acceptable to society and the scientific community [100]. It is hoped that this chapter will promote interactions among the many stakeholders involved in designing and developing biomaterials, including biologists, engineers, surgeons, and physicians. It will also promote future research activities in this rapidly developing area.

References

1. M.S. Kang, J.H. Kim, R.K. Singh, J.H. Jang, H.W. Kim, Acta Biomater. **16**, 103 (2015)
2. F.M. Chen, X. Liu, Prog. Polym. Sci. **53**, 86 (2016)
3. R. Davis, A. Singh, M.J. Jackson, R.T. Coelho, D. Prakash, C.P. Charalambous, W. Ahmed, L.R.R. da Silva, A.A. Lawrence, Int. J. Adv. Manuf. Technol. **120**, 1473 (2022)
4. J. Li, D.J. Mooney, Nat. Rev. Mater. **1**, 16071 (2016)
5. A. Prasad, Mater. Today: Proc. **44**, 2447 (2021)
6. B. Erzen, M. Karataş, R. Orhan, E. Aydoğmuş, J. Macromol. Sci. B, **1** (2025)
7. A. Suwardi, F. Wang, K. Xue, M.Y. Han, P. Teo, P. Wang, S. Wang, Y. Liu, E. Ye, Z. Li, X.J. Loh, Adv. Mater. **34**(1), 2102703 (2022)
8. W. Zafar, S.H. Sumrra, Z.H. Chohan, Eur. J. Med. Chem. **222**, 113602 (2021)
9. G.L. Koons, M. Diba, A.G. Mikos, Nat. Rev. Mater. **5**(8), 584 (2020)
10. Y.H. Jung, J.U. Kim, J.S. Lee, J.H. Shin, W. Jung, J. Ok, T.I. Kim, Adv. Mater. **32**(16), 1907478 (2020)
11. S.P. Nichols, A. Koh, W.L. Storm, J.H. Shin, M.H. Schoenfisch, Chem. Rev. **113**(4), 2528 (2013)
12. S.R. Patel, C.M. Lieber, Nat. Biotechnol. **37**(9), 1007 (2019)
13. M.I. Chimowitz, M.J. Lynn, C.P. Derdeyn, T.N. Turan, D. Fiorella, B.F. Lane, S. Janis, H.L. Lutsep, S.L. Barnwell, M.F. Waters, B.L. Hoh, M. Hourihane, E.I. Levy, A.V. Alexandrov, M.R. Harrigan, D. Chiu, R.P. Klucznik, J.M. Clark, C.G. McDougall, M.D. Johnson, L. Pride, M.T. Torbey, O.O. Zaidat, Z. Rumboldt, H.J. Cloft, N. Engl. J. Med. **365**(11), 993 (2011)
14. S.K. Mulpuru, M. Madhavan, C.J. McLeod, Y.M. Cha, P.A. Friedman, J. Am. Coll. Cardiol. **69**(2), 189 (2017)
15. T. Albrektsson, A. Wennerberg, Clin. Implant Dent. Relat. Res. **21**, 4 (2019)
16. A. Aijaz, M. Li, D. Smith, D. Khong, C. LeBlon, O.S. Fenton, R.M. Olabisi, S. Libutti, J. Tischfield, M.V. Maus, R. Deans, R.N. Barcia, D.G. Anderson, J. Ritz, R. Preti, B. Parekkadan, Nat. Biomed. Eng. **2**(6), 362 (2018)
17. D.P. Tokpah, D. Kavaz, V.H. Sumo, P.D. Tantua, I. Ceesay, World J. Adv. Res. Rev. **13**(3), 371 (2022)
18. S. Kargozar, S. Ramakrishna, M. Mozafari, Curr. Opin. Biomed. Eng. **10**, 181 (2019)
19. M.F. Nazar, M.A. Saleem, S.N. Bajwa, B. Yameen, M. Ashfaq, M.N. Zafar, M. Zubair, J. Phys. Chem. B **121**(2), 437 (2017)
20. S. Agarwal, J. Curtin, B. Duffy, S. Jaiswal, Mater. Sci. Eng. C **68**, 948 (2016)

21. A. Oryan, A. Kamali, A. Moshiri, H. Baharvand, H. Daemi, Int. J. Biol. Macromol. **107**, 678 (2018)
22. D. Shekhawat, A. Singh, M.K. Banerjee, T. Singh, A. Patnaik, Ceram. Int. **47**(3), 3013 (2021)
23. J. Li, G.W. Hastings, in *Handbook of Biomaterial Properties*, ed. by W. Murphy, J. Black, G. Hastings (Springer, New York, 2016), pp. 339–352
24. M. Dziadek, E. Stodolak-Zych, K. Cholewa-Kowalska, Mater. Sci. Eng. C **71**, 1175 (2017)
25. Á. Serrano-Aroca, L. Iskandar, S. Deb, Eur. Polym. J. **103**, 198 (2018)
26. Y. Yan, J. Zhang, L. Ren, C. Tang, Chem. Soc. Rev. **45**(19), 5232 (2016)
27. M.W. Bolger, G.E. Romanowicz, D.H. Kohn, Curr. Opin. Biomed. Eng. **11**, 76 (2019)
28. A. El-Fiqi, J.H. Kim, H.W. Kim, Mater. Sci. Eng. C **110**, 110660 (2020)
29. D. Aycan, N. Alemdar, Carbohydr. Polym. **184**, 401 (2018)
30. R. Gelli, F. Ridi, P. Baglioni, Adv. Colloid Interface Sci. **269**, 219 (2019)
31. A. Ashri, N. Amalina, A. Kamil, S. Fazry, M.F. Sairi, M.F. Nazar, A.M. Lazim, Int. J. Biol. Macromol. **107**, 2412 (2018)
32. P. Roach, D. Eglin, K. Rohde, C.C. Perry, J. Mater. Sci. Mater. Med. **18**(7), 1263 (2007)
33. E. Mariani, G. Lisignoli, R.M. Borzì, L. Pulsatelli, Int. J. Mol. Sci. **20**(3), 636 (2019)
34. K. Anselme, Biomaterials **21**, 667 (2000)
35. H.A. Currie, S.V. Patwardhan, C.C. Perry, P. Roach, N.J. Shirtcliffe, in *Hybrid Materials—Synthesis, Characterisation and Applications*, ed. by G. Kickelbick (Wiley-VCH, Weinheim, 2007), pp. 255–300
36. C.J. Kirkpatrick, K. Peters, M.I. Hermanns, F. Bittinger, V. Krump-Konvalinkova, S. Fuchs, R.E. Unger, Itbm-Rbm **26**(3), 192 (2005)
37. R. Sridharan, A.R. Cameron, D.J. Kelly, C.J. Kearney, F.J. O'Brien, Mater. Today **18**(6), 313 (2015)
38. A.P. Pereira, D. Janela, A.C. Areias, M. Molinos, X. Tong, V. Bento, V. Yanamadala, S.P. Cohen, F.D. Correia, F. Costa, J. Pain Res. **18**, 73 (2025)
39. M.N. Helmus, D.F. Gibbons, D. Cebon, Toxicol. Pathol. **36**(1), 70 (2008)
40. W. Jin, P.K. Chu, in *Encyclopedia of Biomedical Engineering*, ed. by R. Narayan (Elsevier, 2019), pp. 425–439
41. G.I. Im, Biomater. Res. **24**, 7 (2020)
42. B. Huzum, B. Puha, R.M. Necoara, S. Gheorghevici, G. Puha, A. Filip, P.D. Sirbu, O. Alexa, Exp. Ther. Med. **22**(5), 1315 (2021)
43. C. Gautam, J. Joyner, A. Gautam, J. Rao, R. Vajtai, Dalton Trans. **45**(48), 19194 (2016)
44. S. Nagarajan, M. Mohana, P. Sudhagar, V. Raman, T. Nishimura, S. Kim, Y.S. Kang, N. Rajendran, ACS Appl. Mater. Interfaces **4**(10), 5134 (2012)
45. Y.S. Hedberg, B. Qian, Z. Shen, S. Virtanen, I.O. Wallinder, Dent. Mater. **30**(5), 525 (2014)
46. Y. Wang, H. Yu, C. Chen, Z. Zhao, Mater. Des. **85**, 640 (2015)
47. M. Rahmati, M. Mozafari, J. Cell. Physiol. **234**(4), 3321 (2019)
48. R. Taktak, A. Elghazel, J. Bouaziz, S. Charfi, H. Keskes, Mater. Sci. Eng. C **86**, 121 (2018)
49. A. Szcześ, L. Hołysz, E. Chibowski, Adv. Colloid Interface Sci. **249**, 321 (2017)
50. E.M. Elmowafy, M. Tiboni, M.E. Soliman, J. Pharm. Invest. **49**(4), 347 (2019)
51. W. Fan, X. Fu, Z. Li, J. Ou, Z. Yang, M. Xiang, Z. Qin, Materials **14**(20), 6065 (2021)
52. R. Ormsby, T. McNally, P. O'Hare, G. Burke, C. Mitchell, N. Dunne, Acta Biomater. **8**(3), 1201 (2012)
53. C.K. Kuo, J.E. Marturano, R.S. Tuan, BMC Sports Sci. Med. Rehabil. **2**, 20 (2010)
54. N. Eliaz, N. Metoki, Mater. **10**(4), 334 (2017)
55. E. Gibon, L.A. Córdova, L. Lu, T.H. Lin, Z. Yao, M. Hamadouche, S.B. Goodman, J. Biomed. Mater. Res. **105**(6), 1685 (2017)
56. M. Long, H.J. Rack, Biomaterials **19**(18), 1621 (1998)
57. R.S Devi, A.S. Kumar, E. Bhuvanes, C. Kavya, Int. J. Adv. Res. Innov. Ideas Educ. **3**(3), 489 (2017)
58. H. Sheardown, E.A. Hicks, A. Tayab, B. Muirhead, in *Biomaterials Science*, ed. by W.R. Wagner, S.E. Sakiyama-Elbert, G. Zhang, M.J. Yaszemski (Elsevier, Netherlands, 2020) pp. 1135–1152

59. N. Tahhan, T.J. Naduvilath, C. Woods, E. Papas, Cont. Lens Anterior Eye **45**(1), 101525 (2021)
60. J.H. Thean, A.A. McNab, Clin. Exp. Optom. **87**(1), 11 (2004)
61. C.Y. Khoo, J. Chong, U. Rajan, Singapore Med. J. **40**(4), 230 (1999)
62. P.C. Nicolson, J. Vogt, Biomaterials **22**(24), 3273 (2001)
63. C.W. Flowers Jr., P.J. McDonnell, Curr. Opin. Ophthalmol. **5**(4), 81 (1994)
64. J.W. Kwon, Y.K. Han, W.J. Lee, C.S. Cho, S.J. Paik, D.I. Cho, J.H. Lee, W.R. Wee, J. Catar. Refract. Surg. **31**(3), 607 (2005)
65. M.S. Loughnan, D. Robinson, L. Sullivan, Curr. Opin. Ophthalmol. **7**(4), 41 (1996)
66. M.E. Mulet, J.L. Alio, M.C. Knorz, Ophthalmology **116**(8), 1455 (2009)
67. B.H. Feldman, T. Kim, J. Catar. Refract. Surg. **36**(2), 332 (2010)
68. H.D. Rernulla, P.A. Rubin, J.W. Shore, F.C. Sutula, D.J. Toaunsend, J.J. Woog, K.V. Jahrling, Ophthalmology **102**(4), 586 (1995)
69. T. Chen, R. Janjua, M.K. McDermott, S.L. Bernstein, S.M. Steidl, G.F. Payne, J. Biomed. Mater. Res. B Appl. Biomater. **77**(2), 416 (2006)
70. I. Tsui, Surv. Ophthalmol. **57**(3), 253 (2012)
71. M.T. Lam, J.C. Wu, Expert Rev. Cardiovasc. Ther. **10**(8), 1039 (2012)
72. M.W. Curtis, B. Russell, J. Cardiovasc. Nurs. **24**(2), 87 (2009)
73. S.K. Jaganathan, E. Supriyanto, S. Murugesan, A. Balaji, M.K. Asokan, Biomed. Res. Int. **2014**, 459465 (2014)
74. S.S. Khan, A. Trento, M. DeRobertis, R.M. Kass, M. Sandhu, L.S. Czer, C. Blanche, S. Raissi, G.P. Fontana, W. Cheng, A. Chaux, J.M. Matloff, J. Thorac. Cardiovasc. Surg. **122**(2), 257 (2001)
75. T. Kumar, A. Singh, S. Thakre, S. Acharya, S. Shukla, S. Kumar, Cureus **15**(7), e42131 (2023)
76. J. Butany, M.S. Ahluwalia, C. Munroe, C. Fayet, C. Ahn, P. Blit, C. Kepron, R.J. Cusimano, R.L. Leask, Cardiovasc. Pathol. **12**(6), 322 (2003)
77. C. Pan, Y. Han, J. Lu, Micromachines **12**(7), 770 (2021)
78. S.H. Cho, W.I. Jo, Y.E. Jo, K.H. Yang, J.C. Park, D.H. Lee, Neurointervention **12**(1), 31 (2017)
79. W.F. DeForge, J. Vet. Cardiol. **22**, 40 (2019)
80. S. Ravi, E.L. Chaikof, Regen. Med. **5**(1), 107 (2010)
81. E. Salernitano, C. Migliaresi, J. Appl. Biomater. Biomech. **1**(1), 3 (2003)
82. S. Ramakrishna, J. Mayer, E. Wintermantel, K.W. Leong, Compos. Sci. Technol. **61**(9), 1189 (2001)
83. H. Ananth, V. Kundapur, H.S. Mohammed, M. Anand, G.S. Amarnath, S. Mankar, Int. J. Biomed. Sci. **11**(3), 113 (2015)
84. A.S. Praveen, A. Arjunan, A. Baroutaji, in *Encyclopedia of Smart Materials*, vol. 1, ed. by Abdul-Ghani Olabi (Elsevier, 2022), pp. 426–435
85. D. Banoriya, R. Purohit, R.K. Dwivedi, Mater. Today Proc. **4**(2), 3534 (2017)
86. C.K. Chng, K. Foong, N.H. Gandedkar, Y.H. Chan, C.L. Chew, Prog. Orthod. **15**(1), 39 (2014)
87. J. Ohtonen, P.K. Vallittu, L.V.J. Lassila, Eur. J. Orthod. **35**(1), 110 (2013)
88. S. Krishnakumar, T. Senthilvelan, Mater. Today Proc. **46**, 9707 (2021)
89. M.K. Al-Omiri, A.A. Mahmoud, M.R. Rayyan, O. Abu-Hammad, J. Endod. **36**(9), 1439 (2010)
90. A. Sood, A. Gupta, G. Agrawal, Carbohydr. Polym. Technol. Appl. **2**, 100067 (2021)
91. J. Jacob, J.T. Haponiuk, S. Thomas, S. Gopi, Mater. Today Chem. **9**, 43 (2018)
92. B. Dhandayuthapani, D.S. Kumar, in *Biomedical Applications of Polymeric Materials and Composites*, ed. by R. Francis, D.S. Kumar (Wiley, 2016), pp. 1–20
93. X. Zhang, H. Zhang, Z. Wu, Z. Wang, H. Niu, C. Li, Eur. J. Pharm. Biopharm. **68**(3), 526 (2008)
94. K. Vig, A. Chaudhari, S. Tripathi, S. Dixit, R. Sahu, S. Pillai, V.A. Dennis, S.R. Singh, Int. J. Mol. Sci. **18**(4), 789 (2017)
95. H.R. Amaral, J.A. Wilson, R.J. do Amaral, I. Pasçu, F.C. de Oliveira, C.J. Kearney, J.C.C. Freitas, A. Heise, Carbohydr. Polym. **252**, 117201 (2021)

96. A. Przekora, Cells **9**(7), 1622 (2020)
97. G.M. Raghavendra, T. Jayaramudu, K. Varaprasad, S. Ramesh, K.M. Raju, RSC Adv. **4**(7), 3494 (2014)
98. S. Das, A.B. Baker, Front. Bioeng. Biotechnol. **4**, 82 (2016)
99. G.M. Raghavendra, K. Varaprasad, T. Jayaramudu, in *Nanotechnology Applications for Tissue Engineering*, ed. by S. Thomas, Y. Grohens, N. Ninan (Elsevier, Netherlands, 2015) pp. 21–44
100. A. Kashi, S. Saha, J. Long-Term Eff. Med. Implants **19**(1), 19 (2009)

Index

S. J. Ikhmayies (ed.), *Advances in Biomaterials Research*, Advances in Material Research and Technology, https://doi.org/10.1007/978-3-031-97481-6

B

D

F

G

I

J

K

Q

R

S

U

MIX
Papier aus verantwortungsvollen Quellen
Paper from responsible sources
FSC® C105338

If you have any concerns about our products,
you can contact us on
ProductSafety@springernature.com

In case Publisher is established outside the EU,
the EU authorized representative is:
Springer Nature Customer Service Center GmbH
Europaplatz 3, 69115 Heidelberg, Germany

Printed by Libri Plureos GmbH
in Hamburg, Germany